COMPOSITE MATERIALS: TESTING AND DESIGN (SEVENTH CONFERENCE)

A conference
sponsored by
ASTM Committee D-30 on
High Modulus Fibers and
Their Composites
Philadelphia, PA, 2–4 April 1984

ASTM SPECIAL TECHNICAL PUBLICATION 893
James M. Whitney, Wright-Patterson Air Force
Base, editor

ASTM Publication Code Number (PCN)
04-893000-33

1916 Race Street, Philadelphia, PA 19103

Library of Congress Cataloging-in-Publication Data

Composite materials.

(ASTM special technical publication; 893)
"Represents a collection of papers obtained from presentations made at the Seventh Conference on Composite Materials: Testing and Design"—Introd.
"ASTM publication code number (PCN) 04-893000-33."
Includes bibliographies and index.
1. Composite materials—Congresses. I. Whitney, James Martin, 1936- II. ASTM Committee D-30 on High Modulus Fibers and Their Composites. III. Conference on Composite Materials: Testing and Design (7th: 1984: Philadelphia, Pa.) IV. Series.
TA418.9.C6C595 1986 620.1'18 85–28639
ISBN 0-8031-0447-2

Library of Congress Catalog Card Number: 85-28639

NOTE

The Society is not responsible, as a body,
for the statements and opinions
advanced in this publication.

Printed in Baltimore, MD
January 1986

Foreword

The Seventh Conference on Composite Materials: Testing and Design was held 2–4 April 1984 in Philadelphia, Pennsylvania. ASTM Committee D-30 on High Modulus Fibers and Their Composites sponsored the conference. James M. Whitney, Wright-Patterson Air Force Base, served as conference chairman and editor of this publication. Most of the papers presented are included in this volume, which complements the first, second, third, fourth, fifth, and sixth conference publications—*ASTM STP 460, ASTM STP 497, ASTM STP 546, ASTM STP 617, ASTM STP 674,* and *ASTM STP 787, Composite Materials: Testing and Design.*

Related ASTM Publications

Composite Materials: Testing and Design (Sixth Conference), STP 787 (1982), 04-787000-33

Composite Materials: Testing and Design (Fifth Conference), STP 674 (1979), 04-674000-33

Composite Materials: Testing and Design (Third Conference), STP 546 (1974), 04-546000-33

Composite Materials: Testing and Design (Second Conference), STP 497 (1972), 04-497000-33

A Note of Appreciation to Reviewers

The quality of the papers that appear in this publication reflects not only the obvious efforts of the authors but also the unheralded, though essential, work of the reviewers. On behalf of ASTM we acknowledge with appreciation their dedication to high professional standards and their sacrifice of time and effort.

ASTM Committee on Publications

ASTM Editorial Staff

Contents

STRENGTH

DELAMINATION

ANALYSIS AND CHARACTERIZATION

Introduction

This volume represents a collection of papers obtained from presentations made at the Seventh Conference on Composite Materials: Testing and Design held in Philadelphia, PA, 2–4 April, 1984. All seven of the conferences held under this title have been sponsored by ASTM Committee D-30 on High Modulus Fibers and Their Composites (New Orleans, 1969, *ASTM STP 460*; Anaheim, 1971, *ASTM STP 497*; Williamsburg, 1973, *ASTM STP 546*; Valley Forge, 1976, *ASTM STP 617*; New Orleans, 1978, *ASTM STP 674*; Phoenix, 1981, *ASTM STP 787*).

Although this conference has traditionally covered all aspects of composite materials technology, it is interesting to note the emphasis of the papers associated with each of the previous six conferences, as it provides an important historical background to research in the area. In the first two conferences, for example, considerable emphasis was placed on test methods. This was very timely, as in the early days of advanced composites technology there was a tendency to apply metallic materials test methods to fiber-reinforced composites. The heterogeneous, anisotropic nature of composite materials required that many of the test methods borrowed from metallic technology be carefully scrutinized for application to advanced composites.

In the third conference, emphasis began to shift away from the more routine tests and began to focus on such areas as fatigue and creep. The concern over the effects of high humidity on the elevated-temperature properties of epoxy resin matrix composites was also introduced into this conference. This was a prelude to the era of "the moisture problem." Fracture, impact response, and environmental effects were emphasized in the fourth conference. The fifth conference represented a wide spectrum of activity in composite materials. Although a session on environmental effects was presented, the concern over moisture effects was subsiding considerably. Fatigue, time-dependent behavior, and nondestructive evaluation were topics of particular interest in the sixth conference.

The seventh conference reflects the current issue of toughness as related to damage tolerance of advanced composites. Sessions on failure mechanisms, delamination, and strength strongly emphasize toughness. Structural aspects of damage tolerance are also addressed. Many of the papers reflect the maturity to which composites technology has arrived. Most of the easy problems have been solved. Researchers are now faced with characterizing and analyzing the complex failure mechanisms associated with stress concentrations and delaminations. The conference also contains a number of papers on the emerging materials. In particular, multi-dimensionally reinforced materials in the form of woven and braided structure are discussed.

This volume contains 23 of the 31 papers presented at the conference. The five major sections correspond to the conference sessions: Structures, Failure Mechanisms, Strength, Delamination, and Analysis and Characterization.

Since topics of major interest are included in this volume, it is highly recommended to researchers and designers in the field of composite materials. This volume along with the other six volumes from the testing and design conferences provide an excellent background for the beginner who wishes to know details about the development of composite materials technology.

Special thanks go to the session chairman, authors, reviewers, and ASTM Staff who made the conference and this resulting volume possible.

James M. Whitney
Materials Research Engineer, Air Force Wright Aeronautical Laboratories, Wright-Patterson Air Force Base, OH; symposium chairman and editor.

Structures

Ramon A. Garrett[1]

Effect of Manufacturing Defects and Service-Induced Damage on the Strength of Aircraft Composite Structures

REFERENCE: Garrett, R. A., "**Effect of Manufacturing Defects and Service-Induced Damage on the Strength of Aircraft Composite Structures,**" *Composite Materials: Testing and Design (Seventh Conference), ASTM STP 893,* J. M. Whitney, Ed., American Society for Testing and Materials, Philadelphia, 1986, pp. 5–33.

ABSTRACT: This paper describes the effects of manufacturing defects and service-induced damage on the static and fatigue strength of aircraft composite structures.

Seven manufacturing defects associated with mechanical fasteners were investigated; out-of-round holes, delaminations at the exit side of drilled holes, porosity, improper fastener seating depth, tilted countersinks, interference fit, and multiple fastener installation and removal cycles. Both static and fatigue test results are described, along with correlation with analysis techniques. The interaction of the effects of these defects on hole wear, measured in fatigue tests of structural joints, is described.

The effects of two types of service-induced damage are also described; low-energy impact damage and penetration damage. The relative sizes of visible and non-visible damage as determined by visual and nondestructive inspection techniques are compared. An evaluation of stitching and the inclusion of glass or Kevlar fiber buffer strips to improve the damage tolerance of carbon/epoxy structures is included. Results of tests of carbon/epoxy panel structures are discussed. Correlation of experimental results with predicted residual static strength is good.

KEY WORDS: composite materials, composite structures, graphite laminates, epoxy laminates, impact damage, defects, residual strength, static strength, fatigue strength, fatigue (materials), postbuckling

Future aircraft will require airframes that are lighter weight, easier to maintain, and more durable than current construction approaches and materials. The use of composite materials in primary structures offers promise of significant weight savings, due to their greater specific static strength and even larger improvement in fatigue strength. However, proof of the ability of composite structures to be tolerant of both initial manufacturing defects and damage induced by service usage was necessary before their application to aircraft primary structures.

[1] Chief technology engineer, Strength, McDonnell Aircraft Company, St. Louis, MO 63166.

Programs to evaluate the damage tolerance of composite structures have been initiated by several different agencies under the Department of Defense. Each of these programs have been directed towards a different aspect of the subject area; this paper summarizes some of the work performed by the McDonnell Aircraft Company in each area.

The effect of manufacturing defects on the static and fatigue strength of laminates with loaded and unloaded fastener holes was conducted in the Ref *1* program and is described in the next section. The effect of low-energy impact damage on composite structures was included in several programs with the United States Navy (Ref *2, 3,* and *4*). Finally, the effect of penetration damage conducted in conjunction with the Ref *2* program is described. Table 10 is included in the Appendix for converting English Units used throughout this paper to SI Units.

Manufacturing Defects at Fastener Holes

The effects of several types of possible manufacturing defects associated with mechanical fasteners in aircraft structures were investigated under the Ref *1* program. This investigation was performed using laminates with loaded fastener holes, a common critical structural design feature in aircraft composite structures. In many cases, tests were conducted with specimens in which the defect or anomaly was more severe than expected from current manufacturing processes. In all cases, defects that resulted in strength reductions greater than 15% would have been detected using current industry inspection techniques and would have been rejected or repaired to meet current acceptance criteria.

Tests were performed to determine the effects on static strength, fatigue strength, and hole wear.

Static Strength

The effect on static strength was evaluated by comparing static strength of joints with a particular defect with the static strength of specimens with no defect. All static strength test specimens were fabricated from 20-ply laminates, with a per ply thickness of 0.0104 in. (0.208 in. total thickness).

Specimens were tested to failure in tension and compression at three environmental conditions: room temperature dry (RTD), room temperature wet (RTW), and elevated temperature wet (ETW). The ETW tests were conducted at 250°F with specimen moisture contents of approximately 0.80% by weight. Hercules AS/3501-6 carbon/epoxy was used for fabrication of test specimens.

Results are summarized in Table 1. Indicated percentages of increased or decreased strength are based on a comparison with baseline specimens. Detailed results are discussed in the following paragraphs.

Out-of-Round Holes—Effects of out-of-round holes on joint strength were evaluated by drilling a second set of holes offset from the first holes by 0.004 in. as shown in Fig. 1. Test results of specimens from two laminates (50/40/10

TABLE 1—*Strength reduction summary.*

	RTD Tension	Compression RT[a]	Compression 250°F[a]
Out-of-round holes			
50/40/10 laminate	[b]	...	...
30/60/10 laminate	−4.8	...	...
Broken fibers exit side of hole			
Severe	−7.3	−8.4	−9.2
Moderate	−1.4	−3.2	−4.2
Porosity around hole			
Severe	[b]	−10.3	−30.8
Severe with freeze-thaw	...	−11.6	...
Moderate	...	−7.1	−13.3
Moderate with freeze-thaw	...	−8.4	...
Improper fastener seating depth			
80% thickness	−16.4	...	...
100% thickness	−34.3	...	...
Tilted countersinks			
Away from bearing surface	[b]	...	−16.7
Toward bearing surface	−21.4	...	−16.7
Interference fit tolerances (in.)			
50/40/10 @ 0.003	[b]	...	+9.1[c]
@ 0.008	[b]	...	+9.1[c]
30/60/10 @ 0.003	[b]	...	[b,c]
@ 0.008	[b]	...	[b,c]
Fastener removal and reinstallation			
100 cycles	[b]	...	−8.3

[a]Moisture content = 0.86%
[b]Less than 2% change.
[c]Tensile loading
NOTE—Three dots indicate no test. See appendix for SI equivalents.

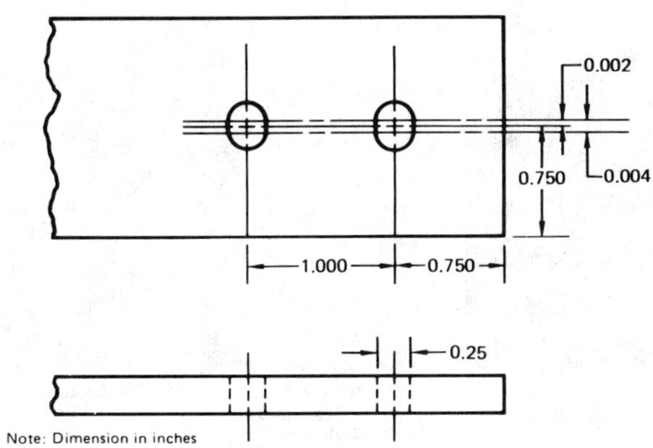

FIG. 1—*Out-of-round holes—specimen.*

and 30/60/10, where %0° plies/%±45° plies/%90° plies are denoted in that order) indicated little sensitivity to this condition.

Delaminations at Exit Side of Hole—Specimens were tested for two conditions; "moderate" delaminations and "severe" delaminations. In order to obtain the various degrees of delamination, drill procedures included use of dull bits without backup material, and improper drill and feed speeds.

Delaminations were detected visually and with ultrasonic C-scan. Delaminations were defined as "moderate" when they extended 10 to 20% (2 to 4 plies deep) into the laminate thickness on the exit side and as "severe" when they extended 20 to 30% (4 to 6 plies). Nondestructive C-scans in the area of the fastener holes with moderate and with severe delaminations are illustrated in Fig. 2 and compared with a nominal condition.

Joint tension strength was evaluated by tests of dry laminates at room temperature. Because of its sensitivity to environment, joint compression strengths were evaluated at RTW and ETW test conditions. Test results are summarized in Table 2 along with baseline strength data.

Porosity—Two levels of porosity were evaluated; "moderate" (1 to 3%) and "severe" (3 to 5%). Desired levels of porosity in the 50/40/10 laminate were obtained by using the altered collation and curing procedures. Specimens were located within panels such that fastener holes occurred in areas of desired porosity levels as indicated by photomicrograph and nondestructive inspections (Fig. 3).

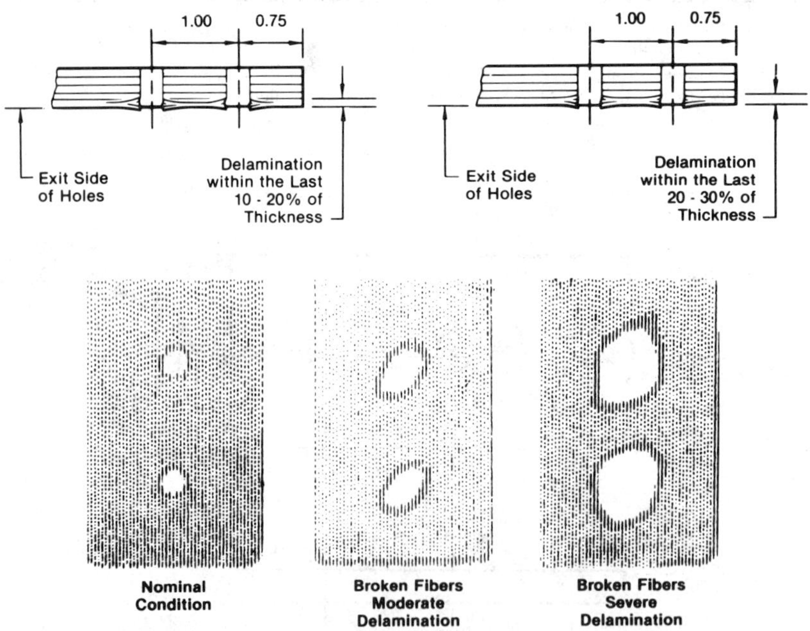

FIG. 2—*C-scans of laminates with delaminations at fastener holes.*

TABLE 2—*Effect of delaminations on joint strength.*

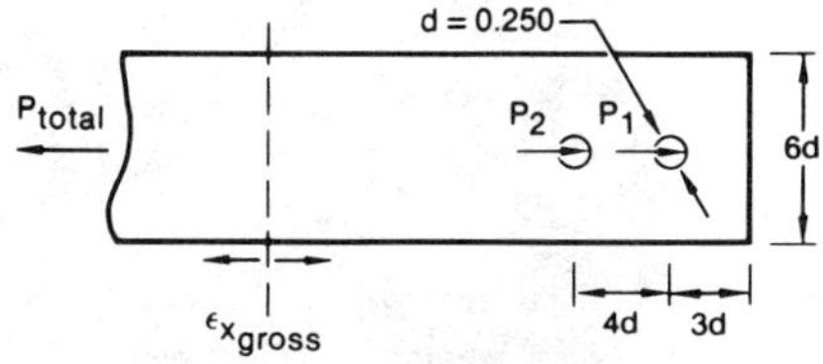

Test Condition (50/40/10 Layup)	Baseline Data F_{x2}^{bru}, ksi	Baseline Data $\epsilon_{x_{gross}}^{u}$, μin./in.	Specimens with Delaminations F_{x2}^{bru}, ksi	Specimens with Delaminations $\epsilon_{x_{gross}}^{u}$, μin./in.	Change, % F^{bru}	Change, % ϵ_{gross}	Loading
			MODERATE				
RTD	+140	+3990	+138	+3790	−1.4	−4.9	tension
RTW	−155	−4740	−150	−4450	−3.2	−6.2	compression
ETW	−120	−3790	−115	−3810	−4.2	+0.6	compression
			SEVERE				
RTD	+140	+3990	+130	+3610	−7.3	−9.5	tension
RTW	−155	−4740	−142	−4580	−8.4	−3.5	compression
ETW	−120	−3790	−109	−3330	−9.2	−12.2	compression

Strengths of baseline specimens and specimens with porosity are compared in Table 3. Little sensitivity to severe porosity was indicated under tensile loading. Under compressive loadings, strength reductions ranged from 7 to 13.3% for specimens with moderate porosity and 10 to 30.8% for specimens with severe porosity. The greatest reductions occurred at 250°F test condition.

Improper Fastener Seating Depth—Effects of excessive countersink depth on joint strength were evaluated by testing composite joint members having fasteners seated too deeply in a typical laminate (50/40/10). Two conditions of countersink depth as shown in Table 4 were evaluated at room temperature in tension.

Resulting joint strengths are compared with baseline strengths in Fig. 5. Strengths for joints with excessive countersink depths (80% and 100%) are compared with strengths of specimens with nominal countersunk depths (52% of laminate thickness). Joint strengths for countersink versus noncountersink laminates indicated that no significant reductions occurred when fastener seating depth was nominal. The relative amount of cylindrical bearing area as compared to countersink bearing area may account for the demonstrated loss in strength. Earlier tests have indicated that the maximum cylindrical bearing capacity is nearly 160 ksi for large edge distances. An analysis of the forces in the region of the countersink indicates an effective bearing capacity of 110 ksi, when friction is accounted for and when sufficient head bearing area still remains. Using these capacities results in predicted reductions of 14% and 30%, to be compared with the demonstrated reductions of 16.4% and 34.3%, respectively.

Tilted Countersinks—Countersink perpendicularity was investigated for two conditions of misalignment. As shown in Fig. 5, the misaligned countersink was

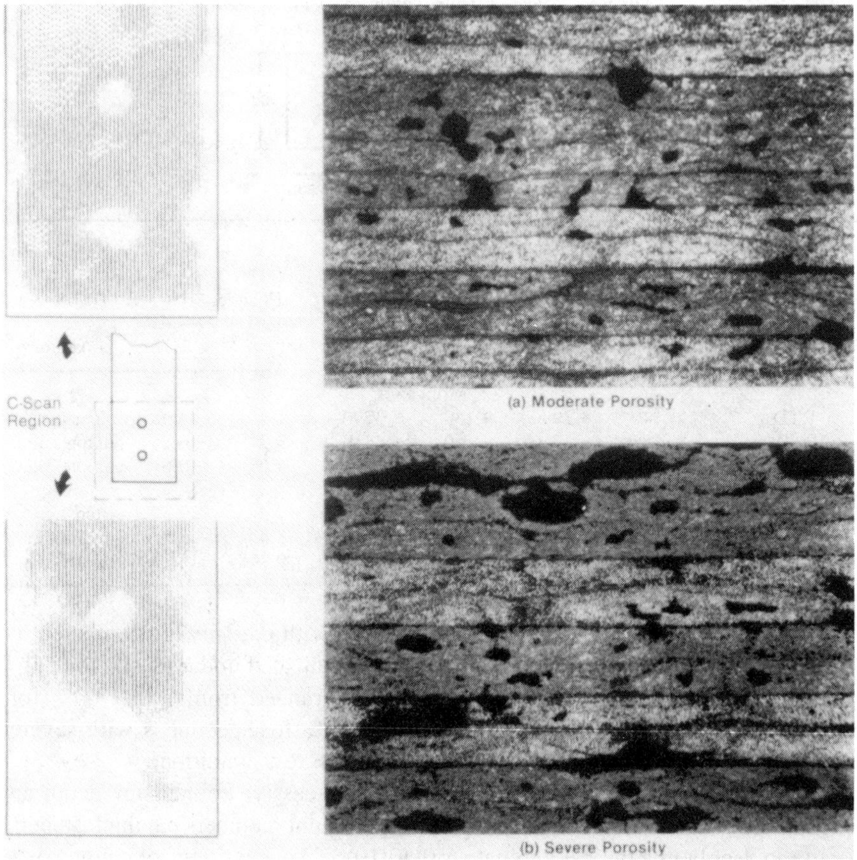

FIG. 3—*Examples of panel porosity.*

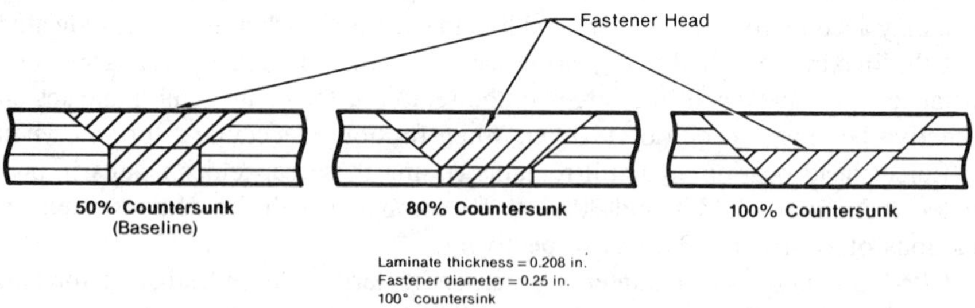

FIG. 4—*Countersink depths evaluated.*

TABLE 3—*Effect of porosity on joint strength.*

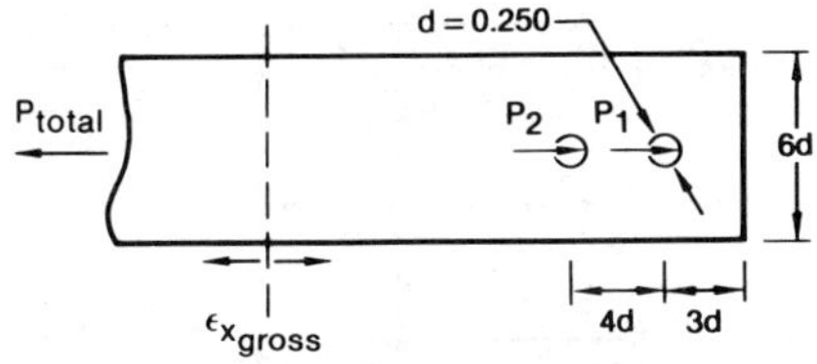

Test Condition (50/40/10 Layup)	Baseline Data F_{x2}^{bru}, ksi	Baseline Data $\epsilon_{x_{gross}}^{u}$, µin./in.	Specimens with Porosity F_{x2}^{bru}, ksi	Specimens with Porosity $\epsilon_{x_{gross}}^{u}$, µin./in.	Change, % F^{bru}	Change, % ϵ_{gross}	Loading
			MODERATE POROSITY				
RTD	+140	+3990	...	...	...	...	tension
RTW	−155	−4740	−144	−4480	−7.1	−5.4	compression
RTW (F-T)[a]	−155	−4740	−142	−4370	−8.4	−7.9	compression
ETW	−120	−3790	−104	−3110	−13.3	−17.9	compression
			SEVERE POROSITY				
RTD	+140	+3990	+140	+3940	0	−1.2	tension
RTW	−155	−4740	−139	−4170	−10.3	−12.1	compression
RTW (F-T)[a]	−155	−4740	−137	−4110	−11.6	−13.3	compression
ETW	−120	−3790	−83	−2550	−30.8	−32.6	compression

[a](F-T) = exposed to freeze-thaw cycles.

TABLE 4—*Effect of countersink depth on joint strength.*

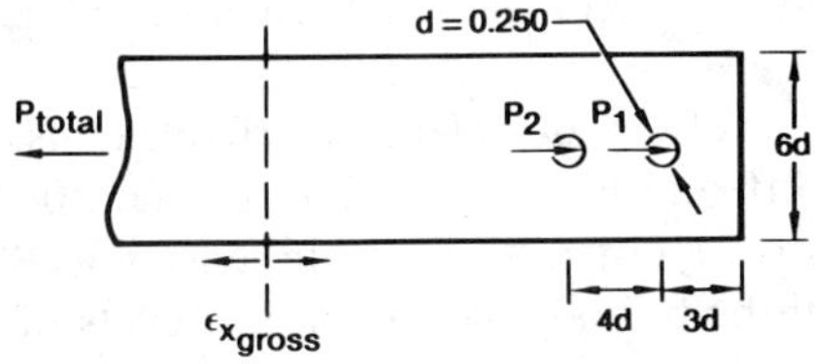

Countersink Depth (50/40/10 Layup) RTD, tension	F_x^{bru}, ksi	$\epsilon_{x_{gross}}^{u}$, µin./in.	Change, % F^{bru}	Change, % ϵ_{gross}
Nominal	140	3990	...	...
80%	117	3240	−16.4	−18.8
100% (knife edge)	92	2540	−34.3	−36.3

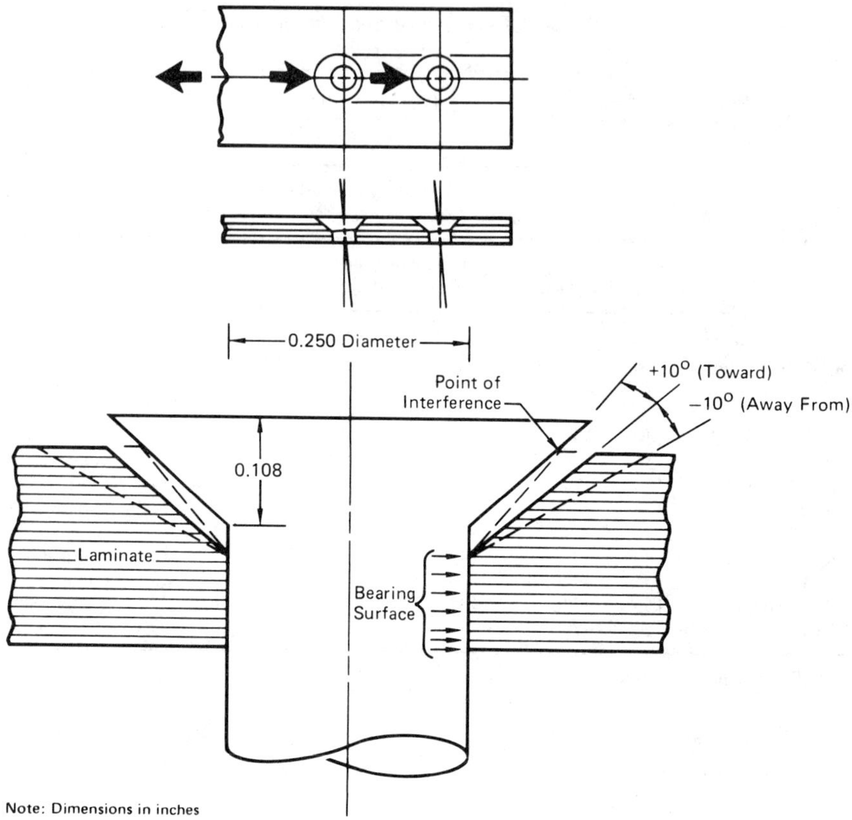

FIG. 5—*Tilted countersink—specimen configuration.*

tilted 10° away from the bearing surface for one condition and tilted 10° toward it for the other. Tests were conducted in tension at RTD and in compression at 250°F after specimen moisture conditioning. Experimental results are summarized in Table 5.

Interference Fit—The effects of fastener interference fits on joint strength were investigated in two different laminates (50/40/10 and 30/60/10). Two-fastener-in-tandem specimens were tested to failure in tension at RTD and ETW conditions. Specimens with both 0.003 and 0.008 in. levels of interference fit were tested.

Results for both layups indicate an insensitivity to interference at room temperature. Joint strength of the more fiber-dominant 50/40/10 layup increased under ETW conditions for both levels of interference. However, joint strengths of the matrix-dominant 30/60/10 layup showed no change at the ETW test condition.

Laminate damage due to fastener installation at interference fits ranging from

TABLE 5—*Effect of tilted countersink on joint strength.*

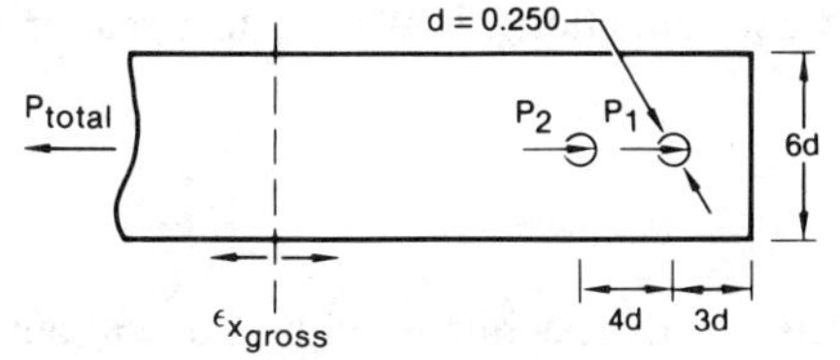

Test Condition (50/40/10 Layup)	Baseline Data $F_{x_2}^{bru}$, ksi	Baseline Data $\epsilon_{x_{gross}}^{u}$, μin./in.	Tilted Countersink $F_{x_2}^{bru}$, ksi	Tilted Countersink $\epsilon_{x_{gross}}^{u}$, μin./in.	Change, % F^{bru}	Change, % ϵ_{gross}	Loading
			AWAY FROM BEARING				
RTD	+140	+3990	+140	+3910	0	−2.0	tension
ETW	−120	−3750	−100	−2990	−16.7	−20.2	compression
			TOWARD BEARING				
RTD	+140	+3990	+110	+3030	−21.4	−24	tension
ETW	−120	−3750	−100	−2898	−16.7	−22.7	compression

0.002 to 0.008 in. were further evaluated for fasteners requiring pull-through installation techniques. Little or no damage resulted from a fastener interference of 0.0035 in. or less; however, damage is indicated at the fastener exit side as well as along the entire fastener length for interference fits from 0.004 through 0.008 in.

Fastener Removal and Reinstallation—These tests were used to evaluate whether repeated installation and reinstallation would locally damage the laminate hole area, or affect joint strength, or both. Fasteners were installed, torqued to 50 in. · lb, and completely removed. This procedure was repeated 100 times prior to strength testing. Specimens were tested to failure in tension at RTD and in compression at ETW.

Visual appearance of all fastener holes after installation and reinstallation cycling was unchanged. Strength data listed in Table 6 indicates little sensitivity

TABLE 6—*Effect of fastener removal and reinstallation on joint strength.*

Test Condition (50/40/10 Layup)	Baseline Data $F_{x_2}^{bru}$, ksi	Baseline Data $\epsilon_{x_{gross}}^{u}$, μin./in.	After 100 Cycles Fastener Installation $F_{x_2}^{bru}$, ksi	After 100 Cycles Fastener Installation $\epsilon_{x_{gross}}^{u}$, μin./in.	Change, % F^{bru}	Change, % ϵ_{gross}	Loading
RTD	+140	+3990	+140	+3900	0	−2.2	tension
ETW	−120	−3790	−110	−3510	−8.3	−7.4	compression

to RTD tension test conditions. Compressive strength values indicated an increased sensitivity (8.3% reduction); however, test scatter for the compression tests was large.

Fatigue Strength and Hole Wear

Tests were conducted with specimens with and without internal porosity to determine the effects on joint fatigue life, hole wear, and failure modes. Emphasis was placed on generation of hole wear data and its relation to joint fatigue life.

A pure bearing, 0.375-in.-diameter, single fastener test specimen was used. Tension-tension ($R = +0.1$) and tension-compression ($R = -1.0$) constant-amplitude testing was performed at room temperature with specimens in the as-manufactured condition. Hercules AS/3501-6 carbon/epoxy was also used for fabrication of all fatigue and hole wear test specimens.

All constant-amplitude fatigue specimens were cycled to failure, or 10^6 cycles, whichever occurred first. Specimens that did not fail in 10^6 cycles were tested to determine residual strength. Constant-amplitude fatigue testing was performed at three stress levels for each specimen type. Selection of the stress levels for fatigue testing was based on load-deflection data obtained from static tests. During fatigue testing, load-deflection data were also obtained each time a specified hole wear level was reached.

The evaluation included three layups; the fiber-dominated layup 50/40/10 (20-ply laminate, 0.0104 in. per ply), and two matrix-dominated layups: 19/76/5 (21-ply laminate, 0.0104 in. per ply) and 30/60/10 (20-ply laminate, 0.0104 in. per ply).

Residual strengths were, in general, equal to or greater than nonfatigued specimen static strengths; however, in most cases, these specimens had acquired hole wear of 0.02 in. or greater during fatigue testing. For structural applications, hole elongations of 0.02 in. exceed the usual yield criteria for metallic joints that may also represent a tentative criteria for composite joints.

Typical results of tension-tension ($R = +0.1$) and tension-compression ($R = -1.0$) cyclic loading for the 50/40/10 laminate at room temperature, dry (RTD) test conditions are summarized in Fig. 6 in terms of fatigue cycles required to produce an 0.02-in. hole wear in the fastener hole. Comparative results for all layups for tension-tension ($R = +0.1$) cycling, is summarized in Fig. 7. For tension-compression ($R = -1.0$), the 19/76/5 and 30/60/10 matrix-dominant layups sustained fewer load cycles prior to developing an 0.02-in. hole wear, compared to the 50/40/10 layup (Fig. 8).

Tests of specimens with moderate porosity were conducted to evaluate the effects of this anomaly on joint durability. Earlier static tests indicated that moderate levels of porosity had a minor effect on static joint strength at room temperature. Specimens with moderate porosity in regions of fastener holes were tested under $R = +0.1$ and $R = -1.0$ fatigue loadings at RTD conditions. No reduction of static strength or joint fatigue life was indicated.

The rate of hole wear in other tests of composite joints without porosity at

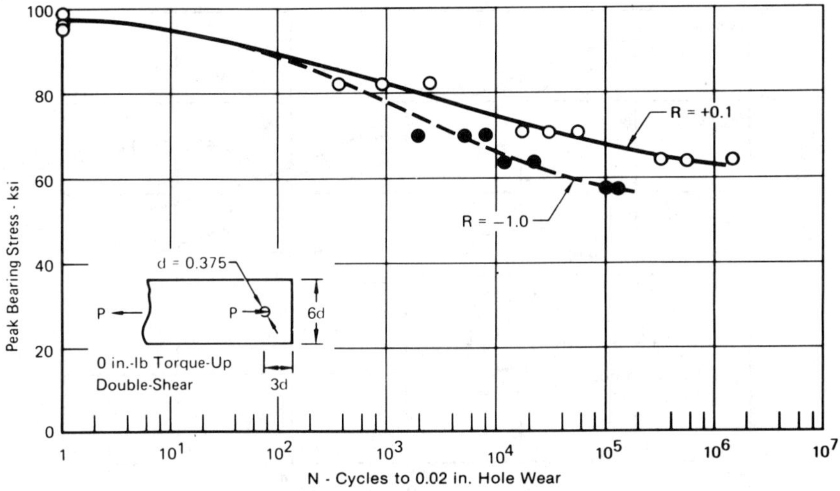

FIG. 6—*RTD baseline joint fatigue life.*

$R = +0.1$ was investigated in terms of number of cycles to produce three levels of wear (0.005, 0.010, and 0.020 in.) These data indicate that the matrix-dominant 19/76/5 layup exhibited earliest initiation of hole wear, but had the most gradual rate of accumulation. Conversely, the fiber-dominant (50/40/10) layup exhibited the most-delayed initiation of hole wear, but had the most rapid accumulation. It was noted that the drop-off of joint spring rates of test specimens with the 50/40/10 correlated with the initiation of hole wear. The 30/60/10 layup exhibited an intermediate performance.

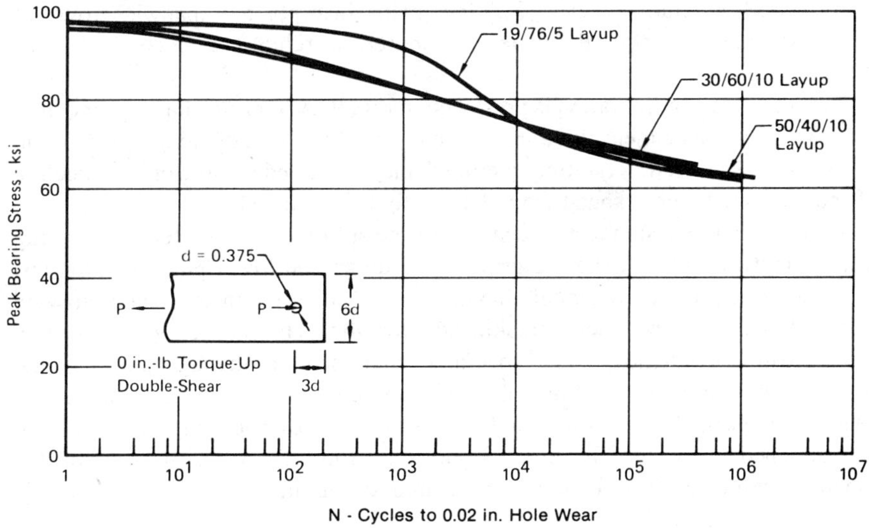

FIG. 7—*Comparison of* R = +*0.1 joint fatigue life.*

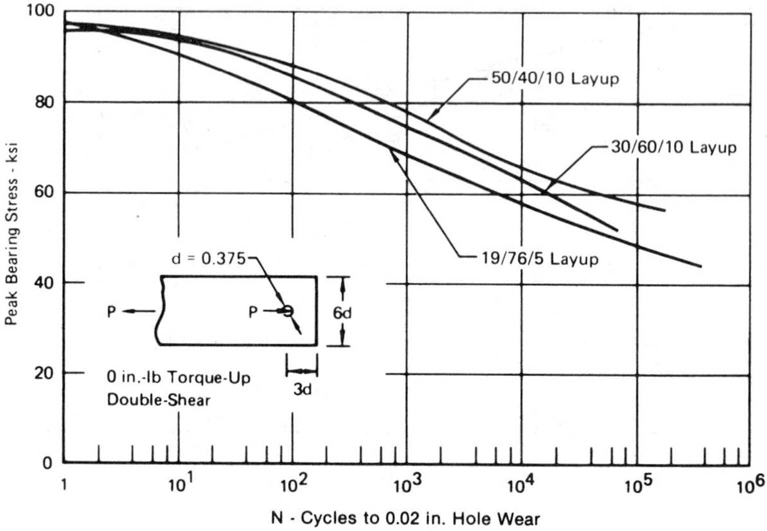

FIG. 8—*Comparison of* R = −*1.0 joint fatigue life.*

Low-Energy Impact Damage

The effect of low-energy impact damage on the static and fatigue strength of carbon/epoxy wing-cover-skin structures and of integrally stiffened panels typical of postbuckling fuselage structures was evaluated.

Wing-Cover-Skin Structural Panels

Effects of low-energy impact to an upper wingskin were evaluated in static compression and fatigue tests [*2*]. Specimens incorporated spanwise rows of Kevlar stitches simulating the patterns proposed for reinforcing cocured skin-to-stiffener joints.

The specimen configuration is presented in Fig. 9. A 0.50-in.-diameter indenter was used while specimens were supported over a 3-by-3-in. opening. An 8-ft · lb energy level was used. Resulting internal damage detected in ultrasonic inspection of the four specimens ranged from 1.2-in. to 1.5-in. diameter.

For these compression tests, each end of the specimen was bolted to a loading adapter that fit within hydraulic grips in the upper and lower platens of the test machine. Back-to-back channels having access holes for instrumentation were clamped onto the specimen for skin stabilization. The area between the two central rows of stitches in the specimen contained the damage and was not stabilized by the channels. The column composed of the specimen with loading adapters on each end was supported at two locations by additional fixturing that was attached to the test machine. Strain data from back-to-back gages on the specimen indicated little bending. Measured strains in the stabilizing channels were negligible.

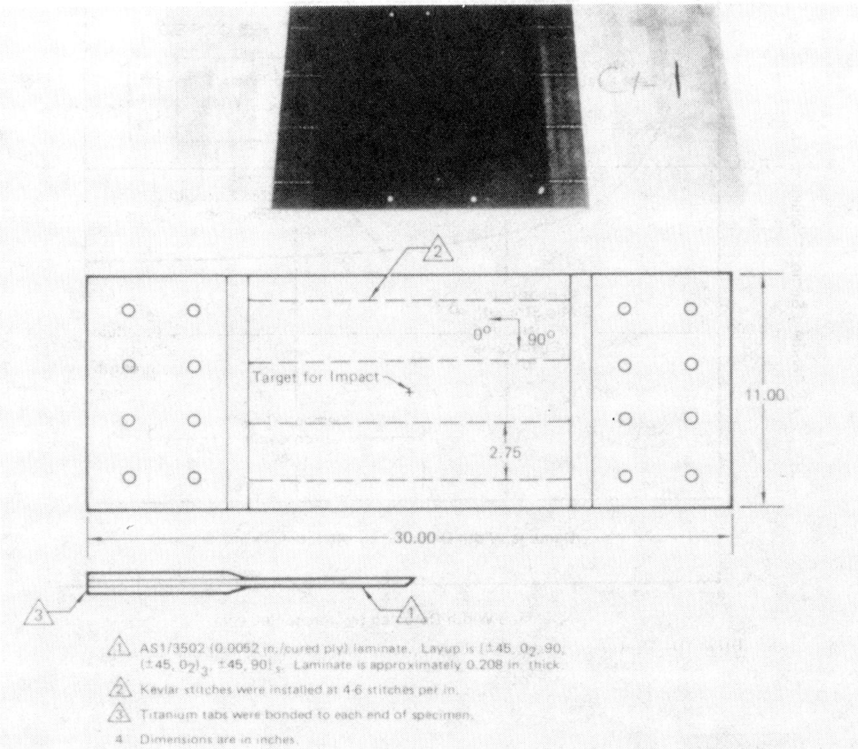

FIG. 9—*Specimen configuration—skins with nonvisible impact damage.*

Results of static compression tests indicated that stitches were not sufficient to prevent damage propagation and overall failure at strain levels above −4500 μin./in. Strain data indicate local bending in the damaged area at low loads, possibly contributing to the failures.

A fatigue test of a fourth specimen was conducted to assess damage containment features of the stitch pattern. A compression-dominated spectrum was used. Damage detected in ultrasonic inspection was initially 1.6-in. long and 1.4-in. wide but grew to 1.7-in. long and 2.7-in. wide after 24 000 equivalent flight hours of spectrum loading. This specimen was tested at a reduced strain level, and damage was contained by parallel rows of stitches that were spaced at 2.75 in. In residual strength tests of this specimen, the far-field strain at failure was −4200 μin./in.

These test results are summarized in Fig. 10 in terms of far-field failure strains as a function of damage sizes detected in ultrasonic inspections. Test results for coupons with a 0.25-in. diameter hole are also shown in Fig. 10. The predicted strengths, presented as a solid line, were determined using the methodology of Ref. *1*. These predictions correlate well with test results for specimens having a

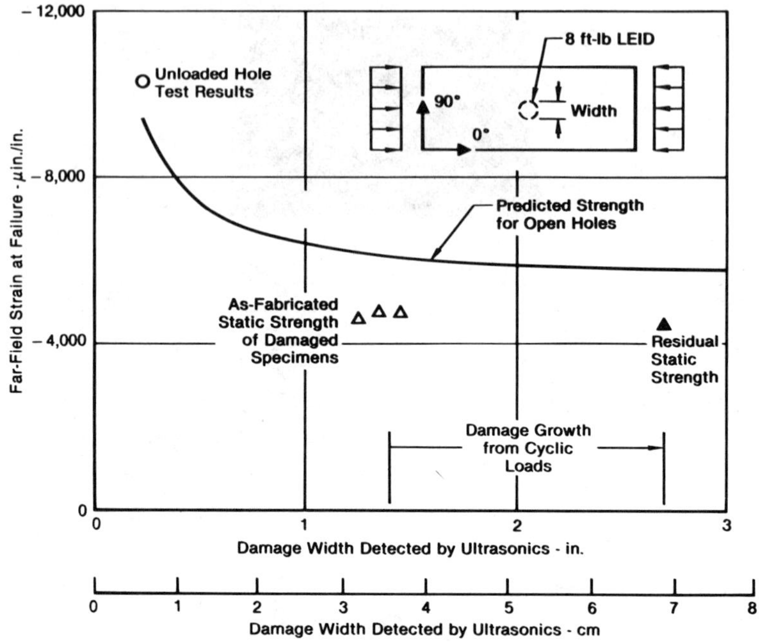

FIG. 10—*Residual strength of specimens with nonvisible impact damage.*

0.25-in.-diameter hole. Predictions for specimens containing impact damage were made for damage modeled as open round holes.

Test results indicate that impact damage produced an effective strain concentration greater in magnitude than a round hole of equivalent size. Strengths predicted for an equivalent hole size were unconservative by approximately 30%, possibly due to local structural instability of delaminate plies within the damaged zone.

Test results are summarized in Table 7, where strain data are shown for the limit load level used for the fatigue test (−20 000 lb) and for the failure load levels. Sizes of nonvisible impact damage determined by ultrasonics are also shown. No fatigue failure occurred in four lifetimes of spectrum fatigue loads.

Integrally Stiffened Postbuckling Fuselage Panels

The effects of low-energy impact on the structural integrity of two types of fuselage panels were evaluated: fuselage compression panels and fuselage shear panels.

Fuselage Compression Panels—Tests were performed to evaluate the effectiveness of Kevlar stitches for containing nonvisible low-energy impact damage in cocured skin-stiffener joints of buckled composite panels loaded in compression.

The baseline behavior of curved stiffened panels under compression postbuckling loads was determined in a previous test program [*3*]. Fatigue failures were precipitated by local disbonds occurring in skin-stiffener joints.

Cyclic load tests were conducted on a curved stiffened panel identical to those previously tested except that each cocured joint was reinforced with two rows of Kevlar stitches. Low-energy impacts were made to produce internal damage in two areas of high peel stress areas where disbonds had initiated in the earlier fatigue tests. The damage was produced by a spherical indenter with 10-ft · lb impact energy.

Disbonds were detected and growth was monitored by periodic ultrasonic inspections. C-scan inspection records in the region of the center stiffener are shown in Fig. 11, where sound attenuation occurring from disbonds and from air trapped under hat stiffeners is indicated by a darkened area. Disbonds in unstitched panels initiated and grew with increasing load cycles. An identical panel with stitching survived 1 000 000 cycles and showed only minor evidence of disbonding.

The damage containment capability of Kevlar stitches was significant. An unstitched, undamaged baseline panel suffered failure after 3500 load cycles where the peak load was −55 kips. A stitched, undamaged panel was cyclically loaded without failure for 280 000 cycles to −50 kips followed by an additional 280 000 cycles to −55 kips. This test sequence using a stitched, damaged panel was repeated; again the reinforced panel survived both groups of 280 000 load cycles. In addition, ultrasonic inspections conducted after each block of 140 000 cycles indicated good containment of the impact damage with no disbond growth. Reinforcement with Kevlar stitches was, in this case, an effective means of assuring durability of cocured joints having nonvisible impact damage.

Fuselage Shear Panels—Tests were also performed to determine the effect of low-energy impact damage on the static and fatigue strength of integrally stiffened shear-loaded panels operating in the postbuckling regime. The as-manufactured ultimate strength of the baseline test panel was 829 lb/in. A typical panel under postbuckling loads is shown in Fig. 12. Maximum mid-panel lateral deflection for the static panels was in order of 0.2 in.

The panel configuration and the locations of low-energy impact damage is shown in Fig. 13. The panel skin was a 7-ply laminate, with thickness per ply or 0.0052 in. (t_{total} = 0.0364 in.). An impact apparatus was designed with a slotted metal impact tube to guide the impact weight to the desired impact point. The ½-in.-diameter round ball impact tool rested against the panel and was centered inside the impact tube by a tool guide. All damaged panels were impacted on the skin side.

Panels were first impacted in the center bay using decreasing energy levels starting at 10 ft · lb. The threshold level at which full penetration was achieved was in the range of 4.0 to 4.5 ft · lb. Impact energy levels up to 4.0 ft · lb have been estimated for fuselage lower surface for foreign object damage such as ice and gravel impacts during landing and take-off and for fuselage sides and corners for ground handling impacts from objects such as tools [*5*].

TABLE 7—*Residual strength of specimens with nonvisible impact damage.*

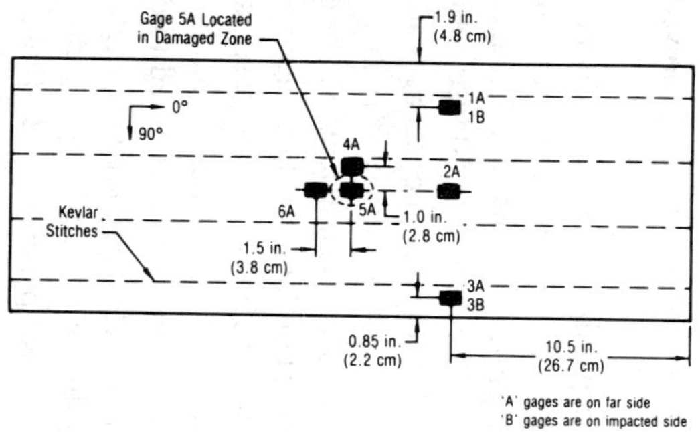

Typical Strain Data[a]

Event[b]	Load, kips (kN)	Specimen Number	Strain Gage[d]							
			1A	1B	2A	3A	3B	4A	5A	6A
Low-load strain survey		C1-2	−809	−809	−952	−939	−976	...	...	...
		C1-3 (0 SFH)[c]	−684	−670	−914	−1044	−1027	−876	1125	−921
	−20 (−89)	C1-3 (24k SFH)	−794	−785	−878	−930	−918	−866	F[e]	−910
		C1-4	−713	−906	−838	−782	−775	−918	−1119	−882
		C1-5	−936	−858	−860	−789	−771	...	...	...

Strains at maximum fatigue load		C1-2	−2916	−2922	−3063	−2924	−3010	...	...	...
		C1-3 (0 SFH)	−2607	−2532	−2953	−3262	−3208	−3092	−5310	−3096
	−64.5 (−287)	C1-3 (24k SFH)	−2914	−2874	−2912	−2957	−2923	−2658	F	−3162
		C1-4	−2346	−3472	−2571	−2793	−2688	−3333	−5544	−3015
		C1-5	−3248	−3182	−2209	−2551	−2471	...	...	...
Ultimate strength	−100 (−446)	C1-2	−4735	−4728	−4765	−4651	−4820	...	...	...
	−94 (−416)	C1-3 (24k SFH)	...	...	...	...	...	...	...	...
	−100 (−466)	C1-4	−3817	−5276	−3656	−4493	−4034	F	F	−4524
	−96 (−426)	C1-5	−4691	−4606	F	−3854	−3746	...	...	...

[a]Strains are in units of μin./in.
[b]All tests conducted at room temperature.
[c]Specimen C1-3 tested in spectrum fatigue to an equivalent of 24 000 Spectrum Flight Hours (SFH).
[d]Specimens C1-2 and C1-5 did not have strain gages 4A, 5A, and 6A. No ultimate strains recorded for Specimen C1-3 because of instrumentation malfunction.
[e]F = Gage failure.

Size of Internal Damage[a]

Specimen	L, in. (cm)	W, in. (cm)
C1-2	1.57 (3.99)	1.35 (3.43)
C1-3 (0 SFH)	1.57 (3.99)	1.36 (3.45)
C1-3 (24k SFH)	1.70 (4.32)	2.70 (6.86)
C1-4	1.79 (4.55)	1.43 (3.63)
C1-5	1.78 (4.52)	1.25 (3.18)

[a]Dimensions given are pretest damage size except where noted.

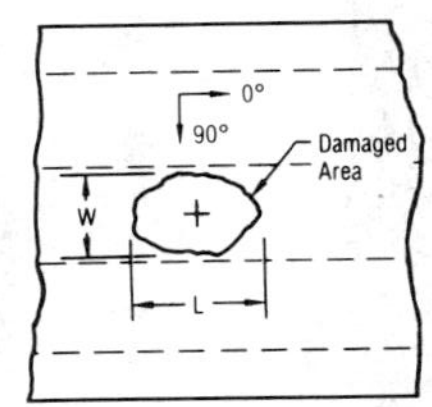

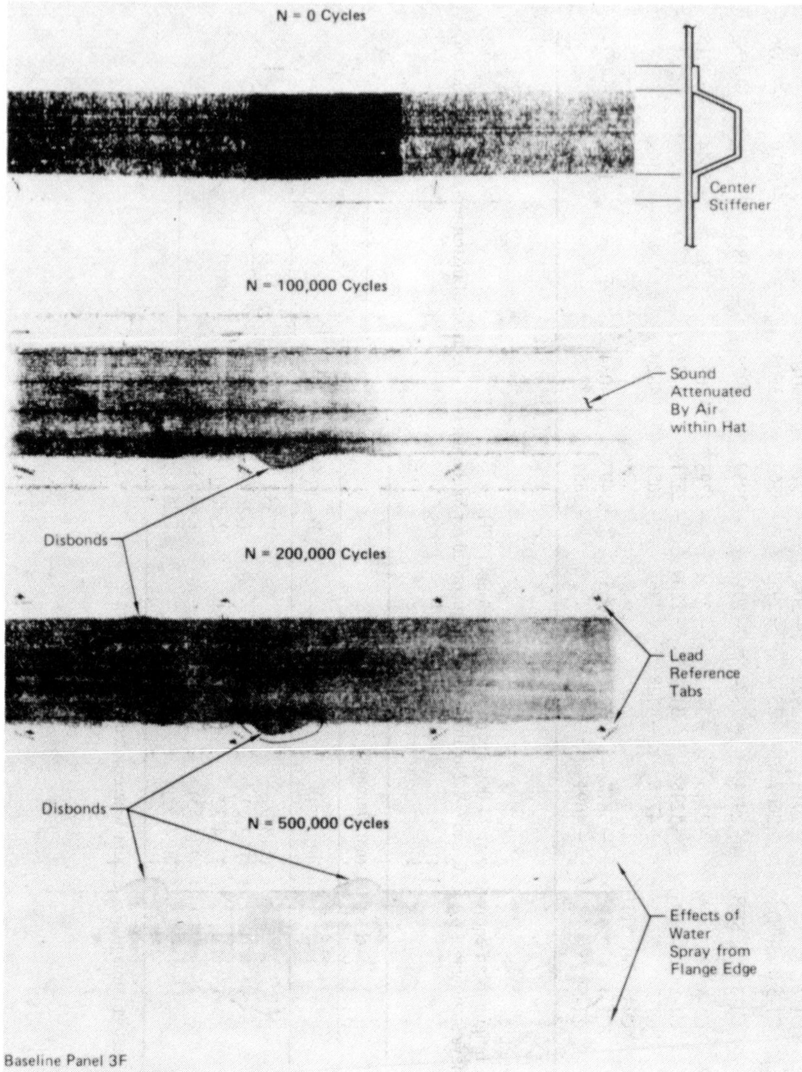

FIG. 11—*Growth of disbonds in flange-to-skin joint.*

A test panel after being subjected to multiple strikes in the range of 4.0 to 4.5 ft · lb is shown in Fig. 14. It was observed that while damage on the skin outer surface appeared different at different energy levels, the damage on the inner surface, in the form of delamination of the outer ±45° ply, was similar for all strikes. A second impacted panel showed similar behavior.

Enhanced radiographic inspections of the strikes at 4.25 and 4.3 ft · lb are shown in Fig. 15. An energy level of 4.30 ft · lb was adequate to achieve broken

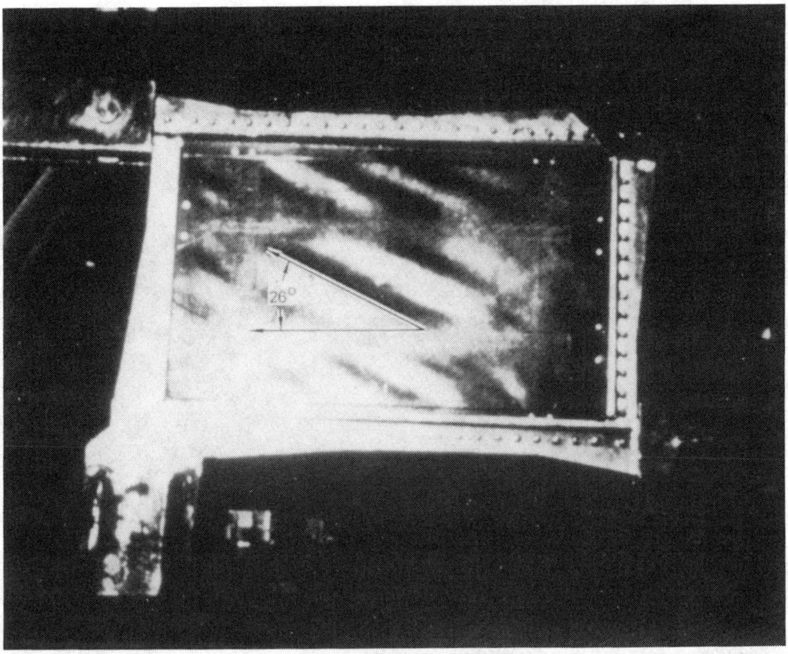

FIG. 12—*Buckled shear panel.*

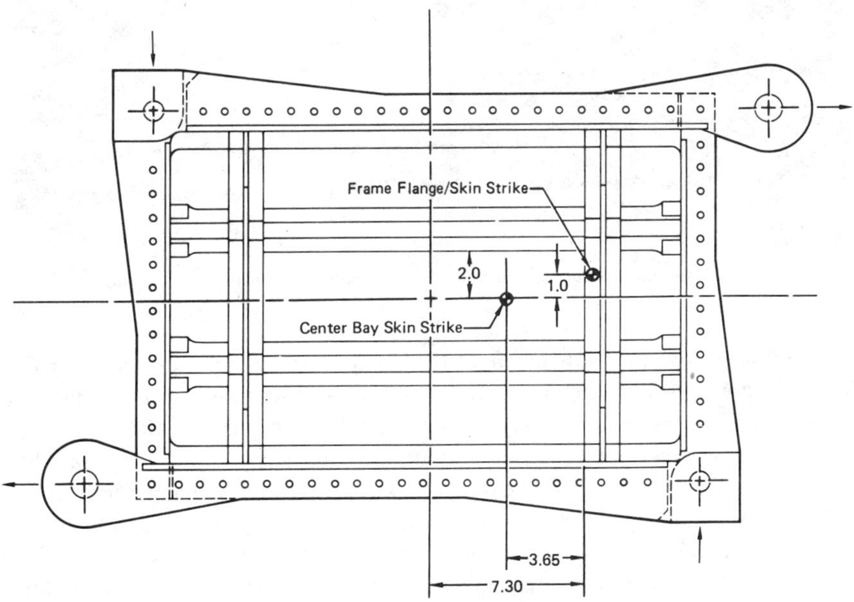

FIG. 13—*Shear panel configuration and impact damage locations.*

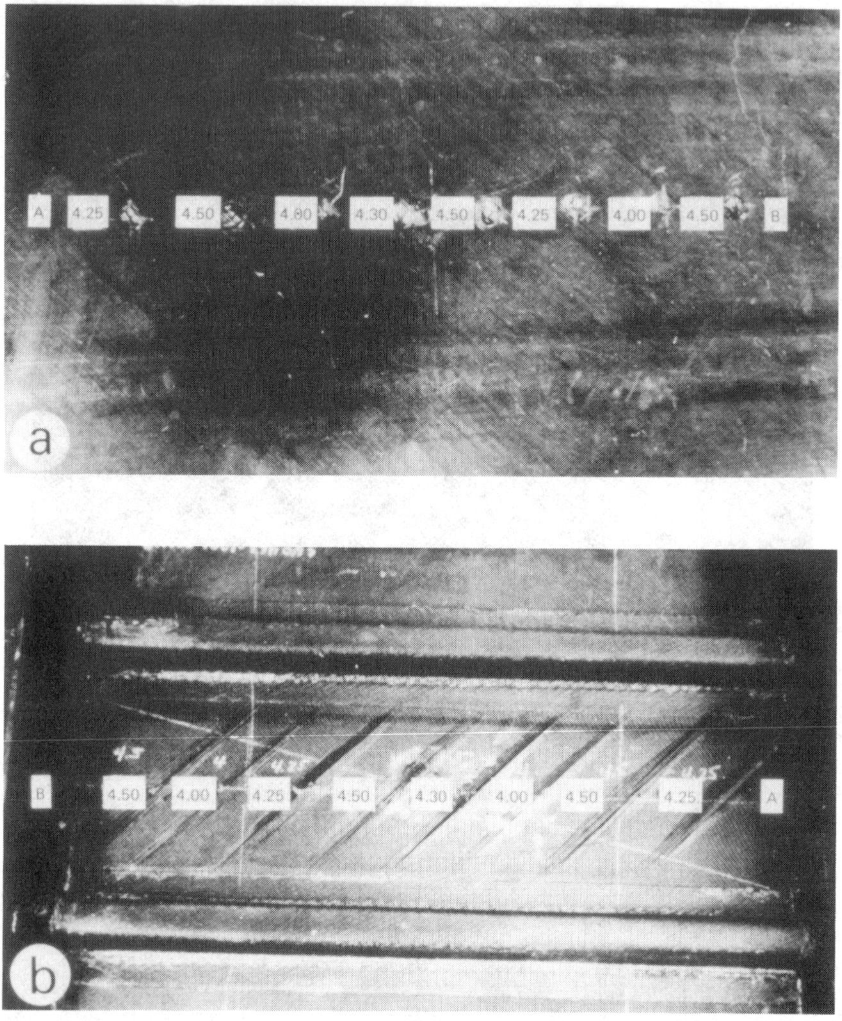

FIG. 14—*Impact damage investigation;* (a) *skin side and* (b) *stiffener side.*

fibers across full ½-in. diameter. Internal damage was diamond shaped, probably due to 0, ±45, 90 ply orientations. Through-the-thickness damage as indicated by radiographic inspection was similar for both energy levels, although visual appearances were different. Delamination on the inner surface of the skin (stiffener side) in the outer +45° ply extended to the adjacent hat flange (both sides).

Five additional panels were damaged using an energy level of 4.30 ft · lb. Three panels were impacted at a buckle crest location in the center bay of the panel. The remaining two panels were impacted at a location where failures occurred during fatigue loading of the baseline panels. These impact locations are shown in Fig. 13. All panels were impacted on the skin side.

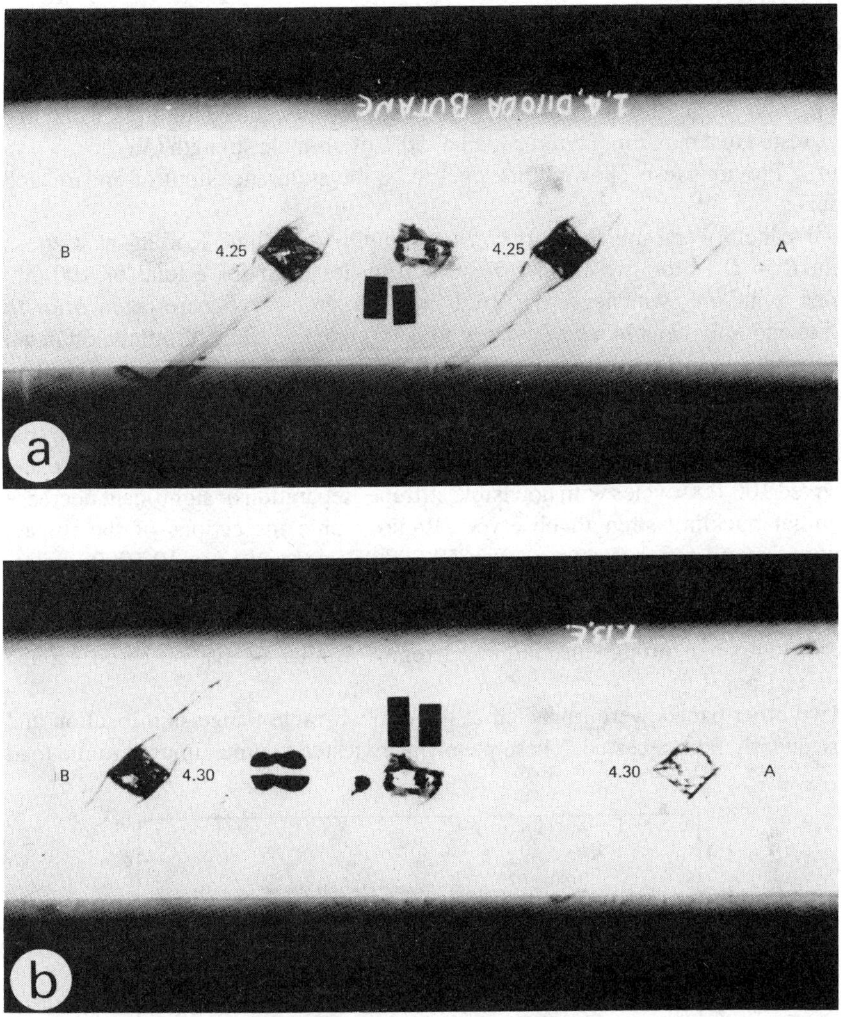

FIG. 15—*Radiographic inspection;* (a) *4.25 ft · lb and* (b) *4.30 ft · lb.*

Of the three panels impacted in the center bay, one panel was statically tested to failure; the remaining two panels were fatigue tested at a load level to preclude stiffener disbonding. The two panels impacted at the critical stiffener/skin interface regions were fatigue tested to a load that produced stiffener disbonding in the baseline panels.

The panel for static testing was subjected to a center bay impact of 4.30 ft · lb. The degree and type of damage was similar to that obtained in the exploratory tests (Figs. 14 and 15). Initial buckling for this panel occurred at 104 lb/in. and panel static failure occurred at 771 lb/in., compared to an average of 829 lb/in. for the undamaged panels. Strain data was similar to that obtained in tests of undamaged panels; however, the maximum strain magnitudes achieved were less due to the lower failing load. This panel failed across the tension diagonal through the impacted region, similar to baseline static test panels.

The two fatigue test panels that were subjected to center bay impact damage were tested to a maximum fatigue load of 50% of ultimate strength (Nxy^{max} = 415 lb/in.). Previous tests showed this level to be the endurance limit of undamaged panels.

All panels were subjected to constant-amplitude fatigue loading at a stress ratio, $R = 0.1$, for two blocks of 50 000 cycles each for a total of 100 000 cycles or failure, whichever occurred first. Strain surveys were taken prior to testing and after each block of cycling to determine the effect of fatigue on panel performance. Panels surviving 100 000 cycles were subjected to a residual strength test. A summary of all fatigue results is presented in Table 8 and in Fig. 16.

The impact damage sustained by the first fatigue test panel was similar to that observed both in the exploratory tests and in the static test panel. This panel survived 100 000 cycles with no visible stiffener separation or significant decrease in initial buckling strength observed. Radiographic inspections of the impact region prior to fatigue testing, after 50 000 cycles, and after 100 000 cycles, indicated that the damaged region did not grow during fatigue loading. Residual strength for the panel was 778 lb/in. with sheet rupture occurring across the tension diagonal through the impacted region similar to that observed for the static test panel.

Two other panels were impacted at the critical frame flange/skin location and subsequently fatigue tested. These panels were tested to a maximum fatigue load

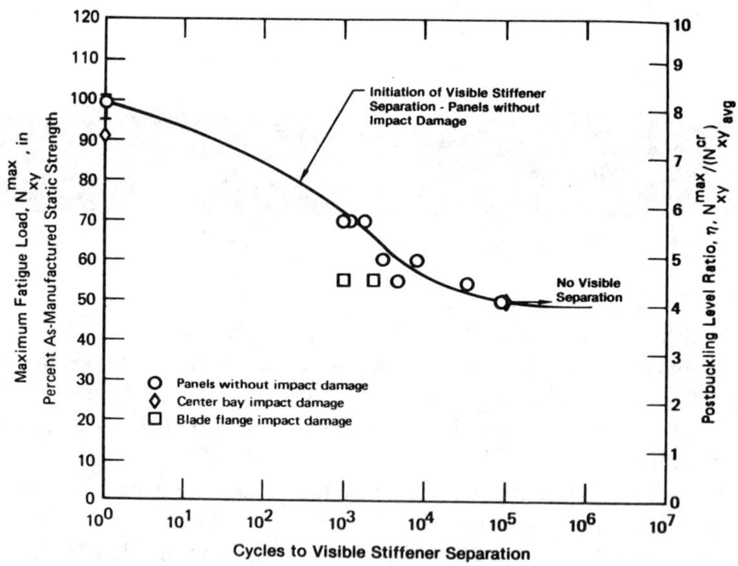

FIG. 16—*Panel fatigue performance.*

TABLE 8—*Summary of test results.*

Panel	N_{xy}^{cr} 1 Cycle, lb/in.	Static Strength, lb/in.	N_{xy}^{max} Fatigue, lb/in.	$\eta = \frac{N_{xy}^{max}}{(N_{xy}^{cr})_{avg}}$	Cycles to Visual Disbonding	Cycles Completed	N_{xy}^{cr} 50 000 Cycles, lb/in.	N_{xy}^{cr} 100 000 Cycles, lb/in.	Residual Static Strength, lb/in.	Comments
1	98	851	...	...	1	1	...	...	...	As-manufactured static test panels
2	80	797	...	...	1	1	...	...	...	
3	93	838	...	...	1	1	...	...	...	
4	97	...	580	5.69	2 100	94 700	...	...	...	
5	102	...	580	5.69	1 000	50 000	...	...	...	
6	96	...	580	5.69	1 200	50 000	...	...	849	
7	99	...	497	4.87	7 500	50 000	38	...	...	As-manufactured fatigue test panels
8	108	...	497	4.87	3 000	50 000	48	...	...	
9	112	...	415	4.07	100 000[a]	100 000	112	112	947	
10	77	...	415	4.07	100 000[a]	100 000	30	30	973	
11	135	...	456	4.47	33 000	100 000	114	90	860	
12	78	...	456	4.47	4 500	100 000	48	48	890	
13	104	754	...	...	1	1	...	...	...	Impact damage—center bay skin 1 static panel 2 fatigue panels
14	102	...	415	4.07	100 000[a]	100 000	102	98	858	
15	124	...	415	4.07	300[b]	100 000	73	65	778	
16	115	...	456	4.47	950	100 000	63	54	918	Impact damage—blade flange/skin 2 fatigue panels
17	120	...	456	4.47	2 300	100 000	77	77	861	

[a]Visual disbonding not observed after 100 000 cycles.
[b]Porosity noted prior to test.

of 55% of ultimate strength to assure frame flange/skin separation. Both panels survived 100 000 cycles with visible stiffener separation occurring at the impact location at 950 and 2300 cycles, respectively. Baseline panels tested to this level separated at 4500 and 33 000 cycles, respectively.

The frame flange/skin separation in the impacted region became extensive for both panels during the first 50 000 cycles. The frame flange along the impacted side of the center bay became nearly totally separated from the skin, causing a shift in the center bay buckling mode. Mid-panel strain response indicated that the majority of separation occurred during the first 50 000 cycles. Buckling strengths of 115, 63, and 53 lb/in. prior to fatigue testing and after 50 000 and 100 000 cycles, respectively, also indicate that the majority of damage occurred during the first 50 000 cycles. Residual strengths for both impacted panels was greater than the baseline panels, with similar sheet rupture across the tension diagonal.

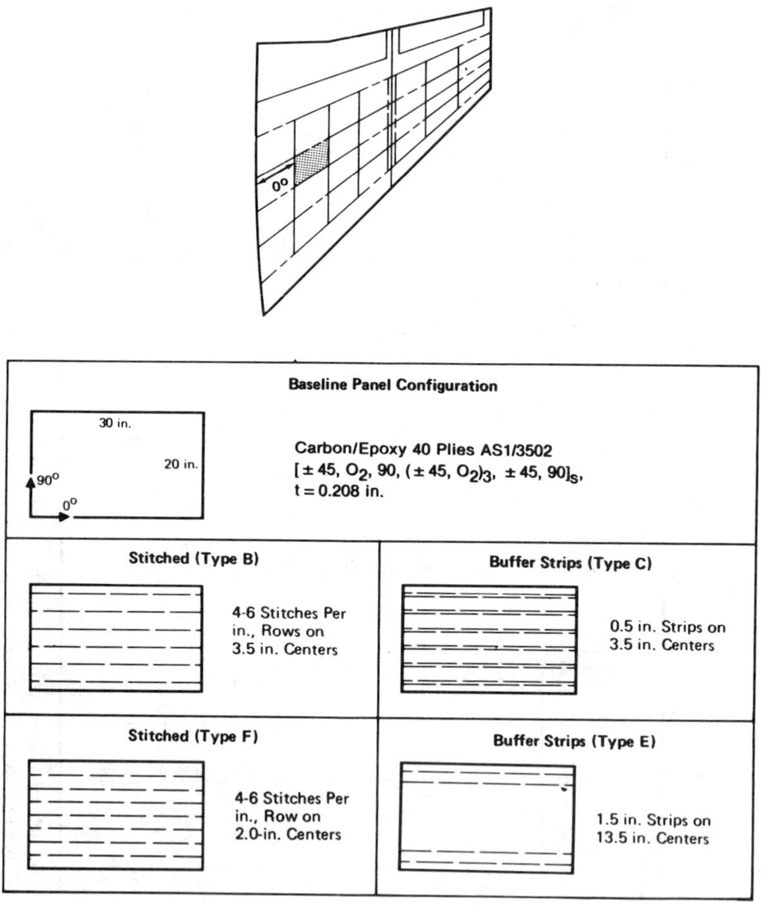

FIG. 17—*Penetration test specimens.*

Penetration Damage

Tests and analyses were performed to evaluate the effects of penetration damage on the residual strength of composite structure incorporating various damage containment features. Flat, unstiffened, carbon/epoxy specimens representative of monolithic wing skins were damaged. Good correlation was obtained between measured residual strengths and predictions made using the maximum strain failure criterion.

Three damage containment features were evaluated. These features, Fig. 17, included parallel rows of Kevlar stitches, closely-spaced parallel rows of imbedded glass/epoxy buffer strips, and wider-spaced parallel rows of wide glass/epoxy buffer strips.

The stitched specimens utilized Kevlar thread having a breaking strength of 120 lb and installed at four to six stitches per inch. Improved durability and damage containment were previously demonstrated in structures where cocured skin-to-substructure joints were reinforced with Kevlar stitches. In the specimens that incorporated glass/epoxy buffer strips, the 0° plies of graphite/epoxy were locally replaced, through the thickness, with 0° plies of glass/epoxy on either 3.5-in. centers or 13.5-in. centers.

Panel Penetration

Damage ranged from a small, relatively clean, hole to a large-diameter hole surrounded by delaminated plies. Delaminated areas were detected through ultrasonic inspections.

Ultrasonic inspection records presented in Fig. 18 illustrate the extent of delamination in specimens incorporating the various containment features. Relative to baseline composite specimens, delaminations in stitched specimens were limited in width to the rows of stitches adjacent to the penetration. Ability of stitches to contain delaminations has been noted in other investigations. Damage in specimens incorporating buffer strips was also limited in width by the imbedded strips; however, some additional laminate damage was noted along the strips.

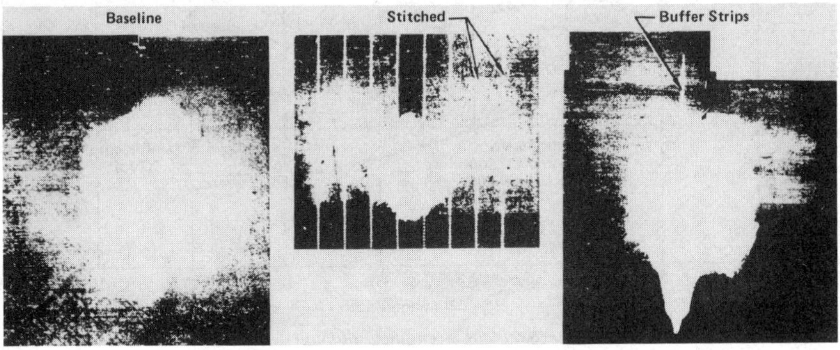

FIG. 18—*Delamination damage detected in ultrasonic inspections.*

The residual tensile strength of damaged specimens was determined in room-temperature static tests and correlated with analytic predictions. Specimens with buffer strips, particularly the wider strips at spar locations, exhibited significant improvement in strength relative to baseline specimens. Stitching had no effect on residual strength.

Residual strengths are tabulated in Table 9, along with the maximum visible damage and the maximum delamination damage detected by ultrasonics. Static failures were sudden with little time after failure initiation, except for those panels with buffer strips. Wide buffer strips stopped cracks from propagating across the width and turned the cracks lengthwise to propagate along the imbedded strips. Narrow, closely spaced buffer strips generally slowed to the propagation across the width.

Residual tensile strengths are presented in Fig. 19 in terms of strain-to-failure for various damage sizes. Strain-to-failure was calculated using the gross cross-sectional area. The damage presented for each specimen ranges from the maximum visible damage to the internal damage detected with ultrasonics. Specimens with buffer strips (shaded) exhibited significant improvement in residual strength relative to strength of baseline specimens (unshaded) having similar damage.

Two analytical procedures for predicting residual strength were evaluated. All predicted strengths were corrected for a finite specimen width (20 in.). Linear elastic fracture mechanics analysis techniques [6] were used to predict a lower

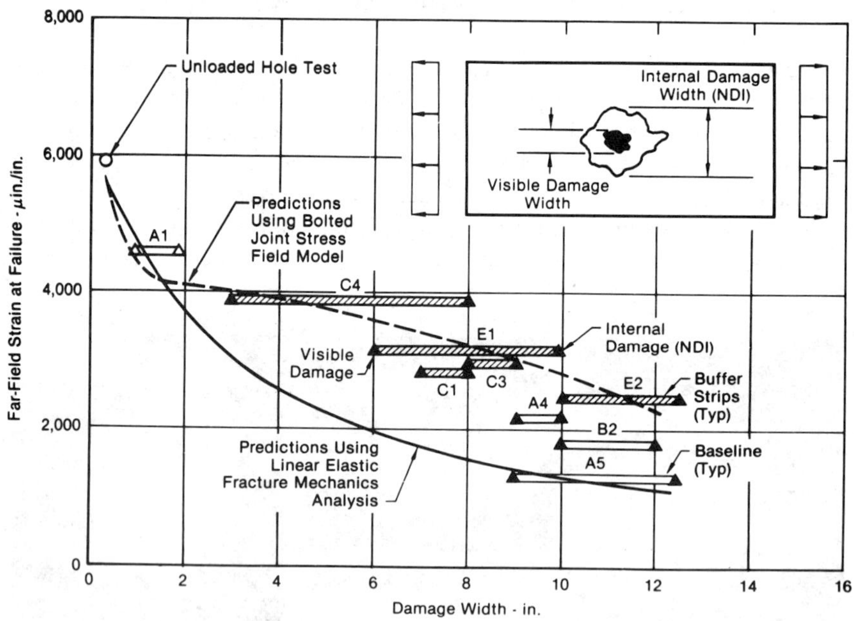

FIG. 19—*Residual strength of damaged specimens.*

TABLE 9—*Test results—penetration damaged specimens.*

Specimen Type	ID	Visible Damage, Diameter, in.	Internal Damage Length (0°) × Width (90°), in.	Residual Tensile Strength, lb
Baseline carbon/epoxy	A1	1	2 × 2	180 250
	A2	2[a]	[a]	65 250
	A3	5	8 × 8	[b]
	A4	9	14 × 10	84 000
	A5	9	14 × 14	50 600
Carbon/epoxy with rows of stitches on 3.5-in. centers	B1	6	6 × 6	[b]
	B2	10	12 × 12	69 850
	B3	4[a]	[a]	[d]
	B4	6	8 × 8	[d]
Carbon/epoxy with 0.5-in. buffer strips on 3.5-in. centers	C1	7	12 × 8	105 000
	C2	2[a]	[a]	74 200
	C3	8	16 × 9	107 500
	C4	3	16 × 8	143 250
Baseline aluminum	D1	4[c]	...	[b]
Carbon/epoxy with 1.5-in. buffer strips on 13.5-in. centers	E1	6	10 × 10	115 000
	E2	10	16 × 13	89 500
Carbon/epoxy with rows of stitches on 2.0-in. centers	F1	6	8 × 8	[d]
	F2	7	9 × 11	[d]

[a]Multiple penetrations over 20 in. by 20 in. area.
[b]Structural tests not planned.
[c]Permanent deformation over 16-in. diameter with multiple penetrations.
[d]Structural tests not complete.

bound. Penetration damage size was assumed to consist of through-the-thickness defects equal in width to the maximum visible and maximum internal damage.

In the second approach, the damage was assumed to be a circular hole in an orthotropic plate, and the methodology of Ref 7 was used to predict strain distributions about the hole. These strains were used in conjunction with the Maximum Strain Failure Criterion to predict far-field strain to failure as a function of damage (hole) size. Good correlation was first obtained between predictions and test results for a 0.25-in.-diameter fastener hole followed by extension of predictions to larger damage sizes. The residual strength of specimens with buffer strips compare well with these predictions.

Conclusions

Several conclusions were drawn from results of the programs described in this paper.

First, it was concluded that manufacturing defects that produced the more significant strength reductions were easily found by current nondestructive eval-

TABLE 10—*English Units and International System of Units (SI) equivalent conversions.*

Quantity	English Unit		SI Equivalent
Energy or Torque	1 ft · lb	=	1.356 N · m
	1 in · lb	=	0.1130 N · m
Force	1 kip	=	4.448 kN
	1 lb	=	4.448 N
Length	1 in.	=	2.540 cm
Pressure or Stress	1 ksi	=	6895 kPa
Running Load	1 lb/in.	=	175.1 N/m
Temperature	°F		°C = 1.8(°F) + 32.0
			K = °C + 273.2

uation techniques and would have been rejected or repaired by current acceptance criteria. Tolerances and controls being used in fabrication and assembly of composite aircraft structures are adequate to assure uniform strength and structural performance.

Second, the propagation of damage from low-energy impact is dependent on type of loading and strain levels. The strength loss of the damaged laminate can be approximated on the basis of an "equivalent" round hole. The propagation under repeated loads is relatively slow and can be confined by relatively simple reinforcement techniques such as stitching.

Third, penetration damage is more significant than low-energy impact damage (LEID). The loss of laminate strength due to penetration damage of a given size is greater than from LEID of the same apparent size. To reduce the strength loss, more significant reinforcement techniques such as buffer strips are required; stitching is not adequate.

APPENDIX

Refer to Table 10 for conversion of English Units to International System of Units (SI).

References

[1] Garbo, S. P. and Ogonowski, J. M., "Effect of Variances and Manufacturing Tolerances on the Design Strength and Life of Mechanically Fastened Composite Joints," Contract F33615-77-C-3140, USAF RPT AFWAL-TR-81-3041, U.S. Air Force, April 1981.

[2] Hinkle, T. V. and Garrett, R. A., "High Strain Composite Wing for Fighter/Attack Type Aircraft—Concept Validation," NADC-80146-60, Naval Air Development Center, Sept. 1982.

[3] Hinkle, T. V. and Garrett, R. A., "Examination of Postbuckled Compression Behavior of Curved Panels," MCAIR/NASC Contract N00019-79-C-0204, MDC RPT A7264, Aug. 1982.

[4] Renieri, M. P. and Garrett, R. A., "Postbuckling Fatigue Behavior of Flat Stiffened Graphite/Epoxy Panels Under Shear Loading," NADC Contract N62269-79-C-0463, RPT NADC-78137-60, Naval Air Development Center, July 82.

[5] Butler, B. M., "Wing/Fuselage Critical Component Preliminary Design (Northrop)," AFFDL-TR-78-174, Air Force Flight Dynamics Laboratory, March 1979.

[6] Avery, J. G. and Bradley, S. J., "Design Manual for Battle Damage Tolerant Fiber Composite Structures," NASC Contract N00019-80-C-0048, Boeing Report D180-26092-1, June 1980.
[7] Ogonowski, J. M., "Effect of Variances and Manufacturing Tolerances on the Design Strength and Life of Mechanically Fastened Composite Joints: Volume 3-Bolted Joint Stress Field Model (BJSFM) Computer Program User's Manual," AFWAL-TR-81-3041, U.S. Air Force, Vol. 3, April 1981.

Ray E. Bohlmann,[1] *Gary D. Renieri,*[1] *and Bob L. Riley*[1]

Bolted Composite Repairs Subjected to Biaxial or Shear Loads

REFERENCE: Bohlmann, R. E., Renieri, G. D., and Riley, B. L., "**Bolted Composite Repairs Subjected to Biaxial or Shear Loads,**" *Composite Materials: Testing and Design (Seventh Conference), ASTM STP 893,* J. M. Whitney, Ed., American Society for Testing and Materials, Philadelphia, 1986, p. 34–47.

ABSTRACT: The methodology to analyze bolted composite repairs subjected to biaxial or shear loads was developed. The methodology was incorporated into an existing McDonnell Aircraft Company (MCAIR) interactive computer program, BREPAIR, which was initially developed for uniaxial loading. The analysis determines bolt loads while accounting for hole clearances and determines patch and laminate strains for doubly symmetric finite regions where the damage is represented as a circular cutout. The analysis was verified by comparing predicted results to industry-recognized solutions for laminates containing circular holes (unrepaired), and to test results of an unrepaired and repaired shear panel. Results of this investigation verified that the methodology developed was a viable analytical tool for determining bolted repair designs for damaged laminates subjected to biaxial or shear loads.

KEY WORDS: composite materials, bolted repair, biaxial loading, circular damage, composite repair, finite plate, infinite plate, metal patch, orthotropic materials, repair methodology, shear loading

Bolted repair methods are being developed concurrently with the increasing use of composites for Navy fleet aircraft. Significant contributions have recently been made to the bolted repair technology of composite structures [*1–8*]. Considerable effort has been directed towards the development of adequate repair methodology that includes computer software to aid in establishing design of repairs and testing to verify such designs.

One of the first bolted repair computer analysis routines, BREPAIR, was developed by McDonnell Aircraft Company (MCAIR) to define repair patch geometry and fastener requirements in far less time and cost than conventional finite-element analysis [*2*]. The initial form of BREPAIR was capable of analyzing repaired laminates subjected to uniaxial loading only.

However, structures such as wing or fuselage skins sustain complex biaxial or shear loads or both. For such loadings, an extension of the initial BREPAIR was necessary. Therefore, the objective of this study was to add biaxial and shear

[1] Technical specialist, lead engineer, and branch chief, respectively, Technology, McDonnell Douglas Corporation, McDonnell Aircraft Company, St. Louis, MO 63166.

loading analytical capability to BREPAIR. A summary of the work performed, including an overview of the theoretical derivation of pertinent equations, comparison of the analysis results to industry-recognized solutions for laminates containing circular holes, test verification, and conclusions are presented.

Theoretical Derivation—Overview

BREPAIR is a methodology that combines numerical boundary value techniques with a two-dimensional compliance method to obtain the bolt loads for two connected elements, the "skin" and the "patch," as shown in Fig. 1. This section is intended to give the reader an overview of the technical aspects of the program. The analytic details of this derivation are contained in Ref 7.

Two different solution techniques to the boundary value problem are used: "boundary collocation" and "boundary integration." Both use a series expansion for their solution. Typically, each term in the series is composed of two parts: one that is dependent on material and geometric properties, and the other is an arbitrary coefficient that reflects loading on the region.

Both solution techniques were employed to find the stresses or displacements at an arbitrary point in a finite region that has a number of loaded holes. The boundary integral method was used to determine the effect of each loaded hole on the stresses at the point for an infinite region. The stresses predicted at the boundary of the actual finite region are also determined as a function of all the loaded holes. The stresses at the boundary are reversed using the boundary collocation technique to effect removing everything outside the boundary. Finally,

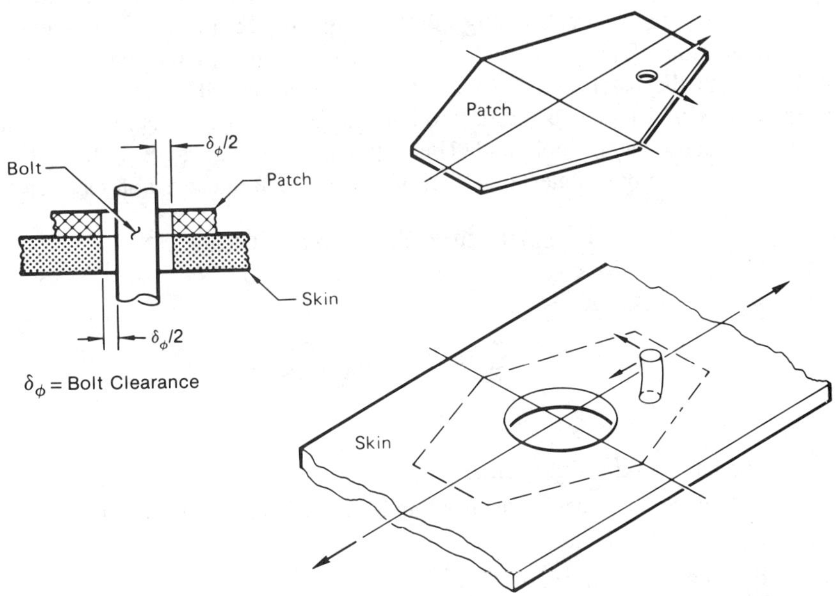

FIG. 1—*Bolted load transfer between patch and skin.*

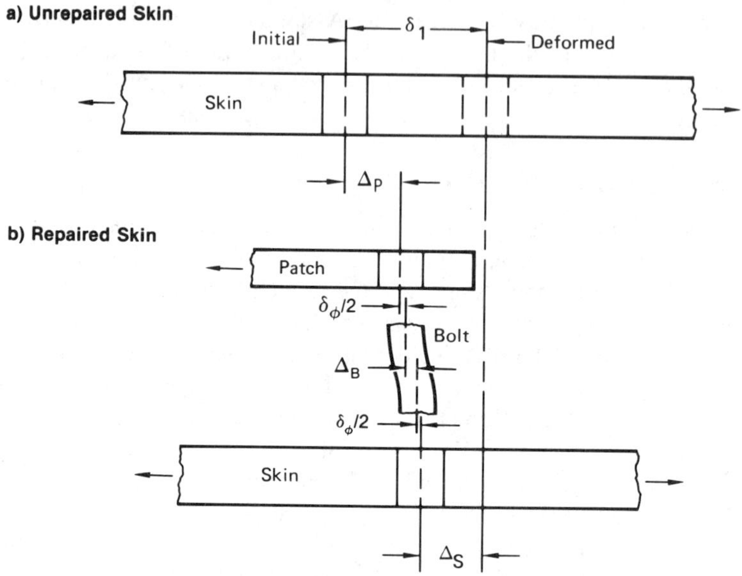

FIG. 2—*Displacements at a bolt BREPAIR program.*

the series from the boundary collocation is used to find the stresses at any interior point.

For explanation purposes, the two-dimensional compliance method is easily described by starting with the one-dimensional situation and extending it to two dimensions. A loaded skin is assumed to deform a certain amount, δ_1, as illustrated in Fig. 2*a*. If a patch is attached, Fig. 2*b*, the skin would deform less; how much less is a function of the magnitude of the load transmitted through the bolt. However, both the patch and the bolt are also deforming. The deformation of the repaired skin is equal to the deformation of the unrepaired skin minus the deformations caused by the bolt load. This in turn must be equal to the deformation of the patch and the bolt under the effect of the bolt load plus the bolt clearance.

$$\Delta(\text{Repaired Skin}) = \Delta(\text{Patch} + \text{Bolt} + \text{Clearance})$$

or in terms of Fig. 2,

$$\delta_1 - \Delta_s = \Delta_P + \Delta_B + \delta_\phi$$

where

δ_1 = unrepaired skin displacement,
Δ_s = differences between unrepaired and repaired skin displacement,
Δ_P = patch displacement,
Δ_B = bolt displacement, and
δ_ϕ = bolt clearance.

Rearranging the terms, those that are dependent on the value of bolt load are set equal to those that are not.

$$\Delta_P + \Delta_B = \delta_1 - \Delta_s - \delta_\phi$$

The magnitude of the bolt load is not known, but the effect on the patch, bolt, and skin of a unit bolt load could be determined. Since the other terms, such as the unrepaired skin deformation, are not a function of the bolt load, there would be one equation with the one unknown. This one-equation/one-unknown system can be expanded to a N-equation/N-unknown system.

Methodology Verification

This section provides comparisons of the BREPAIR analysis to industry-recognized solutions to laminates containing circular holes (unrepaired), and a comparison of the shear analyses to actual test results for unrepaired laminates and laminates containing metal repair patches.

Unrepaired Verification—Comparison to Existing Analyses

Tangential edge-of-hole strain concentrations can significantly affect the failure strength of a laminate containing a circular hole. To ascertain the accuracy of BREPAIR for predicting such strains, edge-of-hole strain distributions were determined for infinite and finite plates containing a typical 63.5-mm (2.5-in.) diameter hole. Isotropic and orthotropic laminates subjected to uniaxial tensile, uniaxial compressive, biaxial tensile-tensile, biaxial tensile-compressive, and shear loading were analyzed. BREPAIR predictions were compared to a finite-element solution (SY5) and a closed-form infinite-plate solution by Greszczuk [9].

Figures 3 and 4 present the tangential edge-of-hole strains for an infinite and finite plate, respectively, resulting from biaxial tensile-compressive loading of a typical carbon/epoxy (AS/3501-6) laminate. Similarly, Fig. 5 shows results for shear loading. Results presented in Figs. 3 through 5 indicated that BREPAIR predictions are accurate for infinite and finite plates containing circular holes subjected to inplane biaxial or shear loading.

Biaxial and Shear Analyses Verification—Shear Test

It has been shown in the previous sections that the newly developed BREPAIR analysis was verified for unrepaired laminates subjected to biaxial and shear loadings by comparison with other analytical techniques. To further verify the new BREPAIR analysis, a test was needed that would introduce biaxial or shear type loads. Therefore, a shear test was performed on an unrepaired specimen followed by a shear test for the specimen with a repair patch. Satisfactory correlation in these tests would verify not only the shear analysis capability but also the biaxial analysis capability.

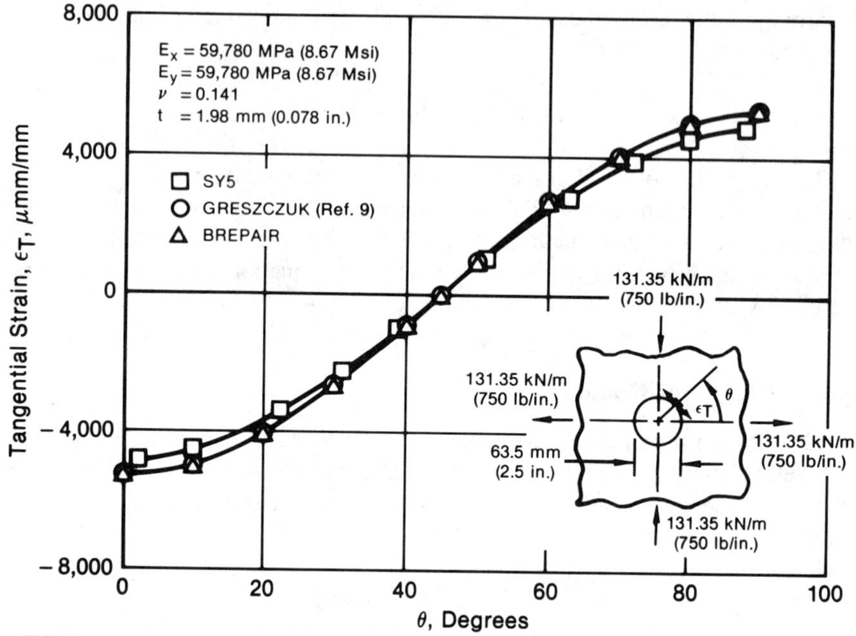

FIG. 3—*Edge-of-hole tangential strain, biaxial tension-compression, orthotropic laminate, infinite plate.*

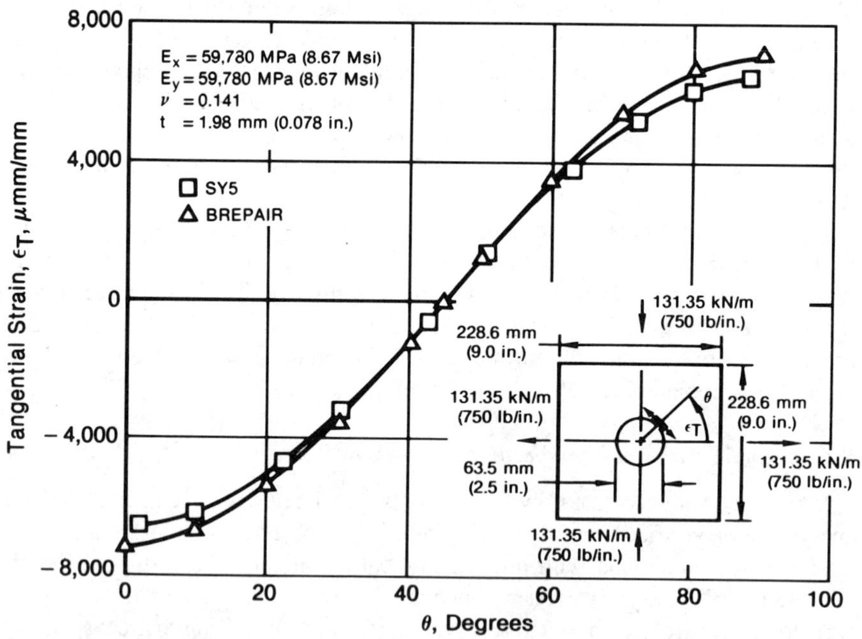

FIG. 4—*Edge-of-hole tangential strain, biaxial tension-compression, orthotropic laminate, finite plate.*

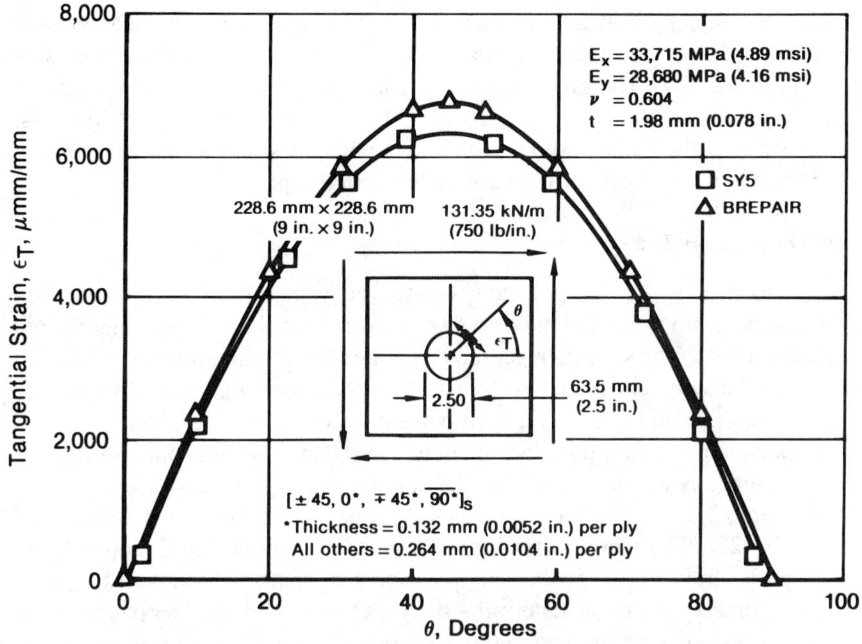

FIG. 5—*Edge-of-hole tangential strain, shear, orthotropic laminate, finite plate.*

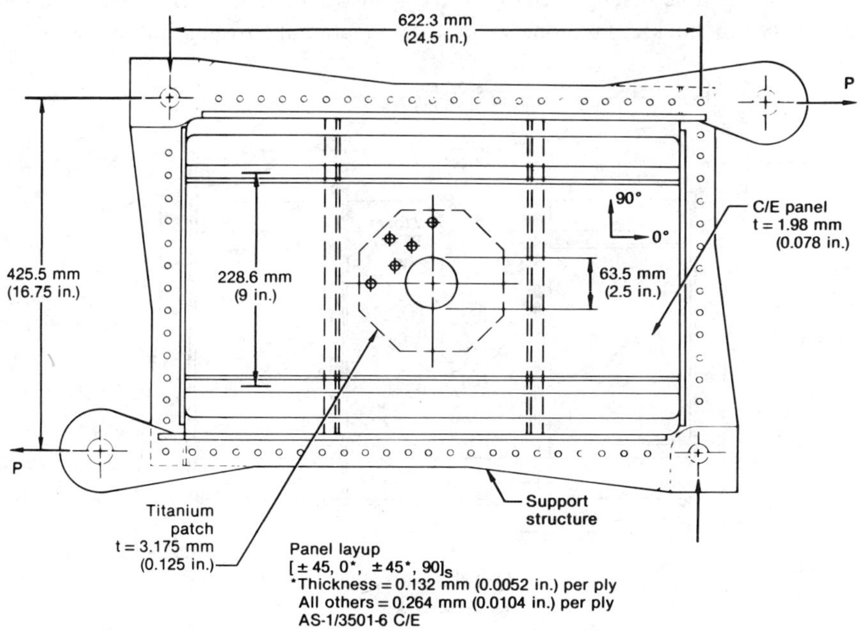

FIG. 6—*Typical bolted repair shear test setup.*

From elementary mechanics of materials, it can be shown that shear loading (N_{xy}) rotated 45° is equivalent to biaxial tensile ($N'_x = N_{xy}$) and compressive ($N'_y = N_{xy}$) loading. Therefore, the shear panel specimen can be considered a biaxial verification test at 45° as well as a shear verification test provided that both cases give the same response such as tangential strain at the edge of the hole or bolt fastener loads in the case of the bolted repair.

Shear Verification Tests

The objective of the test program was to verify the ability of the modified BREPAIR program to predict the response of shear loaded specimens (unrepaired and repaired test). The specimen geometry for the shear verification tests is shown in Fig. 6. The laminate layup is typical of aircraft skins that would be loaded primarily in shear and contains a high percentage of ±45° tape plies.

The specimen was first tested in the unrepaired condition at room temperature with strain-gage readings recorded at the corners of the panel and at the edge of the 63.5-mm (2.5-in.) diameter hole as shown in Fig. 7. The panel was loaded in 8.89-kN (2000-lb) increments. Before each load increment was applied, the back-to-back strain gages on the compression diagonal at the hole edge were compared for divergence to determine the onset of buckling. The panel began to buckle at the edge of the hole along the compression diagonal (Gage 12, 14, 15, and 16, Fig. 8) at just above 35.59 kN (8000 lb). At 40 kN (9000 lb), the test was discontinued to prevent damage to the unrepaired hole. The buckle at the edge of the hole disappeared when the load was removed. In order to stay in the linear (nonbuckled) region, the predicted membrane strains and test strains

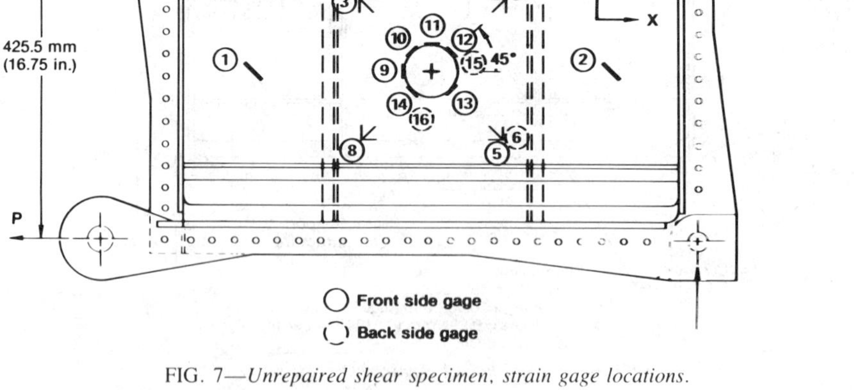

FIG. 7—*Unrepaired shear specimen, strain gage locations.*

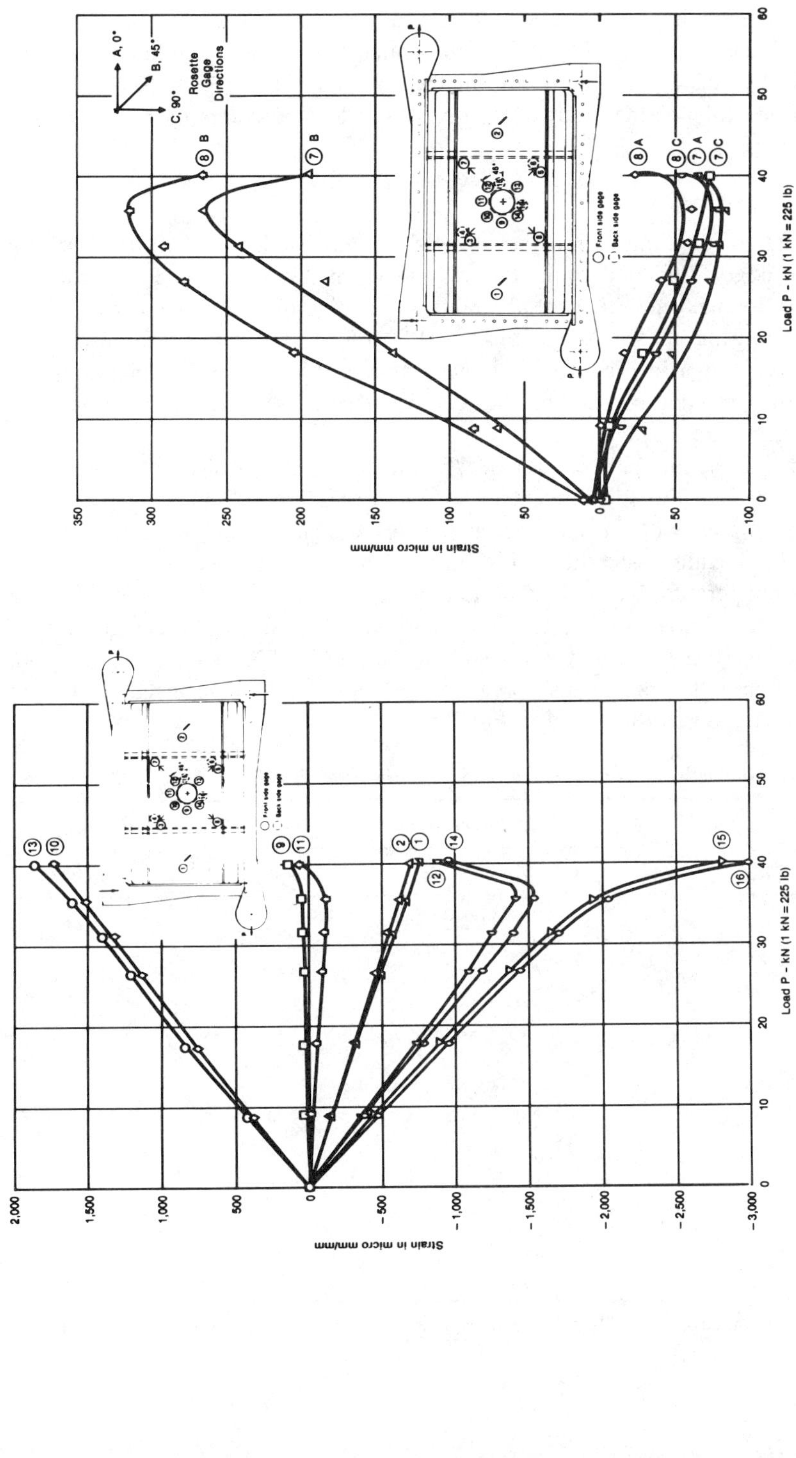

FIG. 8—*Unrepaired shear panel test, strain-gage results.*

in Table 1 are given for the 35.59 kN (8000 lb) level. BREPAIR can only predict membrane strains (no bending); therefore, valid comparisons can only be made for locations having single gages where bending is negligible and back-to-back gages that can be averaged to obtain membrane strains where bending is present. Legs A, B, and C axial strains from the corner rosettes (3, 4, 5, and 6) are transformed to shearing strains for comparison to predictions by the following relationship.

$$\gamma_{xy} = 2\epsilon_B - (\epsilon_A + \epsilon_C)$$

The center of the strain gages at the hole are actually 2.0 mm (0.08 in.) from the hole edge. Predicted strains are also given at this location. The prediction error for the maximum tangential strains at the edge-of-hole ranged from 6.0 to 12.6% along the tension diagonal to −0.5 to 5.3% along the compression diagonals as shown in Fig. 7. The larger error on the tension diagonal could be due to the presence of bending that could not be accounted for since back-to-back gages were not available.

Following the unrepaired test a 3.175-mm (0.125-in.) 6AL-4V titanium patch was installed on the specimen with 4.76-mm (3/16-in.) diameter Big-foot Jo-bolts. Photographs of the specimen before test are shown in Figs. 9 and 10 and strain-gage locations are shown in Fig. 11. The specimen was tested to failure at room temperature with a loading rate of 22.2 kN/min (5000 lb/min). Failure occurred at 221 kN (49 700 lb). This load was well above the design goal of 133.45 kN (30 000 lb) that is a shear flow of 214.5 kN/m (1225 lb/in.). An examination of the specimen showed that the primary failure occurred away from the repair region as illustrated in Fig. 11.

TABLE 1—*Unrepaired shear test results and BREPAIR predictions for 35.59 kN (8000 lb) load.*

Gage Number	Strain, μmm/mm		Error, %
	Test	Prediction[a]	
	SHEAR STRAIN		
3	700	...	...
Avg 3,4	777	912	+17.4
4	853	...	...
5	797	...	...
Avg 5,6	792	912	+15.2
6	787	...	...
	TANGENTIAL STRAIN		
10	1545	1740	+12.6
13	1629	1740	+6.8
12	−1390	...	...
Avg 12,15	−1653	−1740	+5.3
15	−1915	...	...
14	−1522	...	...
Avg 14,16	−1748	−1740	−0.5
16	−1974	...	...

[a]Predictions at edge of hole based on 2 mm (0.08 in.) from edge to allow for gage location.

FIG. 9—*Repaired specimen—patch, front side.*

Test strains are compared to predictions at 214.5 kN/m (1225 lb/in. shear flow) since most gages showed a linear response up to this level. As in the unrepaired test, predictions versus test comparisons were made at single strain-gage locations where only membrane loading is expected and at back-to-back gage locations where bending is expected. The prediction error shown in Table 2 for the corner rosettes (Gages 3, 4, 5, and 6) and the tangential strain at the edge of the repaired hole (Gages 10 and 13) is less than 6%. The back-to-back strain gages on the patch (Gages 15 and 16) show that the eccentricity of the patch offset from the skin results in substantial bending strains. The patch shearing strain prediction in the x, y axes system compares within 7% of test when the back-to-back strains are averaged and transformed to the x, y axes system.

FIG. 10—*Repaired specimen—patch, back side.*

While a majority of the predictions were in the 5 to 10% range and several were within 15%, the correlations were generally very good considering the presence of the various error contributors such as buckling, localized stiffeners, and errors in gage placement. Based on these limited tests, the modified BREPAIR program has been verified to be a viable technique for predicting the behavior of shear (or biaxial) loaded composite skins with a circular hole (unrepaired or repaired).

Conclusions

A methodology has been developed to analyze composite bolted repairs subjected to biaxial or shear loading. The analysis has been incorporated into the

FIG. 11—*Failed repaired specimen—patch, front side.*

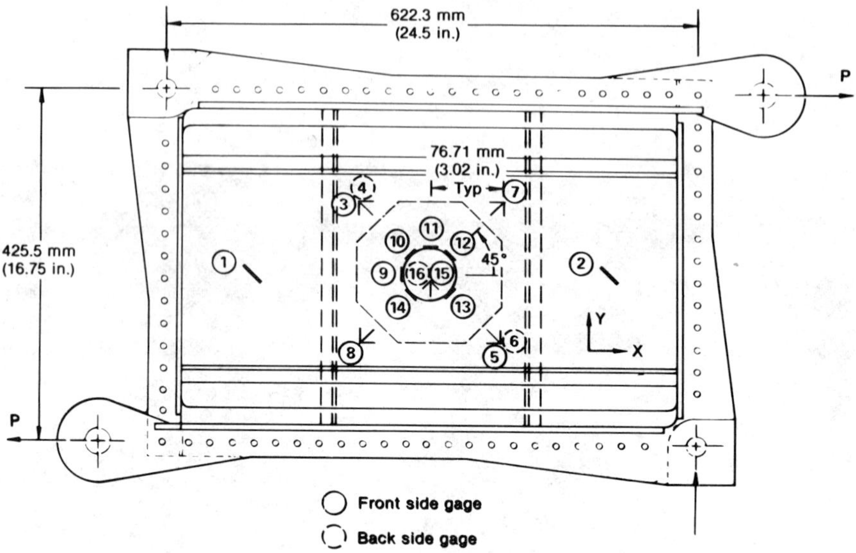

FIG. 12—*Repaired shear specimen, strain gage locations.*

TABLE 2—*Repaired shear test results and BREPAIR predictions for 133.45-kN (30 000-lb) load.*

Gage Number	Strain, μmm/mm		Error, %
	Test	Prediction[a]	
	SHEAR STRAIN, *xy*		
3	3808	...	...
Avg 3,4	3670	3800	+3.4
4	3533	...	...
5	4269	...	...
Avg 5,6	3772	3800	+0.6
6	3274	...	...
	TANGENTIAL STRAIN		
10	4648	4371	−6.0
13	4367	4371	+0.1
	SHEAR STRAIN, *xy*		
15	2466	...	...
Avg 15,16	910	822	−9.7
16	−646	...	...

[a]Predictions at edge of hole based on 2 mm (0.08 in.) from edge to allow for gage location.

BREPAIR computer program. The following conclusions were made regarding the BREPAIR analysis.

1. The analysis provides accurate results when compared to existing industry-recognized analyses for composite laminates containing circular holes.
2. Analysis predictions compare closely with shear panel test results (unrepaired and repaired).

Acknowledgments

This work was sponsored in part by the Naval Air Development Center (NADC) Contract N62269-81-C-0297, Mark Libeskind, project engineer. The analytical development effort of Mr. D. Douglas Wilmarth is greatly appreciated.

References

[*1*] Watson, J. C., ''Bolted Field Repair Composite Structure,'' Naval Air Development Center Report NADC-77109-30, March 1979.
[*2*] Bohlmann, R. E., Glaeser, D. A., and Watson, J. B., ''Field Repair of Composite Structural Components,'' Naval Air Development Center Report NADC-78078-60, Feb. 1980.
[*3*] Bohlmann, R. E., Renieri, G. D., and Libeskind, M. in *Joining of Composite Materials, ASTM STP 749,* K. Kedward, Ed., American Society for Testing and Materials, Philadelphia, 1981, pp. 97–116.
[*4*] Watson, J. C., ''Bolted Repair Verification of Composite AV-8B Wing Skins,'' Naval Air Development Center, Progress Report for Contract N62269-81-C-0222, 3 April 1981 through 12 June 1981.
[*5*] Bohlmann, R. E., Renieri, G. D., and Horton, D. K., ''Bolted Repair Analysis Methodology,'' Preliminary to Final Report, Naval Air Development Center, Contract N62269-81-C-0297, Report MDC A7592, May 1982.
[*6*] Bohlmann, R. E. and Renieri, G. D., ''Bolted Repair Analysis Methodology—Software Report and User's Guide,'' Naval Development Center, Contract N62269-81-C-0297, Report MDC A7654, May 1982.
[*7*] Bohlmann, R. E., Renieri, G. D., and Horton, D. K., ''Bolted Repair Analysis Methodology,'' Naval Air Development Center, Report NADC-81063-60, Dec. 1982.
[*8*] Watson, J. C., ''AV-8B Structural Repair Development,'' presented at Sixth Conference on Fiberous Composites in Structural Design, New Orleans, 24–27 Jan. 1983, MCAIR 82-027.
[*9*] Greszczuk, L. L., ''Stress Concentrations and Failure Criteria for Orthotropic and Anisotropic Plates with Circular Openings,'' MDAC Paper WD 1485, April 1971.

R. L. Ramkumar,[1] G. C. Grimes,[1] and S. J. Kong[1]

Characterization of T300/V-378A Graphite/Bismaleimide for Structural Applications

REFERENCE: Ramkumar, R. L., Grimes, G. C., and Kong, S. J., "**Characterization of T300/V-378A Graphite/Bismaleimide for Structural Applications,**" *Composite Materials: Testing and Design (Seventh Conference), ASTM STP 893,* J. M. Whitney, Ed., American Society for Testing and Materials, Philadelphia, 1986, pp. 48–63.

ABSTRACT: This paper discusses the potential of T300/V-378A graphite/bismaleimide for use in aircraft structures that are likely to be exposed to service temperatures in the 120 to 204°C (250 to 400°F) range. The microstructure of cured T300/V-378A laminates was examined to detect the presence of any process-induced defects; moisture absorption studies were conducted to characterize the hygroscopic behavior of the material; and its basic properties were generated through static tests on $[0]_{16T}$, $[90]_{16T}$, and $[\pm 45]_{4s}$ laminates. These were followed by static tests on plain $[0/\pm 45/90]_{2s}$ T300/V-378A specimens and specimens with unloaded (open and filled) and loaded holes. Tests were conducted under room temperature, dry (RTD) and elevated temperature, wet (ETW) conditions. Plain specimens and specimens with open holes were also subjected to constant-amplitude fatigue tests at minimum-to-maximum cyclic stress ratios (*R*) of 0.05 and −1.0.

Subsequently, the tolerance of $[0/\pm 45/90]_{4s}$ T300/V-378 laminates to low-velocity impact damage by a hard object was evaluated by measuring impact energy levels corresponding to incipient (not necessarily visible) damage and through-penetration. Specimens from impact-damaged panels were then tested to obtain residual static compression strengths. Finally, multibay, hat-stiffened, flat shear panels were fabricated using the T300/V-378A material system, and were subjected to pure-shear static and fatigue tests under RTD and ETW conditions.

Results were compared with available data on T300/5208 and AS/3501-6 graphite/epoxy laminates to assess the relative merits of the T300/V-378A graphite/bismaleimide material system. In general, the RTD matrix-dominated response of T300/V-378A laminates is inferior to that of AS1/3501-6 laminates. But, the T300/V-378A laminates retain most of their RTD strengths under 177°CW (350°FW) (wet) conditions, where conventional epoxy-matrix materials suffer property losses that occur at the glass transition temperature. The V-378A bismaleimide material, therefore, provides a viable alternative to conventional epoxy-matrix materials, for applications in components that experience humid conditions and temperatures in the 120 to 205°C (250° to 400°F) range.

KEY WORDS: composite materials, fatigue (materials), bismaleimide, T300/V-378A, unidirectional tape, moisture absorption, mechanical properties, environmental effects, fastener holes, static tests, fatigue tests, bearing strengths, shear panel, impact damage, damage tolerance

[1] Engineering specialist, senior technical specialist, and manager, respectively, Northrop Corporation, Aircraft Division, Hawthorne, CA 90250.

Take-off and hovering maneuvers of V/STOL aircraft, and aerodynamic heating during supersonic flight, are situations during which parts of aircraft structural surfaces are exposed to service temperatures beyond the 104°C (220°F) limit customarily employed for conventional epoxy-matrix composites. Currently used epoxy-matrix composites undergo glass transition around 104°C (220°F), when they are saturated with moisture [*1*], and they consequently exhibit low bolt bearing strengths under these conditions [*2*]. A need for alternative matrix materials therefore exists, for use at temperatures beyond 104°C (220°F), in the presence of moisture.

Polyimide-matrix materials, such as PMR-15 and LARC-160, have been developed to provide composites that can withstand service temperatures up to 316°C (600°F) [*3*]. The first generation condensation and addition polyimides, however, require processing temperatures (600°F) and pressures (200 psi) that result in fabrication costs that are significantly higher than those for epoxy-based composites. Recently, a class of addition-cured polyimide-matrix materials, called bismaleimides, has been developed for application up to 204°C (400°F). The bismaleimide resin is a fully imidized hot-melt system, supplied as 100% solids on graphite reinforcement. It is curable at 177°C (350°F) under a 689 kPa (100 psi) pressure, like an epoxy-matrix composite, and is generally postcured at 245°C (475°F) in an aircirculating oven. Some of the commercially available bismaleimide materials that have recently been evaluated for structural applications are: U.S. Polymeric's V-378A, Hexcel's F-178, Ferro's CPI 2272, Hercules' 4001, etc. This paper summarizes the results obtained at Northrop Corporation from various tests on T300/V-378A graphite/bismaleimide laminates [*4–7*]. Additional data on the material system are available in Refs *8* to *11*.

Fabrication of Laminates

In general, the T300/V-378A laminates can be processed like graphite/epoxy and are comparable in fabrication quality to graphite/epoxy panels. Initial concerns regarding the T300/V-378A material included an offensive odor from the prepreg, and difficulties encountered during secondary bonding of tabs to the test panels. The cure cycles used in Refs *4* to *7* for the fabrication of T300/V-378A test panels are shown in Figs. 1 and 2. A 30 min dwell at 79°C (175°F), used in Refs *5* and *6*, was eliminated in Refs *4* and *7*. Cured panels in Refs *4*, *5*, and *7* were subjected to an unrestrained postcure at 245°C (475°F) for 4 h, followed by an exposure to 288°C (550°F) for 1 h. In Ref *6*, postcure was restricted to 4 h at 245°C (475°F), without any exposure to 288°C (550°F). All the fabricated panels were determined to be acceptable based on established quality control tests. They were practically void free (<<1% by volume), and their ultrasonic through-transmission records and photomicrographs of their cross-sections attested to it. In 40-ply-thick panels [*7*], however, microscopic cracks were visible in the matrix of the central 90° and ±45° plies, when observed at a magnification of over 25. These were small in size and few in number, and were possibly

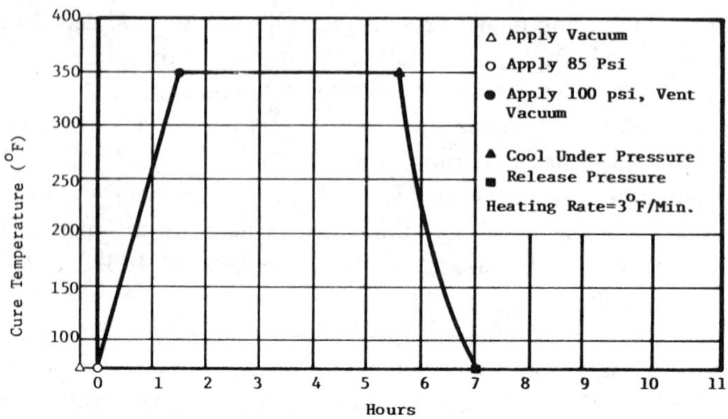

FIG. 1—*Cure cycle used in Refs* 4 *and* 7 *for T300/V-378A panels.*

induced by residual thermal stresses when cooled down after a postcure at 288°C (550°F). These defects adversely affected the matrix-dominated mechanical response of T300/V-378A laminates as will be discussed later. Elimination of the postcure at 288°C (550°F) could possibly lessen these adverse effects.

Moisture Absorption Study

In Ref 5, tests were conducted on $[0/\pm45/90]_s$ specimens to obtain the rate of moisture absorption as a function of time and the maximum moisture absorbed as a function of relative humidity. A single-stage moisture absorption test procedure was adopted. Test temperatures ranged from 60°C (140°F) to 93°C (200°F). Relative humidity levels of 60, 70, and 95% were used. Prior to conditioning, specimens were dried at 82°C (180°F) in vacuum until their weights remained unchanged. To simulate a one-dimensional moisture absorption in the thickness direction, the four edges of the specimens were sealed with aluminum foil tapes.

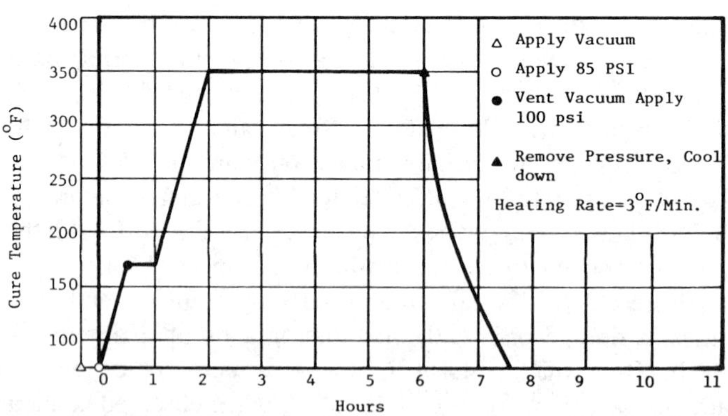

FIG. 2—*Cure cycle used in Refs* 5 *and* 6 *for T300/V-378A panels.*

The maximum moisture content in the T300/V-378A specimens was assumed to vary with relative humidity (RH) as follows [*12*]:

$$M_{max} = a\phi^b \tag{1}$$

where M_{max} is the maximum moisture content, expressed in percentage weight gain, and ϕ is the relative humidity, also expressed in percentage. A least squares fit of the obtained test results to the form in Eq 1 yielded:

$$M_{max} = 0.000705\ \phi^{1.69} \tag{2}$$

The methods presented in Ref *13* were used to determine the diffusivity coefficient for the T300/V-378A material. Based on the observation that the rate of moisture absorption is a function of temperature only, while the maximum moisture content is mainly dependent on the relative humidity, the following expressions were used [*13*]

$$D = \pi(h/4M_{max})^2\ (M_2 - M_1)^2/(\sqrt{t_2} - \sqrt{t_1})^2 \tag{3}$$

or

$$D = (h^2/t)\left[-\frac{\ell n(1 - G)}{7.3}\right]^{4/3} \tag{4}$$

where D is the diffusivity coefficient in mm²/s, h is the laminate thickness in millimetres, M_{max} is the maximum moisture content in percent for a given environment, M_1 and M_2 are percent moisture contents in the laminate after t_1 and t_2 seconds of exposure to the given environment, and $G = M/M_{max}$ is a dimensionless time-dependent parameter ($G_1 = M_1/M_{max}$, etc.). Diffusivity coefficient values were computed using Eqs 3 and 4 for each temperature/relative humidity test condition. Results from five replicates yielded 15 D values that were averaged at each test condition.

The computed diffusivity coefficients for T300/V-378A laminates were expressed in the following form [*12*], assuming that D is only a function of temperature

$$D = D_o e^{-C/T} \tag{5}$$

Where T is the absolute temperature, and D_o and C are constants. An Arrhenius plot of the average D value at each test temperature yielded the following result

$$D = 0.391 e^{-4075/T} \tag{6}$$

The diffusivity coefficient for T300/V-378A laminates was subsequently compared with *D* values for other epoxy-matrix composites [*12*] at various temperatures. The T300/V-378A material was determined to have a *D* value that was approximately ten times the *D* value for the T300/5208 material, and approximately seven times the *D* value for the AS/3501-5 material. The V-378A bismaleimide material therefore absorbs moisture much faster than the 5208 or the 3501-5 epoxy material. Likewise, the desorption rate for V-378A is also expected to be much larger than that for 5208 or 3501-5. Nevertheless, it is seen from Eq 2 that the equilibrium moisture content for the V-378A material under 100% relative humidity conditions is approximately the same as that for the 3501-6 epoxy material (1.7%).

Basic Mechanical Properties

The basic mechanical properties of the T300/V-378A material system were obtained through tension and compression tests of $[0]_{16T}$, $[90]_{16T}$, and $[\pm 45]_{4s}$ specimens [*4*]. Test results obtained under room temperature, dry (RTD) conditions are presented in Table 1. The transverse shear capability of the material

TABLE 1—*Basic mechanical properties for T300/V-378A under RTD conditions* [4].

Test Method	Laminate	Environment	Average Properties
ASTM D 3039-76	$[0]_{16T}$	RTD	E_{11}^{t} = 133 GPa (19.30 Msi)
			σ_{11}^{tu} = 1510 MPa (218.9 ksi)
			ϵ_{11}^{tu} = 11 308 μ
			ν_{12}^{t} = 0.311
ASTM D 3410-75	$[0]_{16T}$	RTD	E_{11}^{c} = 130 GPa (18.79 Msi)
			σ_{11}^{cu} = 1435 MPa (−208.2 ksi)
			ϵ_{11}^{cu} = 11 963 μ
			ν_{12}^{c} = 0.313
ASTM D 3039-76	$[90]_{16T}$	RTD	E_{22}^{t} = 9 GPa (1.32 Msi)
			σ_{22}^{tu} = 34 MPa (4.94 ksi)
			ϵ_{22}^{tu} = 3586 μ
			ν_{21}^{t} = 0.027
Northrop ETL Standard	$[90]_{16T}$	RTD	E_{22}^{c} = 9 GPa (1.27 Msi)
			σ_{22}^{cu} = −191 MPa (−27.7 ksi)
			ϵ_{22}^{cu} = −25 779 μ
			ν_{21}^{c} = 0.01
ASTM D 3518-76	$[\pm 45]_{4s}$	RTD	G_{12} = 5 GPa (0.784 Msi)
			τ_{12}^{u} = 74 MPa (10.7 ksi)
			γ_{12}^{u} = 24 864 μ
Northrop ETL Standard	$[\pm 45]_{4s}$	RTD	G_{12} = 5 GPa (0.707 Msi)
			τ_{12}^{u} = 95 GPa (13.7 ksi)
			γ_{12}^{u} = 35 417 μ

NOTE—Physical properties for the $[0]_{16T}$ specimens were: 62.1% of fiber volume, 30.1% of resin weight, specific gravity of 1.563, and approximately 1% of porosity by volume. Physical properties for the $[\pm 45]_{4s}$ specimens were: 62.2% of fiber volume, 30.3% of resin weight, specific gravity of 1.572, and approximately 0.3% of porosity by volume.

was determined through short-beam shear tests on $[0]_{16T}$ specimens, and flatwise tension tests were conducted on $[\pm45]_{4s}$ specimens. The average short-beam shear strength was measured to be 117 MPa (17.0 ksi), and the average flatwise tensile strength was measured to be 5 MPa (778 psi). A comparison of the properties in Table 1 with those obtained by McKague et al [*10*] (see Table 2) indicates that: (*a*) the 0° tension and the ±45° tension properties are approximately the same; (*b*) the 0° compression properties in Ref *10* are slightly higher; and, (*c*) the 90° tension properties in Ref *10* are much higher. The larger 0° compression and 90° tension properties in Ref *10* are possibly a result of the restricted postcure temperature used in Ref *10*.

Figure 3 presents a comparison of the basic T300/V-378A properties with the corresponding T300/5208 properties [*10*]. Under RTD conditions, the T300/V-378A properties are approximately 10% higher than the T300/5208 properties. Under 350FW conditions, the 0° compression properties of the T300/V-378A material are approximately 30% larger than the corresponding T300/5208 values. The 350FW 90° tensile strength and failure strain values for T300/V-378A are approximately 300% and 115% higher than those for T300/5208, respectively. And, the 350FW shear strength and ultimate shear strain for T300/V-378A are approximately 150% and 30% higher than those for T300/5208, respectively. The T300/V-378A material is far superior to the T300/5208 material under elevated temperature, wet (265FW, 350FW) conditions.

Effect of Open and Filled Holes

Static tests on $[0/\pm45/90]_{2s}$ T300/V-378A specimens, with and without holes, yielded results that are presented in Table 3 [*7*]. Specimens with 6.35 mm diameter central holes (unloaded or open) had a width to hole diameter ratio of 8. Under

TABLE 2—*Basic mechanical properties for T300/V-378A under various environmental conditions* [10].

Test Type	Environment[a]	Ultimate Strain, μmm/mm	Strength, MPa (ksi)
0° tension	RTD	11 710	1741 (252.5)
0° compression	RTD	−14 843	−1759 (−255.1)
0° compression	350FW	−8 103	−1045 (−151.5)
90° tension	RTD	6 087	59 (8.584)
90°tension	350FD	6 327	43 (6.189)
90° tension	265FW	3 399	24 (3.409)
90° tension	350FW	3 905	22 (3.177)
±45 tension[b]	RTD	21 405	77 (11.185)
±45 tension	350FD	48 087	62 (9.049)
±45 tension	265FW	64 000	62 (8.967)
±45 tension	350FW	69 231	54 (7.875)

[a]The "wet" specimens were saturated under 75% relative humidity conditions.

[b]Reported ultimate strain and strength values are the ultimate shear strain and the shear strength, respectively.

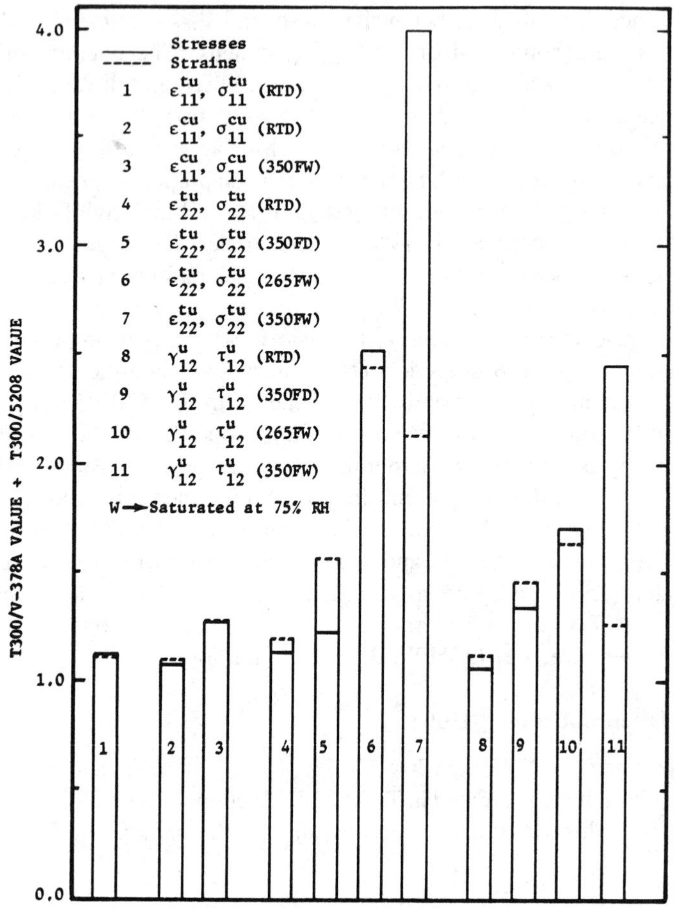

FIG. 3—*Summary of T300/V-378A basic properties* [10].

RTD static tension, the ultimate strain was lowered by 36%, from 10 865 μmm/mm to 6993 μmm/mm, due to the 6.35-mm diameter hole. In AS1/3501-6 laminates [*6*], a 6.35 mm diameter hole reduces the RTD tensile ultimate strain to approximately 6000 μmm/mm, lower than the value realized in T300/V-378A laminates. Under 350FW conditions, a 31% reduction in the ultimate tensile strain was caused by the hole in the T300/V-378A laminates. Under static compression, ultimate strain values of 5850 and 6550 μmm/mm were measured under RTD and 350FW conditions, respectively. Specimens tested under 350FW conditions contained approximately 1% of moisture by weight. In AS1/3501-6 laminates [*6*], a 6.35-mm-diameter hole reduces the RTD ultimate compressive strain value to approximately 7500 μmm/mm. The poorer compressive behavior of the T300/V-378A laminate is partially a result of the 288°C (550°F) postcure-induced defects in the matrix of the test specimens.

TABLE 3—*Static test results on* $[0/\pm45/90]_{2s}$ *T300/V-378A specimens* [7].

Specimen Type[a]	Load Type	Environment	Modulus, GPa (Msi)	Strength, MPa (ksi)	Failure Strain, μmm/mm
Plain	tension	RTD	44.8 (6.50)[b]	487 (70.6)	10 865
Plain	tension	350FW	45.3 (6.57)	483 (70.6)	9 504
6.35-mm-dia open hole	tension	RTD	46.9 (6.80)[b]	328 (47.6)	6 993
6.35-mm-dia open hole	tension	350FW	47.0 (6.82)	338 (49.1)	6 602
Plain	compression	RTD	43.3 (6.29)	−493 (−71.5)	−10 775
Plain	compression	350FW	48.0 (6.96)	−446 (−64.7)	−9 600
6.35-mm-dia open hole	compression	RTD	53.4 (7.75)	−309 (−44.9)	−5 850
6.35-mm-dia open hole	compression	350FW	56.1 (8.13)	−266 (−38.5)	−6 550

[a]All the specimens were 51-mm wide. Static tension specimens were 305-mm long, and static compression specimens were 203-mm long. Specimens with open 6.35-mm central hole had a *W/D* ratio of 8.

[b]Estimated values.

Results from constant amplitude fatigue tests on $[0/\pm45/90]_{2s}$ T300/V-378A specimens are presented in Table 4 [7]. The RTD and 350FW threshold strain levels, below which fatigue failure does not occur for a million cycles of the imposed constant-amplitude fatigue loading, decrease to 2700 and 3500 μmm/mm, respectively, under fully reversed ($R = -1$) loading conditions, in the presence of a 6.35-mm-diameter open hole.

TABLE 4—*Constant-amplitude fatigue test results on* $[0/\pm45/90]_{2s}$ *T300/V-378A specimens* [7].

Specimen Type	Environment	Reference Static Failure Strains, μmm/mm		Constant-Amplitude Fatigue Threshold Strain Level[a] μmm/mm for	
		Tension	Compression	$R = 0.05$[b]	$R = 1$
Plain	RTD	10 865	−10 775	7 661	±6 357
Plain	350FW	9 504	−9 600	6 700	±3 744
6.35-mm-dia open hole	RTD	6 993	−5 850	5 946	±2 662
6.35-mm-dia open hole	350FW	6 602	−6 550	5 612	±3 504

[a]Maximum absolute strain value at which the specimens survived one million cycles of the imposed constant-amplitude fatigue loading without suffering fatigue failure.

[b]*R* is the algebraic minimum-to-maximum cyclic strain ratio.

Static compression tests on $[\pm 45/0_3/90/0_2]_{2s}$ T300/V-378A specimens, with 1.55% of moisture by weight, provided the results in Fig. 4 [*6*]. Filled-hole results correspond to tests in which a protruding head fastener was installed finger tight in the 6.35-mm-diameter hole. It is seen that the compression modulus is slightly larger for specimens with open or filled holes, in comparison to plain specimens. The modulus increases slightly with an increase in the test temperature. The ultimate compressive strain (and the compressive strength) values decrease with test temperature. The filled-hole specimens yielded larger ultimate strains than open-hole specimens. The ultimate strain for plain specimens reduced from 10 000 μmm/mm under RTW conditions to 5685 μmm/mm under 400 FW conditions. For specimens with 6.35-mm-diameter filled and open holes, the ultimate strains reduced from 8631 to 5851 μmm/mm, respectively, under RTW conditions, to 3734 and 3402 μmm/mm, respectively, under 400FW conditions. The AS1/3501-6 graphite/epoxy laminates have RTW ultimate compressive strain values of 12 258 and 7629 μmm/mm for plain and open-hole specimens, respectively [*6*]. But, the ultimate strains reduce to approximately 4400 and 4900 μmm/mm, respectively, under 250FW conditions [*6*]. The epoxy matrix laminates undergo glass transition near 250°F when they contain approximately 1% of moisture by weight. Their compressive strengths are reduced drastically beyond 250FW conditions, while the V-378A matrix composites retain a respectable fraction (0.60) of their RTW compressive strengths even under 400FW conditions.

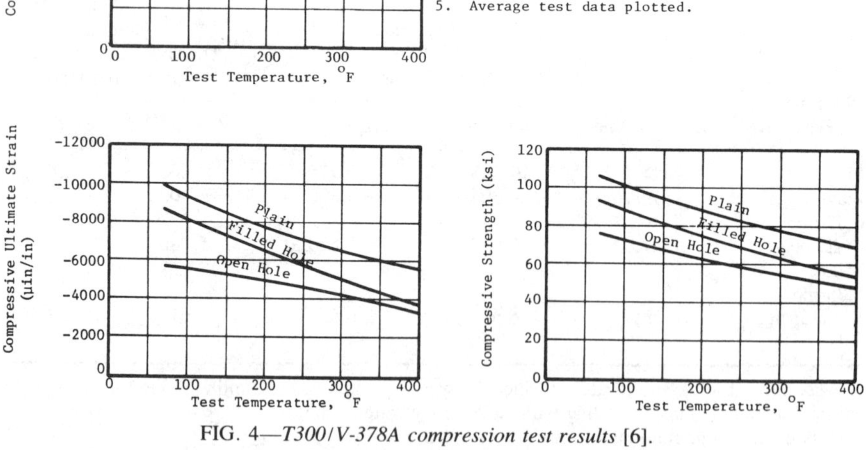

FIG. 4—*T300/V-378A compression test results* [6].

Bearing Strength for T300/V-378A Laminates

The $[\pm45/0_3/90/0_2]_{2s}$ T300/V-378A specimens with 2.36% of moisture by weight and 6.35-mm-diameter central holes, were tested in a double lap configuration, using steel plates, to obtain bearing strengths at various temperatures [*6*]. Figure 5 presents the variation in the bearing yield strength with test temperature. The bearing yield strength corresponded to the proportional limit of the load versus joint deflection curve, and was lower than the stress corresponding to a 4% elongation in the hole diameter (the criterion used in ASTM Test Method for Bearing Strength of Plastics (D 953-80). The bearing yield strengths for T300/V-378A are comparable to those for AS/3501-6 up to 250°F [*2*]. Beyond 250FW conditions, epoxy-matrix laminates undergo glass transition and lose their bearing capability, while the V-378A matrix composites possess a bearing yield strength of approximately 248 MPa (36 ksi) even under 450°FW conditions.

The bearing test specimens were 51-mm wide and had an edge distance that was four times the hole diameter. Still, a combination of bearing and shear-out failure modes was observed in the failed specimens, due to the low percentage (12.5%) of ±45° plies.

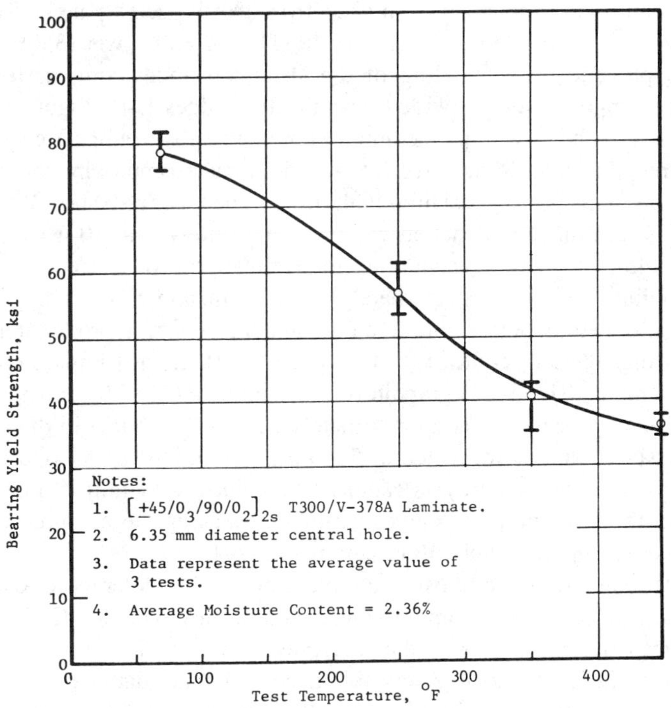

FIG. 5—*Bearing yield strength versus temperature* [6].

Tolerance to Low-Velocity Impact by a Hard Object

The susceptibility of T300/V-378A laminates to low-velocity impact threats like dropped tools, runway debris, etc. was addressed in this study [*14*]. Two 254-mm-square $[0/\pm45/90]_{5s}$ panels were initially centrally impacted in a modified drop tower arrangement to cause through-penetration. A hemispherical-tipped impactor, with a tip diameter of 12.7 mm, was screwed into the bottom platform of the drop weight, below the instrumented tip. The test panel was bolted down along all four edges, to a steel substructure with 25.4-mm-wide flanges. The through-penetration impact tests were instrumented to measure the contact forces as a function of time. Force versus time information from each test was then fed into a post-processor to yield force and energy variations with time and impactor/panel deflection. Incipient damage was manifested by the first "major" unloading phenomenon in the force versus time trace, and occurred at an energy level (E_I) between 5 and 8 N · m (4 and 6 ft · lb). Through-penetration absorbed between 68 and 81 N · m (50 and 60 ft · lb) (E_{TP}). Based on these results, another test panel was impacted at energy levels of 9.5, 16, and 27 N · m (7, 12, and 20 ft · lb), at two locations each, to obtain three damage levels beyond incipient damage. Ultrasonic *C*-scan records of the impact-damaged panels yielded damage sizes corresponding to the various impact energy levels.

Post-impact residual compression strengths were determined using 254-mm-long, 127-mm-wide specimens extracted from the impacted panels. Tests were performed under room temperature, dry (RTD) and 350°F wet (350FW) conditions. To prevent gross buckling of the specimen under compression, lateral knife-edge supports were provided near the free edges [*14*]. Figure 6 presents the variation of the post-impact compression strength with impact energy. Post-impact strengths under RTD and 350FW conditions are approximately the same. The (*C*-scan) damage size variation with impact energy is also shown in the same figure. It is seen that an impact energy level of 13.56 N · m (10 ft · lb), or 2429 N · m/m (45.5 ft · lb/in.) accounting for laminate thickness, reduces the compressive failure strain to approximately 2500 μmm/mm. This energy level and residual failure strain correspond to a *C*-scan damage size approximately 51 mm (2 in.) in diameter. In contrast, a 13.56 N · m (10 ft · lb) impact on a 6-mm (0.25-in.) thick AS1/3501-6 graphite/epoxy laminate (2135 N · m/m) results in a *C*-scan damage size that is approximately 32 mm (1.26 in.) in diameter, and a compressive failure strain value of 5250 μmm/mm [*14*]. And, over 68 N · m (50 ft · lb) of impact energy is required to induce a 51-mm (2-in.) diameter damage in the same graphite/epoxy laminate, reducing the compressive failure strain value to approximately 3050 μmm/mm [*14*].

Bismaleimide matrix composites are therefore far less tolerant to low-velocity impact damage compared to epoxy-matrix composites. The 288°C (550°F) post-cure-induced matrix defects in the fabricated T300/V-378A laminate possibly contribute significantly to this low tolerance to impact damage. Nevertheless, they retain their RTD post-impact strength even under 350FW conditions (see Fig. 6) where epoxy-matrix composites cannot be used.

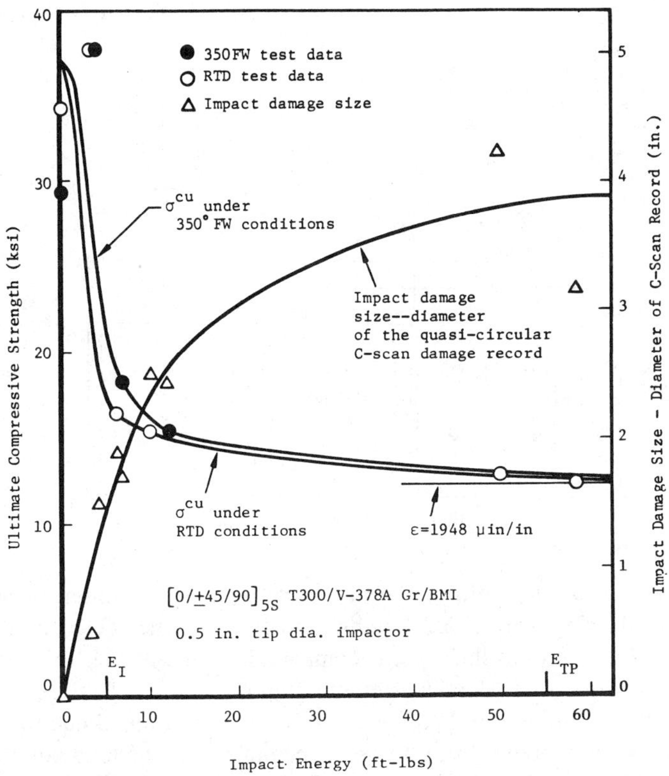

FIG. 6—*Residual compressive strength and impact damage size as a function of impact energy for* $[0/\pm45/90]_{5s}$ *T300/V-378A graphite/bismaleimide laminates* [7].

Tests on Hat-Stiffened, Flat Shear Panels

Hat-stiffened, flat shear panels, representative of an aft fuselage section in a typical V-STOL aircraft, were also fabricated using the T300/V-378A material (see Fig. 7). An "eccentrically loaded cantilever beam" was used to introduce nominally pure shear in the test specimen without any stress concentration or adverse effects. A schematic of this setup is shown in Fig. 8. In this setup, the panel was part of an eccentrically loaded cantilever beam that contained two dummy metal end bays. These end bays helped to distribute the shear load uniformly in the test panel. Metal chords were used to simulate the frames in a typical fuselage panel. External loading was applied through the centroid of the panel to minimize the stressses in the chords, which in turn resulted in the use of realistic chord sizes.

Two static tests were performed, one under RTD conditions and the other under 350FW conditions. Tests under 350FW conditions were conducted by building a heat chamber around the entire panel and the metal chords. Baffles were introduced into the heat chamber to make the circulation of the hot air uniform. Baffle locations were adjusted to achieve a uniform temperature reading

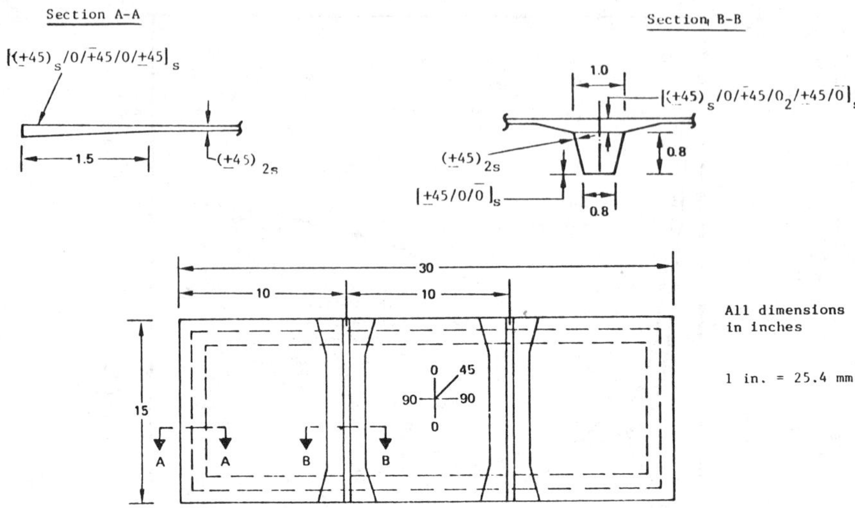

FIG. 7—*Shear panel design* [7].

over the entire panel planform. From the test results, the initial buckling load was obtained using a modified Southwell plot [*15*]. The RTD initial buckling load of 27.7 kN/m (158 lb/in.) was comparable to the 30.5 kN/m (173 lb/in.) value obtained for an 18% thicker graphite/epoxy panel in Ref *15*. Under RTD and 350FW conditions, the T300/V-378A shear panels failed due to the delamination of the hat stiffener from the web, accompanied by a loud popping sound. When the stiffener delaminated, some splitting between the fibers and fiber breakage were also precipitated in the surface (45°) ply of the web, adjacent to the delamination location. The ultimate failure loads under RTD and 350FW conditions were 79.4 kN/m (453 lb/in.) and 75.9 kN/m (433 lb/in.), respectively.

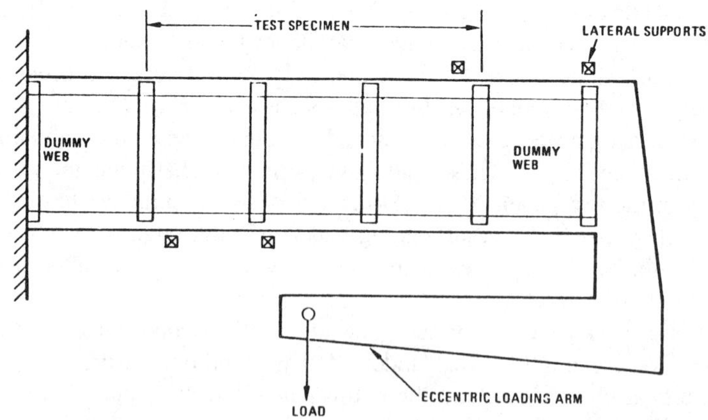

FIG. 8—*Cantilever beam with eccentric loading arm* [7].

The RTD failure value for the T300/V-378A shear panel is less than half the value (159.8 kN/m or 913 lb/in.) measured for the graphite/epoxy panel in Ref *15*. Even if the thickness difference between the compared panel webs is accounted for, the failure load for the T300/V-378A panel is only 59% of the corresponding graphite/epoxy value. The failure mode—delaminating of the stiffeners from the web—is governed by the transverse tensile and interlaminar shear properties of the matrix materials. Again, defects induced by the 288°C (550°F) postcure of the V-378A matrix panel are partially responsible for its inferior behavior compared to the 3501-6 epoxy-matrix panels. Nevertheless, the V-378A matrix panel retains almost all (96%) of its RTD strength under 350FW conditions, establishing its potential where epoxy-matrix panels fail.

Conclusions

Based on the discussed test results, the following conclusions are drawn.

1. The V-378A matrix laminates absorb moisture much faster than the 5208 or the 3501-5 epoxy-matrix laminates. But, the equilibrium moisture content in V-378A matrix laminates under 100% RH conditions is approximately the same as that in 3501-6 epoxy-matrix laminates.
2. Under RTD conditons, the inplane mechanical properties of the T300/V-378A material are comparable to T300/5208 properties. But, the flatwise tensile strength of the V-378A matrix laminate is only 40% of the value for 3501-6 matrix laminates.
3. Under 350FW conditions, matrix-dominated properties (0° compression, 90° tension, and shear) of the T300/V-378A material are much larger than those of the T300/5208 material.
4. The presence of a 6.35-mm-diameter hole in quasi-isotropic T300/V-378A laminates results in an RTD tensile failure strain value of approximately 7000 μmm/mm, larger than the 6000 μmm/mm value realized in AS/3501-6 laminates. Under 350FW conditions, where epoxy-matrix composites cannot be used, the T300/V-378A laminates retain an ultimate tensile strain value of 6600 μmm/mm with the 6.35-mm-diameter hole.
5. A 6.35-mm-diameter hole in quasi-isotropic T300/V-378A laminates lowers its RTD ultimate compressive strain value to 5850 μmm/mm, lower than the 7500 μmm/mm value realized in AS/3501-6 laminates. Under 350FW conditions, though, the V-378A matrix laminates retain an ultimate compressive strain value of 6550 μmm/mm.
6. Filled holes result in higher ultimate compressive strain values than open holes. Under 400FW conditions, T300/V-378A laminates retained compressive failure strain values of 3730 and 3400 μmm/mm with filled and open 6.35-mm-diameter holes, respectively.
7. The $R = -1$ million cycle threshold strain levels for quasi-isotropic T300/V-378A laminates with 6.35-mm-diameter holes are 2700 and 3500 μmm/mm under RTD and 350FW conditions, respectively.

8. The bearing yield strength of T300/V-378A laminates is comparable to that of AS/3501-6 up to 250FW conditions, the limit for epoxy-matrix composites. The V-378A matrix laminates, though, possess a bearing yield strength of approximately 248 MPa (36 ksi) even under 450FW conditions.
9. The V-378A matrix laminates possess a much lower tolerance to low-velocity impact compared to 3501-6 matrix laminates. A 14 N · m (10 ft · lb) impact by a 12.7-mm-tip-diameter object reduces the compressive failure strain of the V-378A matrix laminate to approximately 2500 μmm/mm. This is less than half the value (5250 μmm/mm) retained by 3501-6 matrix laminates.
10. In hat-stiffened flat panels subjected to a pure shear loading, failure is induced by the delamination of the hat stiffener from the web. When similar panels are fabricated using T300/V-378A and AS/3501-6, the RTD failure shear stress resultant for the V-378A matrix panel is less than half the value measured for the 3501-6 matrix panel, though the initial buckling loads are approximately the same. Nevertheless, the V-378A matrix panel retains the RTD shear strength even under 350FW conditions.
11. The processability of V-378A matrix composites is similar to epoxy-matrix composites, and the processing of bismaleimide composites has finally transcended the experimental stage to where the cured laminate microstructure is practically defect-free. The restriction of the postcure temperature to below 245°C (475°F) is expected to improve some of the results obtained in Refs *4* and *7*. The V-378A matrix composites are, therefore, promising candidates for the temperature range between epoxy applications and polyimide applications (120 to 204°C or 250 to 400°F).

Acknowledgment

This paper contains results that were generated at Northrop Corporation on many Independent Research and Development (IRAD) programs and a Navy-sponsored program. The Northrop/Navy program was administered by the Naval Air Development Center with Mr. T. E. Hess as the program technical monitor.

References

[*1*] Ramkumar, R. L., "Effect of Adverse Environments on Composites," *Proceedings,* SESA/JSME Joint Conference on Experimental Mechanics, Honolulu, 23–30 May 1982.
[*2*] Kong, S. J., "Bolt Bearing Strengths of Graphite/Epoxy Laminates," presented at the AIAA/ASME SDM Conference, Atlanta, GA, April 1981.
[*3*] *Graphite/Polyimide Composites,* M. B. Dexter and Davis, Jr., Eds., NASA Conference Publication 2079, National Aeronautics and Space Administration, Aug. 1979.
[*4*] Grimes, G. C., Dusablon, E. G., and Ramkumar, R. L., "Tension, Compression and Shear Properties of Some Basic T300/V-378A Graphite/Bismaleimide Laminates," Report NOR 81-148, Northrop Corp., Dec. 1981.
[*5*] Kong, S. J., "Moisture Absorption Study of Graphite/Polyimide," Report NOR 81-59, Northrop Corp., March 1981.
[*6*] Kong, S. J., "Matrix Dependent Properties of T300/V-378A Graphite/Bismaleimide," Report 82-219, Northrop Corp., Dec. 1982.

[7] Ramkumar, R. L., Grimes, G. C., and Dusablon, E. G., "Structural Evaluation of an Organic Matrix Filamentary Material for Use at Elevated Temperatures," Naval Air Development Center, Report NADC-82069-60, Jan. 1982.
[8] McKague, L., "V378A Polyimide Resin—A New Composite Matrix for the 80's," ASTM Symposium on Composites for Extreme Environments, Bal Harbour, FL, Nov. 1980.
[9] Wilkins, D. J., "A Comparison of the Delamination and Environmental Resistance of a Graphite/Epoxy and a Graphite-Bismaleimide," Naval Air Systems Command, Report NAV-GD-0037, Sept. 1981.
[10] McKague, L., Reynolds, J. D., and Fruit, J. in *Composite Materials: Testing and Design (Sixth Conference), ASTM STP 787,* American Society for Testing and Materials, Philadelphia, 1982, pp. 73–84.
[11] "Mechanical Property Summary-T300/V378A Five Harness Satin Cloth," Fiberite Corporation, 1982.
[12] Loos, A. C. and Springer, G. S., "Moisture Absorption of Graphite/Epoxy Composites Immersed in Liquids and in Humid Air," *Journal of Composite Materials,* Vol. 13, April 1979.
[13] Shen, C. H. and Springer, G. S., "Moisture Absorption and Desorption of Composite Materials," *Journal of Composite Materials,* Vol. 10, Jan. 1976.
[14] Ramkumar, R. L., "Environmental Effects on Composite Damage Criticality," Naval Air Development Center, Report No. NADC-79067-60, Jan. 1982.
[15] Agarwal, B. L., "Flat Stiffened Graphite/Epoxy Tension Field Panels Under Constant-Amplitude Fully-Reversed Fatigue Loads," Naval Air Development Center, Report No. NADC-81169-60, Aug. 1981.

Robert A. Riddle[1] *and Scott W. Beckwith*[1]

Development of Test and Analysis Methods for Thick-Wall, Graphite/Epoxy Filament-Wound Composite Materials Fracture Toughness

REFERENCE: Riddle, R. A. and Beckwith, S. W., "**Development of Test and Analysis Methods for Thick-Wall, Graphite/Epoxy Filament-Wound Composite Materials Fracture Toughness,**" *Composite Materials: Testing and Design (Seventh Conference), ASTM STP 893,* J. M. Whitney, Ed., American Society for Testing and Materials, Philadelphia, 1986, pp. 64–83.

ABSTRACT: Fracture control techniques for full-scale structural articles generally require laboratory test data, analysis of key parameters, and evaluation using fracture mechanics techniques. These parameters, usually fracture toughness or critical strain energy release rate, are often measured for metals and even thin-wall composite materials. However, data on thick-wall composites greater than 25 mm is virtually nonexistent.

The Space Shuttle solid rocket booster case is currently designed using D6AC steel. However, considerable weight savings are achieved by using conventional filament-wound composites to replace the steel case. The Filament Wound Case (FWC) program is aimed at developing graphite/epoxy composite segments for the Space Shuttle.

Each FWC segment is 5.9 to 8.2 m (19.5 to 27.0 ft) in length and 3.7 m (12 ft) in diameter. The thickness of the composite varies from 28 mm (1.12 in.) in the center, or membrane region, to 46 mm (1.8 in.) in the joint region. The layup in the membrane region consists of 11 layers of $\pm 29°$ helical and 90° hoop fibers with the Hercules AS4-12K graphite and HBRF-55A resin system. In the joint region, 0° broadgoods are inserted to increase strength and stiffness. The joint helical and hoop layers are also Hercules AS4-12K graphite fiber with HBRF-55A resin, while the 0° broadgoods are AS4-12K graphite prepreg with 3501-5A resin.

As part of the fracture control program, emphasis was placed on the development of specific test methods to measure pertinent fracture parameters for the thick-wall, filament-wound graphite composite. This paper discusses several of the samples selected and analyzed to date as well as the experimental results derived from full-thickness samples removed from a dissected composite segment.

Test methods currently include both Mode I and Mode II fracture toughness measurements. Interlaminar fracture toughness and strain energy release rates were determined for both modes using a double cantilever beam and modified rail shear test, respectively. Mode I cross-ply fracture toughness was determined using a wide, double-edge notch specimen. Finite-element analyses were conducted on all of the specimens. A J-integral post processor was used to develop corresponding Mode I and Mode II strain energy release rates.

[1] Staff engineer and superintendent, respectively, Hercules Aerospace Division, Materials Technology Department, Magna, UT 84044. Dr. Riddle is now an engineer at Lawrence Livermore National Laboratory, Livermore, CA 94550.

The values determined from the thick-wall, filament-wound composites were compared to other data from autoclaved systems reported in the open literature. Problems encountered in testing these thick-wall samples and existing analysis limitations are also discussed.

KEY WORDS: filament-wound composites, fracture control, graphite fiber, epoxy, fracture mechanics, fracture toughness, critical strain energy release rate, *J*-integral, delamination, mixed-mode fracture, test methods, composite materials

Nomenclature

a	Notch or crack depth
E_{yy}, E_{zz}, E_{xx}	Lamina, laminate extensional moduli
E_L, E_T	Laminate moduli, defined in relation to crack plane (L = longitudinal, perpendicular to crack plane, T = transverse)
G_I, G_{II}	Strain energy release rates (Modes I and II)
G_{xy}	Lamina, laminate shear modulus
G_{TL}	Laminate shear modulus defined in relation to crack plane
J	*J*-integral
K_I, K_{II}	Stress intensity factors (Modes I and II)
W	Specimen width
x, y, z	Cartesian coordinate directions
$\nu_{yz}, \nu_{yx}, \nu_{zx}$	Poisson's ratio
ν_{LT}	Poisson's ratio

The Space Shuttle solid rocket booster case is currently designed using D6AC steel. However, a significant weight savings in the booster rocket assembly is possible by using filament-wound composites to replace portions of the steel case [*1*]. The Filament Wound Case (FWC) design concept for use on the Space Shuttle is shown in Fig. 1.

The FWC segments are filament wound to an internal diameter of 3.7 m (146 in.) through a series of steps. The process generally consists of winding the Hercules AS4-12K graphite fiber to form helical (±29°) patterns (Fig. 2) and hoop (90°) patterns (Fig. 3) on a metal mandrel. The cured composite segments, such as those shown in Fig. 4, are then removed mechanically from the mandrel. The pin holes are drilled in the ends (Fig. 5), and the unit undergoes ultrasonic inspection as shown in Fig. 6.

The objective of applying fracture control techniques to the filament-wound case is to produce a component of high reliability with a minimum number of full-scale burst tests. Translation of the principles of fracture control, which were developed primarily for metallic pressure vessels and aircraft structures, to something applicable to thick-wall composite structures involves consideration of the concepts of initial flaw sizes and types, critical flaw sizes, and subcritical flaw growth characteristics.

Specifically addressed in this paper is the design and analysis of material tests

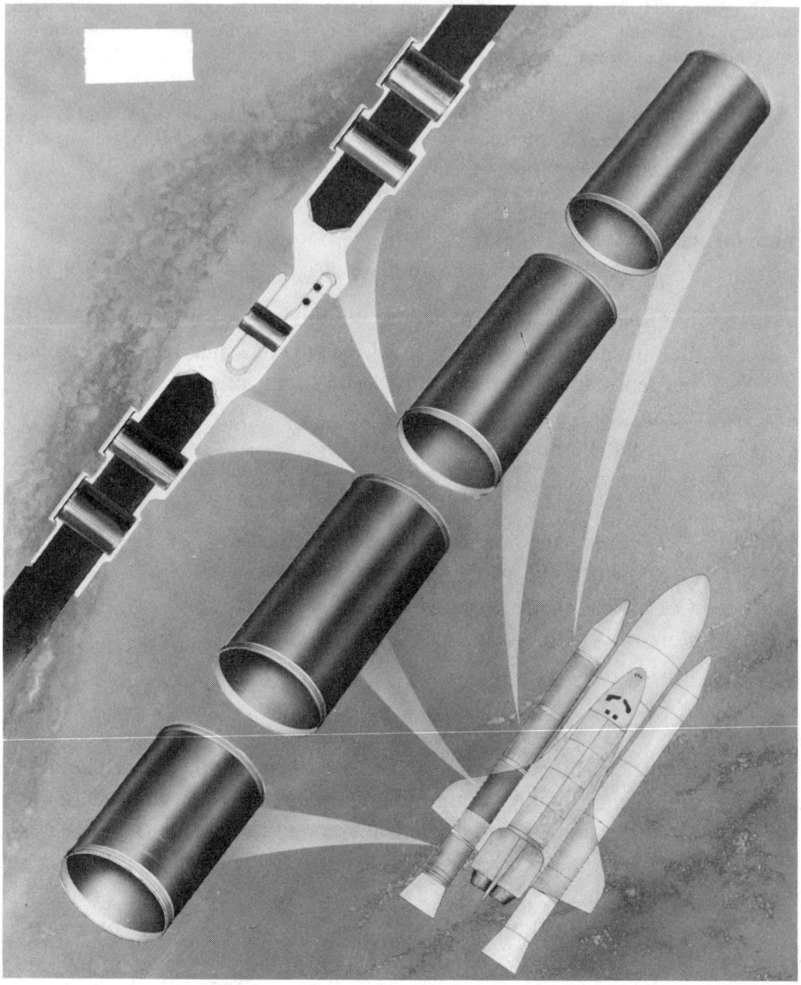

FIG. 1—*Filament Wound Case (FWC) concept.*

used to define the material allowables for critical flaw size calculations. The necessary material properties are defined by the expected failure modes.

Structural Loading Conditions

Potential failure modes for the FWC depend on both the loading conditions and the types of preexisting or initial flaws. During qualification and use, the FWC is subjected to hydroproof, flight, and compressive loading conditions. The initial flaw types of particular interest are delaminations and surface cuts.

To predict delamination growth using a fracture mechanics approach, the crit-

FIG. 2—*Full-scale FWC segment in helical winding phase.*

ical interlaminar strain energy release rates for Modes I, II, and III are necessary. However, since the analysis performed to determine critical delamination sizes is largely two-dimensional and axisymmetric, Modes I and II are the loading modes of primary importance.

In predicting delamination growth, crack growth is typically under mixed-mode loading conditions. The pressure causes the Mode I applied strain energy release rate (G_I) for a delamination to be negative for hydroproof and flight loads. This negative G_I causes compressive normal forces between the delamination surfaces, and hence tends to reduce the propensity for delamination growth. The Mode II applied strain energy release rate (G_{II}) is therefore responsible for delamination growth, and the maximum G_{II} occurs in the regions of maximum interlaminar shear stress. For compressive loading, which occurs just prior to ignition in the FWC of the Space Shuttle solid rocket motors, G_I is positive, and both G_I and G_{II} contribute to delamination growth.

Ideally, mixed-mode failure curves would be available with interlaminar fracture values for both positive and negative G_I plotted against a range of G_{II} values, and with frictional effects accounted for when G_I is negative. However, the tests to determine critical G_I and G_{II} values form a necessary basis for prediction of delamination growth, within analytical limitations.

FIG. 3—*Full-scale FWC segment in hoop winding phase.*

FIG. 4—*Mandrel removal from full-scale FWC forward segment.*

FIG. 5—*FWC full-scale hole drilling operation.*

FIG. 6—*FWC in ultrasonic NDE inspection station.*

The failure mode depends on the relative depth of the crack with respect to the laminate thickness for surface cuts. Delaminations blunt the crack advance through the composite wall thickness for shallow cuts. Failure occurs in this case because of stress redistribution in the remaining laminate material, and additional fiber bending in the region of the cut. Self-similar flaw growth occurs for relatively deep surface cuts, with the resulting catastrophic failure mode.

Whether blunting delamination or self-similar flaw growth occurs depends on the relative magnitudes of the Mode I and Mode II interlaminar critical strain energy release rates, as compared to the Mode I cross-ply critical strain energy release rate. The occurrence of the blunting delamination near the crack tip of a shallow surface cut is due to the relatively more predominant effect of interlaminar shear stresses in the crack tip region, as compared to the in-plane stresses. These interlaminar shear stresses are more pronounced for shallow cuts as compared to deep cuts.

Based on the initial flaw types and expected failure modes, the material parameters necessary to predict flaw sizes are Modes I and II interlaminar and Mode I cross-ply fracture toughness values.

Previous Work

Unfortunately, most of the previous work on developing the pertinent fracture parameters for composites deal with thin-wall, autoclave systems [*2–8*], and only recently have thicker, filament-wound composites been investigated [*9,10*]. The latter studies were conducted on materials with a nominal thickness of 7.1 mm (0.28 in.) for Mode I cross-ply only. A variety of Mode I [*11–16*] and Mode II [*14,17*] tests have recently been developed and applied to thin-wall composites with a good degree of success.

Many of these studies have yielded a range of fracture toughness or strain energy release values for graphite/epoxy composites that provide a starting point for fracture control design and analysis. The work in Ref *9* reviews much of the available fracture work, so an in-depth review will not be given in this paper.

Filament-Wound Case Composite Material

The filament-wound case segments are fabricated with Hercules AS4-12K graphite fiber and two resin systems, HBRF-55A and 3501-5A. The AS4-12K graphite fiber is made from a polyacrylonitrile (PAN) precursor and has a nominal tensile strength of 3.65 GPa (529 ksi), modulus of 235 GPa (34.0 msi), and an elongation of 1.50% as determined by laminate lot acceptance data (295 tests). Tow tests, which may be more typical of filament winding processes, yield values somewhat higher for tensile strength and elongation.

The winding resin is the Hercules HBRF-55A system that is an amine-cure epoxy resin. The resin has ambient temperature mechanical properties of 86.3 kPa (12.5 ksi) tensile strength, 3.17 GPa (460 ksi) tensile modulus, and 7.8% elongation. The 0° broadgoods used in the segment joint areas use the standard

Hercules 3501-5A resin system commonly used for aircraft autoclaved parts. The 3501-5A resin is used because of handling and processing requirements associated with layup of broadgood materials.

The four FWC case segments vary in length from 5.9 to 8.2 m (19.5 to 27.0 ft), and are 3.7 m (12 ft) in diameter. The thickness of the composite varies from 28.4 mm (1.12 in.) in the membrane region to 45.7 mm (1.8 in.) in the joint region. The thickness buildup in the joint region is due to the addition of 0° graphite broadgoods, which are inserted to improve strength and stiffness in this region of bending. Figure 7 shows the layup for both the membrane and joint regions of the first full-scale development segment manufactured. Each of the 11 helical layers (±29°) in both the membrane and joint regions has a thickness of 1.87 mm (0.0736 in.). Likewise, the hoop layers have a thickness of 0.394 mm (0.0155 in.). The 0° broadgoods in the joint region vary in thickness from 2.28 mm (0.0896 in.) near the inside diameter (ID) to 0.650 mm (0.0256 in.) near the outside diameter (OD). In the transition region, the helical layers are feathered in and out to smooth the transition in thickness between the membrane and the joint regions.

Following the standard gel and cure cycles, a thermal stress relief cycle, and then a 24-h postcure cycle at 149 K (300°F), the end sections of a full-scale segment were cut off to provide a nominal 64-cm (25-in.) length ring. The ring, designated a tag-end ring, represents the material found in the joint region of the segments. The specific samples described in later paragraphs were machined using a radial arm saw for the initial rough cut and then an end mill for final finishing. While the first specimens tested and reported in this paper are from the joint region, future tests are planned in which specimens are taken from the membrane region.

All of the subsequent tests were conducted after allowing the materials to dry thoroughly. They were instrumented and tested at 25 K (77°F) and ambient relative humidity.

Interlaminar Fracture Toughness

Two fracture modes were evaluated in the interlaminar loading condition, Mode I and Mode II. These tests are described in the sections that follow.

Mode I Testing

The specimen used to determine Mode I interlaminar fracture toughness is the double-cantilever-beam type shown in Figs. 8 and 9. Pin loading of the specimen is possible because of the specimen thickness, 28.4 mm (1.12 in.), which was machined from the joint region. The notch was machined using a 3-mm (0.125-in.) blade with a 90° taper at the edge.

The interlaminar resistance to flaw growth varies at every ply interface, depending on the fiber ply orientation and the amount of porosity at the fiber interfaces so that with this specimen one gets the minimum fracture toughness

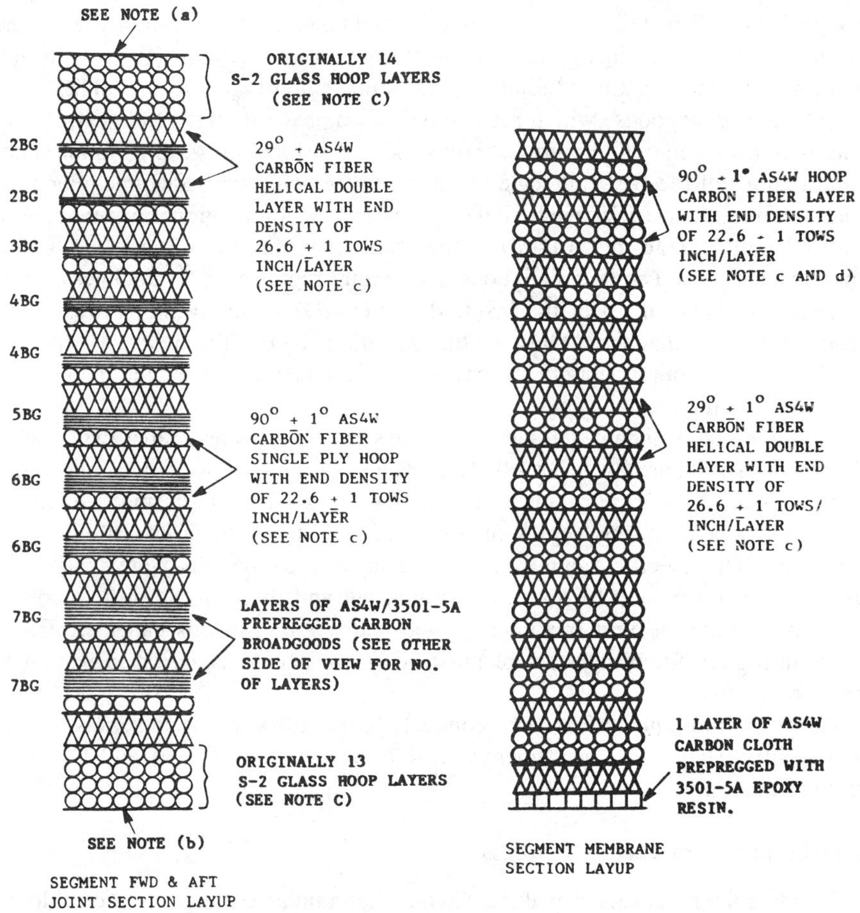

(*a*) Outside diameter glass surface originally consisted of a 14-layer glass hoop buildup. Surface machined to a final OD requirement.
(*b*) Inside diameter glass surface originally consisted of a 13-layer glass hoop buildup. Surface machined to a final ID requirement.
(*c*) Segment carbon and glass fiber hoop and helical layers are wound with HBRF-55A epoxy resin.
(*d*) Two hoop plies equal a layer.

FIG. 7—*Details of segment laminate construction with joint (11 double helicals/10 single-ply hoops/10 broadgoods stacks with 46 plies and OD and ID glass hoop buildups) and membrane (11 double helicals/10 double-ply hoops with 20 hoop plies) details shown.*

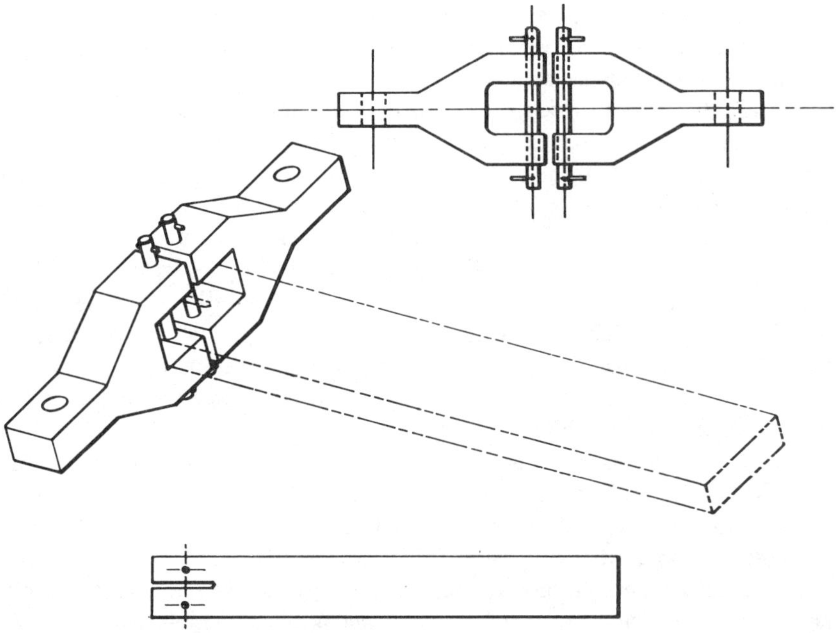

FIG. 8—*Interlaminar Mode I fracture toughness specimen.*

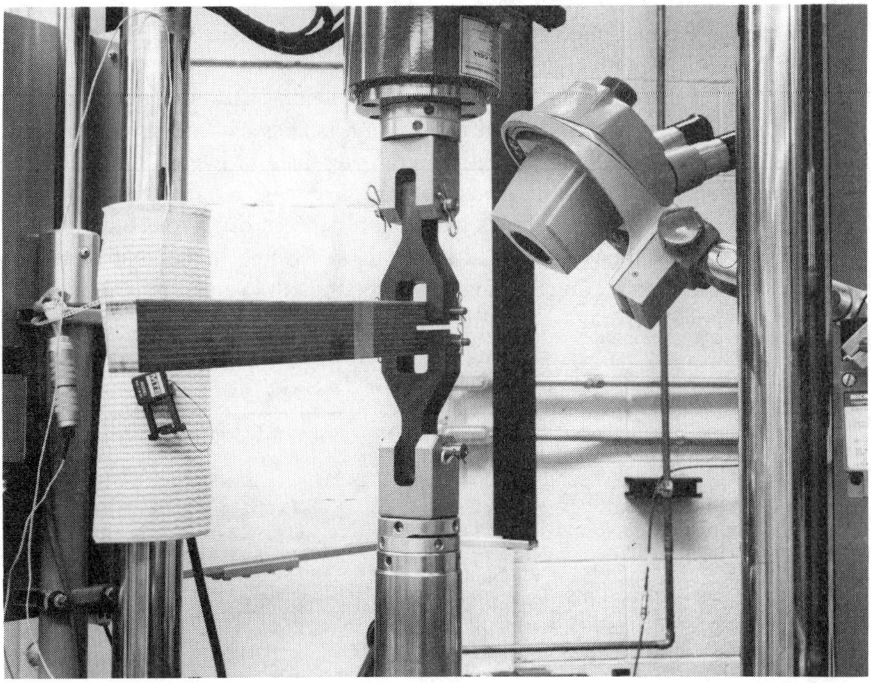

FIG. 9—*Interlaminar Mode I specimen in test fixture.*

value of the ply interfaces near the root of the notch. The ply interface with the minimum fracture strength was consistently either at the helical-to-0° ply interface, or within the helical layer itself.

Although the specimen was originally designed for multiple crack advances, and therefore multiple measurements of the critical interlaminar strain energy release rate, fiber bridging past the first increment of crack growth complicated the failure process. The effort to get a resin-dominated fracture toughness value was negated with the fiber carrying even a small part of the load.

The double-cantilever-beam specimen was modeled on a layer-by-layer basis using a finite-element analysis of the specimen. With the glass hoops at the internal diameter machined away during specimen preparation, the remaining 32 layers were asymmetric with respect to the plane at the midthickness, because of the preponderance of 0° graphite broadgoods near the internal diameter. The lamina properties for the layers are seen in Table 1.

Since the notch was machined as nearly as possible at the specimens midthickness, this material asymmetry with respect to the plane of the notch should cause a condition of mixed-mode loading at the crack tip.

A J-integral post processor was used in conjunction with the finite-element analysis to determine both the applied strain energy release rate and the relative amounts of Mode I and Mode II loading. Since a linear, finite-element analysis was performed, the J-integral is exactly equivalent to the strain energy release rate, and the method of separating the Mode I and Mode II loading components was done as in Ref *17*.

The analysis showed that in spite of the material asymmetry, the loading was nearly pure Mode I, with $G_{II}/G_I = 3.13 \times 10^{-4}$.

For the test results shown in Table 2, the failure load was taken as the maximum specimen load. Both the choice of the maximum load as the failure load and the linear finite-element analysis seemed appropriate since in every case the load-versus-displacement plot for the specimens was very nearly linear to the point of maximum load. Figure 10 shows the fracture surface of a typical sample.

Some of the specimens were load-cycled to determine if the sharp tip of a fatigue crack reduced the interlaminar fracture strength. Since the object of the cyclic loading was to induce a fatigue crack and not to systematically study

TABLE 1—*FWC lamina properties used in analysis.*

		Helical (±29°)	Hoop (90°)	Broadgoods (0°)
E_{yy}	GPa (msi)	8.16 (1.18)	7.76 (1.13)	10.7 (1.552)
E_{zz}	GPa (msi)	9.87 (1.43)	132 (19.2)	10.7 (1.552)
E_{xx}	GPa (msi)	49.4 (7.16)	7.76 (1.13)	118 (17.1)
ν_{yz}	...	0.1886	0.0157	0.3125
ν_{yx}	...	−0.0175	0.0157	0.025
ν_{zx}	...	0.2776	0.2674	0.025
G_{xy}	GPa (msi)	4.5 (0.6643)	3.1 (0.444)	6.6 (0.96)

TABLE 2—*Double-cantilever-beam test results (Mode I interlaminar fracture toughness).*

Specimen Identification	Failure Load, N (lb)	Strain Energy Release Rate, J/m² (in. · lb/in.²)	Fracture Toughness, MPa $\sqrt{m}$ (ksi $\sqrt{in.}$)
TFS-32	1560 (350)	190 (1.08)	1.4 (1.3)
TFS-33[a]	1420 (320)	160 (0.90)	1.3 (1.2)
TFS-34	1560 (350)	190 (1.08)	1.4 (1.3)
TFS-35[a]	1420 (320)	160 (0.90)	1.3 (1.2)
TFS-36[a]	1420 (320)	160 (0.90)	1.3 (1.2)

[a]Fatigue cycled.

fatigue crack growth rates, the maximum load was increased gradually until the fatigue crack grew visibly in less than 5000 cycles. The fatigue crack started to grow at 75% of the subsequent maximum load for two of the samples, and for the third it began at 90% of the maximum load.

The fatigue cracks thus started were unfortunately not a uniform extension of the notch, but tended to follow the helical angle, in the helical ply, at the edge of the specimen. Also, the fatigue cracks did not significantly reduce the load to failure. It was therefore concluded that the interlaminar fracture strength is not very sensitive to the crack tip radius. Also, that unidirectional ply specimens are better candidates for a systematic study of fatigue crack growth rates.

Comparing the average value for G_{Ic} from these tests, 170 J/m² (0.97 in · lb/in.²), with those reported in the literature for similar materials and tests, reasonable agreement is found [*11–16*]. Since the materials in Refs *11–16* were autoclaved, as compared to filament wound in this study, this difference in material processing evidently does not significantly affect the Mode I interlaminar fracture toughness.

The critical stress intensity factors (K_{Ic}) were calculated from the G_{Ic} values using the relationship

$$K_I = \left\{ \frac{\sqrt{2}\, G_I\, E_L}{\left[\sqrt{\dfrac{E_L}{E_T}} - \nu_{LT} + \dfrac{E_L}{2G_{TL}} \right]^{1/2}} \right\}^{1/2} \quad (1)$$

FIG. 10—*Failed surface of interlaminar Mode I specimen.*

Mode II Testing

A modified rail shear specimen was used for measurement of the Mode II interlaminar fracture toughness. The specimen, along with a diagram showing the specimen and loading fixtures assembled, is seen in Figs. 11 and 12. The fixtures are loaded in uniaxial tension through the pin holes at the ends of the fixtures. The ends of the fixtures are adjustable so that the centerline of the applied load passes through the center plane of the specimen, so that pure shear occurs along this plane.

To verify that nearly pure shear did occur in the specimen, the distance between the long vertical parts of the loading fixtures was monitored while load was applied to the specimen. No movement between the fixtures was noted. The fixtures were designed to accommodate specimens within a range of thickness, and also to be able to apply a positive Mode I loading component.

A layer-by-layer linear finite-element analysis was performed on the modified rail shear specimen, similar to that done on the double-cantilever-beam specimen. Using the J-integral post processor on the results of the finite-element analysis, G_{II} was calculated as a function of the applied load.

The G_I component in the specimen loading was neglected because it was small, that is, $G_I/G_{II} = 0.013$.

The results of the testing are seen in Table 3. Once again, the maximum load was used as the failure point. Since the failure was catastrophic, the load drop

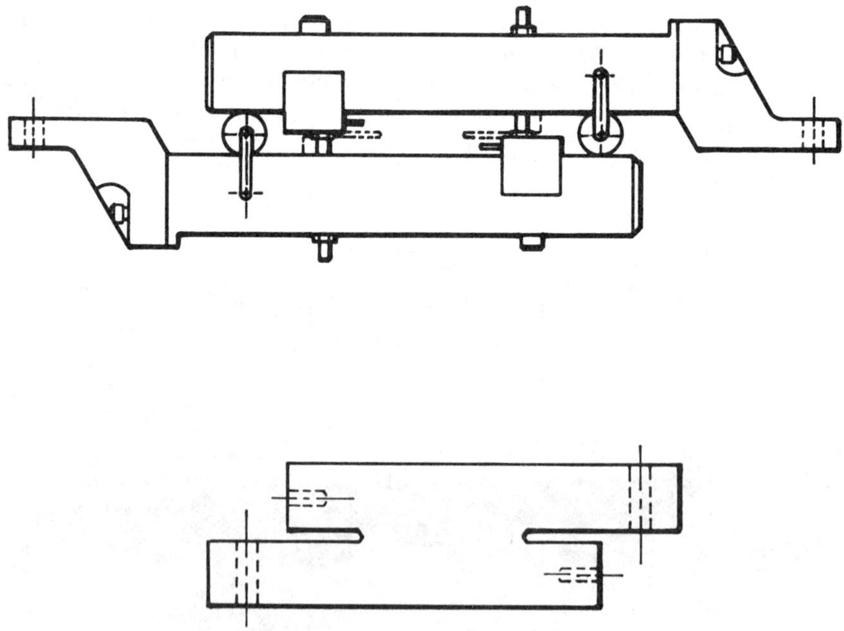

FIG. 11—*Interlaminar Mode II fracture toughness specimen.*

FIG. 12—*Interlaminar Mode II specimen in test fixture.*

off was recorded as a nearly vertical line, and the point of maximum load was clearly defined.

It also appears here, as with the double-cantilever-beam specimen, that efforts to induce a fatigue crack tip at the notch had a negligible effect on the resulting interlaminar Mode II fracture toughness values. The maximum alternating load reached 60% of the average failure load before crack growth occurred. The tendency for nonlinear deformation was more pronounced in this test mode.

Failure occurred in the resin at the interface of the helical and 0° broadgoods layers (Fig. 13). The average shear stress in the specimen at failure was 17.6 MPa (2.55 ksi). This stress level is about 50% of the typically reported shear strengths for this resin and fiber system determined from other test methods. This presents the possibility that the composite material is sensitive to Mode II flaw propagation.

TABLE 3—*Modified rail shear test results (Mode II interlaminar fracture toughness).*

Specimen Identification	Failure Load, N (lb)	Strain Energy Release Rate, J/m² (in. · lb/in.²)	Fracture Toughness, MPa $\sqrt{m}$ (ksi $\sqrt{in.}$)
TFS-37	24 000 (5400)	595 (3.40)	4.92 (4.47)
TFS-39	21 800 (4910)	494 (2.82)	4.48 (4.07)
TFS-40[a]	24 000 (5370)	590 (3.37)	4.90 (4.45)
TFS-41[a]	20 800 (4680)	448 (2.56)	4.27 (3.88)

[a]Fatigue cycled.

Cross-Ply Mode I Fracture Toughness

The specimen used for determining the Mode I cross-ply fracture toughness is the double-edge crack type seen in Fig. 14. This specimen was designed with two specific objectives:

1. to be wide enough to fully load the helical fibers in the remaining ligament, and
2. to be long enough so that the stress concentration at the loading holes does not affect the failure mode.

The specimen is 274 mm (10.785 in.) wide in the hoop direction by 498 mm (19.6 in.) long in the axial direction. Two different notch depths were tested, one of 61.7 mm (2.43 in.) and one of 87.1 mm (3.43 in.), which correspond to $2a/W$ ratios of 0.45 and 0.64, respectively. There was confidence that the helical fibers were fully loaded, particularly with the deeper notch, because of the failure mode, as will be discussed. The two-dimensional, plane-strain, finite-element analysis performed on the specimen indicated that the stress concentration of the

FIG. 13—*Failed surface of interlaminar Mode II specimen.*

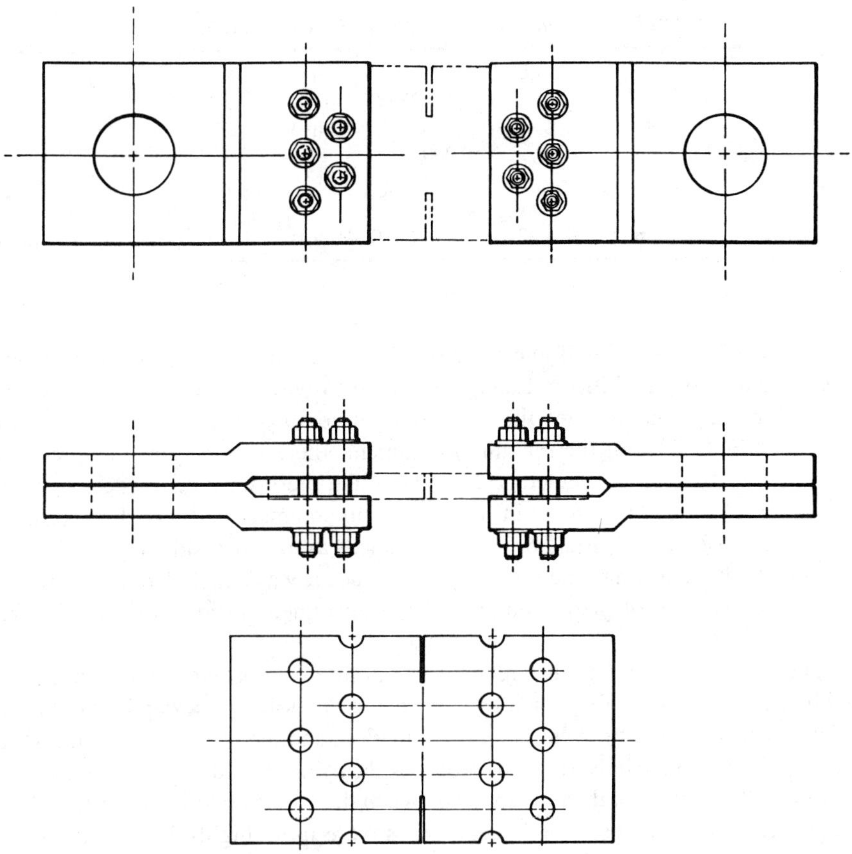

FIG. 14—*Loading diagram for double-edge crack Mode I flaw growth specimen.*

hole did not significantly affect the stress concentration at the notch, and hence did not affect the resulting failure mode.

The laminate properties used for this analysis are seen in Table 4. The in-plane curvature of the sample ($0.547\ m^{-1}$), as these samples were taken from a 3.7-m ring segment, was neglected in the analysis.

The load sharing on the bolt holes was calculated by finite-element analysis by first applying a uniform stress across the remaining ligament, and then determining the resulting forces at the holes. In the final analysis, these forces at the bolt holes were the boundary loading conditions.

The results of the tests are shown in Table 5 and a failed sample is shown in Fig. 15. The applied G_I as a function of load was calculated using the J-integral post processor on the results of the finite-element analysis of the specimen, and also by using the crack closure method, as described in Ref *18*.

TABLE 4—*Laminate properties for double-edge crack specimen.*

E_{yy}	= 24.3 GPa (3.53 msi)
E_{zz}	= 8.62 GPa (1.25 msi)
E_{xx}	= 77.1 GPa (11.1 msi)
ν_{yz}	= 0.300
ν_{yx}	= 0.143
ν_{zx}	= 0.024
G_{xy}	= 14.0 GPa (2.03 msi)

The results reported in Table 5 are based on the crack closure method, though the results from the *J*-integral analysis agreed to within 12%.

Within the limitations of the techniques used to measure the strains in the center of both sides of the remaining ligament, there was reasonable agreement with the analytically predicted strains. Although there were differences between the strain gage readings at the inside and outside diameter sides of the specimen, there was no consistent trend between specimens as to which side, inside diameter or outside diameter, had the greater strain value. It was concluded that there was no significant out-of-plane bending in the specimen, despite the small slight curvature.

There was a significant difference in the failure modes between the specimens with the deep and shallow notches. With the shallow notch depth, the helical fibers near the notch tip were less highly loaded, and therefore interlaminar shear stresses between the helical and 0° broadgoods layers should be greatly increased. As a result, significant delaminations between the layers occurred prior to failure. With the deeper notches, the helical layers were more highly loaded, the fibers were more evenly loaded, and the fibers broke more clearly across the remaining ligament, in a more self-similar type of flaw growth. The resulting predicted G_{Ic} is slightly higher. The values are higher than reported in Refs *5,9,10* but the joint material has a significant number of 0° broadgoods that should increase fracture toughness.

Conclusions

Using the critical Mode I and Mode II strain energy release rates, critical delamination sizes in the FWC segments may be calculated for the various loading conditions using the critical Mode I and Mode II strain energy release rates. Although the effect of the friction between the delamination surfaces is neglected, this is conservative in predicting smaller allowable delaminations.

Although critical flaw sizes could be calculated based on the Mode I critical strain energy release rate, the purpose of this data is beginning to establish that any allowable surface or edge cut would delaminate prior to failure. That is, the most severe surface or edge cuts, where self-similar flaw growth might be expected to occur, and where delamination is minimal, would cause failure in

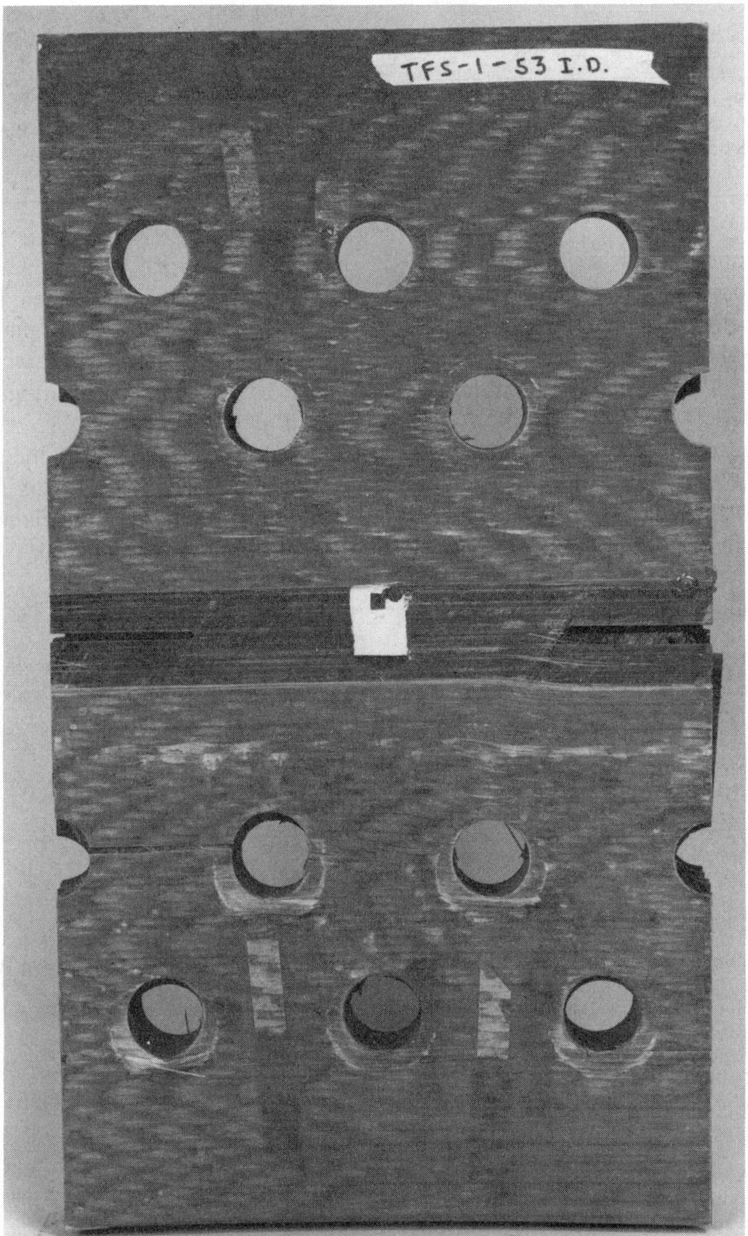

FIG. 15—*Failed surface of Mode I cross-ply specimen.*

TABLE 5—*Results of the double-edge crack (Mode I cross-ply fracture toughness tests).*

Specimen Identification	Failure Load, MN (lb)	2a/W	Strain Energy Release Rate, J/m^2 (in. · lb/in.^{2})	Fracture Toughness, MPa $\sqrt{m}$ (ksi $\sqrt{in.}$)
TFS1-49	2.07 (466 000)	0.64	0.234 (1335)	112 (102)
TFS1-50	2.15 (484 000)	0.64	0.252 (1440)	117 (106)
TFS1-53	2.57 (578 000)	0.45	0.175 (1000)	97.1 (88.3)
TFS1-54	2.59 (583 000)	0.45	0.179 (1020)	97.9 (89.0)

hydroproof. Further tests and analyses are being performed to substantiate this approach. The test method currently in use is to hydroproof the segment to 1.12 times the flight pressure and then conduct four pressure cycles to the maximum expected operating pressure. Any other surface cut of substantial size, where the strength of the case segment might be affected, would delaminate during hydroproof.

On the FWC, this delamination growth would be detected during the before-and-after hydroproof ultrasonic inspections of the case. The criticality of the flaw could then be assessed based on the amount and location of the delamination growth.

Acknowledgments

This paper is based upon studies that were conducted under the Hercules Aerospace Division IR&D effort for advanced filament-wound composite structures on Project RI-52-200. The work was part of a task designed to develop fracture control and accept/reject criteria for thick-wall, filament-wound graphite/epoxy composites such as the Space Shuttle Filament Wound Case (FWC). The assistance of Mr. Joe Medina in testing many of the samples from the full-scale units is appreciated.

References

[*1*] "Solid Rocket Motor (SRM) Filament Wound Case Feasibility Study," Final Report H250-12-1-012, Vol. 1, NASA Contract NAS8-34356, Hercules Aerospace Division, Feb. 1982.

[*2*] Hulsey, R. C., "Delamination Fracture Toughness of a Unidirectional Graphite/Epoxy Composite," M.S. thesis, Texas A&M University Report MM 3724-80-9, 1980.

[*3*] Cruse, T. A., *Journal of Composite Materials*, Vol. 7, April 1973, p. 218.

[*4*] Bathias, C., Esnault, R., and Pellas, J., *Composites*, Vol. 12, No. 3, July 1981, p. 195.

[*5*] Beaumont, P. W. R. and Server, W. L. in *Composite Reliability, ASTM STP 580*, American Society for Testing and Materials, Philadelphia, 1975, p. 443.

[*6*] Ueng, C. E. S., Aberson, J. A., and Lafitte, B. A., *Journal of Composite Materials*, Vol. 11, April 1977, p. 222.

[*7*] Morris, D. H. and Hahn, H. T., *Journal of Composite Materials*, Vol. 11, April 1977, p. 124.

[*8*] Wetherhold, R. C., "Strength of Notched Composite Materials," Center for Composite Materials Report CCM-79-05, University of Delaware, Newark, June 1979.

[*9*] Bau, H. and Beckwith, S. W., "Fracture Mechanics Evaluation of Filament Wound Case Materials Subjected to Operational Environments," *AIAA/ASME/SAE 18th Joint Propulsion*

Conference, 21–23 June 1982, AIAA Paper 82-1070, American Institute of Aeronautics and Astronautics.

[*10*] Beckwith, S. W., "Elevated Temperature and Thermal Postcure Effects on Composite Material Fracture Toughness," *Proceedings,* 28th National SAMPE Symposium and Exhibition, 12–14 April 1983, Society for the Advancement of Materials and Process Engineering.

[*11*] Devitt, D. F., Schapery, R. A., and Bradley, W. L., *Journal of Composite Materials,* Vol. 14, 1980, pp. 270–285.

[*12*] Williams, D. R., "Mode I Transverse Cracking in an Epoxy and a Graphite Fiber Reinforced Epoxy," M.S. thesis, Texas A&M University Report MM 3724-81-16, 1981.

[*13*] Cullen, J. S., "Mode I Delamination of Unidirectional Graphite/Epoxy Composite Under Complex Load Histories," M.S. thesis, Texas A&M University Report MM 3724-81-13, 1981.

[*14*] Vanderkley, P. S., "Mode I/Mode II Delamination Fracture Toughness of a Unidirectional Graphite/Epoxy Composite," M.S. thesis, Texas A&M University Report MM 3724-81-15, 1981.

[*15*] Whitney, J. M., Browning, C. E., and Hoogsteden, W., *Journal of Reinforced Plastics and Composites,* Vol. 1, 1982, pp. 297–313.

[*16*] Nicholls, D. J. and Gallagher, J. P., *Journal of Reinforced Plastics and Composites,* Vol. 2, 1983, pp. 2–17.

[*17*] Riddle, R. A., Streit, R. D., and Finnie, I., in *Transactions,* American Society of Mechanical Engineers, *Journal of Engineering Materials and Technology,* Vol. 105, 1983, pp. 268–272.

[*18*] Rybicki, E. F., Schmueser, D. W., and Fox, J., *Journal of Composite Materials,* Vol. 11, 1977, pp. 470–486.

Failure Mechanisms

John F. Mandell,[1] *Dodd H. Grande,*[1] *Tseng-Hua Tsiang,*[1] *and Fredrick J. McGarry*[1]

Modified Mircodebonding Test for Direct *In Situ* Fiber/Matrix Bond Strength Determination in Fiber Composites

REFERENCE: Mandell, J. F., Grande, D. H., Tsiang, T.-H., and McGarry, F. J., **"Modified Microdebonding Test for Direct *In Situ* Fiber/Matrix Bond Strength Determination in Fiber Composites,"** *Composite Materials: Testing and Design (Seventh Conference), ASTM STP 893,* J. M. Whitney, Ed., American Society for Testing and Materials, Philadelphia, 1986, pp. 87–108.

ABSTRACT: This paper describes a test method for determining the *in situ* fiber/matrix bond strength of composites. In a sequence of steps, a compressive load is applied to individual fibers oriented normal to a polished surface, and the interface is observed microscopically between loading steps, until debonding occurs. The direct result of a test is the force to produce debonding of a particular fiber. The data are reduced to a nominal interfacial shear strength through a finite-element analysis using a simplified model of the fiber, surrounding matrix, and uniform composite properties beyond the matrix. Bond strength results for a variety of composites range from 23 MPa for E-glass/polyester to 86 MPa for T300/5208 graphite/epoxy. The results are consistent with data available from other test methods.

KEY WORDS: composite materials, bond strength, interface, test methods, finite-element analysis

The importance of the fiber/matrix interface has been long recognized in composites technology; often its failure precipitates general cracking under mechanical loading or environmental exposure. However, the absence of a practical test method for the interfacial strength for realistic composites has prevented direct consideration of this property in most studies. While fiber and matrix properties are readily available for common composite systems, the interface quality is usually addressed only indirectly through its influence on the composite transverse and interlaminar shear strength. This paper describes a direct test conducted on the individual fibers of conventional composites that may provide a practical method of interfacial strength measurement for a broad range of composite systems.

[1] Principal research associate, graduate student, graduate student and professor, respectively, Department of Materials Science and Engineering, Massachusetts Institute of Technology, Cambridge, MA 02139. Mr. Tsiang is presently at Rohr Industries, Inc., Chula Vista, CA 92012.

Although no interface test has been available for typical composites, a number of methods using model systems have provided useful information on fiber/matrix adhesion. These include:

1. standard adhesion tests employing the matrix as the adhesive and bulk adherends of the fiber composition [*1*,*2*],
2. fiber pullout [*3*,*4*],
3. single fibers in bulk resin samples loaded so as to produce debonding [*5*,*6*], and
4. single fibers in bulk resin loaded so as to break the fiber into fragments of critical length [*7*,*8*].

Relative to these methods, an *in situ* interface test for realistic composites has the obvious advantages of reflecting actual processing conditions and allowing interface characterization by the user without the need for specialized model construction. It can also be used to determine *in situ* interface degradation during fatigue and environmental exposure of laboratory specimens [9] and, possibly, the monitoring or failure analysis of parts in service.

Experimental Method

The modified microdebonding apparatus is depicted in Figs. 1 and 2. A composite specimen is prepared with some fibers normal to a surface that has been polished using standard metallography techniques. The apparatus applies a compressive load to the end of an individual fiber in a series of steps of increasing force, with the specimen inspected for the onset of debonding between each step. The force necessary to produce debonding is recorded, and a micrograph is taken of the fiber and surrounding region. The nominal interfacial shear strength is calculated from the debonding force using the results of a finite-element solution that incorporates the fiber diameter, spacing to the nearest neighbor fiber, and the elastic (orthotropic) properties of the fiber, matrix, and far-field composite.

The initial version [*9*] of the microdebonding test was similar to the modified (current) version except that a small group of fibers was compressed. This produced debonding, but it was not possible to reduce the debonding force to a nominal interface strength because of the complex details of the contact stresses that were directly involved in the debonding process [*10*]. The original version also was sensitive to the geometry and alignment of the loading probe. The modified version was developed to load only the central part of the diameter of a single fiber, since the finite-element analysis indicated that this would remove the debonding site from the complex contact problem, allowing an approximate debonding stress to be calculated. This change also eliminated the sensitivity to probe alignment and geometry. While the modified version requires precise probe positioning and loading as well as vibration isolation, it has resulted in a practical test giving reproducible data that can be reduced to a nominal interfacial shear strength value. Additional details of the development of the test apparatus may be found in Ref *11*.

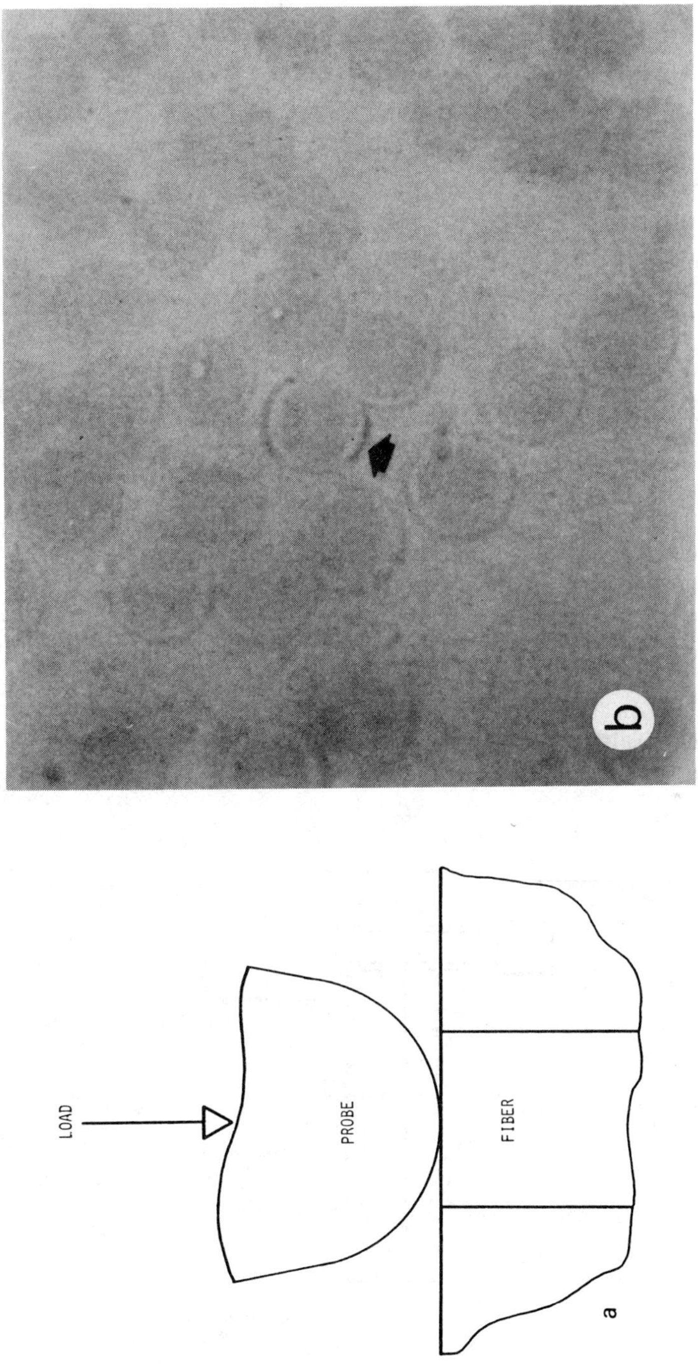

FIG. 1—(a) *Schematic of loading for microdebonding test, and* (b) *micrograph of debonded fiber in S-glass/epoxy (fiber diameter = 10 μm).*

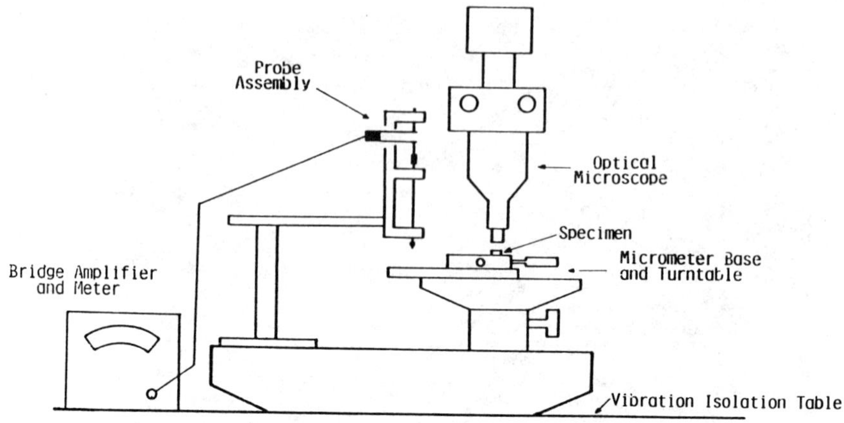

FIG. 2—*Schematic of microdebonding apparatus.*

Figures 2 and 3 show details of the apparatus. Several features are important: (1) the micrometer base allows initial location of the point to be loaded to within 1 μm; (2) the turntable with magnetic stops allows repeated rotation from the loading to the viewing position with an accuracy better than 1 μm; (3) the probe mechanism with a 5 to 15-μm-radius ground diamond tip allows soft loading contact through the Instron Mod. 2620-526 extensometer used as a load transducer, with load measurement to within 0.1 g and frictional sticking of less than 0.1 g (for the results given here, a 6-μm-radius tip was used for the graphite fibers, while a 13-μm tip was more convenient for the larger diameter glass

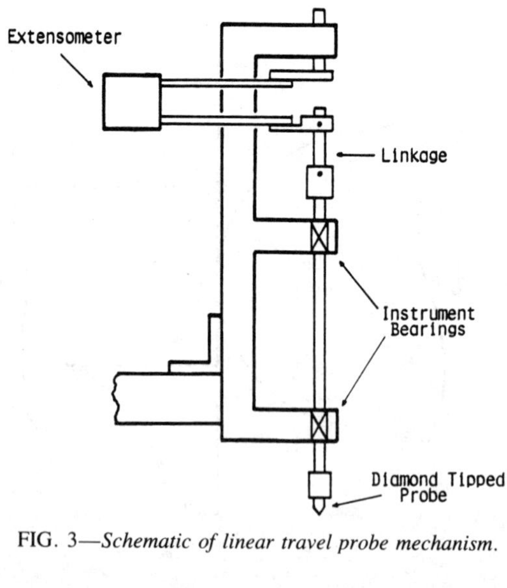

FIG. 3—*Schematic of linear travel probe mechanism.*

fibers); (4) a research quality light microscope capable of detecting debonding for the material of interest, also used to obtain a photomicrograph of the fiber and surroundings from which fiber diameters and spacings can be determined within 0.5 μm; and (5) the apparatus is mounted on a vibration isolation table, which is essential in most locations.

With a transparent matrix, debonding is clearly visible as a dark ring around glass fibers (Fig. 1*a*) or as a reflection at the interface of graphite fibers; unfortunately, the debond is more difficult to demonstrate with a micrograph than in direct viewing.

Specimen Preparation

Specimens for the microdebonding test consist of composite laminates that have been sectioned perpendicular to the reinforcing fibers. They may be either unidirectional or multidirectional laminates or chopped strand materials such as sheet molding compound (SMC), which have been sectioned perpendicular to the fibers in a particular lamina. Specimens are then potted in a polyester mounting compound and polished to a very smooth surface using standard metallographic techniques. Specimen preparation typically consists of the following steps.

1. Section specimen perpendicular to fibers.
2. Pot specimens in polyester compound.
3. Polish specimens using standard metallographic techniques (alumina down to 0.05 μm with rayon bonded to cotton cloth or diamond compound with a silk cloth).
4. Store in dessicator jar until time of test.

Because of difficulties in obtaining a flat surface, tests on single fibers in bulk resin have been unsuccessful, while tests on single strands have been very difficult.

Apparatus Calibration and Alignment

The following steps are taken for apparatus calibration and alignment.

1. Calibrate bridge amplifier and load transducer using analytical weights.
2. Align the cross hair of the ocular lens to correspond with the location of probe contact on the specimen. This is most easily done by scratching an X on the polished surface of the potting compound surrounding the specimen and locating the probe near the intersection of the lines through a combination of adjusting the probe position, adjusting the objective lens, and rotating the ocular lens containing the cross hair.

Test Procedure

The test procedure consists of the following steps.

1. Mount the specimen on the stage using plasticene. (Use a press to level the plasticene on the specimen base.)
2. Locate the region of interest on the specimen using the micrometer drive stage.
3. On a nearby area, check alignment of the optics, which may shift during use of the camera, etc.
4. Align the crosshair with the center of the fiber of interest.
5. Rotate the turntable until the magnetic stop engages in the probe position.
6. Slowly raise the specimen using the microscope coarse focus knob until the probe nearly contacts the specimen.
7. Using the fine adjustment knob, raise the specimen until the desired load is indicated. Hold the load for 5 s, then lower the specimen to unload.
8. Rotate the turntable until the magnetic stop engages with the specimen under the optical axis of the microscope.
9. Inspect the specimen for initiation of debonding or evidence of off-centered loading or probe sliding (the probe leaves a small indentation on the fiber end). In the event of an off-centered loading, the test point should be discarded. If evidence of the probe's sliding is observed, the linkage of the probe mechanism should be realigned and the specimen releveled.
10. Repeat Steps 5 through 9 until interfacial debonding is detected.
11. Record the load at initiation of debonding and photograph the tested fiber for measurement of the fiber diameter and spacing.

Analysis

Two-dimensional and axi-symmetric finite-element analyses were used for two purposes: (1) to better understand the test method, and (2) as a tool in data reduction. Figure 4 shows a typical finite-element mesh and boundary conditions, and Fig. 5 gives a schematic of the model with coordinates. Based on studies of mesh gradation and model geometry [*10*], arrangements such as Fig. 4 were chosen to avoid influences of these parameters on the interfacial shear stress distribution. The two-dimensional (plane-strain) and axi-symmetric analyses both used anisotropic hybrid stress elements; essentially identical results were obtained from each method, and the data given here are from the axi-symmetric model unless otherwise noted. Neither analysis used special elements at the interface to provide continuity of the shear stress, and some discrepancy was observed in the shear stress just inside the fiber compared with that in the matrix. (The interfacial shear stress given in this paper is actually the value for the first element into the matrix in each case, since this gave more consistent results.)

One of the major concerns with the test is the sensitivity of the interface stress to the details of the local probe/fiber contact problem. Details of the probe and

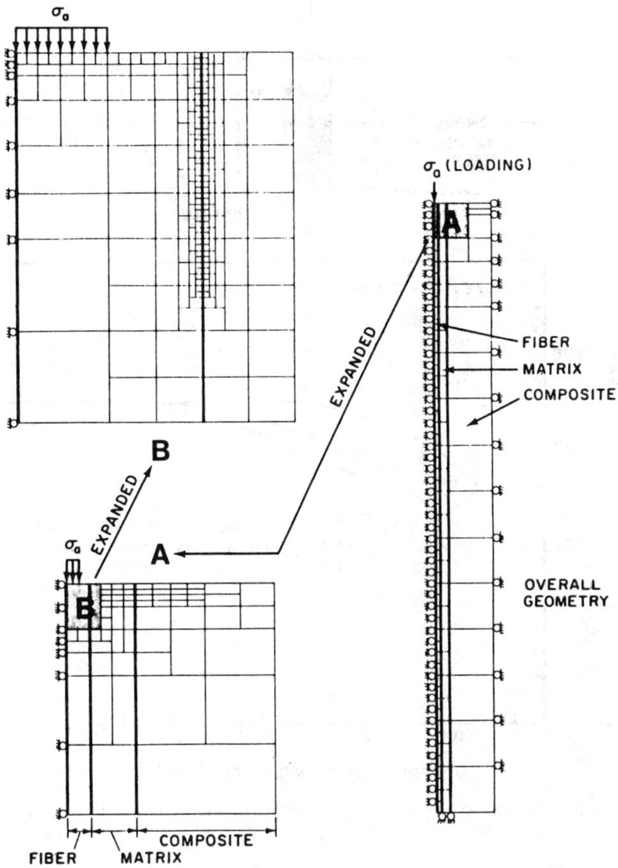

FIG. 4—*Typical finite-element mesh.*

continuum anisotropic body interaction including frictional effects have been investigated separately [*10,12*]. While probe deformations were not included in the results given here, extreme cases were simulated by a uniform pressure (very soft probe) and a step deformation in the fiber (very hard probe). Figure 5 (two-dimensional analysis) shows the distribution of shear stress along the interface for a pressure loading over various fractions of the fiber end, where D_σ is the loaded diameter, D_f is the total fiber diameter, σ_a is the applied pressure, and $\overline{\sigma} = \sigma_a D_\sigma / D_f$ is the average pressure over D_f (the distances, D_σ and D_f are used for the two-dimensional model, while the corresponding areas, A_σ and A_f are used for the axi-symmetric analysis).

Figure 5 indicates that the interfacial shear stress, τ_{xy} normalized by the average pressure, reaches nearly the same maximum value regardless of the fraction of the fiber end that is loaded, over the range studied. The position of the maximum stress is below the free surface by a fraction of D_f that depends on the area loaded

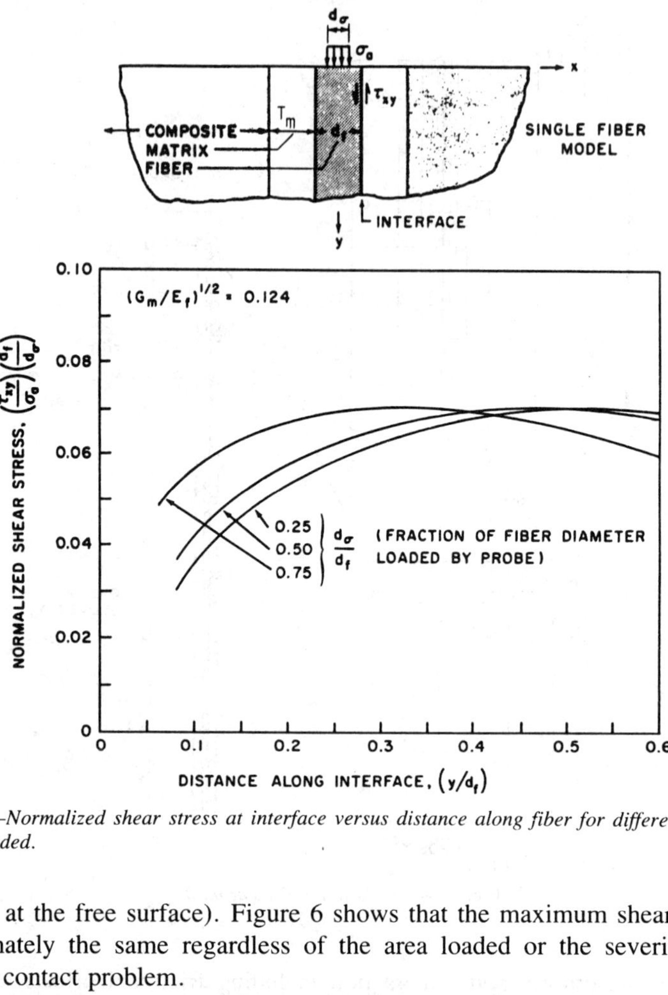

FIG. 5—*Normalized shear stress at interface versus distance along fiber for different fractions of fiber loaded.*

(τ_{xy} = 0 at the free surface). Figure 6 shows that the maximum shear stress is approximately the same regardless of the area loaded or the severity of the assumed contact problem.

The finite-element results [*10*] also show that the shear stress is the dominant stress component at the interface; the radial normal stress component is compressive. Thermal and hygral residual stresses have not been analyzed for this problem. However, literature micromechanics solutions for analogous problems (broken fibers) indicate that the maximum shear stress may not be significantly affected by residual stresses if the fiber is surrounded by a composite of moderate fiber volume fraction [*13*]. Residual stresses could be significant for very low fiber contents or for single fibers in bulk resin. Precise treatment of the mechanical and residual stress problem at the corner of the fiber and matrix on the free surface requires an analysis similar to that for free-edge singularities in laminates [*14*]. While it is clear that a singularity would be found at this point if an appropriate linear elastic analysis were used, the surrounding stresses in this region are very low, and it is believed that failure initiates below the surface as

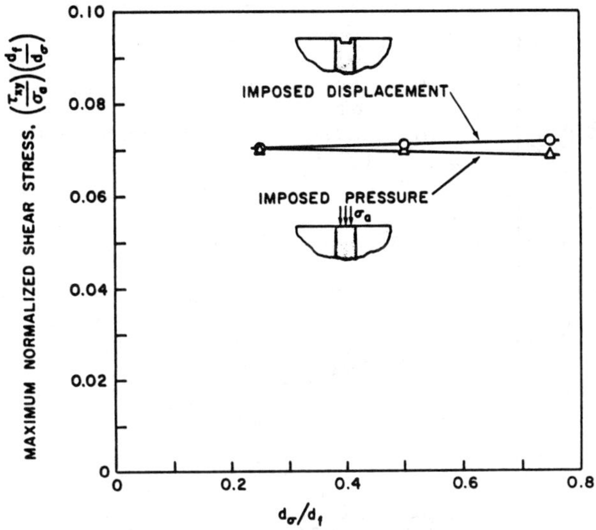

FIG. 6—*Normalized maximum shear stress at interface versus type of loading and fraction of fiber loaded,* $(G_m/D_f)^{1/2}$ = *0.124.*

a result of the shear stress shown in Fig. 5. This stress generates from the probe contact area, outward through the fiber to the interface; it does not generate from the singular point. Neither the stresses near the singularity nor the point of actual debond initiation (surface versus below surface) have been established; it is assumed that debonding initiates due to the maximum stress (Fig. 5) below the surface.

Micromechanics analyses [*15,16*] of similar cases lead to the expectation that the maximum interfacial shear stress will increase as the matrix layer becomes stiffer, either by increasing its modulus relative to that of the fiber, or decreasing its thickness. The expected trends are for the magnitude of the shear stress to vary with $(G_m/E_f)^{1/2}$ and $(T_m/D_f)^{-1/2}$, where G_m is the matrix shear modulus, E_f is the fiber axial Young's modulus, and T_m is the thickness of the matrix layer. Figure 7 gives finite-element results for the effects of varying the values of G_m and E_f, and also for fiber anisotropy. The results (from the two-dimensional analysis) are in good agreement with the $(G_m/E_f)^{1/2}$ trend (the position of the maximum shear stress also moves closer to the free surface as the ratio increases [*10*]). In each case, the assumed far-field composite moduli are consistent (through micromechanics theory) with the values for the fiber and matrix [*10*]. The moduli used for the orthotropic fiber are E_L = 276 GPa, E_r = 13.8 GPa, G_{LT} = 13.8 GPa, and ν_{LT} = 0.20. The maximum interfacial shear stress is not sensitive to the orthotropicity of the fiber (Fig. 7). Figure 8 gives data from the axi-symmetric analysis for the maximum shear stress as a function of both $(G_m/E_f)^{1/2}$ and T_m/D_f for three material systems of interest. The finite-element results show less

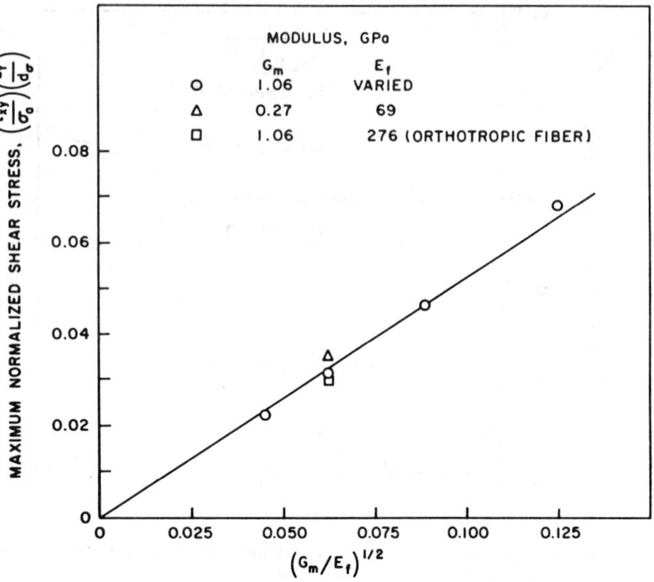

FIG. 7—*Maximum shear stress at interface versus* $(G_m/E_f)^{1/2}$ *for* $d_\sigma/d_f = 0.50$, *imposed pressure*, $T_m/D_f = 1.0$.

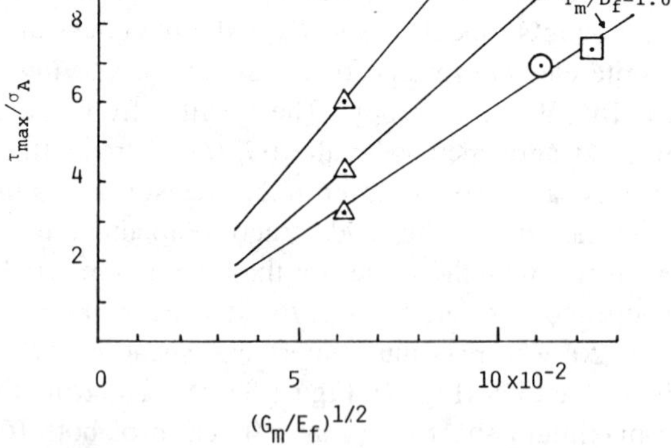

FIG. 8—*Variation of normalized maximum shear stress with* $(G_m/E_f)^{1/2}$ *for three fiber spacings.*

sensitivity to T_m/D_f than that predicted by micromechanics shear lag theory, as will be discussed later.

In summary, the finite-element results show that the maximum interfacial shear stress occurs at a distance that is a fraction of a fiber diameter below the free surface for most cases analyzed. It is assumed that debonding initiates at this point due to the shear stress; however, there is no experimental confirmation of the position where debonding initiates, as it quickly spreads over an area of the circumference and up to the surface. The maximum shear stress for a given probe force is not affected by the details of the probe-fiber contact interaction nor the size of the contact area out to at least 75% of the fiber diameter. The maximum shear stress is significantly affected by the relative elastic constants of the fiber and matrix as well as the thickness of the matrix layer. In practice, the elastic constants are known to within a good approximation under most conditions, and can be included in the data reduction. The thickness of the matrix layer is an approximation in the finite-element model intended to represent the local fiber arrangement and spacing. The actual matrix layer is always nonsymmetrical, with significantly different distances to neighboring fibers in different directions. The effect of actual local fiber arrangement cannot be included in a simple data reduction scheme. Test results indicate that the most important aspect of fiber arrangement is the distance to the nearest neighbor fiber, and only this parameter is included in the stress calculation.

Test Results and Data Reduction

Effects of Fiber Spacing

Figures 9 and 10 give results for the S-glass and graphite/epoxy composite systems described in Table 1. Each figure is plotted as the average stress applied to the end of the fiber at debonding, $\overline{\sigma}$ (force on the probe/fiber cross-sectional area) against T_m/D_f, the normalized distance to the nearest neighbor of the particular fiber tested. The solid lines represent a least squares fit to a power law equation for those points with $T_m/D_f > 0.1$; the curve fit parameters for each material are listed in Table 2. In most composites, there are many fibers with a very close nearest neighbor, essentially in contact. Very small fiber spacings cause difficulty with the data reduction methods, and were not included in the curve fit. Results for values of $T_m/D_f < 0.05$ are shown with standard deviation bars next to the coordinate. Although difficult to analyze, closely spaced fibers give no experimental problems or added scatter.

The raw data in Figs. 9 and 10 need to be reduced to an interfacial shear strength to be most useful. For a given T_m/D_f ratio this is easily accomplished through the finite-element results with the appropriate G_m/E_f ratio. However, there is a problem with the effects of fiber spacing on the results. Figure 11 gives the average applied stress at failure, $\overline{\sigma}$, normalized by the value at $T_m/D_f = 0.40$, $\overline{\sigma}_{.40}$, plotted against the fiber spacing. The three curves represent: (1) the curve fit to the experimental data for S-glass/epoxy, (2) the finite-element prediction,

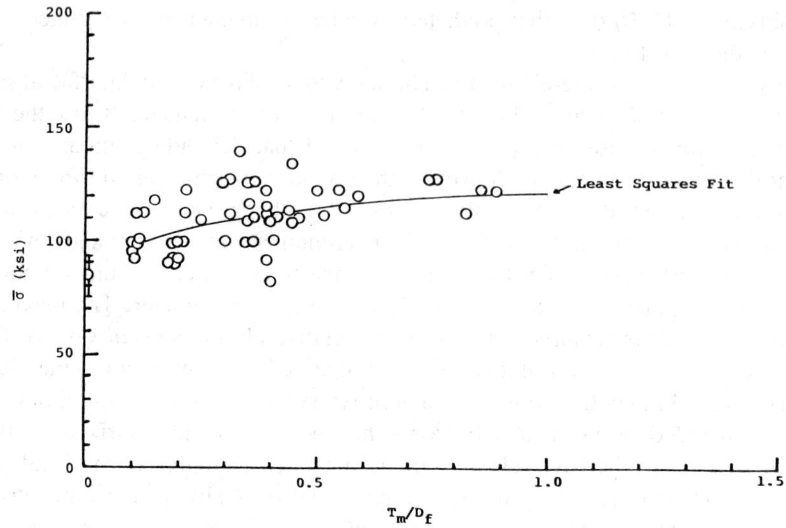

FIG. 9—*Microdebonding results and least squares fit of data for S-glass/epoxy (1 ksi = 6.895 MPa).*

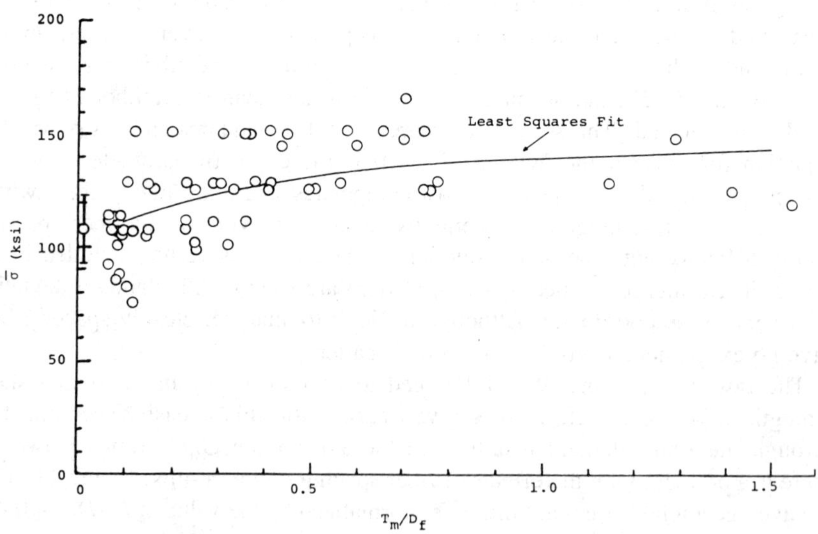

FIG. 10—*Microdebonding results and least squares fit of data for graphite/epoxy (1 ksi = 6.895 MPa).*

TABLE 1—*Materials (unidirectional composites).*

	S-Glass/Epoxy	Graphite/Epoxy
Fiber	S-2 glass (OCF)	Celion 6000 (Celanese)
Matrix	SP 250 (3M)	System F (Celanese) (100 parts Epi-Res 508 with 22.5 parts Epi-Cure 841)
Coupling agent	463 (OCF)	[a]
Nominal fiber volume fraction	0.50	0.60
Typical fiber diameter, μm	9 to 10	6 to 7
Fiber axial modulus, E_f, GPa	86	241
Matrix shear modulus, G_m, GPa	1.07	1.21
$(G_m/E_f)^{1/2}$	0.111	0.071

[a]This system is described in Ref *8* as C-6K-U, unsized fiber with a composite short-beam shear strength of 85 MPa.

and (3) a micromechanics prediction based on a shear lag analysis of fiber pullout discussed later [*16*]. For each material investigated in detail to date (two glass/epoxy materials and one graphite/epoxy), the curves were ordered as shown in Fig. 11. The qualitative trends are all in the same direction, but the experimental results show less variation with T_m/D_f than do the predictions. The predicted trends are in even greater error for the fibers that are nearly in contact.

The reason for the inability to accurately predict the effects of fiber spacing may derive from the approximation used in the theoretical models. The models assume that the fiber is surrounded by a uniform matrix layer of thickness, T_m, while in the experimental procedure T_m is the distance to the nearest neighbor fiber, with other fibers typically positioned at much greater distances. Thus, the fiber is exposed to three main geometric factors not included in the analysis: (1)

TABLE 2—*Curve fit parameters for raw data.*

	S-Glass/Epoxy	Graphite/Epoxy
CURVE FIT TO $\bar{\sigma} = A\,(T_m/D_f)^B$, $T_m/D_f > 0.10$		
A, MPa	855	965
B	0.104	0.096
Coefficient of correlation	0.498	0.481
$\bar{\sigma}_{.40}$, MPa	772	883
DATA FOR T_m/D_f = 0 to 0.05		
$\bar{\sigma}_o$, MPa	585	696
Standard deviation, MPa	63	103
Coefficient of variation, %	10.8	14.9
$\bar{\sigma}_0/\bar{\sigma}_{.40}$	0.755	0.787

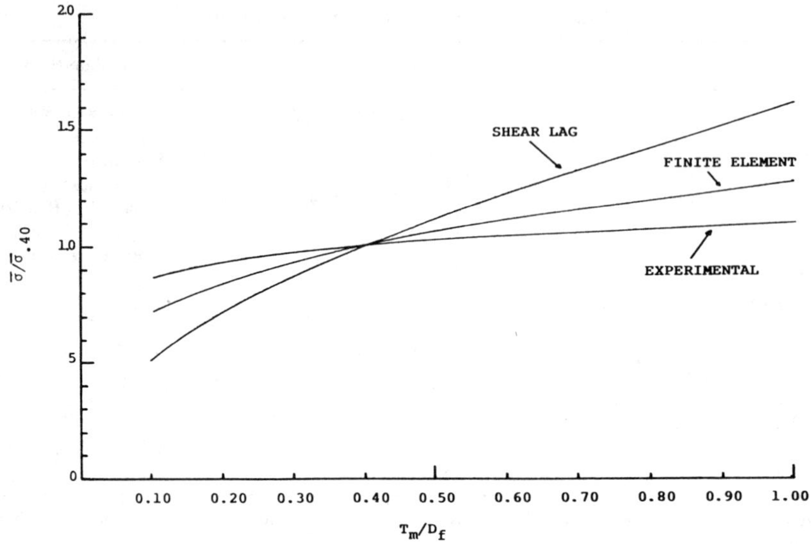

FIG. 11—*Variation of* $\overline{\sigma}$ *versus* T_m/D_f *as predicted by finite-element analysis, shear-lag analysis, and for experimental S-glass/epoxy results (all results normalized by* $\overline{\sigma}_{.40}$*).*

T_m refers to only one position on the circumference, (2) the stress field around the fiber is not symmetrical, and (3) the neighboring fibers are actually discrete, although modeled as a continuum. Additionally, the analysis is linear elastic, while the matrix may be significantly nonlinear, particularly for high ratios of τ_{DEB} to matrix shear yield strength. Despite these complications, the nearest neighbor fiber is clearly the dominant parameter, as all theoretical models imply maximum stresses at this position on the circumference, and debonding is experimentally observed to initiate at this point. No practical method has been found for including the overall local fiber arrangement in the data reduction, and so this discrepancy remains, and the experimental curve fit trends with T_m/D_f are used to reduce the data. As shown in Table 2, the exponents from the least squares curve fits are nearly identical for the two material systems, but the extension of this observation to other systems is uncertain.

Data Reduction Scheme

The data reduction scheme adopted here assumes that there should be no effect of T_m/D_f on the bond strength, and that any observed differences result from inadequacies in the analysis; it is not known whether this assumption is valid. Use of the experimental trends as the basis for data reduction requires a choice as to a value of T_m/D_f where the finite-element results will be assumed accurate. While logically this would be isolated fibers with large values of T_m/D_f, few such can be found in most composites. As a compromise, the finite-element results have been applied at $T_m/D_f = 0.40$, where adequate experimental data are available.

The power law curve fit relationship was used as a basis for shifting all of the data points with $T_m/D_f > 0.1$ in Figs. 9 and 10 to an adjusted value equivalent to a spacing $T_m/D_f = 0.40$, termed $\overline{\sigma}_{.40}$. The curve fit parameters in Table 2 for both materials are very similar, with an exponent, B, close to 0.10. Analysis of the sensitivity of the data reduction method to various parameters led to the conclusion that a power law with an exponent of 0.1 for each material was an adequate basis for shifting of the data (a variety of other methods were also attempted). The adjusted value $\overline{\sigma}_{.40}$ was obtained for each data point from the experimentally determined values of $\overline{\sigma}$ and T_m/D_f for that point through the relationship

$$\overline{\sigma}_{.40} = \overline{\sigma}[.40/(T_m/D_f)]^{0.1} \tag{1}$$

The interfacial shear strength for the adjusted data then is obtained from the finite-element analysis assuming that failure occurs by a maximum stress criterion for the shear stress along the interface direction, τ_{xy}, in Fig. 5. The interfacial debonding shear strength is calculated from

$$\tau_{DEB} = \overline{\sigma}_{.40}\,(\tau_{max}/\overline{\sigma}_{.40})_{FEM} \tag{2}$$

where $(\tau_{max}/\overline{\sigma}_{.40})_{FEM}$ is the ratio of the maximum value of τ_{xy} to the average applied stress on the fiber end from the finite-element analysis at $T_m/D_f = 0.40$, using the appropriate elastic constants. The value of $(G_m/E_f)^{1/2}$ often differs from the value used in the FEM analysis (Fig. 8); actual values are listed for each material tested, and the value of $(\tau_{max}/\overline{\sigma}_{.40})$ is linearly interpolated along the lines in Fig. 8.

Pooling of the data from different values of T_m/D_f by this method yields the interfacial shear strength values given in Table 3. Although the graphite/epoxy material debonds at a slightly higher average fiber axial stress, the data reduction including the effect of the higher fiber modulus results in a lower interfacial shear

TABLE 3—*Test results for interfacial strength.*

	S-Glass/Epoxy	Graphite/Epoxy
Pooled $\bar{\sigma}_{.40}$, MPa	779	889
$(\tau_{max}/\bar{\sigma}_{.40})$ from FEM	0.0892	0.0498
τ_{DEB} from pooled data, MPa	69	44
Standard deviation, MPa	7.3	5.9
Coefficient of variation, %	10.5	13.3
τ_{DEB} from $\bar{\sigma}_{.40}$ of curve fit (not pooled), MPa	70.3	48.0
τ_{DEB} from $\bar{\sigma}_0$ (T_m/D_f = 0 to 0.5), MPa	69.6	46.2
Standard deviation, MPa	7.5	6.8
Coefficient of variation, %	10.8	14.9
$(\tau_{max}/\bar{\sigma}_{.40})$ from shear lag	0.0877	0.0561
τ_{DEB} from pooled data, MPa	68.3	50.0

strength than for the S-glass/epoxy. The value of τ_{DEB} calculated directly from the curve fit value at $T_m/D_f = 0.40$ is similar to that from the data shifting method; however, the shifting method can be applied to a new material where too few data points are available for a curve fit, if the power law exponent of 0.1 is assumed. Figure 12 gives τ_{DEB} for each S-2 glass/epoxy data point calculated by the shifting method, plotted at its experimental value of T_m/D_f. As intended (in the assumption that the bond strength is independent of spacing), the debonding data calculated in this way show no apparent trend with fiber spacing. The shifting method of data reduction is convenient, allowing the use of a limited number of data points distributed over a range of T_m/D_f. More results are required to determine whether the exponent of 0.1 can be used for other composites.

The interfacial strength based on the fibers with $T_m/D_f < 0.05$, including many in apparent contact, can also be calculated using a variation in the shifting method. Figures 9 and 10 indicate that the data for closely spaced fibers are consistent with the trend of the remaining data. Table 2 indicates that the mean $\overline{\sigma}_0$ ($\overline{\sigma}$ for spacings close to zero) is approximately 75% of the curve fit value at $T_m/D_f = 0.4$. Assuming that $\overline{\sigma}_0 = 0.75\overline{\sigma}_{.40}$, the interfacial shear strength for these points is calculated as

$$\tau_{DEB} = (\overline{\sigma}_0/0.75)\,(\tau_{max}/\overline{\sigma}_{.40})_{FEM} \tag{3}$$

As shown in Table 3, the values of τ_{DEB} obtained from $\overline{\sigma}_0$ are similar to those obtained from shifting of the data for $T_m/D_f > 0.1$, and the scatter is also very

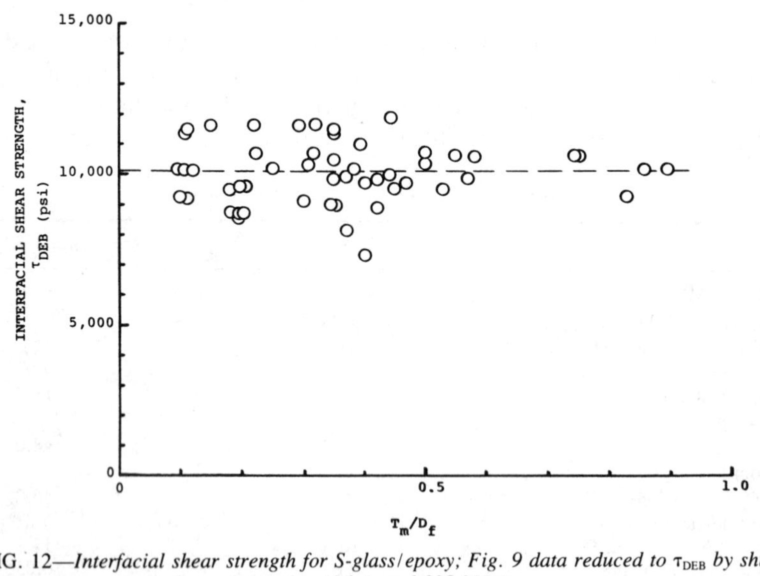

FIG. 12—*Interfacial shear strength for S-glass/epoxy; Fig. 9 data reduced to* τ_{DEB} *by shifting to* $\overline{\sigma}_{.40}$ *according to power law relationship (1 ksi = 6.895 MPa).*

similar. The value of $\overline{\sigma}_0$ for three material systems studied in detail all were between 0.75 and 0.79 of $\overline{\sigma}_{.40}$. If this is the case for other systems, then testing only closely spaced fibers and applying Eq 3 is the most convenient method for obtaining τ_{DEB}.

A final method for reducing the data is to use a closed-form micromechanics solution in place of the finite-element results. As indicated previously, the finite-element results conform to micromechanics solutions in their trend with the elastic constants (Figs. 7 and 8), while both methods disagree with the trend of the experimental results with fiber spacing (Fig. 11). Rearrangement of a shear-lag model for fiber pullout based on a rigid far field with a layer of matrix (or interface) next to the fiber (from Ref *16*) gives a variation of the shear stress along the interface as

$$\tau_{xy} = \frac{\overline{\sigma}}{2}(G_m/E_f)^{1/2}(D_f/T_m)^{1/2}[\sinh(\alpha x) - \cosh(\alpha x)] \tag{4}$$

where $\alpha = [(G_m/(T_m D_f E_f)]^{1/2}$.

The maximum shear stress is then

$$\frac{\tau_{max}}{\overline{\sigma}} = \frac{1}{2}(G_m/E_f)^{1/2}(D_f/T_m)^{1/2} \tag{5}$$

Figure 13 compares Eq 4 with the finite-element results for S-glass/epoxy. The shear-lag model is not accurate at the free surface where the shear stress should

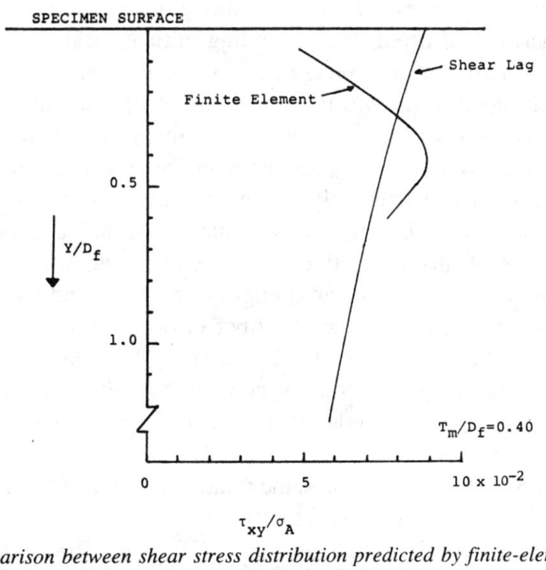

FIG. 13—*Comparison between shear stress distribution predicted by finite-element analysis and that predicted by shear-lag micromechanics analysis for S-glass/epoxy.*

go to zero, and the limited area of probe contact is not considered; however, the maximum shear stress is not greatly different from that predicted by the finite-element analysis. The shear-lag model disagrees much more for close fiber spacings as shown in Fig. 11, but this is not a problem with the data reduction method applied at $T_mD_f = 0.40$. When the data are reduced using the shear-lag model, the values of τ_{DEB} are not greatly different from those calculated from Eq 2, as shown in Table 3. Thus, while the finite-element analysis is thought to give a more accurate data reduction, approximate results can be obtained with the shear-lag equations as long as they are applied in the range of $T_m/D_f = 0.3$ to 0.5.

Discussion and Test Limitations

At this stage of development the microdebonding test is considered to be a useful tool for materials development and for observing the effects of environmental exposure, load history, and other factors. There are a number of concerns in the interpretation of the bond strength data and in the practical operation of the apparatus that need further study and development before it should be used as a standard method to establish absolute quantitative bond strength values.

Bond Strength Interpretation

The most immediate concern is the extent to which τ_{DEB} as calculated actually represents the quantitative local stress condition responsible for debonding. Compared with macroscopic tests that can be used for interface sensitive composite properties, only the microdebonding test clearly provides a result related solely and directly to separation of the interface. (This was the observation that originally led to development of the test.) However, among other problems to be discussed, there is no experimental proof that debonding initiates at the point of maximum shear stress, rather than at the free surface, since it spreads to cover the whole area immediately upon initiation. Bond strength results obtained for a variety of composites are an indication of how data from this test compare with that from other tests. Average values of τ_{DEB} are listed in Table 4 for several systems. The relative values for epoxy versus polyester matrix are consistent with expectations and with early model tests [*5,6*]. However, there is scant basis for knowing what to expect in terms of absolute values. The Celanese graphite/epoxy system is reported to have a short beam shear strength of 85 MPa and an interfacial shear strength value of 67 MPa in the single fiber critical length based test [*8*]. The microdebonding test gives a value of approximately 44 MPa, lower than the other two. The epoxy and polyester strand specimens in Table 1 have τ_{DEB} values of 57 and 23 MPa, while the single fiber critical length test is reported to give corresponding values of 78 and 28 MPa [*17*]. Thus, the microdebonding test gives absolute bond strengths in the same relative order as does the critical length model test, but the values are consistently lower.

The much higher bond strength in Table 4 for the T300/5208 as compared with the Celanese graphite/epoxy may be due to two factors. First, the Celanese

TABLE 4—*Results for several unidirectional composites.*

Material	τ_{DEB} from Critical Length Test [7], MPa[a]	$(G_m/E_f)^{1/2}$	τ_{DEB}, MPa[b]	Coefficient of Variation, %
Celanese graphite/ epoxy (Table 1)	67 [8]	0.071	44	13.3
T300/5208 graphite/ epoxy[a]	...	0.071	88[d]	7.6
S-glass epoxy (Table 1)	...	0.111	69	10.5
S-glass epoxy (Similar to Table 1, but OCF 449 Coupling Agent)	...	0.111	68	5.4
E-glass epoxy (3M Type 1009)	...	1.21[e]	57[d]	
E-glass epoxy[f] (single strand specimens, OCF)	79 [17]	1.21[e]	56[d]	6.4
E-glass/polyester[g] (single strand specimens, OCF)	28 [17]	1.21[e]	23	8.2

[a]Data from the single fiber critical length test.
[b]Calculated from Eq 2 unless otherwise noted.
[c]Oval fiber with slightly irregular surface.
[d]Closely spaced fibers, calculated from Eq 3.
[e]Assumes G_m = 154 000 psi.
[f]Epon 828/EMPOL 1052 resin with epoxy compatible coupling agent.
[g]OCF E701 (20%) + OCF E650 (80%) resin with polyester compatible coupling agent (*f*) and (*g*) specimens were supplied by OCF [*17*].

fiber in this case is unsized, while sized versions of the same fiber give much higher bond strengths in the critical length model test [*8*]. Second, the T300 fiber is oval or bimodal in cross section, with a slightly irregular surface. It is treated as a smooth, circular fiber of equivalent cross-sectional area in calculation of τ_{DEB}.

The data reduction scheme could give significantly different values for τ_{DEB} if it were based on a value of T_m/D_f other than 0.4, which was chosen as a compromise of available experimental data and uncertainties in the analyses. At this time, the major problem with the analyses is that they fail to model the effect of T_m/D_f (assuming that τ_{DEB} should be spacing insensitive). This problem should be minimized at large values of T_m/D_f, where the data and predictions both tend to flatten out (Fig. 11). Although there are few experimental points in this range, τ_{DEB} can be calculated from the curve fit using the FEM results at T_m/D_f = 1.0 (Fig. 8). Compared with the value of τ_{DEB} calculated directly from the curve fit in Table 3 at T_m/D_f = 0.4, the results at 1.0 are reduced from 70 to 59 MPa for S-glass/epoxy and from 48 to 38 MPa for graphite/epoxy; they would be significantly increased relative to the Table 3 values if based on T_m/D_f = 0.10. The values based on large spacings are expected to involve less error due to the data reduction method, so they are thought to be more accurate. Thus, it is anticipated that the results in Table 3 could be too high by 25% or more from this one aspect of data reduction. This factor affects each material tested in the same direction, but not necessarily to the same extent. The other major variable in the data reduction is $(G_m/E_f)^{1/2}$; while no experimental investigation of the accuracy with respect to this factor has been carried out, the FEM and micromechanics results are in close agreement as to its effect, and it is not anticipated to be a major source of error.

The data reduction would change significantly if it was assumed that the matrix yielded prior to debonding. General yielding would eliminate the effect of $(G_m/E_f)^{1/2}$ in the data reduction, which is responsible for the relatively low calculated bond strength for graphite/epoxy, as compared with E and S-glass composites. The debonding strength calculated for the graphite/epoxy is below that usually cited for the shear strength of epoxies of this type [*18*]. If the matrix shear yields before debonding, then τ_{DEB} should be approximately equal to the *in situ* matrix shear yield strength, in the absence of strain hardening. The *in situ* matrix shear yield strength appears to be an upper bound for τ_{DEB} in the microdebonding test. However, local yielding could result in significant errors if the usual data reduction scheme is applied; in particular, closely spaced fibers could appear to have a very high τ_{DEB} if local yielding reduced the calculated effect of the nearby neighboring fiber.

Sources of Experimental Error

Significant scatter is introduced by limitations of the current apparatus and test procedures. In the current results, the greatest sources of scatter are the size of the loading steps and the measurement of the fiber diameter. The size of the steps is a matter of convenience, since in this work they are greater than the minimum possible. This is particularly noticeable with the graphite/epoxy, where the loading steps were 0.5 g, representing 17% of the debonding force of approximately 3 g. The effects of these relatively large steps can be seen in the horizontal banding of the data in Fig. 10. Using the current test procedure, large loading steps systematically raise the calculated τ_{DEB}. Another source of scatter is the fiber diameter, which could only be measured within ± 0.25 μm, representing $\pm 7\%$ of the fiber area used in the calculation of $\overline{\sigma}$; this leads to a random error, as does a similar inaccuracy in determining T_m. It is estimated that improvements in equipment and load increments could reduce the scatter given in Tables 3 and 4 to approximately half of the present values.

Recommendations for Test Improvement

The current test method is inconvenient and the data reduction method needs improvement, as noted in the preceeding. Nevertheless, consistent results can be obtained with different samples of material run at widely varying times with different probes and variations in the apparatus. The relative results such as for epoxy versus polyester matrices are consistent with expectations, and the scatter, while substantial, is tolerable. It is thought that the apparatus and data reduction scheme in their present form could provide a useful tool for those interested in interface properties; more details are available in Ref *11*. The authors are aware of two other facilities that have set up similar equipment and are using it successfully. One of these derives from the earlier multifiber apparatus [*9*], used for aramid composites [*19*]. The current single fiber version is not useful for aramid composites due to splitting of the fibers by the probe prior to debonding. It is possible that a modified, larger radius probe could be developed for single fiber debonding of aramid fibers. The other facility [*20*] uses a modified version of the current single fiber microdebonding apparatus, and was developed following Ref *11*.

Several recommendations are in order for potential users of this test. It is inconvenient to switch from use of the test to conventional use of the microscope, due to the cumbersome realignment procedure. Either the equipment should be redesigned for easier initial alignment, or else the microscope should be dedicated to the test. A microscope with the best possible optics is both convenient and gives improved accuracy in determining D_f and T_m; for similar reasons, polishing equipment to provide the flattest, smoothest possible surface is recommended. Finally, a skilled, careful operator is essential. It is anticipated that a new generation of the apparatus could be significantly more practical and accurate with a modest investment.

When operating properly, a typical test on a well-polished surface takes 5 to 10 min per fiber. While this is not prohibitive, the procedure could be simplified and automated if the debonding event could be detected without the step-by-step observation in the microscope. No extensive study of possible detection methods has been made yet by the authors.

To date, specimens have been dried and stored in a dessicator prior to testing at ambient conditions. Since debonding occurs near the surface, the temperature and moisture content there will quickly approach equilibrium with the surroundings. Conducting the test under a variety of temperature and humidity conditions would require local environmental control during the test. It would also be interesting to run the test in a cyclic fatigue mode, and efforts are currently underway to attempt this. The time scale of loading has been standardized to 5 s in tests to date, but the effects of time under load should be studied. Lastly, it is evident that the data reduction scheme would be enhanced by a more thorough analysis of the effects of the local fiber arrangement. Analysis of the effects of residual stresses, matrix yielding, and the possible significance of the elastic free-edge singularity at the corner of the fiber and matrix on the free surface would be helpful in interpreting the results.

Conclusions

The modified (single fiber) microdebonding test provides a quantitative measure of the *in situ* fiber/matrix bond strength in typical glass and graphite fiber reinforced plastics. The apparatus is inexpensive when used as a stage in a research quality optical microscope. Reduction of the data to obtain an interfacial shear strength from the force at debonding through a finite-element analysis is approximate, requiring a curve fit to experimental data over a range of fiber spacings. The absolute value of the resulting bond strength is uncertain but relative differences can be determined. The coefficient of variation in the results is less than 15% for the materials tested, and could be further reduced by improvements in the apparatus and test procedure.

Acknowledgments

This research was supported in part by the Army Materials and Mechanics Research Center (Contract DAAG-46-77-C-0070) and by the Owens Corning Fiberglass Corp.

References

[*1*] Anderson, G. P., Bennett, S. J., and DeVries, K. L., *Analysis and Testing of Adhesive Bonds*, Academic Press, New York, 1977, pp. 1–21.
[*2*] Plueddemann, E. P. in *Proceedings*, 29th Conference of R.P./Composites Institute, Society of the Plastics Industry, 1974, Paper 24A.
[*3*] McGarry, F. J. and Marshall, D. W. in *Symposium on Standards for Filament-Wound Reinforced Plastics, ASTM STP 327*, American Society for Testing and Materials, Philadelphia, 1963, p. 133.
[*4*] Broutman, L. J., in *Interfaces in Composites, ASTM STP 452*, American Society for Testing and Materials, Philadelphia, 1969, p. 27.
[*5*] Mooney, R. D. and McGarry, F. J. in *Proceedings*, 14th Conference of R.P./Composites Institute, Society of the Plastics Industry, 1959, Paper 12E.
[*6*] Broutman, L. J., *Polymer Engineering and Science*, Vol. 6, 1963, p. 263.
[*7*] Drzal, L. T., Rich, M. J., Camping, J. D., and Park, W. J., in *Proceedings*, 35th Conference of R.P./Composites Institute, Society of the Plastics Industry, 1980, Paper 20–C.
[*8*] Ying, L. in *Proceedings*, 38th Conference of R.P./Composites Institute, Society of the Plastics Industry, 1983, Paper 12–F.
[*9*] Mandell, J. F., Chen, J. H., and McGarry, F. J. in *Proceedings*, 35th Conference of R.P./Composites Institute, Society of the Plastics Industry, 1980, Paper 26D.
[*10*] Tsiang, T. -H., "Damage Development in Fiber Composites Due to Bearing," ScD. thesis, Department of Materials Science and Engineering, Massachusetts Institute of Technology, 1983.
[*11*] Grande, D. H., "Microdebonding Test for Measuring Shear Strength of the Fiber/Matrix Interface in Composite Materials," M.S. thesis, Department of Materials Science and Engineering, Massachusetts Institute of Technology, 1983.
[*12*] Tsiang, T. -H. and Mandell, J. F., "Bearing/Contact for Anisotropic Materials," to be published.
[*13*] Chamis, C. C. in *Composite Materials*, Vol. 6, L. J. Broutman, R. H. Krock, and E. P. Plueddemann, Eds., Academic Press, New York, 1974, p. 53.
[*14*] Wang, S. S. and Choi, I., *Journal of Applied Mechanics*, Vol. 49, 1982, p. 541.
[*15*] MacLaughlin. T. F. and Barker, R. M., *Experimental Mechanics*, Vol. 12, 1972, p. 172.
[*16*] Greszczuk, L. B. in *Interfaces in Composites, ASTM STP 452*, American Society for Testing and Materials, Philadelphia, 1969, p. 42.
[*17*] Eckstein, Y., personal communication, Owens-Corning Fiberglass Corp., Technical Center, Granville, OH, 1984.
[*18*] Penn, L. S. and Chiao, T. T. in *Handbook of Composites*, G. Lubin, Ed., Van Nostrand Reinhold, New York, 1982, p. 57.
[*19*] Elkink, F., personal communication, AKZO Research, Arnheim, Holland, 1984.
[*20*] Tse, M. K., personal communication, Dow Chemical Co., Freeport, TX, 1984.

David J. Nicholls[1]

Effect of Stress Biaxiality on the Transverse Tensile Strain-to-Failure of Composites

REFERENCE: Nicholls, D. J., "**Effect of Stress Biaxiality on the Transverse Tensile Strain-to-Failure of Composites,**" *Composite Materials: Testing and Design (Seventh Conference), ASTM STP 893*, J. M. Whitney, Ed., American Society for Testing and Materials, Philadelphia, 1986, pp. 109–114.

ABSTRACT: Thermoplastic matrices are now available with neat resin strain-to-failure values of up to 100%. These high strain-to-failure values could be expected to result in dramatically higher transverse strain-to-failure values in thermoplastic composites than epoxy composites. However, measurements of the transverse strain-to-failure of some thermoplastic composites have revealed only a modest increase in transverse strain capability. The objective of this work was to show that the matrix is biaxially stressed in a transverse tension test and that this significantly reduced strain-to-failure values.

To accomplish this objective, a specimen configuration was designed to introduce biaxial stresses into the neat resin. Data was generated on polyetheretherketone and polycarbonate, both ductile thermoplastics. Test results showed that biaxial stresses reduced the strain-to-failure of these materials by well over an order of magnitude. Thus, this work shows that, while a thermoplastic matrix's ductility may translate into composite toughness and delamination resistance, it does not necessarily translate into dramatically increased composite transverse strain-to-failure values.

KEY WORDS: composite materials, transverse testing, transverse strain-to-failure values, biaxial stresses, thermoplastic, polyetheretherketone, polycarbonate

It is well known that the strain-to-failure of a unidirectional composite tested in tension transverse to the fibers is a fraction of the strain-to-failure of the neat matrix. Historically, the three explanations for this behavior were strain magnification, stress magnification, and residual stresses [*1,2*]. Strain magnification occurs because most of the volume of a composite is relatively inextensible fiber. Therefore, the actual strain of the matrix between the fibers is larger than the bulk strain measured on the composite. Stress magnification results because the fibers cause localized stress concentrations in the matrix. Finally, residual stresses reduce the transverse strain-to-failure since the matrix has essentially been stressed prior to the test. The residual stresses are due to thermal coefficient of expansion

[1] Captain, AFA/DFEM, U. S. Air Force Academy, Colorado Springs, CO 80840.

mismatches between the fiber reinforcement and the matrix. These three explanations are adequate to explain the reduction of strain-to-failure from 3 to 5% for a typical epoxy resin to 0.5% for a transverse tension test on a typical graphite/epoxy composite.

Since transverse strain-to-failure is a matrix-dominated property, the introduction of thermoplastic matrices, with neat resin strain-to-failure values on the order of 90%, was expected to have a dramatic effect on composite transverse strain-to-failure. However, experimental results show only a modest increase to 0.88% in transverse strain-to-failure for a composite with a matrix such as polyetheretherketone (PEEK) [*3*]. The purpose of this paper is to discuss the biaxial stress state placed on the matrix during a transverse tension test and to show that this can have a significant effect on the strain-to-failure of the composite.

Procedure

The first task in this work was to develop a test procedure to produce biaxial loads in a neat resin specimen. One popular test specimen, the maltese cross, was rejected because it has sharp corners. These sharp corners make measurement of strain-to-failure inaccurate since they are crack initiation sites. Another commonly used specimen that produces biaxial stress is the internally pressurized cylinder. However, this specimen was rejected because of our equipment and material limitations.

The testing approach we chose was based on the idea that in a very short, wide specimen (as illustrated in Fig. 1) the grips would restrain lateral contraction. Thus, we would get a biaxial stress state. Unfortunately, the materials we used were so ductile that they necked through the thickness away from the grips so that the test specimen was no longer constrained. At first, we thought that we could still produce the effect we wanted if we made the gage section thinner than the section in the grips. However, the specimens still necked through the thickness.

In order to prevent the specimen from necking away from the constraint, we decided to imbed the constraint in the neat resin. The first attempt at this was to imbed two steel rods 0.635 cm (0.25 in.) in diameter. However, the interface

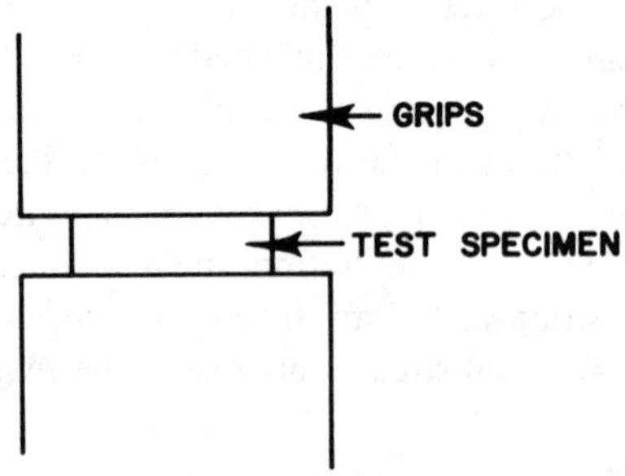

FIG. 1—*Distance between grips is so short, specimen cannot neck.*

between the steel rods and the matrix failed before the matrix did so that the strain-to-failure data could not be interpreted as accurate.

In our final and successful attempt, we made a specimen with composite ends separated by neat resin in the middle as shown in Fig. 2. To prevent failure of the interface between the composite and the neat resin regions, the neat resin was the same material as the composite matrix. The function of the composite sections was to introduce stresses in the lateral direction into the neat resin section through the Poisson's ratio difference. The specimen as a whole was tested in uniaxial tension.

Originally, it was planned that a number of composite matrix materials be included in this work. However, it was found that the type of test specimen just described could only be fabricated if a high viscosity polymer was used. Otherwise, fiberwash into the neat resin area made the test results meaningless. The two materials ultimately tested were the injection molded grade of polyetheretherketone (PEEK) and polycarbonate.

The PEEK specimens were fabricated from a neat resin plaque and APC-1 unidirectional graphite prepreg supplied by Imperial Chemical Industries. First, the APC-1 prepreg was molded to make a unidirectional composite in a matched die metal mold at 1.38 MPa (200 psi) and 382°C (720°F). Then the composite was cut into two pieces, each 4.42 cm by 15.2 cm (1.74 in. by 6.0 in.). A neat resin strip, 1.27 cm by 15.2 cm (0.5 in. by 6.0 in.), was also cut. These pieces were assembled in a 10.2-cm by 15.2-cm (4.0-in. by 6.0-in.) mold and molded together at 382°C (720°F) and 1.38 MPa (200 psi). Three 4.45-cm (1.75-in.) wide test specimens could be cut from such a panel after the panels had been trimmed. All specimens were end tabbed.

The procedure for the polycarbonate test specimens was similar, but since prepreg was not available, film stacking was used to make the composite. Alternate layers of 0.0121-cm (0.005-in.) polycarbonate film and UC-307, eight

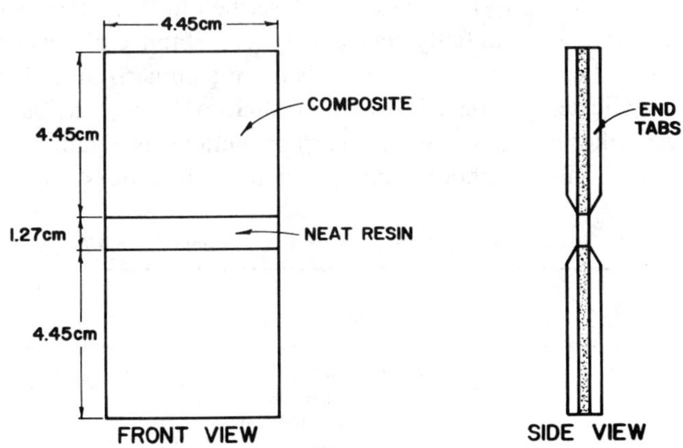

FIG. 2—*Final specimen configuration.*

harness satin graphite cloth were molded in a matched die metal mold. This composite was cut into two pieces and assembled with a neat resin strip in the same way that the PEEK panel was assembled. The whole panel was then molded together at 1.38 MPa (200 psi) and 200°C. These specimens were also end tabbed. All biaxial test specimens were strain gaged with micro-rosette strain gages and tested in tension on an MTS machine at a load rate of 44.5N/s (10 lb/s).

Neat resin dogbone specimens of both materials were tensile cut from neat resin plaques. These were also tested in tension on an MTS machine. However, strain was measured by crosshead travel. Crosshead travel, normally considered inaccurate, was used in this instance because the strains measured were on the order of 90% so that any error caused by machine compliance would be insignificant.

Experimental Work (Results)

PEEK and polycarbonate biaxial test specimens were fabricated and tested as described in the preceding section. Initially, micro-rosette strain gages were placed in the center of the neat resin section, at the edge of the specimen but located halfway between the composite regions, and at the corner formed by the neat resin section, the composite section and the edge of the specimen. Since the readings from all strain gages were the same, later tests used only the strain gage located at the center of the neat resin section.

Using these strain rosettes, we measured the strain of the neat resin parrallel and perpendicular to the applied load. We found that the stress-strain curve was linear to failure and that the Poisson's ratio was 0.033. Since the Poisson's ratio of PEEK and polycarbonate is normally 0.3, it can be concluded that biaxial stresses were introduced into the neat resin. Furthermore, since the Poisson's ratio measured is close to that measured for a composite (0.025), the biaxial stresses introduced are probably similar to those found in a composite.

PEEK and polycarbonate dogbones were fabricated and tested in order to generate uniaxial strain-to-failure data as described in the previous section. The stress-strain curves were initially elastic and then exhibited almost perfect plastic behavior. Both the uniaxial and biaxial data are summarized in Table 1. There were three replicates per data point. This data confirms that, as stated in the literature for other materials [*4,5*], sharp reductions in strain-to-failure occur when these materials are under biaxial tension-tension stress. Tested specimens

TABLE 1—*Strain-to-failure for specimens tested in uniaxial and biaxial tension.*

Material	Stress State	Strain to Failure, %
PEEK	uniaxial tension	95.0
PEEK	biaxial tension	1.1
Polycarbonate	uniaxial tension	73.0
Polycarbonate	biaxial tension	1.0

FIG. 3—*Failed uniaxial and biaxial specimens.*

are shown in Fig. 3 to illustrate the brittle appearance of the fracture of the biaxial specimen as compared to the uniaxial specimen where considerable necking occurred.

A problem was encountered in some of the polycarbonate biaxial test specimens because the fracture occurred in the composite rather than in the neat resin section of the test specimen. This did not occur in the PEEK biaxial specimen and may be attributed to poor bonding between the polycarbonate and the graphite fibers. Data was only recorded where failure took place in the neat resin.

Discussion

It has been demonstrated elsewhere in the literature [*6*] that in a uniaxial transverse tension test, the matrix is actually in a biaxial stress state. Therefore, I will only briefly describe why biaxial stresses are developed. In a neat resin uniaxial tension test, the test specimen elongates parallel to the load application direction. At the same time, it contracts in the direction perpendicular to the load application direction. A typical ratio of the strains perpendicular and parallel to the load application direction (Poisson's ratio) is 0.3. In contrast, a typical Poisson's ratio for a composite tested in tension transverse to the fiber direction is 0.025. Since the matrix retains its tendency to deform in accordance with its own Poisson's ratio, it must be restrained by the fibers. This is illustrated in Fig.

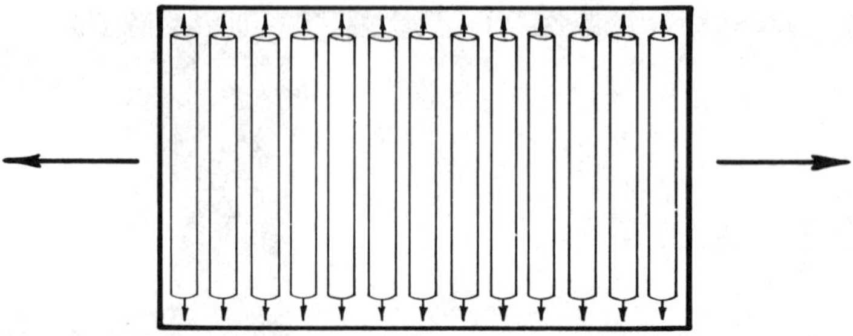

FIG. 4—*Mismatch of Poisson's ratio stresses matrix biaxially.*

4. Thus, the matrix is stressed both by the applied load and, perpendicularly to the applied load, by the fiber reinforcement. This is a state of biaxial, tension-tension stress.

It has been reported previously in the literature that the stress-strain behavior of a number of materials is very dependent on whether the applied stresses are biaxial or uniaxial. In particular, dramatic reductions in strain-to-failure occur when a material is stressed by a biaxial, tension-tension load. This work confirms that the strains-to-failure of the matrix materials used in this program are sharply reduced under biaxial stress conditions. Therefore, the biaxial load conditions present in a transverse tension test of a composite are important in determining the measured strain-to-failure.

Conclusion

The use of high-strain, thermoplastic matrices increases the toughness and damage tolerance of composite materials. However, the existence of biaxial stresses in the transverse tension test limits the strain capability of the thermoplastic matrix. Therefore, the improvement in the strain-to-failure of a thermoplastic composite versus a thermosetting composite is much less than would be expected from a comparison of the neat resin stress-strain behavior.

References

[1] Hull, D., *An Introduction to Composite Materials*, Cambridge Solid State Science Series, pp. 145–154.

[2] Chamis, C. C., in *Fracture and Fatigue*, L. J. Broutman, Ed., Academic Press, 1974, pp. 126–128.

[3] Hartness, J. T., "Polyetheretherketone Composite Fabricated from Unidirectional Tape," SAMPE Symposium Exposition, Society for the Advancement of Materials and Process Engineering, 3–5 April 1984.

[4] Streit, R. D., "Development of a Simple Biaxial Strain Test," Joint Conference on Experimental Mechanics, May 1982.

[5] Goel, V. S., Moet, J. D., and Harris, R., Tech Report #C 7–13 4, American Society for Metals, 1967.

[6] Greszczuk, L. B., *AIAA Journal*, American Institute of Aeronautics and Astronautics, Vol. 9, No 7, pp. 1274–1280.

H. Thomas Hahn[1] *and Jerry G. Williams*[2]

Compression Failure Mechanisms in Unidirectional Composites

REFERENCE: Hahn, H. T. and Williams, J. G., "**Compression Failure Mechanisms in Unidirectional Composites,**" *Composite Materials: Testing and Design (Seventh Conference), ASTM STP 893,* J. M. Whitney, Ed., American Society for Testing and Materials, Philadelphia, 1986, pp. 115–139.

ABSTRACT: The present paper examines compression failure mechanisms in unidirectional composites. Possible failure modes of constituent materials are summarized and analytical models for fiber microbuckling are reviewed from a unified viewpoint. Due to deficiencies in available models, a failure model based on nonlinear material properties and initial fiber curvature is proposed.

The effect of constituent properties on composite compression behavior was experimentally investigated using two different graphite fibers and four different epoxy resins. The predominant macroscopic-scale failure mode was found to be shear crippling. In a soft resin, shear crippling was in the form of buckling of fibers on a microscopic scale. However, for stiff resins, failure was characterized by the formation of a kink band. For unidirectional laminates, compressive strength, and compressive modulus to a lesser extent, were found to increase with increasing magnitude of resin modulus. The change in compressive strength with resin modulus was predicted using the proposed nonlinear model.

KEY WORDS: composite materials, fatigue (materials), compression failure, delamination, unidirectional composites

The inherent weakness of current graphite-epoxy composites to impact and delamination has prompted the development of improved material systems through the use of tougher resins and higher strain fibers [*1,2*]. Present technology, however, is limited in that improved toughness has in general been accompanied by a sacrifice in other properties such as strength and stiffness at elevated temperature. The longitudinal compressive strength of unidirectional composites depends on many factors including the stiffness and strength of the matrix and fiber. Better understanding of compressive failure mechanisms is needed to more accurately predict strength changes resulting from the use of different resins and fibers.

Over the past two decades, much effort has been concentrated on understanding the failure mechanisms and predicting the strength of compression loaded lam-

[1] Professor, Department of Mechanical Engineering and Materials Research Laboratory, Washington University, St. Louis, MO 63130.

[2] Aerospace engineer, Structures and Dynamics Division, National Aeronautics and Space Administration, Langley Research Center, Hampton, VA 23665.

inates. Since fracture of composites is usually instantaneous and catastrophic, identification of critical failure modes is not easily accomplished. The problem is complicated because changes in material properties or the presence of defects can lead to completely different failure modes. Thus, an analytical model accurate for one material system may not predict failure for another material system. Development of a unified model that can be applied to various material systems and failure modes is not within grasp at present. The current approach is to identify critical failure modes for each material system and to develop a unique model for each failure mode.

The present paper reviews failure modes in and analytical models for unidirectional composites subjected to longitudinal compression loading. The results of an experimental investigation on the effect of constituent material properties on compression behavior are discussed and compared with other available data. A possible sequence of failure initiation and propagation is proposed based on observations made during testing and from the examination of failed specimens.

Background on Failure Modes

Failure Modes of Fibers

The compressive failure characteristics of a fiber can be studied by embedding it in a resin casting [*3*]. The transparency available in an epoxy resin and the high axial stiffness of a large cross-section of resin relative to the fiber makes it possible to monitor the controlled failure of the fiber during loading. Compression failure modes characteristic of a fiber bundle have also been successfully observed using transparent resins to encapsulate the fibers [*4*].

Fibers fail differently depending on their internal structure, Fig. 1. High-modulus graphite fibers fracture in shear along a maximum shear plane [*3*]. The

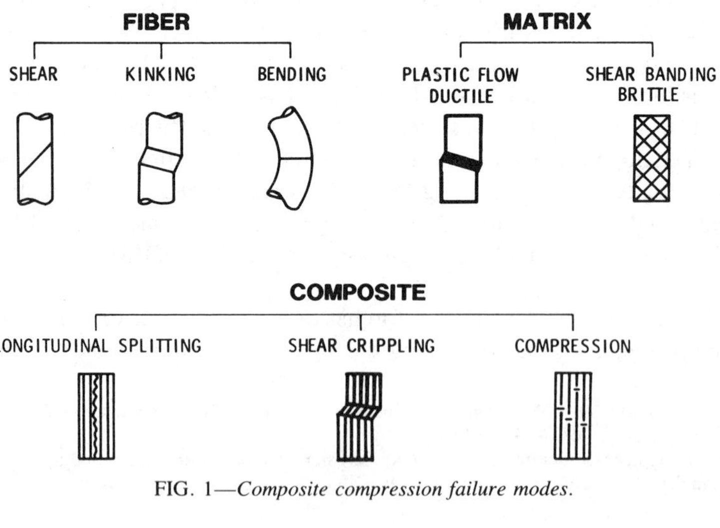

FIG. 1—*Composite compression failure modes.*

initiation of fracture is usually in the form of a hairline crack around the fiber circumference. The crack grows inward from the surface layer fairly normal to the fiber axis. However, final failure is along a maximum shear plane. Higher modulus is the result of better alignment of internal structure, and is associated with lower compressive failure strain. However, PAN Type II and Type A graphite fibers have been observed not to fracture due to compression loading at strains up to 3% [*3*]. Kevlar fibers, on the other hand, fail in a kink mode because of Kevlar's characteristic weak bond in the radial direction that permits individual fibers to split into fibrils [*5*].

Both shear failure and fiber kinking are characteristic failure modes for fibers with well-aligned fibrillar structure. The basic mechanisms for the two failure modes seem to be the same; that is, they are both the result of aligned fibrillar structure and a weak radial bond. However, the low ductility for graphite fibers leads to fracture while the development of fibrils for Kevlar fibers results in kinking.

Brittle fibers with amorphous structure such as glass do not usually fail in the aforementioned failure modes. These fibers can fail in bending, starting from the tension side. Medium to high-strength fibers may also fail in bending.

The failure modes strongly depend upon the lateral support provided the fiber during loading. In the absence of a strong lateral support, all fibers would fail by buckling. As the support stiffness increases, buckling is suppressed and the fiber begins to fail in shear. However, no shear failure has yet been observed for glass fibers or for high-strength graphite fibers such as T300 and T700.

Failure Modes of Resins

Compression tests on bulk resins reveal two types of failure. For ductile resins, plastic flow is frequently observed in a broad band oriented ~45° to the loading axis. For brittle resins, however, shear banding, namely narrow zones of shear yielding, can precede ultimate failure.

Failure Modes of Composites

Because of the weakness of the matrix and the fiber/matrix interface compared with the strength of the fibers, unidirectional composites can fracture along the fibers even when loaded by compression, Fig. 1. Transverse tensile stresses develop in the matrix due to Poisson's ratio differences between the matrix and fiber, and stress concentrations caused by voids can initiate fracture in the fiber/matrix interface [*6*].

If a fiber buckles, the matrix/fiber interface may fracture in shear and lead to ultimate failure. However, if the matrix is ductile and the interface is strong, the fiber can bend without matrix failure and eventually fracture in bending. The eccentricity introduced by such fiber fracture may lead to longitudinal splitting with continued compression loading.

A more likely failure mode of composites associated with fiber buckling and

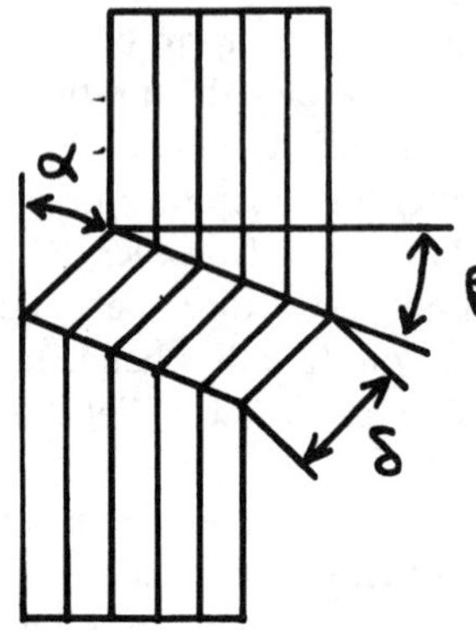

FIG. 2—*Kink-band geometry.*

kinking is shear crippling. Macroscopically, shear crippling looks like a shear failure on a plane at an angle to the direction of loading. Microscopic inspection, however, indicates shear crippling is frequently the result of kink-band formation, Fig. 2.

Some of the reported values for the kink-band boundary angle and the segment length are listed in Table 1 [*7–9*]. The kink orientation angle is not well defined because it will depend on the load history. The angle α can be shown to be twice the angle β if no volume change is assumed within the kink band [*9*].

In graphite fiber composites, fiber breaks are usually observed at the kink-band boundaries [*8,10*], while fiber kinking and extreme fiber bending without fracture are typical for Kevlar and glass composites, respectively [*7–9*].

The kink bands are most clearly observed when failure is gradual and also when longitudinal splitting is prevented by application of hydrostatic pressure [*8*]. Gradual failure is also observed when there is a gradient in the stress field, for example, near a hole [*1*] or in bending, and when fiber tows are surrounded by matrix as in carbon/carbon composites.

The third failure mode of composites is associated with pure compression failure of fibers. In this case, the fracture surface is likely to be at an angle, about 45°, to the loading. Post-failure examinations of fracture surfaces of graphite/epoxy composites alone are usually inadequate to distinguish between fiber kinking and fiber compression failure because the broken fiber segments resulting

TABLE 1—*Geometric parameters for kink band.*

Material	β	δ/d_f	Reference
Gr/Ep	22 to 26°	~3	[*7*]
	45°	. . .	[*10*]
Carbon/Carbon	24 to 28°	~7	[*9*]
Gl/Ep	20 to 30°	. . .	[*8*]
Kv 49/Ep	55 to 60°	. . .	[*6*]
	$\alpha \approx 2\beta$	. . .	[*8*]

from kinking failure are randomly displaced during the catastrophic failure event [*11*].

The available data in the literature strongly suggest that the most likely failure mode in graphite/epoxy composites for strength-critical applications is shear crippling involving fiber kinking. Compression failure of a composite starts with kinking of a few fibers. The kinked fibers disrupt the stability of the neighboring fibers so that the neighboring fibers also fail in the kinking mode. This damage propagation process continues until the composite completely fails. In some cases, fiber kinking may be initiated at several different locations and proceed to converge. The transverse tensile stress in the region where the two advancing kink bands meet may be sufficiently high to cause longitudinal splitting.

The failure modes discussed thus far depend on various material properties and geometrical parameters. Some properties and parameter values may promote one failure mode while other values may favor another. The effect of material properties and geometrical parameters is qualitatively discussed in the following section.

Background on Analytical Models

General Formulation

Consider a fiber embedded in a continuum. A free-body diagram for an infinitesimal segment of the fiber is shown in Fig. 3. The parameters in the figure are defined as follows:

P = axial compressive force,
Q = transverse shear,
M = bending moment,
p = applied distributed axial force,
q = applied distributed transverse force, and
m = applied distributed bending moment.

Assuming small deflection and the initial fiber axis to be along the x axis, we can write the equilibrium equations as [*12*]

$$q + \frac{dQ}{dx} + P\frac{d\omega}{dx} = 0 \tag{1}$$

$$p - \frac{dP}{dx} + Q\frac{d\omega}{dx} = 0 \tag{2}$$

$$\frac{dM}{dx} - Q + m = 0 \tag{3}$$

where ω is the slope of the deflected fiber axis.

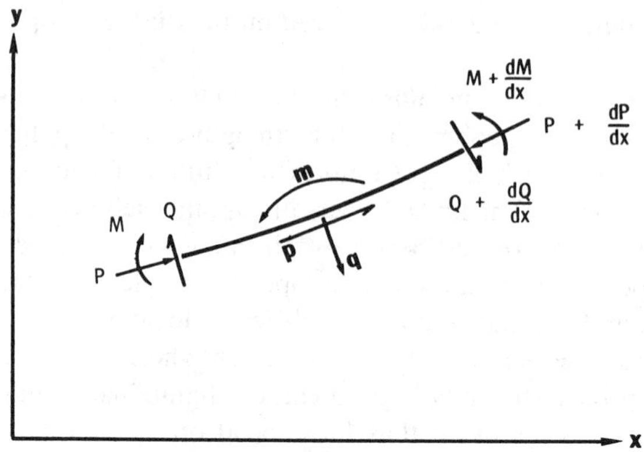

FIG. 3—*Free body diagram for a fiber segment.*

The exact determination of the forces and moments p, q, m, and M is not possible without the knowledge of the displacement fields in the fiber and the surrounding continuum. That is, one must solve the foregoing equations in conjunction with the equilibrium equations for the continuum using the appropriate constitutive relationships. Thus, the final solution for the failure load will depend on properties of the constituent materials.

The problem can be simplified if certain assumptions are made for the forces and moments p, q, m, and M. Various simplifying assumptions proposed in the literature are reviewed in the following section.

Single Fiber Embedded in Infinite Matrix

The model of a fiber surrounded by infinite matrix has been solved within the framework of elasticity [*12–14*]. This model approximates the behavior of a composite with a low fiber volume content. The salient features of this model can be studied by way of the following simplified analysis.

The combination of Eqs 1 and 2 with $m = 0$ and $P =$ constant leads to

$$\frac{d^2M}{dx^2} + q + P\frac{d^2v}{dx^2} = 0 \tag{4}$$

where v is the transverse deflection of the fiber and $\omega = dv/dx$. Noting that

$$M = E_f I_f \frac{d^2v}{dx^2} \tag{5}$$

and assuming that

$$v = f \cos \frac{\pi}{\ell} x \tag{6}$$

$$q = Kv, \; K = constant \tag{7}$$

gives the buckling load as [*12*]

$$P_{fc} = 2(KE_f I_f)^{1/2} \tag{8}$$

Here, E_f and I_f are the fiber modulus and moment of inertia, respectively. The corresponding buckle half wavelength is

$$\ell = \pi \left(\frac{E_f I_f}{K} \right)^{1/4} \tag{9}$$

According to the more rigorous solutions [*12,14*], K is not a constant but depends on ℓ as well as on material properties. However, K is approximately proportional to the matrix modulus E_m [*15*]. Thus,

$$P_{fc} \propto (E_m E_f I_f)^{1/2} \tag{10}$$

$$\ell \propto \left(\frac{E_f I_f}{E_m} \right)^{1/4} \tag{11}$$

If the interaction between fibers can be neglected, for example, in composites with low fiber volume content, the resulting buckling strain ϵ_c of composite is given by

$$\epsilon_c = \frac{P_{fc}}{E_f A_f} \propto \left(\frac{E_m}{E_f} \right)^{1/2} \tag{12}$$

where A_f is the cross-sectional area of the fiber.

Fiber Buckling in Composites

Since it is difficult to include the exact interaction effect between fibers, a variety of approximations have been attempted for the forces and moments p, q, m, and M based on simplified displacement fields. Furthermore, fibers have frequently been idealized as plates so that a two-dimensional analysis could be applied.

If two neighboring fibers are assumed to buckle out of phase with each other, the foundation constant, K, may be assumed to be [*16,17*]

$$K = 4\frac{E_m}{h_m} \tag{13}$$

where h_m is the thickness of the matrix layer. The corresponding buckling load in the fiber is obtained by substituting Eq 13 into Eq 8. The composite failure stress then follows from the rule of mixtures, that is,

$$\sigma_c = v_f \sigma_{fc} = v_f \frac{P_{fc}}{A_f} \tag{14}$$

where v_f is the fiber volume content.

Another simple displacement field is associated with fibers buckling in phase. Since all fibers deform the same way, one can assume [*16,17*] (see Appendix for m)

$$p = q = 0 \tag{15}$$

$$m = -\frac{h_f}{v_f}\frac{G_m}{1 - v_f}\frac{dv}{dx} \tag{16}$$

where h_f is the thickness of the fiber layer. The resulting equation of equilibrium is then

$$\frac{d^2M}{dx^2} + \frac{dm}{dx} + P\frac{d^2v}{dx^2} = 0 \tag{17}$$

Substitution of Eqs 5, 6, and 16 into Eq 17 yields

$$P_{fc} = \frac{h_f}{v_f}\frac{G_m}{1 - v_f} + E_f I_f \left(\frac{\pi}{\ell}\right)^2 \tag{18}$$

The corresponding composite buckling stress is thus (see Eq 14)

$$\sigma_c = \frac{G_m}{1 - v_f} + v_f \frac{E_f I_f}{h_f}\left(\frac{\pi}{\ell}\right)^2 \tag{19}$$

If $h_f << 1$, Eq 19 reduces to

$$\sigma_c = \frac{G_m}{1 - v_f} \tag{20}$$

Equation 19 is generally known to yield a much higher strength than is realized experimentally. Many improvements and modifications have been suggested over the years [*7,18–23*]. Most improvements try to incorporate initial curvature of fibers and material nonlinearity. These improvements will be discussed in the following development of a nonlinear model.

Nonlinear Model

A nonlinear model including initial fiber curvature and matrix material nonlinearity is developed using the distributed moment, *m,* calculated from the average shear stress field. Assuming an initial deflection of the form

$$v_o = f_o \cos \frac{\pi}{\ell} x \tag{21}$$

one can calculate *M, m,* and *q* as (see Appendix)

$$M = E_f I_f \frac{d^2}{dx^2} (v - v_o) \tag{22}$$

$$m = -\pi \frac{d_f^2}{4} \tau_{LT} = -\pi \frac{d_f^2}{4} G_{LT} \frac{d}{dx} (v - v_o) \tag{23}$$

$$q = K(v - v_o) \tag{24}$$

Here, G_{LT} is the composite shear modulus, and τ_{LT} and γ_{LT} are the average shear stress and strain, respectively. The fiber diameter is denoted by d_f. In the case of nonlinear stress-strain relationships, G_{LT} is taken as the secant modulus at a given γ_{LT}.

The derivation starts with the following equation of equilibrium,

$$\frac{d^2M}{dx^2} + \frac{dm}{dx} + q + P \frac{d^2v}{dx^2} = 0 \tag{25}$$

Substitution of Eqs 22 to 24 together with Eqs 6 and 21 into Eq 25 yields

$$\sigma_c = v_f \left[G_{LT} + \frac{E_f I_f}{A_f} \left(\frac{\pi}{\ell} \right)^2 + \frac{K}{A_f} \left(\frac{\ell}{\pi} \right)^2 \right] \left(1 - \frac{f_o}{f} \right) \tag{26}$$

Using the condition that σ_c be a minimum at fixed f_o/f and expressing f in terms of γ_{LT}, one obtains

$$\sigma_c = v_f \left[G_{LT} + \sqrt{\frac{KE_f}{\pi}} \right] \frac{\gamma_{LT}}{\gamma_{LT} + \pi f_o/\ell} \tag{27}$$

If K is neglected, as is frequently done, σ_c is given by

$$\sigma_c = v_f G_{LT} \frac{\gamma_{LT}}{\gamma_{LT} + \pi f_o/\ell} \qquad (28)$$

Furthermore, if no initial defect is assumed, that is, $f_o/\ell = 0$, one has

$$\sigma_c = v_f G_{LT} \qquad (29)$$

Even when G_{LT} is approximated by $G_m/(1 - v_f)$, Eq 29 is different from Eq 20 by the factor v_f. The difference derives from the selection of the free body. If the free body includes both a fiber and matrix, Eq 29 without the fiber volume content, v_f, results [*18*]. If only a fiber is taken as the free body, Eq 29 follows. Since Eq 20 is known to overpredict the strength and because applying the equilibrium forces and moments to the fiber is a reasonable model. Eq 29 is preferred to Eq 20.

The importance of initial fiber curvature has been pointed out and incorporated in the model by many investigators [*19–23*]. The results are all similar to Eq 28 without the factor v_f.

Since G_{LT} is in general a function of γ_{LT}, Eq 29 should be solved for a maximum value of σ_c over the allowable range of γ_{LT}. For example, if the shear stress-strain relationships can be approximated as that of an elastic-perfectly plastic material with yield stress τ_y, Eq 28 reduces to

$$\sigma_c = v_f \frac{\tau_y}{\gamma_y + \pi f_o/\ell} \qquad (30)$$

If fibers are weak, then fibers may fail before the matrix and interfaces fail in shear. The factor, $1 - (f_o/f)$, in Eq 26 can be related to the flexural strength, σ_f, of the fibers. The result is then

$$\sigma_c = v_f \frac{G_{LT}}{1 + \frac{\pi}{2}\frac{E_f}{\sigma_f}\frac{d_f}{\ell}\left(\frac{\pi f_o}{\ell}\right)} \qquad (31)$$

Equation 31 is likely to be applicable to Kevlar/epoxy composites, because the flexural strength of Kevlar fibers is very low.

The effect of a partial bond can be included by introducing a bond efficiency factor for m in Eq 16 [7]. The interfacial strength and voids reduce the composite shear strength, that is, the maximum value of γ_{LT} in Eq 28. Thus, Eqs 28 and 31 incorporate most of the important parameters.

The models discussed so far are essentially one dimensional. A two-dimensional analysis is required to predict the kink-band geometry [*21*]. The analytical

TABLE 2—*Constituent materials.*[a]

Fiber	Matrix	Fiber Weight Content, %	Laminate Nominal Thickness, mm
T300	Narmco 5208	59	3.6
	American Cyanamid BP 907	60	3.9
	Union Carbide 4901/MDA	68	3.0
	Union Carbide 4901/mPDA	71	3.0
T700	BP 907	60	3.6
	4901/MDA	66	3.6
	4901/mPDA	65	3.6

[a]Identification of commercial products and companies in the report is used to describe adequately the test materials. The identification of these commercial products does not constitute endorsement, expressed or implied, of such products by the National Aeronautics and Space Administration.

results of Ref *21* indicate that the kink-band boundary angle becomes zero as σ_c approaches G_{LT}.

Experimental Procedure

Two different fibers were combined with four different resins to make seven unidirectional $[0]_{24}$ graphite/epoxy composites. Specific constituent materials as well as respective fiber weight contents and nominal thicknesses of cured panels are listed in Table 2. Nominal properties of the constituent materials available in the literature are listed in Table 3. The panels were fabricated at the NASA Langley Research Center according to manufacturers' suggested cure cycles.

Compression testing was done on an Instron machine at a crosshead speed of 1 mm/min. Specimens were prepared for testing in an IITRI (Illinois Institute of Technology Research Institute) compression fixture according to procedures described in ASTM Test Method for Compressive Properties of Unidirectional or Crossply Fiber-Resin Composites (D 3410-82).

TABLE 3—*Nominal properties of constituent materials.*

Material	Diameter, μm	Modulus,[a] GPa	Failure[a] Stress, MPa	Failure[a] Strain, %
T300	7.0	230	3310	1.4
T700	5.1	238	4550	1.9
5208	...	4.00	57.2	1.8
BP907	...	3.10	89.5	4.8
4901/MDA	...	4.62	103.4	4.0
4901/mPDA	...	5.46	115.1	2.4

[a]In tension.

Specimens were monitored during test for indications of failure through a stereo optical microscope at magnifications up to ×100. Some specimens had one edge polished to monitor failure initiation.

Two specimens from each material system were tested with a strain gage mounted on one surface. When an anomaly was detected in the stress-strain curve, the specimen was unloaded and examined for damage. Failed specimens were examined on optical microscopes at magnifications up to ×300.

Results and Discussion

Compressive Properties

Strength data for the graphite/epoxy composites are shown in Figs. 4 and 5, where N is the total number of specimens and i is the ordinal number of strength. Of all the composites, scatter is highest for the T300/5208 material, however, this variation is consistent with that reported for this material in Ref *24*. The strength of the other composite materials show much less variation.

The graphite/epoxy composites are compared with one another in Fig. 6 for average strength, failure strain, and modulus. In the figure, E is the initial tangent modulus and E_t is the tangent modulus at failure. For the properties other than average strength, each point represents one specimen since only two specimens were strain-gaged for each composite material system. The properties measured for T300/5208 are comparable with those reported in Refs *24* and *25*. The highest strengths and initial moduli are recorded for the T300/4901/MDA and T300/4901/mPDA specimens. The tangent modulus at failure varies much less from material to material than does the initial tangent modulus.

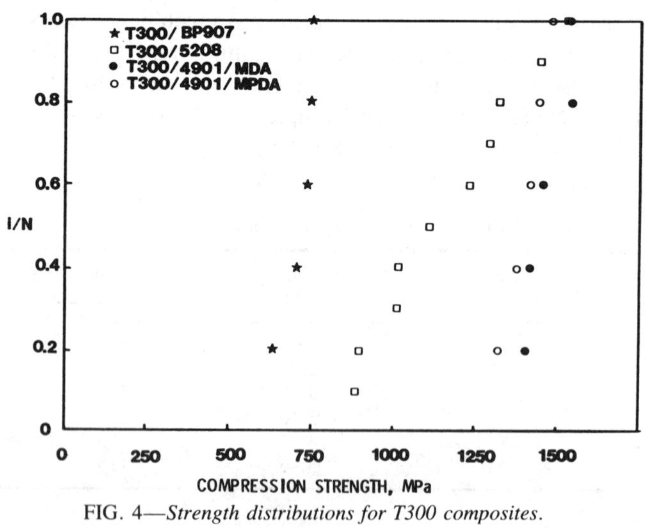

FIG. 4—*Strength distributions for T300 composites.*

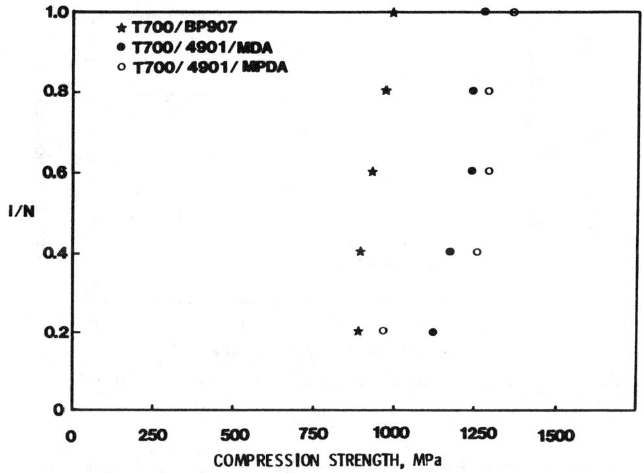

FIG. 5—*Strength distributions for T700 composites.*

The effect of resin tensile modulus on the composite strength and axial modulus is shown in Fig. 7. According to the rule of mixtures, the composite modulus, E_c, is given by

$$E_c = v_f E_f + v_m E_m \tag{32}$$

The rule-of-mixtures prediction based on the maximum variation of constituent properties from Table 1 is shown in Fig. 7. For a laminate with a 60% fiber volume fraction, a change of resin modulus from 3 to 5.5 GPa will result in an

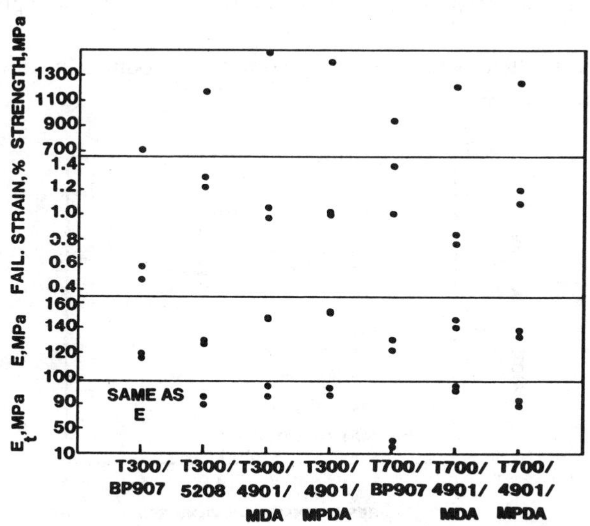

FIG. 6—*Comparison of unidirectional laminate properties.*

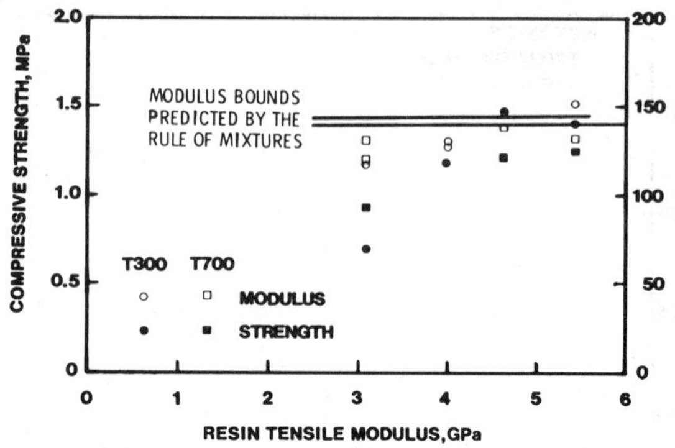

FIG. 7—*Effect of resin tensile modulus on modulus and strength of T300 and T700 composites under compression.*

increase in composite modulus of 1 GPa. This increase is insignificant in comparison to the original composite modulus magnitude, which is on the order of 100 GPa. Thus, the observed modulus increase in Fig. 7 is much higher than predicted by Eq 32.

To ascertain the dependence of compressive modulus of a composite on the resin tensile modulus, the experimental results of Ref *26* for tensile and compressive moduli for a large number of material systems are plotted in Fig. 8. Linear regression analyses were performed on the data and the results are also shown in the figure while the appropriate parameters are listed in Table 4.

One of the two features that stand out in Fig. 8 is that the compressive modulus of a composite is more dependent on the resin modulus than is the tensile modulus. The other feature is that the composite tensile modulus is greater than the composite compressive modulus for resin modulus values less than approximately 5 GPa.

Precise theories that explain the lower composite modulus in compression than

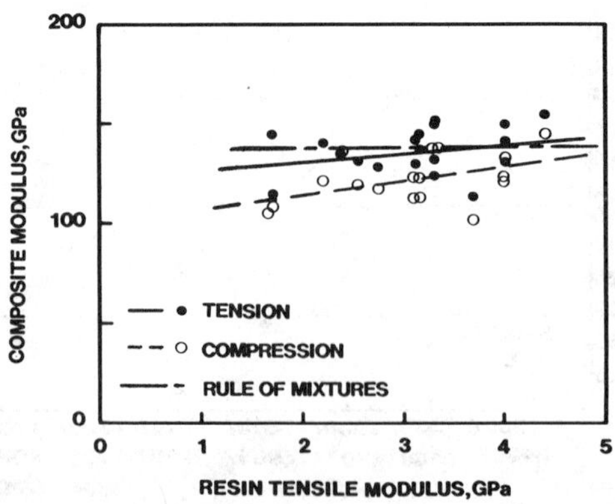

FIG. 8—*Effect of resin tensile modulus on composite moduli (data from Ref* 25).

TABLE 4—*Linear regression analysis*[a] *for the data of Ref* 25.

y	*a*	*b*	Correlation Coefficient
Compressive modulus	7.091	101.0	0.4674
Tensile modulus	4.392	123.4	0.2898
Compressive strength	0.433	−0.1685	0.9021
Tensile strength	0.068	1.586	0.2592
Flexural strength	0.423	0.600	0.7764

[a] $y = ax + b$; and x = resin tensile modulus.

in tension are not available. One possible explanation is that the fibers themselves are less stiff in compression than in tension. The strain hardening observed for graphite/epoxy composites under tension is attributed to improved alignment of internal structure of the graphite fibers during loading. Conversely, initially imperfect alignment of the fiber structure may grow in amplitude during compression loading. However, the change of fiber stiffness alone does not explain the dependence of the composite compressive modulus on the resin modulus.

Another reason may be that the fibers themselves are not perfectly aligned within the lamina. Under compression, the initial misalignment and initial curvature of the fibers may impose disproportionately high stress on the matrix, locally pushing the matrix into the nonlinear range at an early stage of loading. This same argument can also be used to explain strain hardening in tension. A fiber bundle embedded in resin is like a beam column supported on an elastic foundation that can buckle in uniform wavelength when the resin is soft [*4*].

Stress-strain curves for the seven materials studied in this investigation are shown in Fig. 9. All of the materials exhibit strain softening, with T700/BP907 showing the most nonlinearity. The T300/BP907 specimen that shows an abrupt drop in strain while under constant stress was unloaded before final failure and was examined to reveal a shear crippling failure on one of the edges. The shear crippling occurred outside the region where the strain gage was attached. The sudden decrease in the measured strain is believed to be the result of local microbuckling triggering the initiation of shear crippling.

Stress-strain curves of all the other materials are smooth and provide no indication of subcritical local failure. It seems that the shear crippling in T300/BP907 was subcritical because it occurred at a low stress level. In the other composites, local shear crippling did not occur until the laminate was highly stressed and led to the immediate catastrophic failure of the specimen.

The average compressive strengths for the materials tested from Fig. 6 are replotted in Fig. 7 as a function of the resin tensile modulus. The general trend is that compressive strength increases with increasing resin modulus for the range

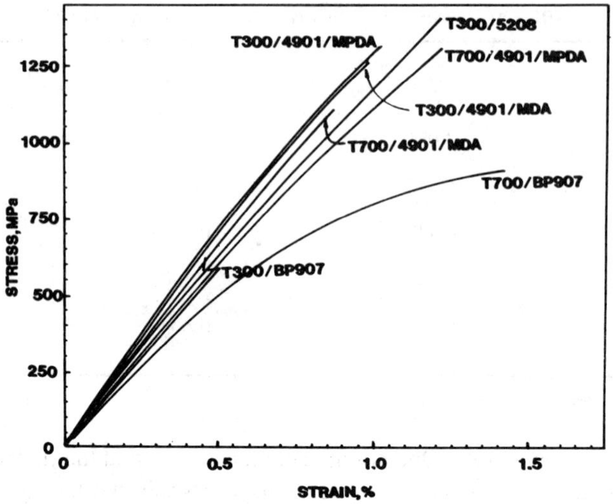

FIG. 9—*Compressive stress-strain curves for T300 and T700 composites.*

of values studied. The increase in compressive strength with resin tensile modulus is also characteristic of the data taken from Ref *26*, Fig. 10. As expected, tensile strength is not appreciably affected by the resin modulus. A linear regression analysis conducted on the data from Ref *26* is presented in Table 4. The regression parameters confirm a strong correlation between compressive strength and resin modulus but a weak correlation between tensile strength and resin modulus.

The current theory represented by Eq 28 was used to predict compressive strength as a function of the resin modulus. To use Eq 28 to predict compressive strength requires the parameter f_o/ℓ to be known. Since the full stress-strain

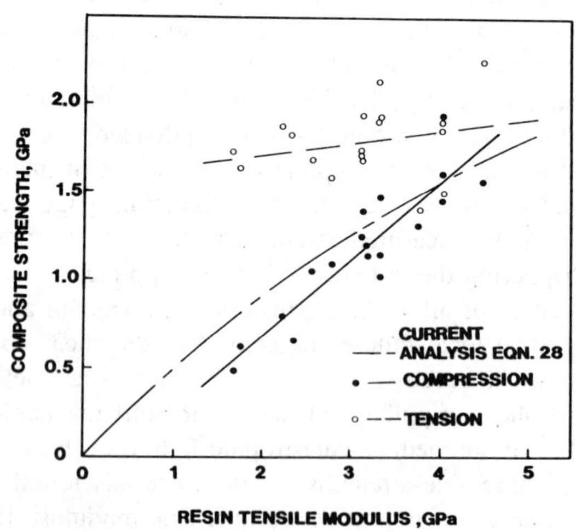

FIG. 10—*Effect of resin tensile modulus on composite strengths (data from Ref* 25).

relationships are not known for most of the materials in Fig. 10, G_{LT} is taken to be the initial modulus for the composite and f_o/ℓ is calculated using representative properties in Eq 28. The initial shear modulus, G_{LT}, is calculated by [27]

$$\frac{1}{G_{LT}} = \frac{v_f}{G_f} + \frac{\eta(1 - v_f)}{G_m} \tag{33}$$

$$\eta = \frac{1}{2}\left(1 + \frac{G_m}{G_f}\right) \tag{34}$$

where G_f is the longitudinal shear modulus of the fiber. For T300/5208 the necessary shear moduli are

$$G_f = 12.4 \text{ GPa and } G_m = 1.48 \text{ GPa}$$

Using $\sigma_c = 1.56$ GPa for $v_f = 0.6$, one calculates

$$\frac{f_o/\ell}{\gamma_{LT}} = 0.295$$

Thus, if $\gamma_{LT} = 1\%$, f_o/ℓ is only 0.295%, which is very difficult to measure.

For the other resins, $(f_o/\ell)/\gamma_{LT}$ was also assumed to be 0.295 and G_m was calculated from E_m under the assumption of an isotropic material with a Poisson's ratio of 0.35. The resulting compression strengths predicted by Eq 28 are presented in Fig. 10. Under these assumptions, Eq 28 is seen to overestimate the composite compressive strength when the matrix resin tensile modulus is less than 3 GPa, but in general is in reasonable agreement with observed experimental data.

Lower compressive strength resulting from lower resin modulus also leads to lower flexual strength, as shown in Fig. 11 by the data taken from Ref *26*. The reason is that, when a composite is weaker in compression, failure will be initiated on the compression side in a flexure test, and hence the flexural strength reflects the compressive strength. Flexural strength does not show as good a correlation with resin modulus as does compressive strength. However, the slope of the linear regression line for flexural strength (Fig. 11) is almost the same as for compressive strength, Table 4.

Compressive failure strains obtained by dividing the failure stresses by the corresponding moduli are presented in Fig. 12. The curve in the figure presents a prediction for buckling strain by the Euler equation for a column with clamped ends, that is, [28]

$$\epsilon_c = \frac{\pi^2}{3}\frac{1}{(L/h)^2} \tag{35}$$

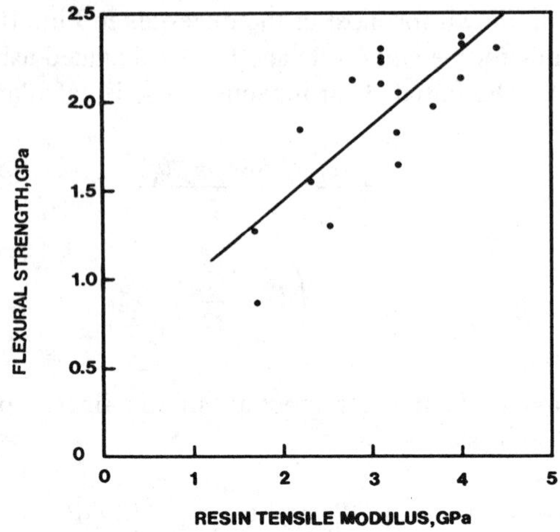

FIG. 11—*Effect of resin tensile modulus on flexural strength (data from Ref* 25).

where L and h are the length and thickness, respectively, of the column. The T300/BP907 and T700/BP907 specimens had tapered tabs, and hence the entire range of possible gage length is shown. All the experimental compressive failure strains are far below the predicted Euler buckling strains.

Failure Modes

With the exception of one T300/5208 specimen and two T300/BP907 specimens, all compression specimens failed suddenly without warning. Monitoring of the polished edge during the test at magnifications up to $\times 50$ did not reveal any sign of damage before ultimate failure. Several specimens were loaded in a

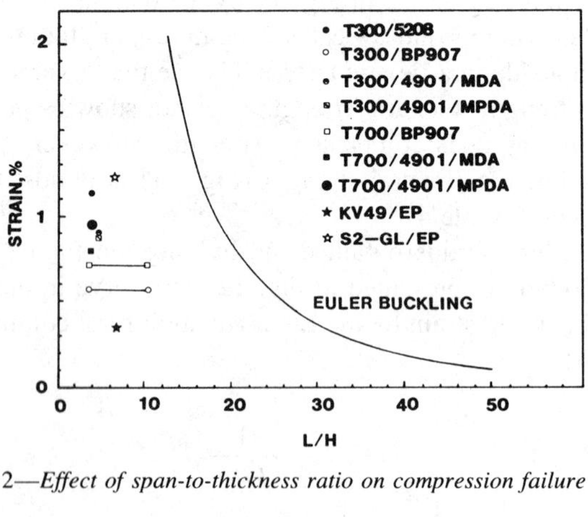

FIG. 12—*Effect of span-to-thickness ratio on compression failure strain.*

stepwise manner and examined under a constant load at each step. Other specimens were loaded and then unloaded for examination. No damage was detected following any of these procedures.

The T300/5208 specimen for which failure was arrested is shown in Fig. 13. The shear crippling did not occur across the entire width since the opposite edge remained intact. Details of the right end of the upper failure are shown in Fig. 13*b*. Most of the broken fiber segments in the kink band are gone, but the few remaining ones are seen to be of short length and tilted. The curvature in the fibers below the secondary kink band is required for kinematic compatibility.

The failure mode observed for a T300/BP907 specimen is shown in Fig. 14. The buckle mode shape is well defined and multiple fractures of the fiber are observed within the buckle region. This is in contrast to the kink-band failure mode in which two fractures occur per fiber and the short broken fibers align in a regular parallel pattern rotated relative to the direction of the applied compression load, Fig. 13. In Fig. 14, the band of microbuckling appears to have started at the left edge and grown into the interior of the specimen. The pattern of failure in Fig. 14*c* suggests that the microbuckling of fibers may start with the buckling of a single fiber and progressively involve additional fibers as the damage prop-

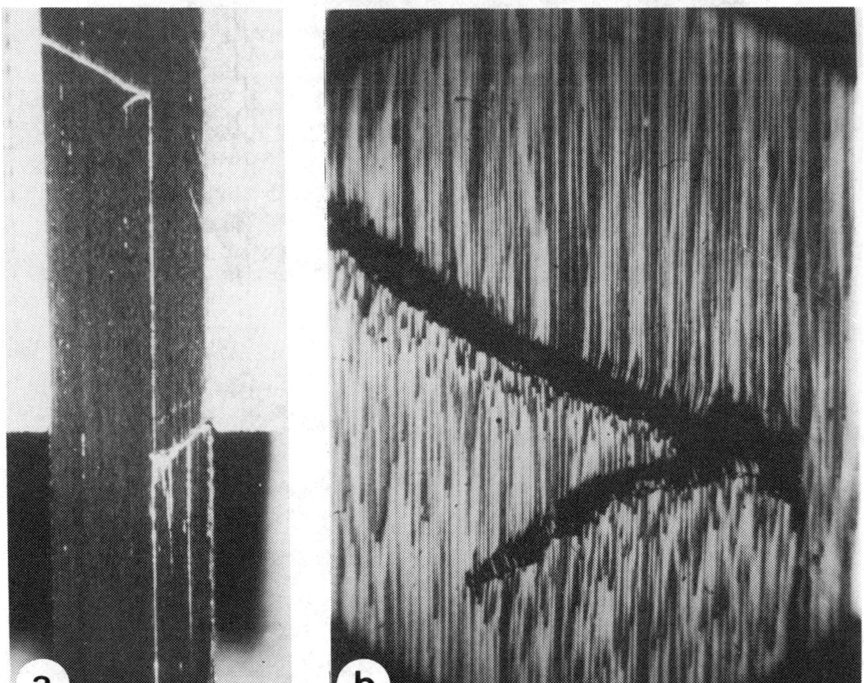

FIG. 13—*Partial failure in a T300/5208 specimen loaded to 1028 MPa;* (a) *overall edge view and* (b) *tip of kink bands.*

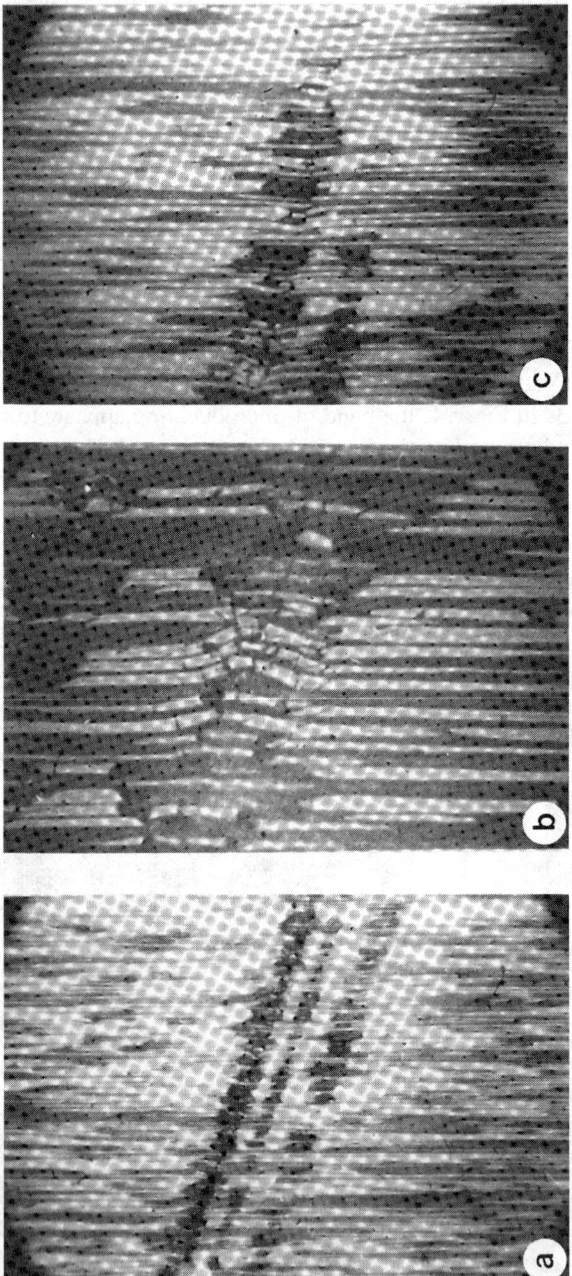

FIG. 14—*Partial failure in a T300/BP907 specimen; (a) overall edge view, (b) microbuckling of fibers in a shear crippling zone, and (c) tip of the microbuckling band.*

agates. Microbuckling causes a tensile stress to develop in the matrix between the buckled and the adjacent straight fiber that reduces the applied load at which the adjacent straight fiber buckles. Microbuckling is more likely to initiate at a free edge than in the interior because the lateral support to the fiber is lower in this region. Microbuckling, however, can also develop in the interior of a laminate in regions of voids or where the fiber-to-matrix bond is defective.

Fracture surfaces of failed specimens did not show much variation for the material systems studied. Shear crippling was observed both through the width, Fig. 15*a* and through the thickness, Fig. 15*b*. Partial shear crippling sometimes caused longitudinal splitting, Fig. 13. Shear crippling could occur on several planes just like slip lines in elasto-plastic materials, Fig. 15*a*.

The failure sequence for unidirectional composites is proposed as follows. As the compression load is increased, the weakest fibers or the fibers that have the least lateral support because of a free boundary, poor fiber to matrix bond, or voids fail first. The failure initiation may also be due to the stress concentrations introduced by test hardware. For the graphite/epoxy composites studied, the failure takes the form of kinking if the matrix is stiff, or of microbuckling if the matrix is soft. Fiber kinking is distinguished from fiber microbuckling in that the former involves the formation of a regular kink-band pattern oriented at less than 90° to the direction of the applied load while microbuckling failure involves large post-buckling deformations of the fiber in which multiple fractures may occur due to the high bending strains. In both cases, however, failure observed on the macroscopic scale for unidirectional composites, that is, looking at a free edge, normally is characterized by shear crippling involving narrow failure zones oriented at less than 90° relative to the applied load. Multiple shear crippling zones may develop in a specimen and the transverse tensile stress that develops

FIG. 15—*Modes of shear crippling;* (a) *across-the-width and* (b) *through-the-thickness.*

in the matrix, particularly as failed fibers become wedged between fibers, may cause matrix fractures to propagate parallel to the fibers.

Conclusions

The compression behavior of unidirectional graphite/epoxy composites was studied using two different fibers and four different resin systems. The fibers were T300 and T700 whereas the resins were 5208, BP907, 4901/MDA, and 4091/mPDA.

The predominant macroscopic failure mode has been identified as shear crippling. When the matrix resin is stiff, shear crippling is the result of fiber kinking on a microscopic scale. Microbuckling can replace kinking if the resin is soft, for example, in T300/BP907. Microbuckling failure is a result of high bending strains in the fiber in the post-buckled state. A partial shear crippling may lead to longitudinal splitting between fibers at the tip of the shear crippling zone because of the required kinematic compatibility. Shear crippling in composites resembles slip lines in metals.

For the materials, loading rates, and test configurations studied, failure was almost always catastrophic, and detecting partial failure was successful in only a very few tests. Little variation in failure mode could be seen from material to material.

The higher tensile strain property of the T700 fiber compared to the T300 fiber did not translate into improved unidirectional laminate strength. The T700 fiber has a smaller diameter that may be a factor. While the resin modulus has a small effect on the composite modulus, it has a strong effect on the composite compressive strength. Tensile properties of the composite are not affected significantly by the resin modulus. The compressive strength can at least qualitatively be predicted by a nonlinear model incorporating initial fiber curvature.

As discussed in the Appendix, the factor, v_f, in Eq 28 is the result of considering the equilibrium of the forces and moments acting on the fiber itself rather than on the representative volume element consisting of the fiber and the matrix. The resulting equation yields a better estimate even when nonlinear material properties and initial curvature of fibers are not included.

Acknowledgments

The work of Mr. Hahn was supported by the NASA Langley Research Center under Grant NAG-1-295. The authors would like to thank Mohsen Sohi for his help in conducting experimental measurements on the graphite/epoxy composites.

APPENDIX

Calculation of Distributed Moment, *m*, Applied to a Buckled Fiber

Consider the cross section of a circular fiber in a uniform shear stress field, τ_{yx}, as shown in Fig. 16*a*. The resultant force, *f*, on an infinitesimal area, $r_f d\theta$, on the fiber surface is given by

$$f = \tau_{yx} r_f \sin\theta \, d\theta \tag{36}$$

The resulting moment, *dm*, about the *z* axis is given by

$$dm = -r_f \sin\theta \, f = -\tau_{yx} r_f^2 \sin^2\theta d\theta$$

Thus, the total moment, *m*, per unit length of the fiber becomes

$$m = -\int_o^{2\pi} \tau_{yx} r_f^2 \sin^2\theta d\theta = -r_f^2 \tau_{yx} \int_o^{2\pi} \sin^2\theta d\theta = -\pi r_f^2 \tau_{yx} \tag{37}$$

In unidirectional composites, τ_{yx} is replaced by τ_{LT}, and hence

$$m = -\pi \frac{d_f^2}{4} \tau_{LT} \tag{38}$$

where d_f is the fiber diameter, that is, $d_f = 2r_f$. Equation 38 is the same as Eq 23.

If the representative area is taken to be a rectangle as shown in Fig. 16*b*, and if the shearing stress, τ_{yx}, is applied on the top and bottom surfaces, the moment, *m*, is simply

$$m = -(h_f + h_m)\tau_{yx} \tag{39}$$

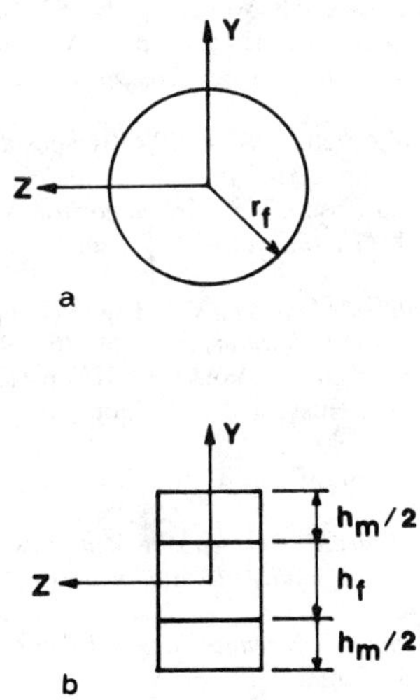

FIG. 16—*Representative cross-sectional area;* (a) *a circular fiber and* (b) *a rectangular fiber.*

In terms of the fiber volume fraction, v_f, where

$$v_f = \frac{h_f}{h_f + h_m} \tag{40}$$

and the composite shearing stress, τ_{LT}, Eq 39 can be written as

$$m = -\frac{h_f}{v_f}\tau_{LT} \tag{41}$$

Equation 41 is the same as Eq 16 when τ_{LT} is approximated by

$$\tau_{LT} = \frac{G_m}{1 - v_f}\frac{dv}{dx} \tag{42}$$

Since Eqs 1 to 3 are to be solved for the buckling of a fiber, Eq 38 is preferred to Eq 41.

References

[1] Williams, J. G. and Rhodes, M. D., "The Effect of Resin on the Impact Damage Tolerance of Graphite/Epoxy Laminates," NASA TM 83213, National Aeronautics and Space Administration, Oct. 1981.

[2] Starnes, J. H., Jr., and Williams, J. G., "Failure Characteristics of Graphite-Epoxy Structural Components Loaded in Compression," *Mechanics of Composite Materials—Recent Advances*, Hashin and Herakovich, Ed., Pergamon Press, 1983.

[3] Hawthorne, H. M. and Teghtsoonian, E., *Journal of Materials Science*, Vol. 10, 1975, pp. 41–51.

[4] Hahn, H. T., "Effects of Constituent Properties on Compression Failure Mechanisms," presented at the Tough Composite Materials Workshop, NASA Langley Research Center, May 1983, NASA CP-2334, National Aeronautics and Space Administration, 1984.

[5] Chiao, C. C. and Chiao, T. T. in *Handbook of Composites*, G. Lubin, Ed., 1982, pp. 272–317.

[6] Greszczuk, L. B. in *Proceedings*, 2nd USA-USSR Symposium on Fracture of Composite Materials, G. C. Sih, Ed., 1981, pp. 231–246.

[7] Kulkarni, S. V., Rice, J. R., and Rosen, B. W., *Composites*, Vol. 6, 1975, pp. 217–225.

[8] Weaver, C. W. and Williams, J. G., *Journal of Materials Science*, Vol. 10, 1975, pp. 1323–1333.

[9] Chaplin, C. R., *Journal of Materials Science*, Vol. 12, 1977, pp. 347–352.

[10] Evans, A. G. and Adler, W. F., *Acta Metallurgica*, Vol. 26, 1978, pp. 725–738.

[11] Hancox, N., *Journal of Materials Science*, Vol. 10, 1975, p. 234–242.

[12] Sadowsky, M. A., Pu, S. L., and Hussain, M. A., *Journal of Applied Mechanics*, Vol. 34, 1967, pp. 1011–1016.

[13] Herrman, L. R., Mason, W. E., and Chan, T. K., *Journal of Composite Materials*, Vol. 1, 1967, pp. 212–226.

[14] Lanir, Y. and Fung, Y. C. B., *Journal of Composite Materials*, Vol. 6, 1972, pp. 387–401.

[15] Greszczuk, L. B., *Microbuckling of Unidirectional Composites*, AFML-TR-71-231, Air Force Materials Laboratory, Jan. 1972.

[16] Rosen, V. W., *Mechanics of Composite Strengthening in Fiber Composite Materials*, American Society for Metals, 1965, pp. 37–75.

[17] Schuerch, H., *AAIA Journal*, Vol. 4, 1966, pp. 102–105.

[18] Greszczuk, L. B., *Journal*, American Institute of Aeronautics and Astronautics, Vol. 13, 1975, pp. 1311–1318.

[19] Davis, J. G., Jr., in *Composite Reliability, ASTM STP 580,* American Society for Testing and Materials, Philadelphia, 1975, pp. 364–377.
[20] Wang, A. S. D., "Non-Linear Microbuckling Model Predicting the Compressive Strength of Unidirectional Composites," ASME Paper 78-WA/Aero-1, American Society of Mechanical Engineers, 1978.
[21] Budiansky, B., *Computers & Structures,* Vol. 16, 1983, pp. 1–4.
[22] Argon, A. S., "Fracture of Composites," *Treatise of Materials Science and Technology,* Vol. 1, Academic Press, 1972.
[23] Hanasaki, S. and Hasegawa, Y., *Journal of Composite Materials,* Vol. 8, 1974, pp. 306–309.
[24] Sinclair, J. H. and Chamis, C. C. in *Compression Testing of Homogeneous Materials and Composites, ASTM STP 808,* R. Chait and R. Papirno, Eds., American Society for Testing and Materials, Philadelphia, 1983, pp. 155–174.
[25] Adsit, N. R. in *Compression Testing of Homogeneous Materials and Composites, ASTM STP 808,* R. Chait and R. Papirno, Eds., American Society for Testing and Materials, Philadelphia, 1983, pp. 175–186.
[26] Palmer, R. J., "Investigation of the Effect of Resin Material on Impact Damage to Graphite/Epoxy Composites," NASA CR 165677, National Aeronautics and Space Administration, March 1981.
[27] Tsai, S. W. and Hahn, H. T., *Introduction to Composite Materials,* Technomic Publishing Co., Lancaster, PA, 1980.
[28] Popov, E. P., *Mechanics of Materials,* 2nd ed., Prentice-Hall, 1976.

Ramesh J. Kar,[1] *Robert E. Herfert,*[1] *and Robert T. Kessler*[1]

Fractographic and Microstructural Examination of Compression Failures in Wet Compression Graphite/Epoxy Coupons

REFERENCE: Kar, R. J., Herfert, R. E., and Kessler, R. T., **"Fractographic and Microstructural Examination of Compression Failures in Wet Compression Graphite/Epoxy Coupons,"** *Composite Materials: Testing and Design (Seventh Conference), ASTM STP 893*, J. M. Whitney, Ed., American Society for Testing and Materials, Philadelphia, 1986, pp. 140–157

ABSTRACT: AS1-graphite/3501-6 epoxy composites exhibit degradation in mechanical properties during testing at elevated temperatures, with laminate moisture contents of 1% or greater. To investigate the effects of environmental variables on the compressive properties, a fractographic and microstructural study was carried out on multi-ply graphite/epoxy coupons overloaded in compression at different temperatures and with varying moisture contents. The analytical techniques used consisted of macroscopic and microscopic fracture examination, optical metallography, and quantitative image analysis (stereology). It was determined that coupons tested at room temperature and low moisture had macroscopic and microscopic fracture characteristics that differed from the characteristics of similar coupons tested at elevated temperatures and higher moisture contents. Microstructural analysis of the fractured coupons helped establish the failure progressions through the laminates, and in conjunction with image analyses, indicated that there was excess resin in 0° orientation plies of the coupons examined. It was concluded that (1) at room temperature, the matrix did not appear to play a role, that is, catastrophic fracture initiated in the outer plies of the composite due to gradual loss of stiffness during overload; and (2) softening of the resin (matrix), due to testing close to the glass transition temperature, was instrumental in controlling the fracture behavior at elevated temperatures, and the resin-rich areas provided increased susceptibility to moisture absorption that led to general weakening of the composite.

KEY WORDS: composite materials, graphite/epoxy, AS1/3501-6 composites, fracture (materials), compression, elevated temperature, microstructure, image analysis

AS1-graphite/3501-6 epoxy composites are known to exhibit degradation in mechanical properties during tests at elevated temperature and conditions of high

[1] Engineering specialist, Aircraft Division; formerly, senior technical specialist, Aircraft Division, now, manager, Materials and Processes Laboratories, Advanced Systems Division; and manager, Microstructural and Chemical Analyses Laboratories, Aircraft Division, respectively, Northrop Corporation, Hawthorne, CA 90250.

humidity such that the laminates have moisture contents of 1% or greater. For example, it has been reported [1] that graphite/epoxy laminates tested in compression at 121°C and under conditions of high humidity fail at load levels that are typically one third of the load values for similar coupons tested at room temperature and low humidity. These differences in bulk properties of such complex coupons may be related to the individual response of each constituent of the lamina to the applied load and test environment, as well as the total synergistic response of all the components acting as a single load-carrying structural member. Therefore, for a complete evaluation of the failure mechanism(s) involved in the performance of laminated coupons, it is necessary to understand both the macro- as well as the micromechanisms involved in the failure.

The objective of the present research program was to investigate whether the effects of elevated temperature and moisture on the compressive strengths of complex graphite/epoxy structural laminates could be understood through a fractographic and microstructural study of laminates that had been tested to failure under different conditions of humidity and temperature.

Experimental Procedure

The specimens chosen for examination in this investigation consisted of multi-ply graphite/epoxy coupons that had been tested to failure under compressive loads at various temperatures and with varying moisture contents. Each coupon was midplane symmetric in nature and consisted of a total of 56 AS1-graphite plies in a 3501-6 epoxy matrix. Table 1 enumerates the locations and orientations of the plies in each of the test coupons. These can broadly be referred to as 17 outer plies (0°, ±45°, 90° orientations) on either side of a core of 22 central plies (0° and +45° orientations). The individual coupon labels, test temperatures, moisture contents (indicative of the humidity conditions), and compressive mechanical properties determined for each coupon are listed in Table 2.

The analytical procedures used in this investigation were similar to those used in the past [2] to analyze graphite/epoxy composite specimens of simple orientation and consisted of methods to study the macro- and micromechanisms associated with the failures.

These procedures included the following:

1. Visual examination and optical photomacrography of failed regions in the coupons. These techniques permitted determination of fracture initiation sites, direction of crack propagation, and broad features.
2. Scanning electron microscopic (SEM) examination of representative fracture sections from the failed laminates. This technique facilitated the evaluation of microscopic fracture characteristics associated with the failures, such as microbuckling and fiber-wetting by resin. In each case, the sections

TABLE 1—*Ply location and orientation in graphite/epoxy test coupons.*

Ply Number	Orientation	Ply Number	Orientation
OUTER PLIES		CENTRAL PLIES (CONT.)	
1	+45	29	0
2	−45	30	0
3	90	31	0
4	0	32	+45
5	+45	33	0
6	−45	34	+45
7	90	35	0
8	0	36	+45
9	+45	37	0
10	−45	38	+45
11	90	39	0
		OUTER PLIES	
12	0	40	+45
13	90	41	−45
14	0	42	+45
15	+45	43	0
16	−45	44	90
17	+45	45	0
CENTRAL PLIES			
18	0	46	90
19	+45	47	−45
20	0	48	+45
21	+45	49	0
22	0	50	90
23	+45	51	−45
24	0	52	+45
25	+45	53	0
26	0	54	90
27	0	55	−45
28	0	56	+45

NOTE—The plies are symmetrically located in the test coupons.

were prepared for SEM examination by vacuum depositing thin layers of gold (approximately 60 nm thick) on the fracture faces to make them electrically conductive. The specimens were then examined in an Etec R-1 Autoscan SEM using standard secondary electron imaging techniques.

3. Optical metallography of representative sections of the coupons in both the longitudinal and transverse directions of the laminates. These procedures were primarily used to supplement information generated from the fracture analysis and were necessary to evaluate the occurrence of resin-rich/fiber-rich regions, void areas, etc.
4. Image analysis (stereology) of representative metallographic sections of the laminates to quantitate fiber and resin contents of the plies in the coupons. Analysis was carried out using an Omnicon FAS II image analysis system in conjunction with a Leitz MM5 optical metallograph.

TABLE 2—*Mechanical properties of test coupons.*

Specimen Label	Specimen Number	Test Temperature °C/°F	Moisture Content, %	Failure Loads, kg/lb	Failure Stress, MPa/ksi	Avg Strain @ Failure
RT/Dry	23	21/70	0.10	$11.9 \times 10^3/26.18 \times 10^3$	437.3/63.4	−0.008672
RT/Dry	46	21/70	0.10	14.6/29.29	489.0/70.9	−0.008475
104C/Wet	65	104/220	1.39	7.5/16.62	277.3/40.2	−0.005060
121C/Wet	57	121/250	1.40	3.48/7.67	128.3/18.6	−0.002195
121C/Wet	71	121/250	1.41	4.10/9.03	151.1/21.9	−0.002305

Results

Macrographic Observations

Preliminary visual examination of the failed coupons revealed significant differences in the overall fracture appearance of coupons tested at elevated temperatures and with moisture contents of 1% or greater (hereafter referred to as 104C/Wet, 121C/Wet, etc.) in comparison with those of coupons tested at room temperature and low moisture (hereafter referred to as RT/Dry). Compression failures in 104C/Wet or 121C/Wet coupons were characterized by one or more cracks that had propagated across the cross sections of the laminates at an angle of 45° to the longitudinal axis (see Fig. 1). In contrast, the RT/Dry coupons were unique in that the central core of plies had separated from the outer plies (see Fig. 2) and were associated with cracks that appeared to change direction as these propagated across the thickness of the coupons. In general, the cracks were oriented at oblique angles ranging from 30° to 45° to the longitudinal axes of the coupons while propagating along the outer plies, whereas in the central ply regions, the cracks were generally oriented normal to the longitudinal axis (see Fig. 2).

Closer examination of the fractures associated with the cracks (exposed by applying controlled tensile loads to the laminates in the laboratory) again revealed significant differences in the fracture features of RT/Dry, 104C/Wet, and 121C/Wet laminates. Figures 3 and 4 illustrate fracture features observed in RT/Dry and 104C/Wet coupons. (Note that the features observed in 121C/Wet coupons were similar to those illustrated for 104C/Wet coupons.)

Whereas the RT/Dry specimens were associated with fractures oriented 45° to the loading direction (see Fig. 3), the 104C/Wet and 121C/Wet coupons were associated with fractures that were transverse to the loading direction in the central plies (see Fig. 4) and a combination of transverse and 45° orientations in the

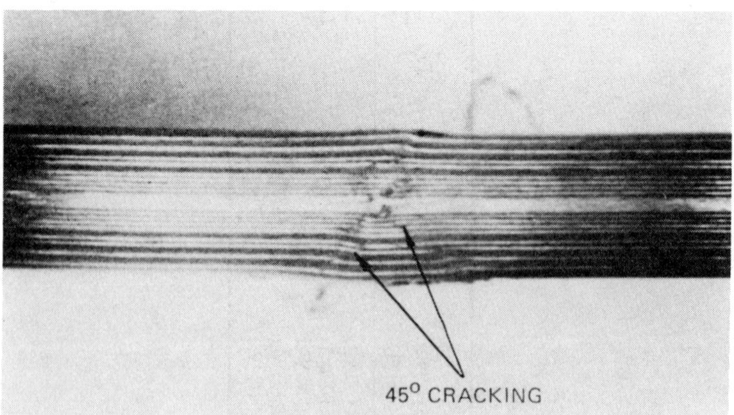

FIG. 1—*Photomacrograph illustrating a 104C/Wet coupon tested to failure in compression.*

FIG. 2—*Photomacrograph illustrating an RT/Dry coupon tested to failure in compression.*

outer plies (again see Fig. 4). It is believed that the ±45° plies were predominantly in the 45° fracture mode whereas the 0° and 90° plies were in the transverse fracture mode.

Scanning Electron Microscopy

Scanning electron microscopic (SEM) examination of representative sections of the fractures revealed that microscopic fracture features such as secondary cracking, fiber "clumping," and "chopping" (illustrated in Fig. 5), characteristic of compression failures and previously reported in simple orientation coupons [*3,4*], were present in all the laminates. Of greater significance, it was observed that fractured fibers in the RT/Dry coupons were associated with extensive

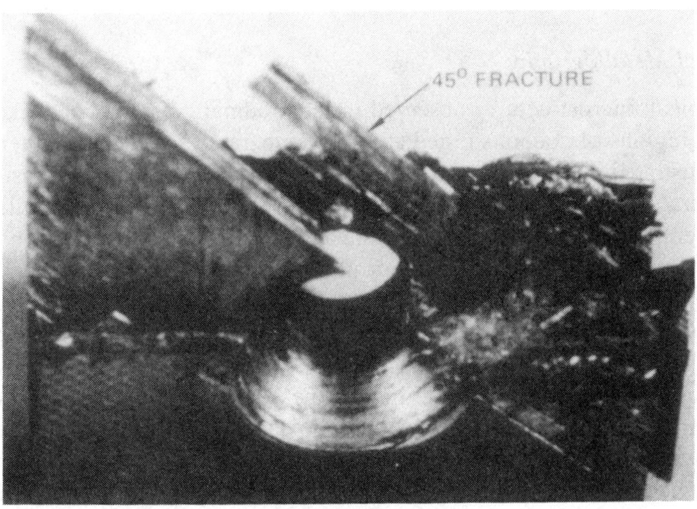

FIG. 3—*Photograph illustrating macroscopic fracture features on an RT/Dry coupon.*

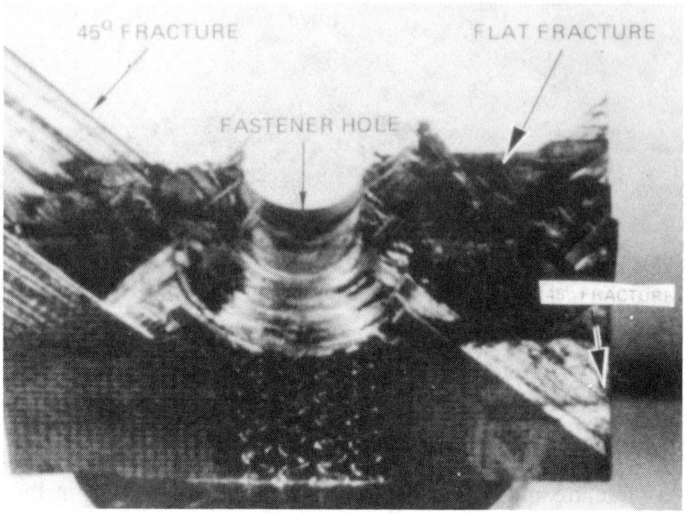

FIG. 4—*Photograph illustrating macroscopic fracture features on a 104C/Wet coupon.*

"wetting" by the resin with the resin-rich areas exhibiting brittle fracture features in the form of small "river-like" cleavage markings (see Fig. 6). In contrast, the 104C/Wet and 121C/Wet coupons were associated with clean fiber surfaces and fiber-resin "pull-away" [*3,4*] (illustrated in Figs. 7*a* and *b*). The resin-rich regions appeared considerably more ductile in nature as suggested by the presence of "flow lines" (see Fig. 8) that were not observed in resin-rich areas of RT/Dry coupons.

Optical Metallography

Typical microstructures observed in longitudinal cross sections taken through failed regions of coupons tested at room temperatures and elevated temperatures are illustrated in Figs. 9 and 10. As shown in Fig. 9, the laminates tested at room temperature, namely RT/Dry, were characterized by two separate failure modes. The central plies appeared to have failed at orientations of 45° to the compressive loading directions in the thickness planes, whereas the outer plies had failed predominantly by delamination with accompanying interply cracking. It was also observed that many of the cracks had occurred in regions of resin separating the plies.

In contrast, as illustrated in Fig. 10, the 104C/Wet (and 121C/Wet) coupons had failed primarily by buckling. Failure consisted of primary fractures oriented approximately at 45° to the compressive axis in the thickness directions, with large secondary cracks (and associated fractures) forming at 90° angles to the

FIG. 5—*SEM photomicrographs illustrating typical microscopic fracture features observed in all the coupons.*

primary cracks in the same plane. As shown in Fig. 10, a large number of plies near the primary fractures were observed to be curved, indicating that buckling had occurred in these regions with a small degree of delamination. Although the extent of primary cracking that had occurred in resin-rich regions could not be estimated, it was observed that a large number of secondary cracks followed the resin-rich areas between plies. These observations suggest that the failures may have occurred along areas of high resin content.

Metallographic examination was also carried out on transverse cross sections taken through failed regions of representative RT/Dry, 104C/Wet, and 121C/Wet coupons. Typical microstructures observed in all the samples were similar (the microstructure observed in a RT/Dry coupon is illustrated in the composite series of photomicrographs shown in Fig. 11). It was determined that the laminates

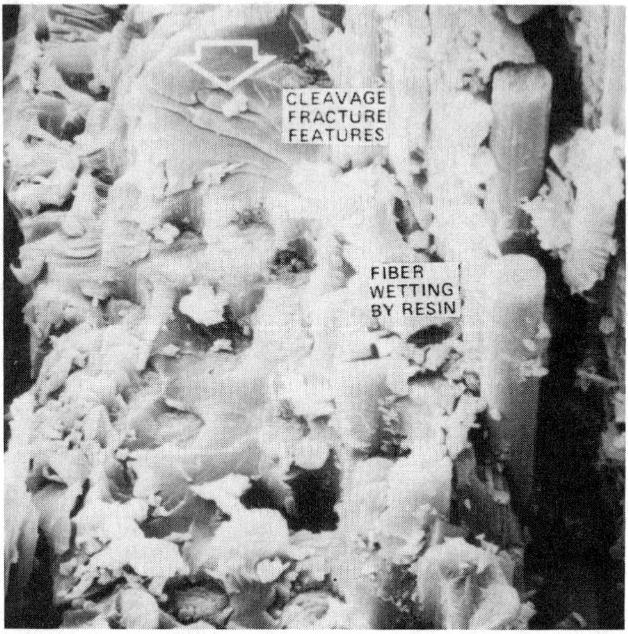

FIG. 6—*SEM photomicrograph illustrating "fiber-wetting by resin" and cleavage fracture features observed in resin-rich areas of an RT/Dry coupon.*

were generally free of voids or irregularly shaped fibers. More significantly, resin-rich areas were observed in all three ply directions (0°, ±45°, and 90°) as are illustrated in Fig. 11. However, there appeared to be a greater preponderance of resin-rich areas in the 0°-ply orientations of all the coupons examined. It was thought that the occurrence of such areas was of significance since they would provide an increased susceptibility to moisture absorption, poor fiber distribution, and weakening of the composite.

Image Analysis

To quantify the resin-rich areas observed metallographically, image analysis was carried out on the center 0° plies of representative RT/Dry and 121C/Wet coupons. Histograms illustrating the fiber contents of these coupons are shown in Figs. 12 and 13. Based on these results, the resin contents of the 0° plies of the laminates were estimated to vary between 56 to 59%. Such high resin contents would certainly be detrimental to the mechanical properties of the laminates.

Discussion

RT/Dry Coupon Failures

Table 3 describes the failure progression through the laminate thickness of a typical RT/Dry coupon. In an analysis of this failure, it was assumed that the

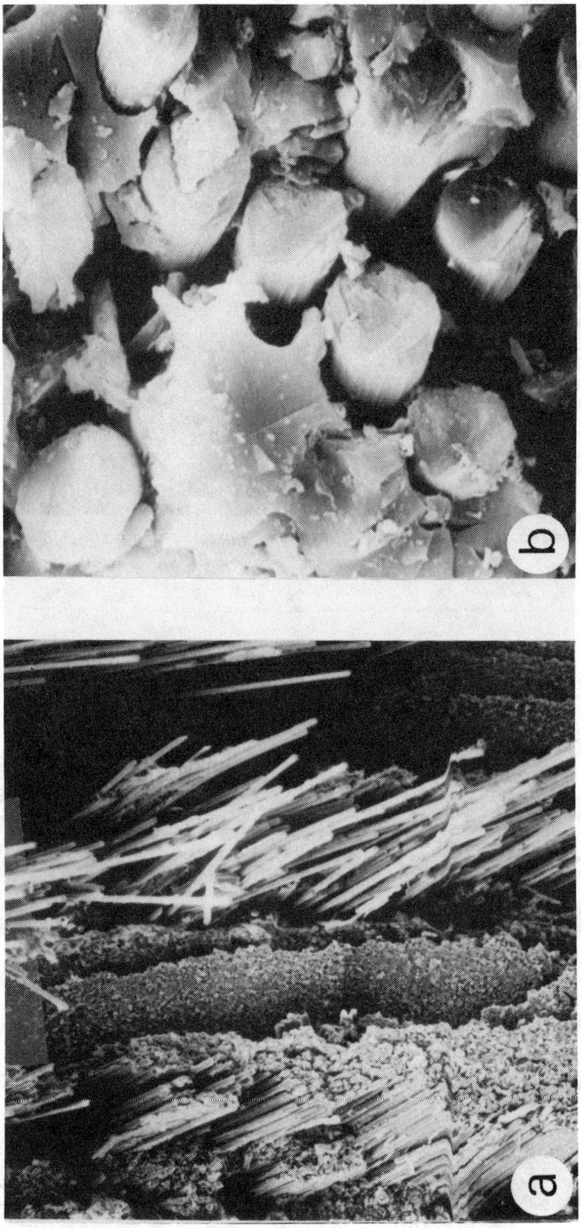

FIG. 7—*SEM photomicrographs illustrating clean fibers* (a) *and fiber-resin pull-away* (b) *in a 104C/Wet coupon. Similar features were observed in 121C/Wet coupons.*

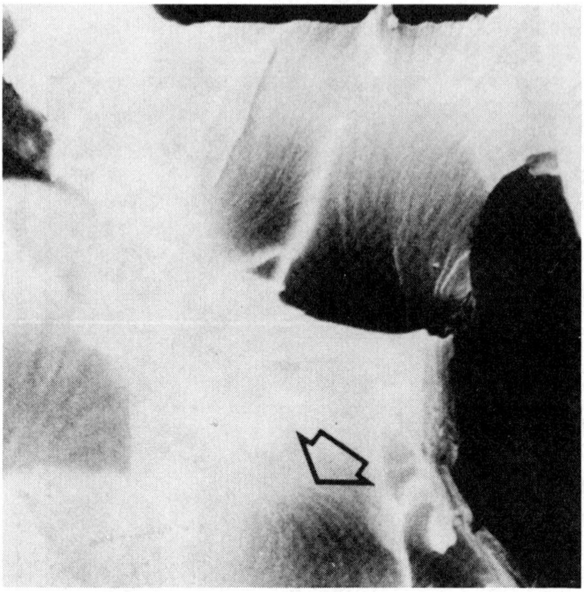

FIG. 8—*SEM photomicrograph illustrating "flow lines" observed in a resin-rich area of a 104C/Wet coupon.*

FIG. 9—*Composite series of optical photomicrographs illustrate failure observed in a longitudinal section taken through a failed RT/Dry coupon.*

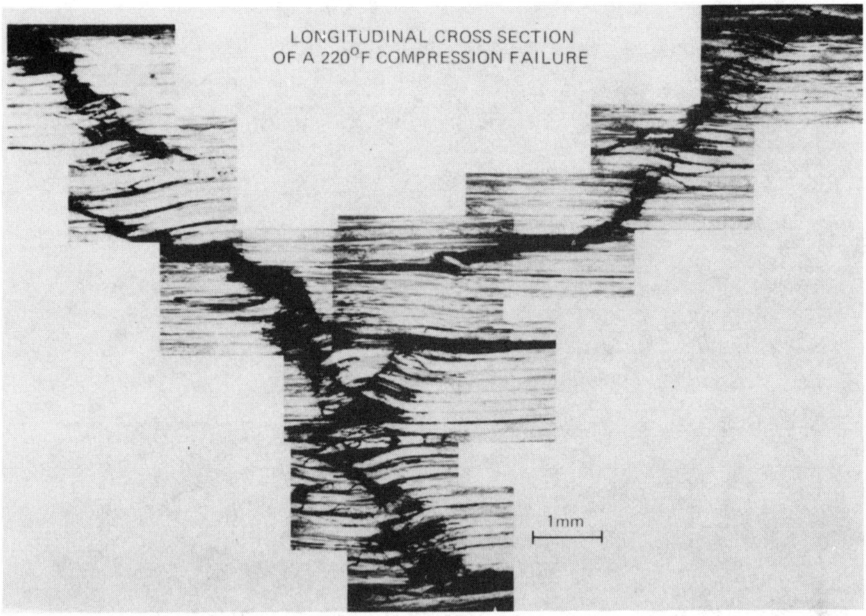

FIG. 10—*Composite series of optical photomicrographs illustrate failure observed in a longitudinal section taken through a failed 104C/Wet coupon.*

specimen acted as a continuum under load with monotonic increasing applied strain at any given cross section and that the shear strains resulting from the unbalanced nature of the laminate were small and did not affect the failure progression. Grimes et al [*3,5*] have estimated proportional-limit lamina strain values for ±45° and 90° orientations as −0.0049 and −0.0094, respectively, and −0.015 as the ultimate strain value for the completely linear 0°-direction lamina. Based on the ultimate strain values measured for RT/Dry composites (which range from −0.0080 to −0.0094 as shown in Table 2), it appears that during compressive loading of these composites, as soon as the applied strain value reached the proportional limit for ±45° plies (−0.0049), plies of this orientation in the outer regions of the composite suffered a gradual loss of stiffness. As the applied compressive strain was further increased to the proportional limit value for 90° plies (approximately −0.0094), they started yielding. As can be seen from Table 1, the outer regions consist of nine ±45° plies, four 90° plies, and four 0° plies. Therefore, when the applied compressive strain approached the ultimate failure strain value for the composite, 13 of the 17 plies in the outer regions had lost their stiffness, resulting in overloading of the remaining four 0° plies in the outer region. As a result, these plies became unstable and buckled, causing delamination as is illustrated in Fig. 9. This loss of load-carrying capacity in the outer plies resulted in the central core of plies (22 in all) being loaded to their ultimate value, resulting in catastrophic failure across the

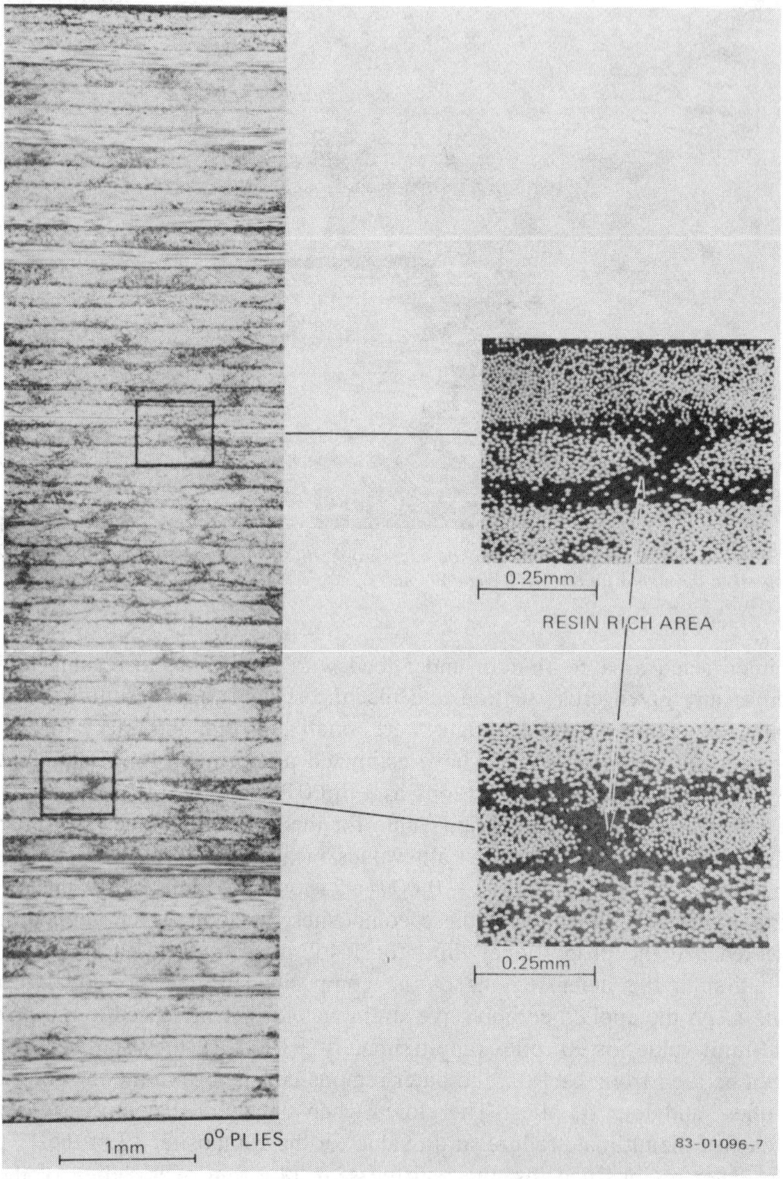

FIG. 11—*Composite series of optical photomicrographs illustrate resin-rich areas observed in an RT/Dry coupon. Similar features were observed in 104C/Wet and 121C/Wet coupons.*

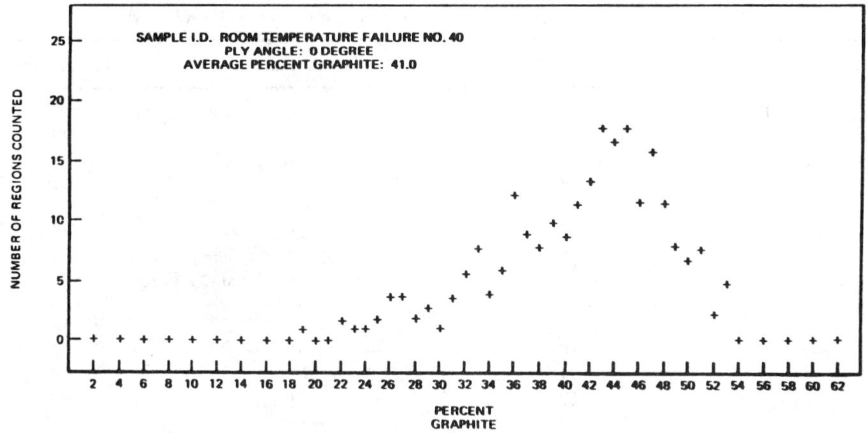

FIG. 12—*Histogram illustrating the volume percent of graphite fibers measured in 0° plies of RT/Dry composites.*

remaining nonfractured thickness of the laminate. The overload fracture surfaces were found to exhibit typical fractographic characteristics such as "clumping," "chopping," and "fiber wetting by resin"[*2*].

104C/Wet and 121C/Wet Coupon Failures

Tables 4 and 5 describe the failure progressions through representative 104C/Wet and 121C/Wet coupons. As can be seen from Table 2, the measured failure strain values for these coupons was much lower than that for RT/Dry coupons.

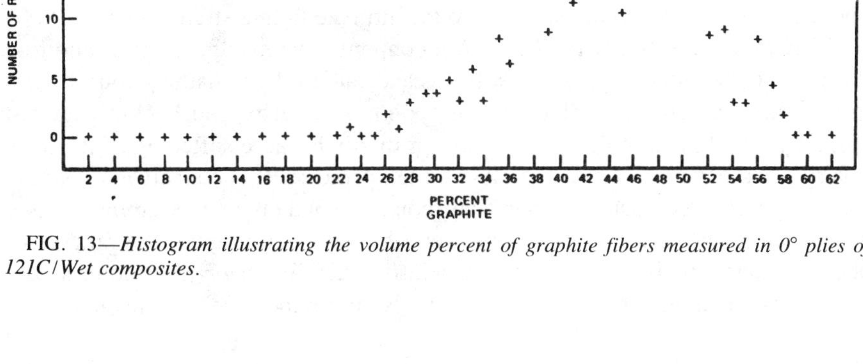

FIG. 13—*Histogram illustrating the volume percent of graphite fibers measured in 0° plies of 121C/Wet composites.*

TABLE 3—*Graphite/epoxy specimen orientation/failure mode RT/Dry.*

Orientation	Ply Number	Orientation	Ply Number
PRIMARY FAILURE[a]		SECONDARY FAILURE (CONT.)	
+45	1	0	29
−45	2	0	30
90	3	0	31
0	4[b]	+45	32
+45	5	0	33
−45	6	+45	34
90	7	PRIMARY FAILURE[a]	
0	8[b]	0	35
+45	9	+45	36
−45	10	0	37
90	11	+45	38
0	12	0	39[b]
90	13[b]	+45	40[b]
0	14	−45	41[b]
+45	15	+45	42
−45	16[b]	0	43
+45	17	90	44
SECONDARY FAILURE[c]		0	45[b]
0	18	90	46
+45	19	−45	47[b]
0	20	+45	48
+45	21	0	49
0	22	90	50[b]
+45	23	−45	51
0	24	+45	52
+45	25	0	53
0	26	90	54[b]
0	27	−45	55
0	28	+45	56

[a]0° plies buckling causing bending, compression and compression/shear fracture followed by delamination between plies shown.
[b]Extensive delamination.
[c]Compression fracture of 0° plies (primary) accompanied by compression/shear fracture of 45° plies (no delamination).

Again, based on the proportional strain limit values reported by Grimes et al [*3,5*] for ±45° and 90° orientation plies tested at elevated temperatures and humidity, it appears that on application of the compressive loads, as soon as the applied compressive strain was close to the ultimate failure strain value (−0.005 for 104C/Wet or −0.002 for 121C/Wet coupons), the nine ±45° plies and four 90° plies in the outer regions lost stiffness and caused the remaining four 0° plies in the outer regions to buckle. At this point, a small plastic buckle occurred spanning all 22 plies of the central core, primarily because softening of the resin (matrix) had occurred due to the elevated temperature. It is conceivable that the softening was accentuated by the high moisture contents of the laminate. As a consequence, the matrix was no longer capable of supporting the central core of plies (primarily 0° fibers), resulting in plastic buckling *prior* to fracture across the entire remaining unbroken cross section. Note that the softening characteristics

TABLE 4—*Graphite/epoxy specimen orientation/failure mode—104C/Wet.*

Orientation	Ply Number	Orientation	Ply Number	Orientation	Ply Number
PRIMARY FAILURE[a]		FOURTH FAILURE[e]		PRIMARY FAILURE (CONT.)	
+45	1[b]	0	26[b]	−45	51
−45	2	0	27	+45	52
90	3	0	28	0	53
0	4	0	29	90	54
+45	5	0	30	−45	55
−45	6	0	31	+45	56
90	7	SECONDARY FAILURE[f]			
0	8	+45	32		
+45	9	0	33[b]		
−45	10	+45	34		
90	11	0	35		
0	12	+45	36		
90	13	0	37		
0	14	+45	38		
+45	15	0	39		
−45	16[b]	PRIMARY FAILURE[a]			
SECONDARY FAILURES[c]		+45	40[b]		
AND		−45	41		
TERTIARY FAILURES[d]		+45	42		
+45	17	0	43[b]		
0	18	90	44		
+45	19	0	45[b]		
0	20	90	46		
+45	21	...	...[b]		
0	22	−45	47		
+45	23[b]	+45	48		
0	24	0	49		
+45	25[b]	90	50[b]		

[a] 0° plies buckling causing bending fracture followed by delamination between plies.
[b] Extensive delamination.
[c] Buckling of both 0° and 45° plies followed by delamination between plies.
[d] Delamination, compression, and compression/shear fracture.
[e] 0° plies failed in compression and compression shear.
[f] Buckling of both 0° and 45° plies followed by delamination as indicated.

of the matrix (resin) were evident as suggested by the microfracture mechanisms observed in these composites such as "lack of fiber wetting by resin" and "flow lines in resin-rich regions."

Conclusions

A fractographic and microstructural study of AS1/3501-6 multi-ply graphite-epoxy laminates tested at various temperatures and with varying moisture content supports the following conclusions.

1. Compression failure in RT/Dry laminates were associated with the outer plies failing primarily by delamination and associated buckling of the central core of plies.

TABLE 5—*Graphite/epoxy specimen orientation/failure mode—121C/Wet.*[a]

Orientation	Ply Number	Orientation	Ply Number
+45	1	0	29
−45	2	0	30
90	3	0	31
0	4	+45	32[b]
+45	5	0	33[b]
−45	6	+45	34
90	7	0	35
0	8	+45	36
+45	9	0	37
−45	10	+45	38
90	11	0	39
0	12	+45	40
90	13	−45	41
0	14[b]	+45	42
+45	15	0	43
−45	16	90	44[b]
+45	17[b]	0	45
0	18	90	46
+45	19	−45	47
0	20	+45	48
+45	21	0	49
0	22	90	50
+45	23	−45	51
0	24	+45	52
+45	25	0	53
0	26	90	54
0	27[b]	−45	55
0	28	+45	56

[a]Plastic yielding and ultimate strength cracking of matrix in plies, extensive delamination within plies, and joggle type failure. Results from the matrix's loss of most of its elastic strength and stiffness.

[b]Extensive delamination.

2. The macroscopic failure mechanism observed in 104C/Wet and 121C/Wet coupons consists of fiber-buckling *prior* to catastrophic fracture.
3. The SEM fracture examination indicated differences in the ductility of the resin areas in 104C/Wet and 121C/Wet coupons and RT/Dry coupons. The 104C/Wet and 121C/Wet coupons were associated with resin that exhibited more ductile fracture characteristics than the resin areas of the RT/Dry coupons.
4. The RT/Dry coupons were associated with microscopic fracture features such as "fiber-wetting by resin" whereas the 104C/Wet and 121C/Wet coupons were associated with pronounced "lack of fiber-wetting by resin" and "graphite fiber pull-away." This observation suggested breakdown in the fiber-resin bond under conditions of elevated temperature and high humidity.
5. Microstructural examination of transverse cross sections of RT/Dry, 104C/Wet, as well as 121C/Wet laminates showed concentrations of excess resin

in the 0° plies of all the coupons. These observations were confirmed by subsequent image analysis. This preponderance of excess resin may have affected the compressive properties of composites tested at elevated temperatures, since these would provide an increased ability for moisture to be absorbed and create a reduction in strengths of the composites as well as poor fiber distribution.

Acknowledgments

This work was performed as part of Northrop's Independent Research and Development Program. The authors would like to acknowledge the experimental assistance of M. H. Ransick and G. C. Grimes.

References

[*1*] Grimes, G. C., unpublished research, Northrop Corporation, Hawthorne, CA, 1982.

[*2*] Herfert, R. E., "Fractography of Graphite/Epoxy Composites," *SAMPE Quarterly*, July 1982, pp. 7–12.

[*3*] Grimes, G. C. and Adams, D. F., "Investigation of Compression Fatigue Properties of Advanced Composites," Northrop Corporation Technical Report NOR 79-17, Oct. 1979.

[*4*] "Fractography of Composites," *Advanced Materials and Processes Manual*, Northrop Corporation, Hawthorne, CA, 1981.

[*5*] Grimes, G. C. in *Test Methods and Design Allowables for Fibrous Composites, ASTM STP 734*, C. Chamis, Ed., American Society for Testing and Materials, Philadelphia, 1981, pp. 281–337.

Strength

Paul A. Lagace[1]

Notch Sensitivity and Stacking Sequence of Laminated Composites

REFERENCE: Lagace, P. A., "**Notch Sensitivity and Stacking Sequence of Laminated Composites,**" *Composite Materials: Testing and Design (Seventh Conference), ASTM STP 893*, J. M. Whitney, Ed., American Society for Testing and Materials, Philadelphia, 1986, pp. 161–176.

ABSTRACT: The effect of laminate stacking sequence and lamination angle on the notched strength of 6-ply graphite/epoxy laminates is investigated. Data is presented for the three different stacking sequences of a 6-ply laminate comprised of two unidirectional plies and two pairs of angled plies: $[\pm\theta/0]_s$, $[0/\pm\theta]_s$, and $[+\theta/0/-\theta]_s$ for θ equal to 15°, 30°, 45°, 60°, 75°, and 90°. Notch geometry is limited to circular holes of which four different diameters were used: 3.175 mm, 6.35 mm, 9.525 mm, and 12.7 mm. Five specimens of each type (hole diameter and laminate) were tested monotonically to failure in tension. In addition, five unnotched specimens of each laminate were also tested. The fracture stresses of the notched specimens are correlated using the method proposed by Mar and Lin and a modified version of the point stress criterion of Whitney and Nuismer. The data shows that there is a distinct stacking sequence effect on notched fracture stress. The fracture stresses of the notched $[+\theta/0/-\theta]_s$ laminates differ significantly from those of the $[\pm\theta/0]_s$ and $[0/\pm\theta]_s$ laminates that are generally the same for a given laminate and hole diameter. It is suggested that these differences are caused by out-of-plane effects of interlaminar stresses (delamination) or the elastic constraint of neighboring plies or both. These factors change as the stacking sequence and lamination angle are varied.

KEY WORDS: composite materials, notch sensitivity, fracture strength, delamination, composite fracture parameter, stacking sequence

The notch sensitivity of composite laminates in the presence of holes, as well as cracks, is a well-documented phenomena from early investigations [*1,2*] to present day work [*3*]. A number of different approaches have been developed and proposed to correlate the fracture stresses of composites with notches [*4–11*]. A common shortcoming of these methods is their lack of consideration of out-of-plane effects, for example, interlaminar stresses, and thus an inability to account for failure at holes due to delamination. The existence of interlaminar stresses at straight free edges was shown by Pipes and Pagano [*12*] and has been the subject of much research. Some work has been conducted on the calculation

[1]Draper Assistant Professor of Aeronautics and Astronautics, Technology Laboratory for Advanced Composites, Massachusetts Institute of Technology, Cambridge, MA 02139.

of interlaminar stresses around holes [for example, Refs *13–16*]. These analyses are generally limited to rather simple laminate construction. Due to these interlaminar stresses, among other possible phenomena, at a hole, a change in laminate stacking sequence, which affects the interlaminar stress state around the hole, may change the fracture mode from in-plane to out-of-plane as is observed with unnotched laminates [*17*]. Several researchers [for example, Refs *18–21*] have looked at the effects of stacking sequence on the fracture of notched laminates. These notch types include both holes and cracks. The work that has been conducted to date has been inconclusive, ranging from showing no stacking sequence effect [*18,21*] to some effect [*19,20*] based on a small amount of data.

In this paper, a systematic experimental investigation of the effect of stacking sequence and lamination angle on the fracture characteristics of notched graphite/epoxy laminates is conducted. The notches considered are limited to centrally located circular holes. Three different possible symmetric stacking sequences of a 6-ply laminate consisting of two unidirectional plies and two sets of angled plies are considered. Results are also compared with unnotched fracture characteristics of the same laminates and correlated by two methods previously proposed [*7,22*].

Experimental Procedure

The material used throughout the program was Hercules AS1/3501-6 graphite/epoxy unidirectional preimpregnated tape supplied in 300-mm-wide rolls and stored at temperature below −18°C. The prepreg was cut using Stanley knives and laid up with the use of a jig to assure proper fiber angles from ply to ply. The material was cured as 300-mm by 350-mm plates in an autoclave under 0.59 MPa and 762 mm Hg vacuum in a two-step process: a 1-h hold at 116°C and a 2-h hold at 177°C. The cured plates were postcured in an oven at 177°C for 8 h. The completed laminates were cut into 50-mm-wide strips on a milling machine with a specially designed table using a water-cooled diamond wheel. Those specimens requiring holes were drilled with diamond-coated drill and reamer sets using a standard drill press. Water cooling was used in the drilling process. This procedure resulted in smooth-edged holes with no back ply delaminations. The 8-ply glass/epoxy loading tabs made of Scotchply Type 1002 were bonded onto each end of the graphite/epoxy strip using film adhesive FM-123-2 supplied by American Cyanamid. This process resulted in a specimen 350 mm long and 50 mm wide as depicted in Fig. 1.

Longitudinal strain gages (Type EA-09-125AD-120 manufactured by Micro Measurements) were bonded onto all specimens in order to obtain data to measure longitudinal modulus. For unnotched specimens, the gage was placed at the specimen center while for notched specimens, the gage was located away from the hole, as pictured in Fig. 1, in order to properly measure the far-field strain. This measurement was used as a means of quality control.

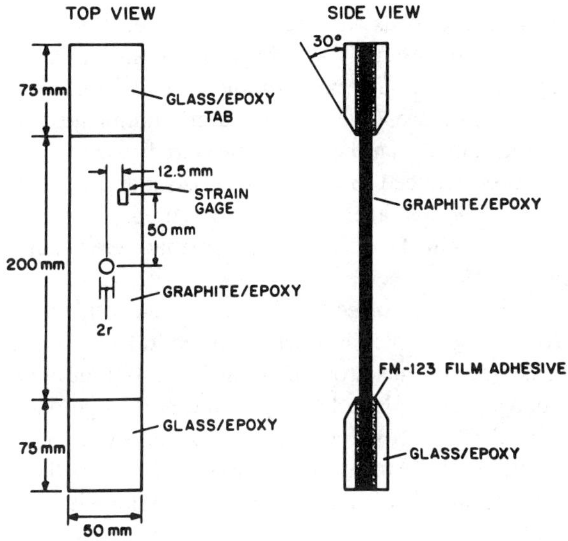

FIG. 1—*Physical characteristics of the coupon specimen.*

A total of 450 test specimens was manufactured and tested as shown in Table 1. A compromise was made between the number of specimens to be run to obtain each data point and the scope of the test program. It was decided that five specimens would be used to obtain each point. This was considered an absolute minimum in order to obtain reasonable confidence in the mean values obtained.

Testing was accomplished on an MTS 810 Material Test System with the aid of hydraulic grips. The specimens were loaded at a constant stroke rate of 1 mm/min yielding an approximate strain rate of 5000 microstrain/min. Load and strain data were recorded through an automated data acquisition system using a PDP-

TABLE 1—*Experimental test program.*

	Laminates		
Hole Diameter, mm	$[\pm\theta/0]_s$[a]	$[0/\pm\theta]_s$	$[+\theta/0/-\theta]_s$
Unnotched	5[b]	5	5
3.175	5	5	5
6.35	5	5	5
9.525	5	5	5
12.7	5	5	5

[a] θ = 15°, 30°, 45°, 60°, 75°, and 90°.
[b] Number indicates number of specimens tested for each lamination angle.

11/34 computer. Fracture loads were recorded after each test and a photograph was taken of each specimen after fracture. The specimens were also monitored by ear throughout the test for any noises indicative of damage.

In the calculation of stresses, the manufacturer's nominal per ply thickness of 0.134 mm was used rather than the actual measured laminate thicknesses. This allowed the data to be reduced to a common base independent of slight manufacturing variations. The average corrected measured per ply thickness for all 450 specimens is 0.135 mm. Thickness measurements were taken at nine locations on each coupon.

An isotropic, finite-width correction factor was not utilized as suggested by other authors [*18*]. The isotropic finite-width correction does not properly account for the differing degrees of anisotropy exhibited in different laminates, nor does it account for changes in in-plane loading at the hole edge due to interlaminar stresses. As a bound to the possible error introduced by not considering finite-width effects, consider the stress distribution away from the hole on a line perpendicular to the applied load. The approximate solution is provided by Konish and Whitney [*23*]

$$\frac{\sigma_x}{\overline{\sigma}} = \frac{1}{2}\left\{2 + \left(\frac{r}{y}\right)^2 + 3\left(\frac{r}{y}\right)^4 - (K_T^\infty - 3)\left[5\left(\frac{r}{y}\right)^6 - 7\left(\frac{r}{y}\right)^8\right]\right\} \tag{1}$$

for y greater than r, where

σ_x = stress parallel to the applied load,
$\overline{\sigma}$ = applied far-field stress,
r = hole radius,
y = distance from the hole center in the transverse direction, and
K_T^∞ = anisotropic stress concentration factor for an infinitely wide plate.

The anisotropic stress concentration factor is determined from the laminate engineering constants as follows

$$K_T^\infty = 1 + \left[2\left(\frac{E_{11}^L}{E_{22}^L}\right)^{1/2} - 2\nu_{12}^L + \frac{E_{11}^L}{G_{12}^L}\right]^{1/2} \tag{2}$$

By integrating the stress expression across the width of the specimen, the total load carried by the net section of the specimen at the hole can be determined. For an infinitely wide plate, this load will be equal to the far-field applied load. In the case of a finite-width plate, the percentage difference of this load with the applied load will give an indication as to the possible error.

For the current program, the maximum possible error will occur for the largest hole diameter of 12.7 mm. The minimum possible value of K_T^∞ for the laminates considered is three. Using this and integrating the stress across the specimen width, it is found that 97.2% of the total applied load is carried for this "worst" case. This implies that a maximum error of 2.8% is introduced by neglecting finite-width effects. This is well within the experimental scatter observed in fracture stresses and can therefore be neglected. This is possible only due to the fact that the hole diameter to specimen width ratio is kept relatively low in this investigation.

Experimental Results

The average measured longitudinal moduli for each laminate, as well as the respective coefficient of variation (CV), are presented in Table 2. The theoretical moduli are also presented for comparison. These moduli were calculated using Classical Laminated Plate Theory and the basic unidirectional ply elastic constants for Hercules AS1/3501-6 of

$$E_L = 130 \text{ GPa} \qquad E_T = 10.5 \text{ GPa}$$
$$G_{LT} = 6.0 \text{ GPa} \qquad \nu_{LT} = 0.28$$

where

L = fiber direction, and
T = transverse direction.

It can readily be seen that the experimentally determined longitudinal moduli are independent of stacking sequence, within experimental error, and agree well with the predicted values.

Unnotched and notched failure stresses were calculated from the maximum applied load using the measured width and calculated nominal thickness. The notched fracture stresses are correlated using two methods.

The first is that suggested by Mar and Lin [*22,24*]. Previous researchers [*25*] have suggested that the basic Fracture Mechanics equation be adapted for composites by changing the value of the exponent 1/2 to yield an equation of the form

$$\sigma_f r^m = \text{constant} \tag{3}$$

TABLE 2—*Theoretical and experimental longitudinal moduli.*

		Experimental Values					
		$[\pm\theta/0]_s$		$[0/\pm\theta]_s$		$[+\theta/0/-\theta]_s$	
θ	Calculated E_L, GPa	E_L, GPa	CV	E_L, GPa	CV	E_L, GPa	CV
15°	116	121	3.8%	118	4.7%	114	3.6%
30°	79.7	81.6	6.5%	82.6	4.3%	80.0	5.1%
45°	57.7	58.7	6.0%	59.7	4.4%	57.2	2.3%
60°	51.7	54.0	6.8%	54.2	2.7%	52.1	2.0%
75°	50.7	52.6	4.9%	53.5	2.7%	50.8	4.0%
90°	50.6	50.3	15.6%	52.5	3.3%	51.0	2.6%

Mar and Lin formalized this concept and proposed the following equation for the fracture stress of notched composites

$$\sigma_f = H_c(2r)^{-m} \tag{4}$$

where σ_f is the fracture stress and $2r$ is the notch length. Mar and Lin proposed that m is the value of the stress singularity at the tip of a discontinuity lying at a fiber/matrix interface. The value of such a singularity was determined by Fenner [*26*] and is dependent upon the Poisson's ratios and shear moduli of the matrix and filament. The value of m, taken from the results generated by Fenner, for AS1/3501-6 graphite/epoxy is 0.28.

The other parameter in the equation is H_c, which Mar and Lin refer to as the "composite fracture toughness." However, this is merely a parameter used to fit the experimental data and is referred to as the "composite fracture parameter" herein.

The Mar-Lin correlation of Eq 4 is applied to the data by setting m to 0.28 and determining the value of H_c for each hole size from the experimental failure stress. An average value of this composite fracture parameter can be determined for the specific laminate.

The second correlation used is the point stress criterion proposed by Whitney and Nuismer [*7,8*]. They proposed that failure would occur when the stress at some characteristic distance, d_o, ahead of a notch reached the unnotched failure stress, σ_o. Using the approximation of Eq 1 for the stress ahead of a hole, they derived the following equation for the notched fracture stress, σ_f

$$\frac{\sigma_f}{\sigma_o} = 2/\{2 + \xi_1^2 + 3\xi_1^4 - (K_T^\infty - 3)(5\xi_1^6 - 7\xi_1^8)\} \tag{5}$$

where

$$\xi_1 = r/(r + d_o) \tag{6}$$

Their work suggested that the characteristic length is constant for a specific material. They found d_o to be 1.02 mm for typical graphite/epoxy systems.

The form of this equation suggests that the notched fracture of various stacking sequences, with the same value of K_T^∞, should be ranked according to their unnotched failure stress. The experimental data shows that this is not so, as out-of-plane effects are neglected. To generalize the Whitney-Nuismer correlation, rather than using the unnotched fracture stress, a "notched fracture stress parameter," σ_p, is proposed. In much the same way that H_c was determined for the Mar-Lin relationship, σ_p can be determined using a d_o of 1.02 mm and the experimental fracture stress. An average value for this notched fracture stress parameter can then be found for each laminate.

The average values for H_c and σ_p and the respective coefficients of variation are reported in Tables 3, 4, and 5 for the $[\pm\theta/0]_s$, $[0/\pm\theta]_s$, and $[+\theta/0/-\theta]_s$

TABLE 3—*Experimental results for* $[\pm\theta/0]_s$ *laminates: unnotched fracture stress and notched fracture parameters.*

θ	σ_o, MPa	CV	H_c, MPa(mm)$^{0.28}$	CV	σ_p, MPa	CV
15°	998	7.3%	1004	7.0%	1007	9.1%
30°	855	5.1%	785	9.9%	815	5.2%
45°	732	7.0%	619	4.5%	655	7.2%
60°	698	3.9%	775	4.9%	819	7.0%
75°	672	4.1%	793	5.9%	829	8.4%
90°	679	8.9%	888	8.5%	912	10.1%

TABLE 4—*Experimental results for* $[0/\pm\theta]_s$ *laminates: unnotched fracture stress and notched fracture parameters.*

θ	σ_o, MPa	CV	H_c, MPa(mm)$^{0.28}$	CV	σ_p, MPa	CV
15°	1038	6.1%	978	8.9%	981	10.0%
30°	945	6.2%	711	2.5%	739	2.6%
45°	787	3.2%	606	1.1%	640	2.6%
60°	814	4.3%	716	4.5%	756	2.9%
75°	733	4.8%	799	3.5%	835	7.6%
90°	732	5.5%	934	10.5%	964	16.3%

TABLE 5—*Experimental results for* $[+\theta/0/-\theta]_s$ *laminates: unnotched fracture stress and notched fracture parameters.*

θ	σ_o, MPa	CV	H_c, MPa(mm)$^{0.28}$	CV	σ_p, MPa	CV
15°	999	8.1%	1257	1.9%	1263	8.4%
30°	918	2.6%	912	4.8%	948	2.1%
45°	730	13.6%	603	1.6%	637	3.6%
60°	587	16.2%	589	3.0%	622	4.1%
75°	549	7.0%	538	2.3%	563	2.0%
90°	561	5.2%	492	1.2%	505	5.4%

laminates, respectively. The average unnotched fracture stresses are also included for comparison.

These average fracture parameters are used in the respective equations to generate predicted notched fracture curves. These are presented in Figs. 2 through 7 along with the experimental data. The horizontal lines represent the fact that the notched strength cannot exceed the unnotched (experimental) strength.

Discussion

Comparison of Correlations

A quick examination of the results presented in Tables 3, 4, and 5 as well as Figs. 2 through 7 suggest that the two correlations provide relatively similar results. This is especially true for lamination angles of 45° and greater where the stress concentration factor, K_T^∞, is near three. However, even for these values of lamination angle, the point stress criterion is too "steep" at small hole diameters (approximately 2 to 6 mm) in that the fracture stress predictions overestimate the experimental result. In these cases, the Mar-Lin correlation provides a better estimate.

This difference is accentuated as the hole diameter approaches zero. In all cases, the Mar-Lin correlation line intersects the measured unnotched fracture stress at a point corresponding to, approximately, a 1 to 2-mm-diameter hole. This implies that there is a notch size below which the composite is insensitive to the presence of the notch. This is an attractive quality of the correlation and

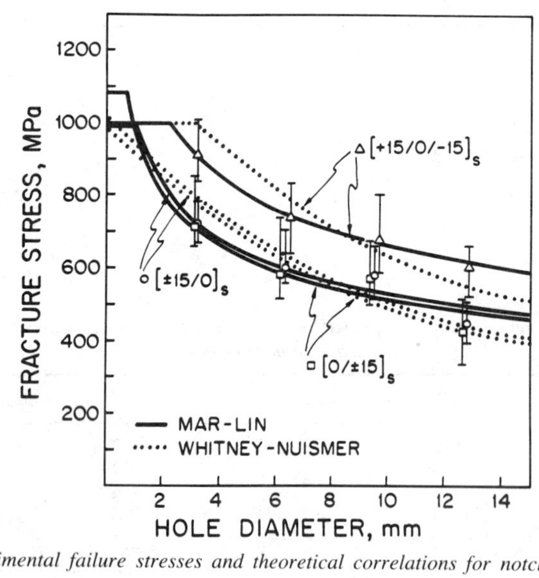

FIG. 2—*Experimental failure stresses and theoretical correlations for notched laminates with* θ = *15°.*

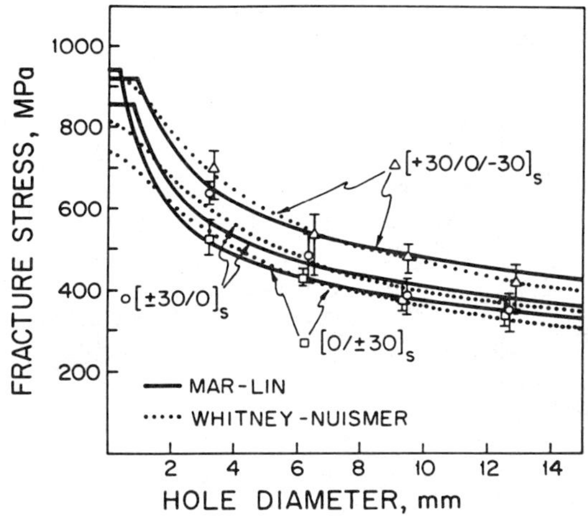

FIG. 3—*Experimental failure stresses and theoretical correlations for notched laminates with* $\theta = 30°$.

can be attributed to the form of Eq 4. The point stress criterion, as applied here, does not even reach the unnotched fracture stress for half of the laminates considered and thus underestimates the fracture stress for holes of diameters less than 1 to 2 mm. This result is due to the fact that a notched fracture stress parameter, σ_p, is used in Eq 5 rather than the unnotched fracture strength as originally proposed by Whitney and Nuismer. However, as can be seen from Tables 3, 4, and 5 by comparing σ_o and σ_p, use of the unnotched fracture stress would generally yield results significantly different from the experimental data.

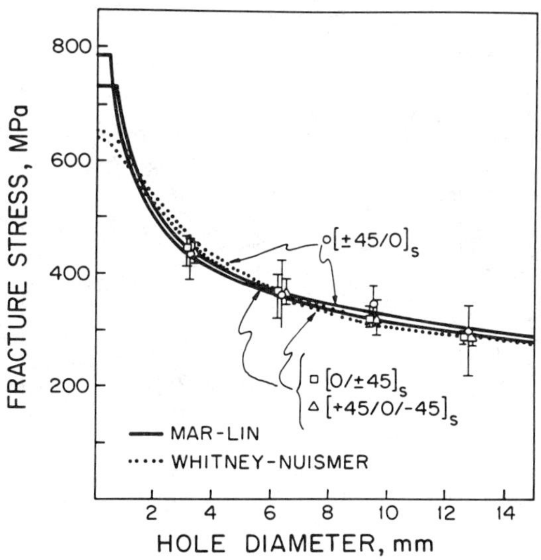

FIG. 4—*Experimental failure stresses and theoretical correlations for notched laminates with* $\theta = 45°$.

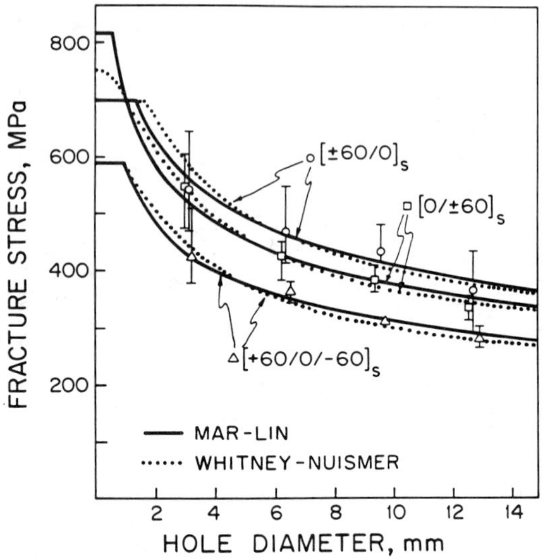

FIG. 5—*Experimental failure stresses and theoretical correlations for notched laminates with* θ = *60°.*

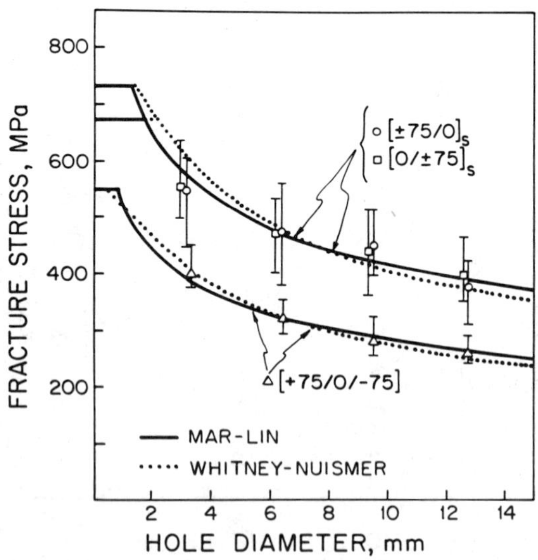

FIG. 6—*Experimental failure stresses and theoretical correlations for notched laminates with* θ = *75°.*

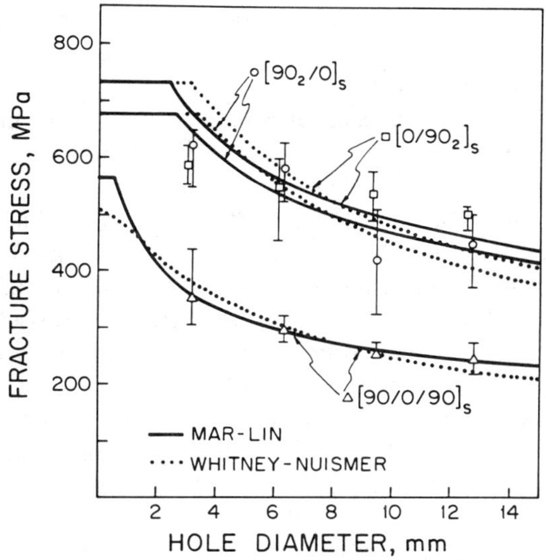

FIG. 7—*Experimental failure stresses and theoretical correlations for notched laminates with* θ = *90°*.

The two cases of θ equal to 15° and 30° show greater difference between the two prediction techniques. For the case of a lamination angle of 15°, the Mar-Lin equation provides a better fit to the data. For θ equal to 30°, the point stress criterion yields a slightly better correlation for two of the three stacking sequences.

Overall, the Mar-Lin equation provides slightly better correlation of the data than the modified point stress criterion of Whitney and Nuismer. The coefficients of variation reported in Tables 3, 4, and 5 are generally lower for the composite fracture parameter, H_c, than for the notched fracture stress parameter, σ_p. These two parameters are relatively of the same value for the same laminate with σ_p being generally larger by about 5%. Neither method is able to properly correlate the $[90_2/0]_s$ and $[0/90_2]_s$ laminate data. Reasons for this are discussed in the following paragraphs.

Stacking Sequence Dependence

The experimental results clearly show that there is an important stacking sequence effect on the strength of notched composites. The dependence is not the same as the dependence of the unnotched tensile strength on the stacking sequence. For the cases of θ equal to 15° and 30°, the $[+\theta/0/-\theta]_s$ laminate has the highest notched fracture stresses (by about 20%) although not possessing the highest unnotched fracture strength. This trend is reversed for lamination angles of 60°, 75°, and 90°. For θ equal to 45°, the three stacking sequences do not have significant differences in notched or unnotched tensile strength. The observed post-mortem fracture modes are also nearly identical.

The $[+\theta/0/-\theta]_s$ laminates are the "best behaved" in that Eq 4 correlates the data well and the composite fracture parameters reported in Table 5 decrease rather smoothly with increasing lamination angle. The observed post-mortem fracture modes for these laminates reveal no or insignificant delaminations in all cases. This is especially illustrated in Fig. 8 where the fracture modes of a $[90_2/0]_s$ and $[90/0/90]_s$ specimen each with a 12.7-mm-diameter hole are pictured. The $[90/0/90]_s$ specimen shows a very clean fracture surface emanating from each side of the hole while the $[90_2/0]_s$ specimen exhibits significant delamination of the outer plies in the vicinity of the hole. This delamination was evident in the post-mortem fracture modes of the $[\pm\theta/0]_s$ and $[0/\pm\theta]_s$ laminates for θ equal to 60°, 75°, and 90°. The notched fracture stresses, and thus the composite fracture parameter, did not change or actually increased with lamination angle for these two laminates. This may be attributable to the occurrence of delamination.

In addition to exhibiting delamination in the vicinity of the hole, the $[\pm\theta/0]_s$ and $[0/\pm\theta]_s$ laminates with θ equal to 60°, 75°, and 90° show a stress-strain behavior indicative of damage long before final fracture. A typical stress-strain plot for a $[90_2/0]_s$ specimen with a 6.35-mm hole is shown in Fig. 9. The plot is initially linear but then the strain becomes discontinuous and nonlinear at a stress less than 50% of the ultimate fracture stress. This stress-strain behavior is typical of all the laminates just mentioned. A similar behavior was observed for these unnotched laminates [*17*], which was indicative of post-first-ply-failure behavior. The present data indicates that similar phenomena occur when the specimens are notched. Conversely, the $[+\theta/0/-\theta]_s$ laminates show only a linear stress-strain curve.

Consideration of the results leads to two possible explanations for the observed dependence of notched fracture stress on stacking sequence, both of which deal with out-of-plane effects. The first is the existence of interlaminar stresses that may cause out-of-plane failure (namely, delamination) to occur in some laminates. The $[\pm\theta/0]_s$ and $[0/\pm\theta]_s$ laminates exhibit virtually the same notched fracture stresses for the cases considered and also show similar fracture modes upon post-mortem examination. This includes delamination for lamination angles of 60°, 75°, and 90°, which does not occur in the $[+\theta/0/-\theta]_s$ laminates. This may account for the increased value of the composite fracture parameter in the former cases.

However, the $[90_2/0]_s$ and $[0/90_2]_s$ laminates did not conform to the correlation equations. The data shows that in the case of the second laminate, the fracture stress does not change significantly for the first three hole diameters: 3.175 mm, 6.35 mm, and 9.525 mm. An examination of the failure modes of these specimens offers a possible explanation for this phenomenon. Photographs of the post-mortem failure modes of two $[0/90_2]_s$ specimens, one with a 3.175-mm-diameter hole and one with a 12.7-mm-diameter hole, are shown in Fig. 10. These photographs are of the edge of the specimen. In the first photograph of the specimen with a 3.175-mm-diameter hole, significant delamination is apparent throughout the laminate. This delamination was also evident at the edge of the hole. In direct

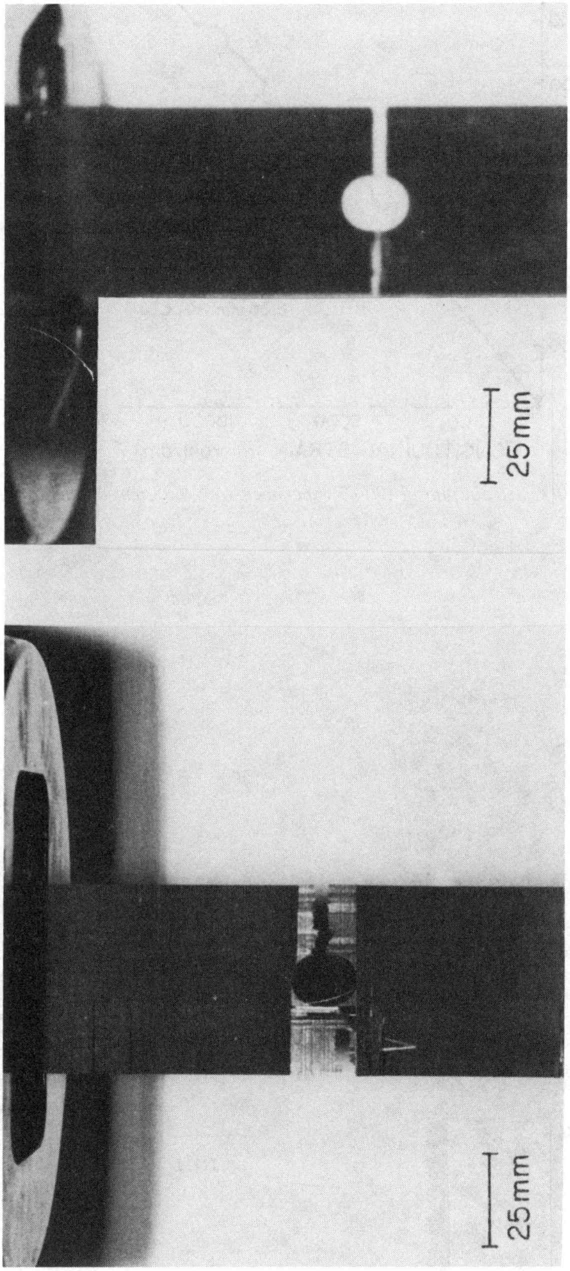

FIG. 8—*Post-mortem fracture mode of specimens with 12.7-mm-diameter hole,* (left) $[90_2/0]_s$ *laminate, and* (right) $[90/0/90]_s$ *laminate.*

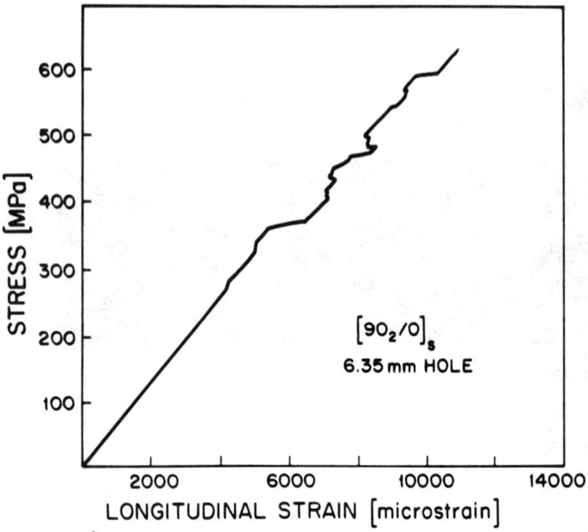

FIG. 9—*Stress-strain plot of* $[90_2/0]_s$ *specimen with 6.35-mm-diameter hole.*

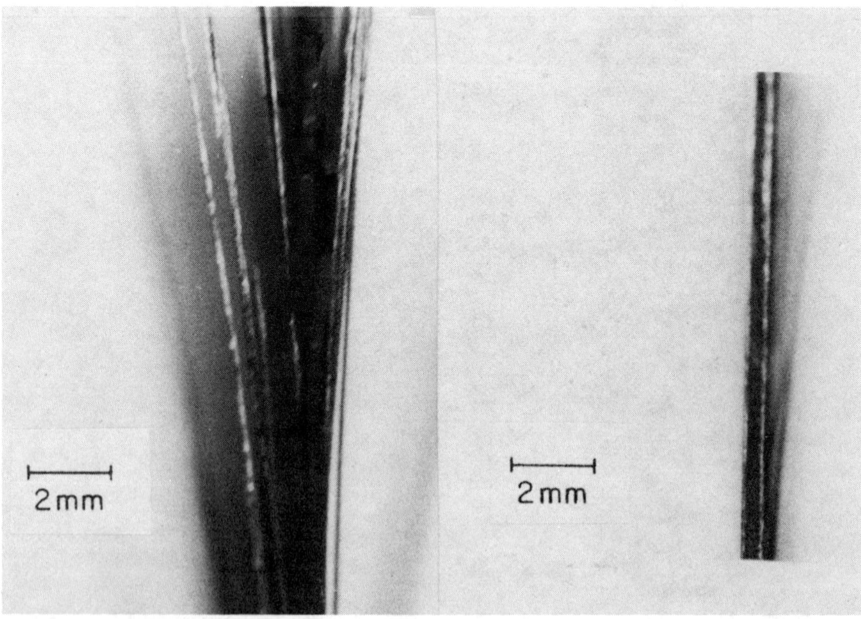

FIG. 10—*Side view of post-mortem fracture mode of* $[0/90_2]_s$ *specimens with* (left) *3.175-mm-diameter hole, and* (right) *12.7-mm-diameter hole.*

contrast with this, the failure of the specimens with a 12.7-mm-diameter hole showed a very clean fracture surface with no evidence of delamination as shown in the second photograph of Fig. 10. It is apparent that the fracture mode changes from out-of-plane to in-plane as the hole diameter increases. The fracture process can be viewed as a competition between out-of-plane and in-plane effects to first cause fracture. The post-mortem fracture observations indicate that the out-of-plane effects are dominant for small holes, while the in-plane notch sensitivity effects are dominant for the larger hole sizes for this particular laminate.

Recent work on the calculation of interlaminar stresses at holes [*16*] shows that the sign of the interlaminar stresses often change around the edge of the hole and the magnitude of these stresses is dependent on the hole radius to laminate thickness ratio, and will thus decrease with increased hole diameter for a given laminate. Caution must thus be exercised in applying in-plane failure analysis methods, such as Eqs 4 and 5, to any laminate data. If failure is controlled by out-of-plane effects, in-plane equations cannot be expected to properly predict the results. More work is needed to better understand the interlaminar stress field around the hole and subsequent failure via delamination.

All the differences in notched fracture stress noted here cannot be attributed to delamination. In the case of the laminates with lamination angles of 15 and 30°, there was still a stacking sequence dependence despite a lack of delamination in the fracture modes. The explanation offered deals with the *in situ* strength of individual plies [*27*]. The strength of any ply changes as the lamination angles of the neighboring plies change. This is due to the differing elastic constraints applied by plies by various angles [*28*]. By changing the stacking sequence, the *in situ* strength of the individual plies may be changed resulting in a change of the overall laminate notched strength. This can interact with delamination to cause the important differences observed in the composite fracture parameters.

Summary

Experimental data has been presented that shows that there is a definite stacking sequence dependence on the notched strength of graphite/epoxy laminates with holes. It is proposed that this dependence can be explained by delamination caused by interlaminar stresses or through-the-thickness constraints of neighboring plies that change the *in situ* in-plane strength of individual plies. In addition, it has been shown that the Mar-Lin equation correlates notched fracture strength data well for laminates that fail in an in-plane fashion and, in fact, generally provides a better correlation than the oft-used point stress criterion.

Acknowledgments

This work was supported by the Air Force Office of Scientific Research under Grant No. AFOSR-82-0071 and Contract No. F49620-83-K-0015. Dr. Anthony K. Amos was the contract monitor.

References

[*1*] Vanderveldt, H. and Liebowitz, H., *Journal of Composite Materials,* Vol. 2, 1968, pp. 254–260.

[2] Wu, E. M. in *Composite Materials Workshop,* Technomic Publications, Lancaster, PA, 1968, pp. 20–43.

[*3*] Lagace, P. A., Mar, J. W., and Dugundji, J., "Fracture, Longevity (Fatigue), Dynamics, and Aeroelasticity of Composite Structures," Air Force Office of Scientific Research Report AFOSR-84-0001TR, June 1983.

[*4*] Greszczuk, L. B. in *Composite Materials: Testing and Design (Second Conference), ASTM STP 497,* American Society for Testing and Materials, Phildelphia, 1971, pp. 363–381.

[*5*] Waddoups, M. E., Eisenmann, J. R., and Kaminski, B. E., *Journal of Composite Materials,* Vol. 5, 1971, pp. 446–454.

[*6*] Cruse, T. A., *Journal of Composite Materials,* Vol. 7, 1973, pp. 218–229.

[*7*] Nuismer, R. J. and Whitney, J. M. in *Fracture Mechanics of Composites, ASTM STP 593,* American Society for Testing and Materials, Philadelphia, 1975, pp. 117–142.

[*8*] Whitney, J. M. and Nuismer, J. M., *Journal of Composite Materials,* Vol. 8, 1974, pp. 253–265.

[*9*] Pipes, R. B., Wetherhold, R. C., and Gillespie, J. W., Jr., *Journal of Composite Materials,* Vol. 13, 1979, pp. 148–160.

[*10*] Kanninen, M. F., Rybicki, E. F., and Griffith, W. I. in *Composite Materials: Testing and Design (Fourth Conference) ASTM STP 617,* American Society for Testing and Materials, Philadelphia, 1976, pp. 52–69.

[*11*] Lo, K. H., Wu, E. M., and Konishi, D. Y., *Journal of Composite Materials,* Vol. 17, 1983, pp. 384–398.

[*12*] Pipes, R. B. and Pagano, N. J., *Journal of Composite Materials,* Vol. 4, 1970, pp. 538–548.

[*13*] Rybicki, E. F. and Hopper, A. T., Air Force Technical Report AFML-TR-73-100, Air Force Materials Laboratory, June, 1973.

[*14*] Tang, S., *AIAA Journal,* Vol. 15, 1977, pp. 1631–1637.

[*15*] Raju, I. S. and Crews, J. H., *Composites Technology Review,* Vol. 4, 1982, pp. 116–124.

[*16*] Carlsson, L., *Journal of Composite Materials,* Vol. 17, 1983, pp. 238–249.

[*17*] Lagace, P. A. in *Proceedings* Sixth Conference on Fibrous Composites in Structural Design, AMMRC MS 83-2, Army Materials and Mechanics Research Center, Nov. 1983.

[*18*] Whitney, J. W. and Kim, R. Y. in *Composite Materials: Testing and Design (Fourth Conference), ASTM STP 617,* American Society for Testing and Materials, Philadelphia, 1977, pp. 229–242.

[*19*] Daniel, I. M., Rowlands, R. E., and Whiteside J. B., *Experimental Mechanics,* Vol. 14, 1974, pp. 1–9.

[*20*] Karlak, R. F. in *Failure Modes in Composites IV,* The Metallurgical Society of the American Institute of Mining, Metallurgical and Petroleum Engineers, 1977, pp. 105–117.

[*21*] Poe, C. C., Jr., *Engineering Fracture Mechanics,* Vol. 17, 1983, pp. 153–171.

[*22*] Mar, J. W. and Lin, K. Y., *Journal of Composite Materials,* Vol. 11, 1977, pp. 405–421.

[*23*] Konish, H. J. and Whitney, J. M., *Journal of Composite Materials,* Vol. 9, 1975, pp. 157–166.

[*24*] Mar, J. W. and Lin, K. Y., *Journal of Aircraft,* Vol. 14, 1977, pp. 703–704.

[*25*] Corten, H. T. in *Fracture, Vol. VII,* H. Liebowitz, Ed., 1972, pp. 676–767.

[*26*] Fenner, D. N., *International Journal of Fracture,* Vol. 12, 1975, pp. 705–721.

[*27*] Flaggs, D. L. and Kural, M. H., *Journal of Composite Materials,* Vol. 16, 1982, pp. 103–115.

[*28*] Stinchcomb, W. W., Reifsnider, K. L., Yeung, P., and Masters, J., *Fatigue of Fibrous Composite Materials, ASTM STP 723,* American Society for Testing and Materials, Philadelphia, 1981, pp. 64–84.

Charles E. Harris[1] and Don H. Morris[2]

Effect of Laminate Thickness and Specimen Configuration on the Fracture of Laminated Composites

---

REFERENCE: Harris, C. E. and Morris, D. H., "**Effect of Laminate Thickness and Specimen Configuration on the Fracture of Laminated Composites,**" *Composite Materials: Testing and Design (Seventh Conference), ASTM STP 893*, J. M. Whitney, Ed., American Society for Testing and Materials, Philadelphia, 1986, pp. 177–195.

ABSTRACT: The effect of laminate thickness on the fracture behavior of laminated graphite/epoxy (T300/5208) composites has been studied. The predominantly experimental research program included the study of the $[0/\pm45/90]_{ns}$ and $[0/90]_{ns}$ laminates with thicknesses of 8, 32, 64, 96, and 120 plies and the $[0/\pm45]_{ns}$ laminate with thickness of 6, 30, 60, 90, and 120 plies. The research concentrated on the measurement of fracture toughness utilizing the center-cracked tension, compact tension, and three-point bend specimens. The development of crack-tip damage prior to fracture was also studied.

Test results showed fracture toughness to be a function of laminate thickness. The fracture toughness of the $[0/\pm45/90]_{ns}$ and $[0/90]_{ns}$ centercracked laminates decreased with increasing thickness and asymptotically approached lower bound values of 1043 MPa $\sqrt{\text{mm}}$ (30 ksi$\sqrt{\text{in.}}$) and 869 MPa$\sqrt{\text{mm}}$ (25 ksi$\sqrt{\text{in.}}$), respectively. The fracture toughness of the $[0/\pm45]_{ns}$ center-cracked laminate increased with increasing thickness but reached an upper plateau value of 1390 MPa$\sqrt{\text{mm}}$ (40 ksi$\sqrt{\text{in.}}$). The fracture toughness of all laminates was independent of crack size except the $[0/90]_{2s}$ laminate that split extensively. The fracture surface of all thick laminates was uniform in the interior and self-similar with the starter notch. With the exception of the $[0/\pm45]_{ns}$ laminate, the fracture toughness of the thicker laminates was relatively independent of specimen configuration.

KEY WORDS: composite materials, graphite/epoxy, fracture toughness, thick laminate, center-cracked tension, compact tension, three-point bend

Fiber-reinforced composite materials have a superior strength-to-weight ratio as well as other mechanical and thermal property advantages over isotropic metals such as steel or aluminum. This has led to increased utilization of these materials as major structural components in many practical applications. Examples can be found in the automotive and aerospace industries, such as the payload bay doors of the space shuttle.

[1] Assistant professor, Aerospace Engineering Department, Texas A&M University, College Station, TX 77843; formerly at Virginia Polytechnic Institute and State University, Blacksburg, VA 24061.

[2] Professor, Department of Engineering Science and Mechanics, Virginia Polytechnic Institute and State University, Blacksburg, VA 24061.

The mechanics of laminated composite structures is theoretically highly developed. However, experimental studies confirming these theories have been conducted almost exclusively on thin laminates generally less than 2.54 mm (0.1 in.) in thickness. In contrast, many structural components are substantially thicker; often five or ten times thicker. As an example, the horizontal stabilizers on the U. S. Air Force F-16's are approximately 160 plies thick.

In general, geometry effects such as thickness variations do not affect the determination of material properties and are easily addressed by applied mechanics theories. This is not necessarily the case in the area of fracture mechanics. For isotropic metals, the state of the art of fracture mechanics is highly developed and the plane-strain fracture toughness is treated as a material property. However, no such fracture toughness material property has been identified for laminated composite materials. Also the measured fracture toughness of composites has shown variations with specimen configuration and specimen geometry such as width and thickness. Furthermore, the correlation between thin laminate fracture characteristics and the fracture behavior of thick laminates has not been established.

Most composite laminates are notch sensitive. Therefore, when a structural component is damaged such as by a dropped tool or has a designed cutout such as a bolt hole, the composite laminate loses much of its original strength. The designer must be cognizant of the fracture mechanics considerations and must know the fracture toughness of the material to properly design a "damage tolerant structure."

The relationship between fracture toughness and specimen thickness has been established for isotropic materials. Fracture toughness is a decreasing function of increasing specimen thickness. A lower bound toughness value corresponding to a condition of plane strain at the crack-tip is asymptotically approached. Discussions of this relationship and the implications thereof can be found in references such as *ASTM STP 410*[3] and *ASTM STP 463*[4]. The crack-tip state of stress is directly related to the relationship between the size of the plastic zone at the crack-tip and the specimen thickness. If the specimen thickness is large relative to the size of the crack-tip plastic zone, sufficient constraint in the interior will be present to produce plane-strain conditions. Plane-stress conditions prevail when the specimen is too thin to provide constraint in the interior. However, no such relationship between fracture toughness and laminate thickness has been established for fiber-reinforced laminated composites.

The first specimen configurations utilized for fracture toughness testing of metals and composites were the center-cracked and double-edge notched tension specimens. Problems associated with testing these configurations began to surface when test results identified the dependence of fracture toughness on specimen

[3] Brown, W. F., Jr., and Srawley, J. E., *Plane Strain Crack Toughness Testing of High Strength Metallic Materials, ASTM STP 410*, American Society for Testing and Materials, Philadelphia, 1966.

[4] Brown, W. F., Jr., *Review of Developments in Plane Strain Fracture Toughness Testing, ASTM STP 463*, American Society for Testing and Materials, Philadelphia, 1970.

thickness. The thick specimens required higher loads that exceeded the load capacity of many testing machines. Also, problems developed with regard to gripping the specimens. (The gripping problem is especially acute for laminated composites.) The three-point bend and compact tension specimens were proposed as alternative configurations and were verified as being acceptable for plane-strain fracture toughness testing by the American Society for Testing and Materials (ASTM) (see Footnotes 3 and 4). Primarily because most laminated composites testing has been confined to thin sheets, no such verification of the similarity of test results from the three specimen configurations has occurred for composite materials.

This paper presents the results of a predominantly experimental investigation of the effects of laminate thickness on the fracture behavior of notched, laminated graphite/epoxy composites. Utilizing center-cracked tension, compact tension, and three-point bend specimens, fracture toughness was determined as a function of crack size and laminate thickness. The role of crack-tip related damage that develops prior to final fracture was also investigated using nondestructive enhanced X-ray radiography and the laminate deply destructive examination techniques.

Experimental Program

The laminate panels were prepared from graphite/epoxy, T300/5208, using a standard tape layup and autoclave curing process. The basic lamina tensile data was determined experimentally from $[0]_8$, $[90]_8$, and $[\pm45]_{2s}$ stiffness coupons. Using the average of four or five replicate tests, the lamina stiffness properties are: E_{11} = 138.7 GPa (20.11 × 10^6 psi), E_{22} = 10.77 GPa (1.56 × 10^6 psi), G_{12} = 5.98 GPa (0.867 × 10^6 psi), and ν_{12} = 0.318, where Subscript 1 refers to the direction parallel to the fibers and Subscript 2 refers to the transverse direction. These property values are typical of those reported in the literature for T300/5208.

The laminate stacking sequences selected for study were $[0/\pm45/90]_{ns}$, $[0/90]_{ns}$, and $[0/\pm45]_{ns}$ where the Subscript ns means multiple layers of the repeated stacking sequence symmetric about the laminate midplane. The $[0/\pm45/90]_{ns}$ and $[0/90]_{ns}$ laminates were tested at thicknesses of 8, 32, 64, 96, and 120 plies, where the individual ply thickness was approximately 0.127 mm (0.005 in.). The thicknesses for the $[0/\pm45]_{ns}$ laminate were 6, 30, 60, 90, and 120 plies. In all cases, the 0° fiber orientation was perpendicular to the machined notch.

The effect of laminate thickness on fracture behavior was primarily studied using the center-cracked tension specimen. The specimen coupon was 50.8 mm (2 in.) by 203.2 mm (8 in.) with a center notch cut by an ultrasonic cutting tool. The notch tip radius was 0.203 (0.008 in.); no attempt was made to sharpen the notch tips. For a thin (6 or 8 plies) and thick (≧90 plies) laminate of each stacking sequence, the crack length-to-width values ($2a/W$) were 0.25, 0.375, 0.50, and

0.625. For other thicknesses, $2a/W$ was 0.50. Also, 50.8 mm (2.0 in.) compact tension specimens and 25.4 mm (1.0 in.) by 127.0 mm (5.0 in.) three-point bend specimens with a/W at 0.50 were tested at the three highest laminate thicknesses. In most cases, four replicate tests were conducted.

All fracture tests were conducted at a constant crosshead displacement rate of 0.02 mm/s (0.05 in./min). The center-cracked tension specimens were held in 51-mm (2-in.) wide wedge-action friction grips such that the specimen length between grip ends was approximately 127 mm (5 in.). Specimen end-tabs were not used. The thin center-cracked specimens, 6 or 8 plies, were tested with an antibuckling support to prevent out-of-plane motion. The compact tension specimens were loaded through 9.5-mm (3/8-in.) loading pins. The three-point bend specimens were supported on 25.4-mm (1.0-in.) diameter cylindrical rods positioned 102 mm (4.0 in.) apart and loaded via a 19-mm (3/4-in.) radius ram. The supports were spring loaded to allow proper initial positioning and free movement once the test commenced. X-ray examinations of the loading regions of the compact tension and three-point bend specimens did not reveal any measurable amount of local bearing or cracking damage. Also, because of the thickness of the compact tension and three-point bend specimens ($\geqq$60 plies), no out-of-plane buckling occurred.

Crack-tip damage that formed in the laminates prior to fracture was investigated using X-ray radiography and the laminate deply technique[5]. Zinc iodide-enhanced X-ray examinations were used to establish the load levels where significant damage developed. Also the progression of damage as the load approached the failure load was documented using X-ray radiography. X-ray examinations were taken at load intervals of approximately 10% of the anticipated failure load or after discontinuities in the load–COD record. Using a gold chloride agent to mark delamination regions, the through-the-thickness damage was studied using the destructive laminate deply technique. This technique allowed the study and documentation of the damage in each individual ply. The deply technique was especially useful for studying the details of broken fibers and for establishing the extent of delaminations as well as their exact interface location.

Determination of Fracture Toughness

Values of stress intensity factor were computed from a finite-element analysis for each of the three specimen configurations and for the four $2a/W$ values. The finite-element code[6] computed the Mode I stress intensity factors and strain-energy release rates based on linear elastic fracture mechanics. The laminate material behavior was considered to be homogeneous and orthotropic. The dif-

[5] Freeman, S. M. in *Composite Materials: Testing and Design (Sixth Conference), ASTM STP 787*, I. M. Daniel, Ed., American Society for Testing and Materials, Philadelphia, 1982, pp. 50–62.

[6] Chu, C. S., Anderson, J. M., Batdorf, W. J., and Aberson, J. A., "Finite Element Computer Program to Analyze Cracked Orthotropic Sheets," NASA CR-2698, National Aeronautics and Space Administration, 1976.

ference between the isotropic and orthotropic values of stress intensity factor were 0.1%, 2.2%, and 7.7% for the center-cracked tension specimens of the $[0/\pm45/90]_{ns}$, $[0/\pm45]_{ns}$, and $[0/90]_{ns}$ laminates, respectively.

Fracture toughness was defined as the value of the linear elastic stress intensity factor at the maximum test load. (This corresponded to catastrophic failure of the center-cracked tension specimens.) This definition of fracture toughness may be somewhat questionable. Because of the formation of crack-tip damage, the laminates of this study were not, generally, linear to failure. One method of accounting for the crack-tip damage zone is to make a "plastic-zone" type adjustment to the fracture toughness value as follows

$$K_Q = Y\sigma_f\sqrt{\pi(a + \rho)}$$

where

K_Q = fracture toughness,
Y = finite width correction factor,
σ_f = applied tensile stress at fracture,
a = original machined notch length,
$\rho = \frac{1}{\pi}\left(\frac{K_Q}{\sigma_0}\right)^2$, and
σ_0 = unnotched laminate strength.

This type of adjustment would increase the value of fracture toughness by about 10% in the extreme case for the smallest crack size and would have little influence on the largest crack sizes. Consequently the adjustment would not affect the fracture toughness versus laminate thickness trend curves that were established for a $2a/W$ value of 0.50. Therefore, the values of fracture toughness reported herein are not adjusted for the damage zone.

Results and Discussion

The $[0/\pm45/90]_{ns}$ Laminate

The fracture behavior of the $[0/\pm45/90]_{ns}$ laminate was very similar to that of an isotropic metal. Fracture toughness, referring to Fig. 1, decreased with increasing laminate thickness and asymptotically approached a lower bound value of 1043 MPa$\sqrt{\text{mm}}$ (30 ksi$\sqrt{\text{in.}}$). There was little difference (less than 10%) in the fracture toughness obtained from the three specimen configurations at 64, 96, and 120 plies. All plotted data in Fig. 1 are the average of four replicate tests (data scatter was minimal and decreased with increasing laminate thickness) and are for a crack length-to-width ratio of 0.50. In addition, the influence of crack length on the fracture toughness of the $[0/\pm45/90]_s$ and $[0/\pm45/90]_{15s}$ center-cracked laminates is shown in Fig. 2. Fracture toughness is practically independent of crack length.

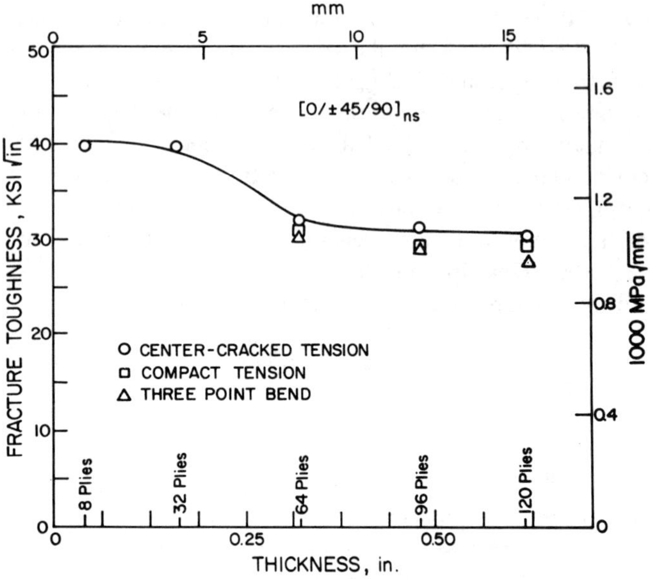

FIG. 1—*Fracture toughness versus thickness for the* $[0/\pm45/90]_{ns}$ *laminate.*

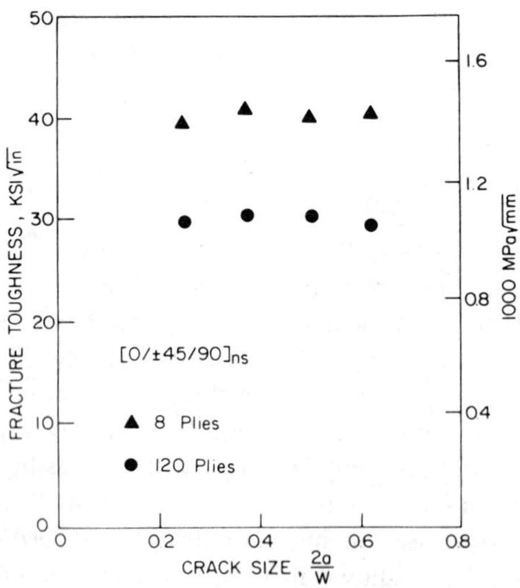

FIG. 2—*Fracture toughness versus crack size of a* $[0/\pm45/90]_{ns}$ *center-cracked laminate.*

A primary characteristic of the fracture behavior of the $[0/\pm45/90]_{ns}$ laminates was the formation of damage at the crack tip prior to fracture. The damage consisted of matrix cracks, broken fibers and delaminations. A typical damage zone in a $[0/\pm45/90]_s$ center-cracked laminate is shown in Fig. 3. This type of damage provides stress relief in the crack-tip region and results in elevated crack growth resistance. The photograph of the fracture surface of a typical $[0/\pm45/90]_s$ center-cracked specimen shown in Fig. 4 illustrates the influence of the damage zone on the final fracture of the thin laminates. While the fracture might be termed as self-similar, macroscopically, the evidence of delaminations that allow ply uncoupling to occur is visible. The surface 0° plies failed by broken fibers more-or-less colinear with the original starter notch while the adjacent 45° plies exhibited substantial matrix cracking parallel to fibers.

The X-ray and deply examinations of the thick laminates revealed that the size of the damage zone was much smaller than in the thin laminates. Furthermore, delaminations were confined to a boundary layer at the surfaces of the thick specimens. As seen in the photographs of broken $[0/\pm45/90]_{15s}$ specimens shown in Fig. 5, the interior of the thick laminates fractured in a nearly uniform manner that was coplanar with the original machined notch surface.

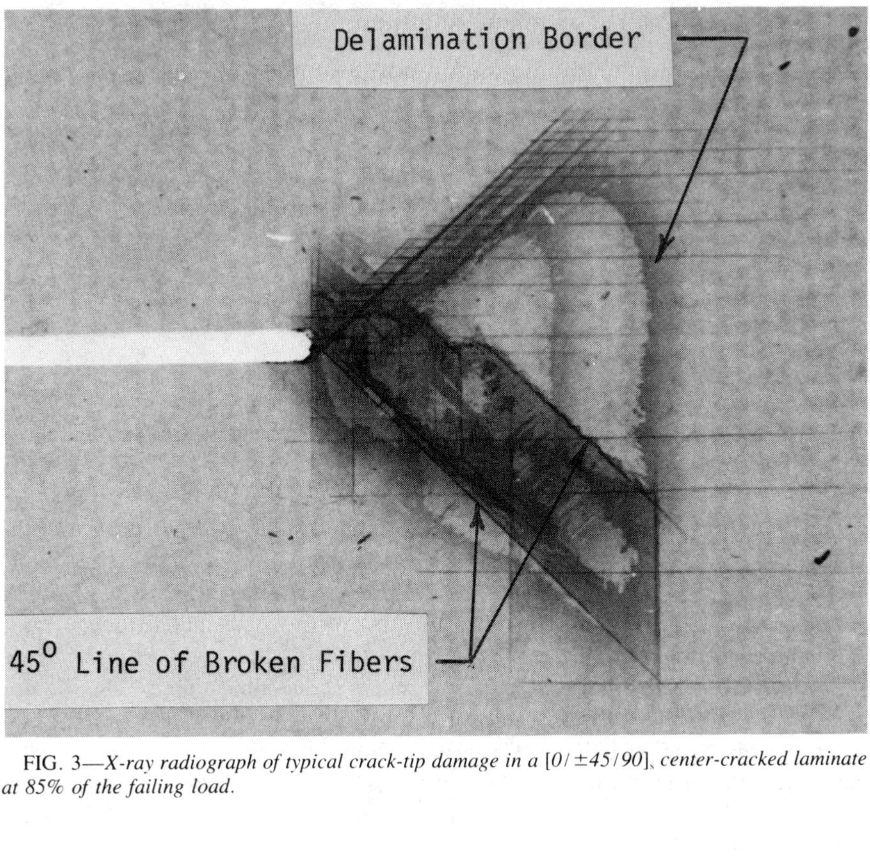

FIG. 3—*X-ray radiograph of typical crack-tip damage in a* $[0/\pm45/90]_s$ *center-cracked laminate at 85% of the failing load.*

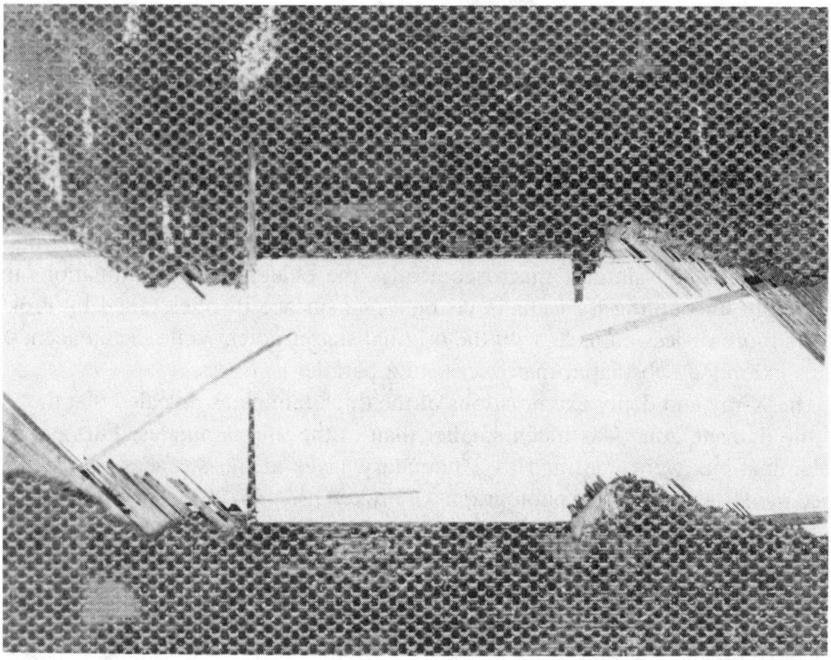

FIG. 4—*Photograph of the fracture surface of a* $[0/\pm45/90]_s$ *center-cracked specimen.*

Therefore, the effect of laminate thickness on the fracture of $[0/\pm45/90]_{ns}$ laminates was to provide constraint in the laminate interior that prevented delaminations from occurring and minimized the size of the crack-tip damage zone. The smaller damage zone offered less crack growth resistance; so the fracture toughness of thick laminates was lower than that of the thin laminates.

The $[0/90]_{ns}$ *Laminate*

As with the quasi-isotropic laminate, the $[0/90]_{ns}$ center-cracked laminate also exhibited fracture toughness values that decreased with increasing laminate thickness. As shown in Fig. 6, the fracture toughness of the $[0/90]_{ns}$ center-cracked laminate asymptotically approaches a lower bound value of 869 MPa$\sqrt{\text{mm}}$ (25 ksi$\sqrt{\text{in.}}$). Also, there was little difference (less than 10%) in the fracture toughness obtained from the three different specimen configurations. The plotted data is the average of four replicate tests. Data scatter was substantial for the 8-ply laminates and minimal for the thick laminates.

This dramatic reduction in fracture toughness is due almost entirely to the role of matrix cracks that form in the 0° plies parallel to the fibers and extend in the direction of the applied load. These axial splits in the $[0/90]_{2s}$ center-cracked laminate are illustrated quite clearly in the radiograph shown in Fig. 7. When

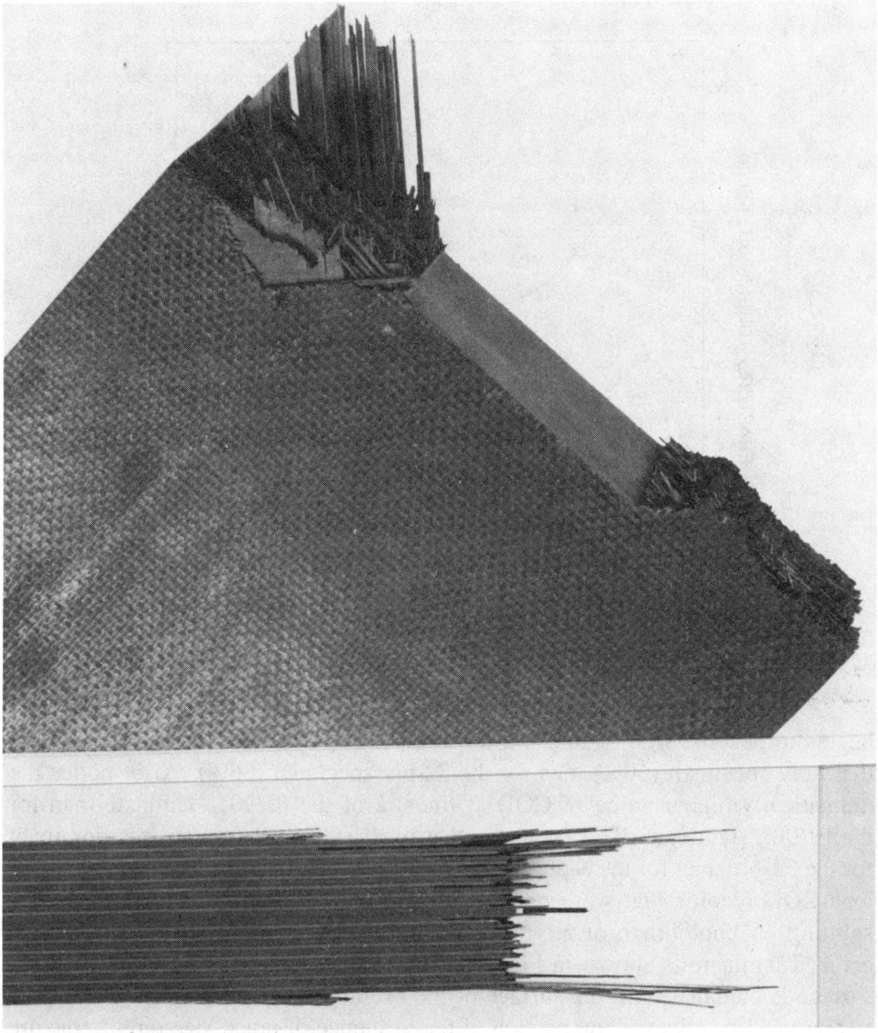

FIG. 5—*Photographs of the fracture surfaces of two* $[0/\pm 45/90]_{15s}$ *center-cracked specimens.*

an axial split forms at the crack-tip, it provides stress relief and load redistribution. The strength of the crack-tip singularity (or stress concentration) is greatly reduced, thereby elevating the notched laminate fracture strength. Oftentimes, these axial splits will extend completely into the grips as can be seen from the photograph of Fig. 8 that shows a typical fractured $[0/90]_{2s}$ center-cracked specimen.

As the laminate thickness was increased, the axial splitting decreased. Figure 9 shows a comparison of the load-COD test records of typical 8-ply and 32-ply center-cracked specimens. The critical load, P_c, where damage development

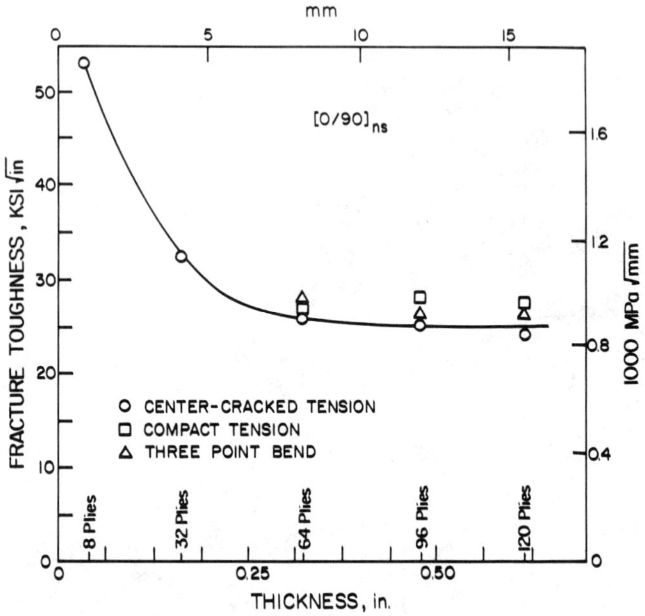

FIG. 6—*Fracture toughness versus thickness for the* $[0/90]_{ns}$ *laminate.*

begins to become significant, is a lower percentage of the failing load, P_m, for the 8-ply specimen (60%) than for the 32-ply specimen (86%). Also, notice the dramatically higher value of COD at fracture of the $[0/90]_{2s}$ laminate than for the $[0/90]_{8s}$ laminate. This indicates that axial splitting is much less dominant for the 32-ply than for the 8-ply laminate. In fact, the $[0/90]_{30s}$ laminate exhibited load-COD records that were linear to failure and showed no evidence of axial splitting. A photograph of a typical fractured specimen of the $[0/90]_{30s}$ center-cracked laminate is shown in Fig. 10. The fracture is uniform and the fracture surface is coplanar with the surface of the original machined notch.

Figure 11 provides a comparison of the influence of crack size on the fracture toughness of the $[0/90]_{ns}$ center-cracked laminate. It was found that the $[0/90]_{2s}$ specimens with larger cracks were much more susceptible to axial splitting than were specimens with the smaller cracks. Therefore, fracture toughness increased with increasing crack length. The fracture toughness of the thick laminates (96 plies) that did not split was independent of crack length.

It is interesting to note that when the fracture toughness of the $[0/90]_{2s}$ laminate was computed at the onset of splitting, K_c, toughness was nearly independent of crack size, as shown in Fig. 11. The value of the critical load taken at the first discontinuity in the load-COD record was used to compute fracture toughness.

In summary, the effect of laminate thickness on the $[0/90]_{ns}$ laminate was to provide constraint against the formation of axial splits in the 0° plies. The axial

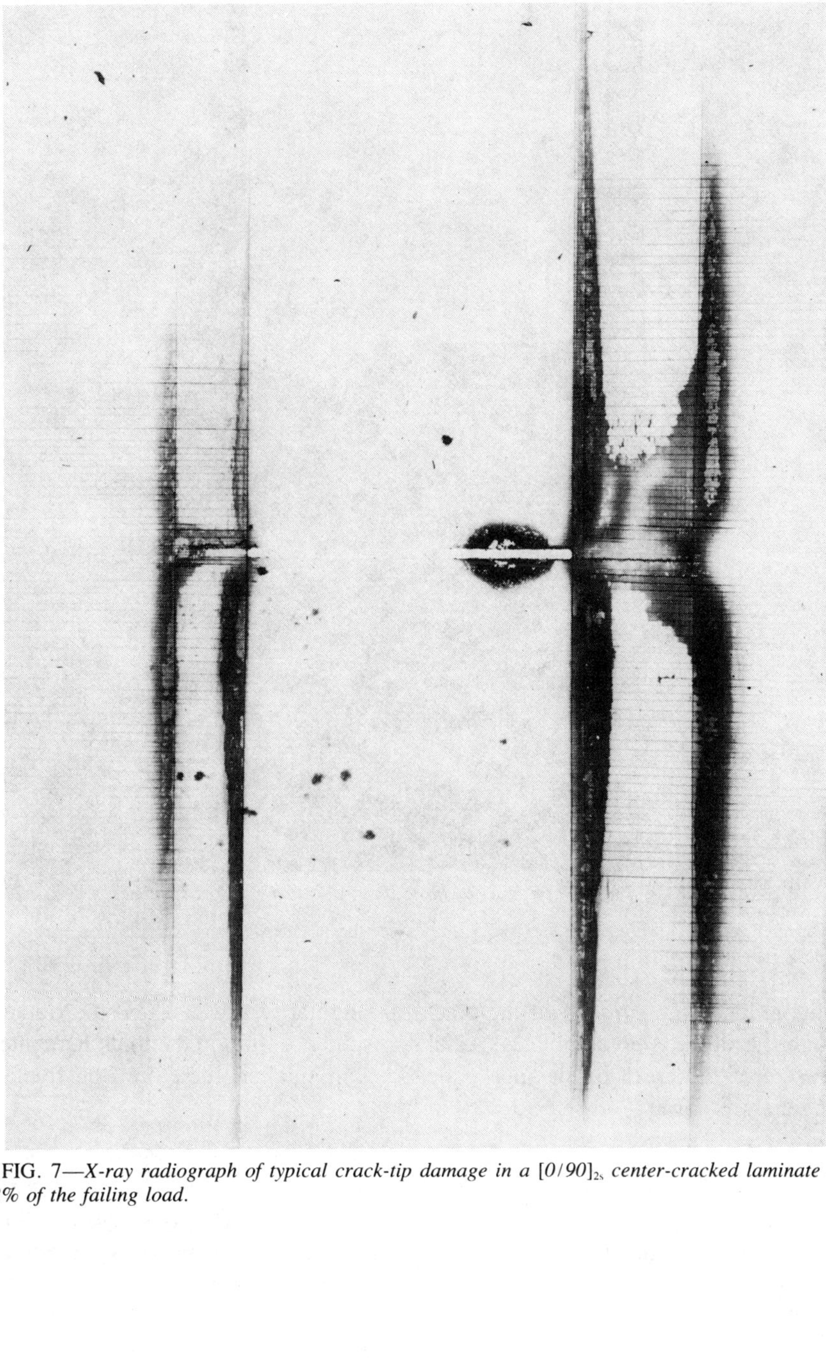

FIG. 7—*X-ray radiograph of typical crack-tip damage in a* $[0/90]_{2s}$ *center-cracked laminate at 92% of the failing load.*

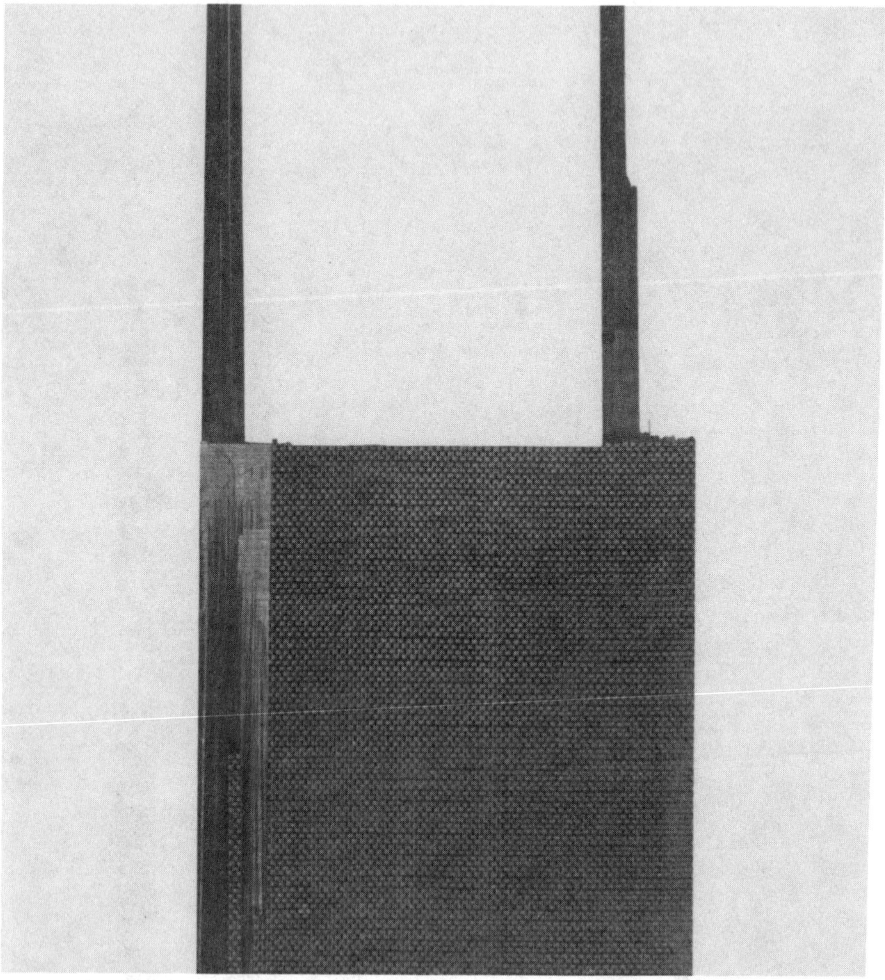

FIG. 8—*Photograph of a fractured* $[0/90]_{2s}$ *center-cracked specimen.*

splits reduced the strength of the crack-tip singularity and elevated the fracture toughness of the 8-ply laminates. Axial splits did not form in the thick laminates; the size of the crack-tip damage zone is much smaller; therefore, the fracture toughness is lower.

The $[0/\pm 45]_{ns}$ *Laminate*

In sharp contrast to the behavior of the first two laminate types, the $[0/\pm 45]_{ns}$ center-cracked laminate exhibited fracture toughness that increased with increasing laminate thickness. Referring to Fig. 12, the fracture toughness of the

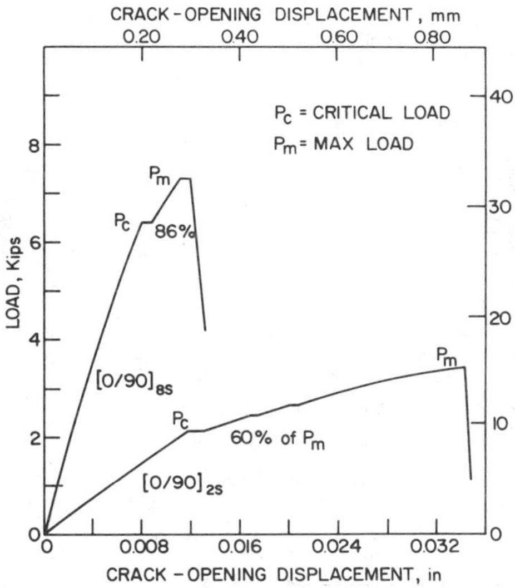

FIG. 9—*Load versus COD for two* $[0/90]_{ns}$ *center-cracked laminates.*

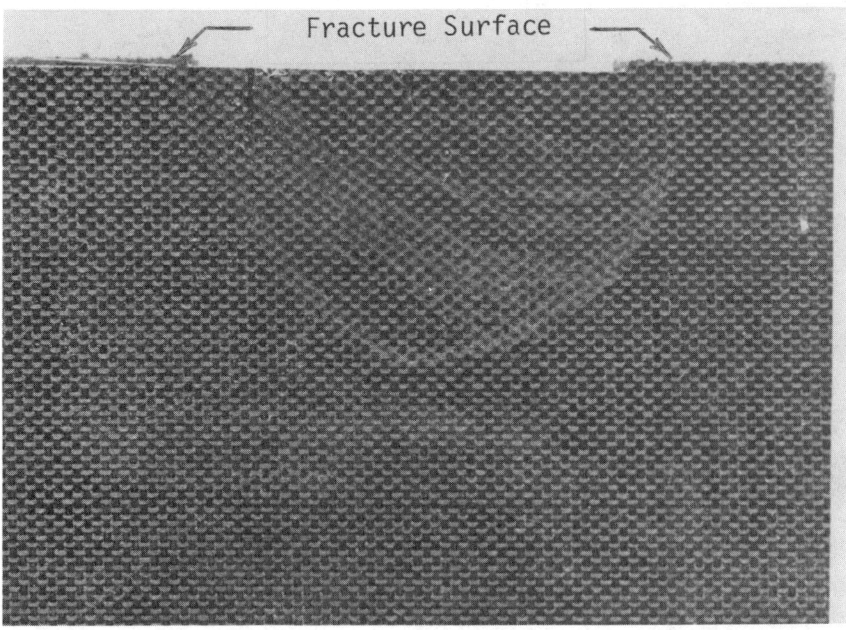

FIG. 10—*Photograph of a fractured* $[0/90]_{30s}$ *center-cracked specimen.*

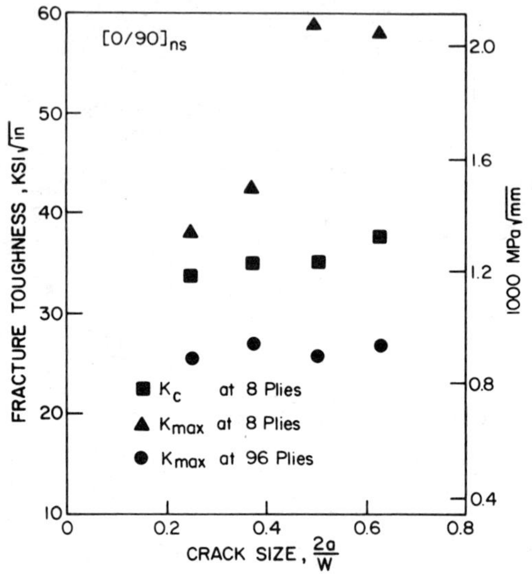

FIG. 11—*Fracture toughness versus crack size of a* $[0/90]_{ns}$ *center-cracked laminate.*

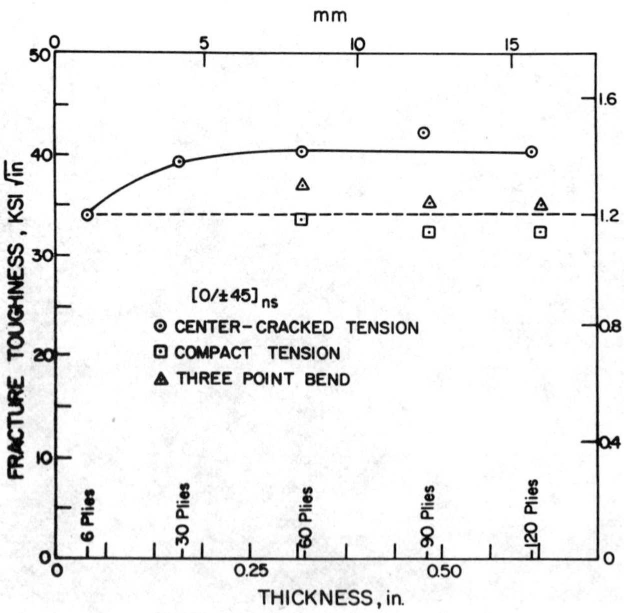

FIG. 12—*Fracture toughness versus thickness for the* $[0/\pm45]_{ns}$ *laminate.*

$[0/\pm45]_{ns}$ center-cracked laminate increases sharply and reaches an upper plateau value of 1390 MPa$\sqrt{\text{mm}}$ (40 ksi$\sqrt{\text{in.}}$) at 30 plies. Beyond 30 plies, fracture toughness is more-or-less independent of laminate thickness. The differences in the fracture toughness (about 15%) determined from the three specimen configurations is greater than was the case with the $[0/\pm45/90]_{ns}$ and $[0/90]_{ns}$ laminates. The data shown in Fig. 12 is the average of four replicate tests and data scatter was quite minimal for all laminate thicknesses. Finally, as can be seen in Fig. 13, fracture toughness is practically independent of crack-length for both the $[0/\pm45]_s$ and $[0/\pm45]_{15s}$ center-cracked laminates.

An examination of the fracture modes of the thin and thick $[0/\pm45]_{ns}$ specimens indicates why the fracture toughness was higher for the thick center-cracked laminates. The X-ray radiograph in Fig. 14 shows the type of damage that formed at the crack tip of the $[0/\pm45]_s$ laminate prior to fracture. Broad regions of delamination are bonded by long matrix cracks that extend in +45° and −45° directions from the notch tip. Dark jagged lines extending at +45° from the notch tip represent broken fibers in the 0° plies. (This was confirmed by deply examinations.) At fracture, this damage state in the 6-ply specimens extended completely to the specimen edge. The photograph of a typical fractured $[0/\pm45]_s$ specimen is shown in Fig. 15. A delamination forming between the +45° and −45° plies allowed complete ply uncoupling to occur. As seen in Fig. 15, the interior −45° plies failed by a matrix crack extending from the notch tip to the specimen edge at a −45° direction parallel to the fibers. The outer +45° ply and 0° surface ply failed together. The +45° ply failed by a matrix crack that extended from the notch tip to the specimen edge in a +45° direction. The 0°

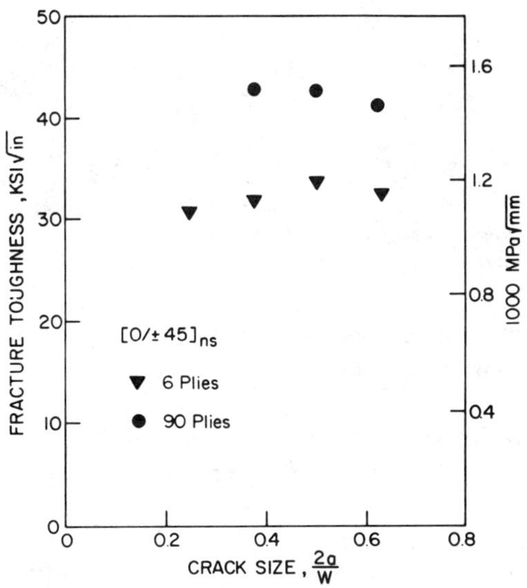

FIG. 13—*Fracture toughness versus crack size of a* $[0/\pm45]_{ns}$ *center-cracked laminate.*

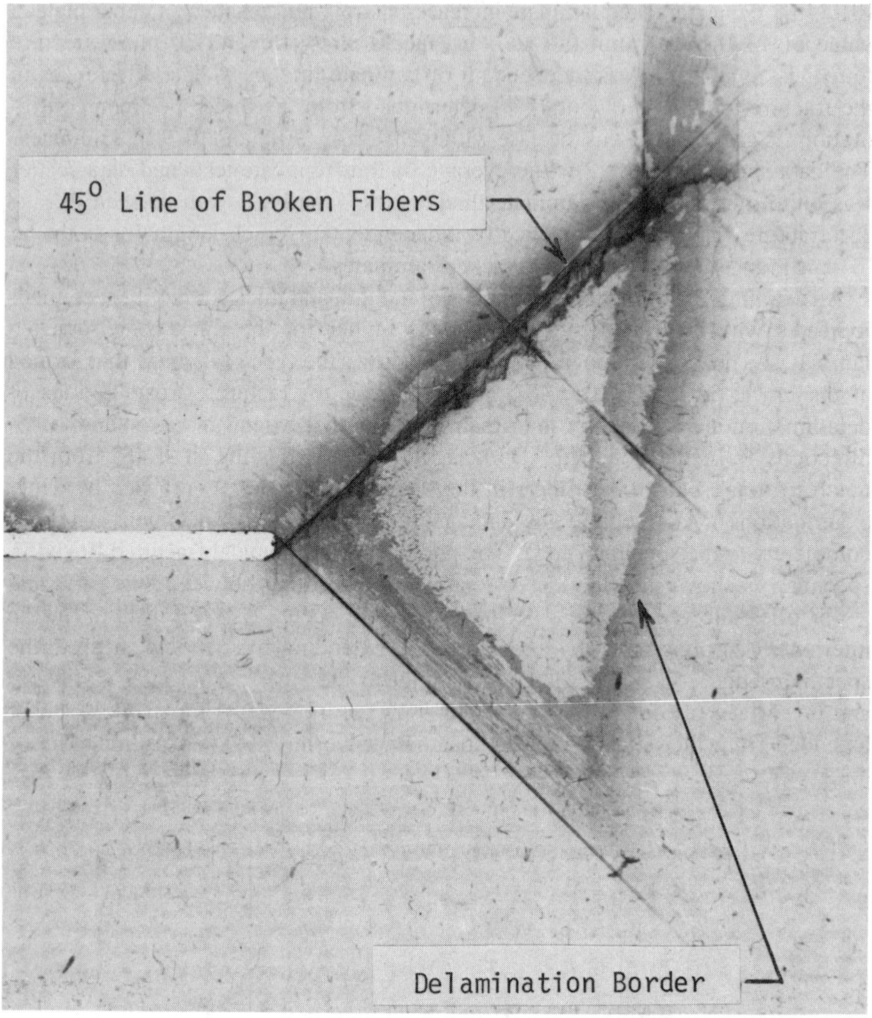

FIG. 14—*X-ray radiograph of typical crack-tip damage in a* $[0/\pm45]_s$ *center-cracked laminate at 99% of the failing load.*

surface ply failed by broken fibers that extended in a +45° line from the notch tip. Therefore, for complete laminate fracture, broken fibers were required in only the two 0° plies.

Because of the thickness constraint, delaminations did not occur in the interior of the thick laminates. This prevented ply uncoupling from taking place. Figure 16 shows the fracture surfaces of several $[0/\pm45]_{20s}$ specimens. The fracture surface in the interior of the laminate is relatively uniform and coplanar with the surface of the machined notch. (Notice the faces of the $[0/\pm45]_{20s}$ specimens exhibit a boundary layer region where delamination and ply uncoupling has

FIG. 15—*Photography of a fractured* $[0/\pm45]_s$ *center-cracked specimen.*

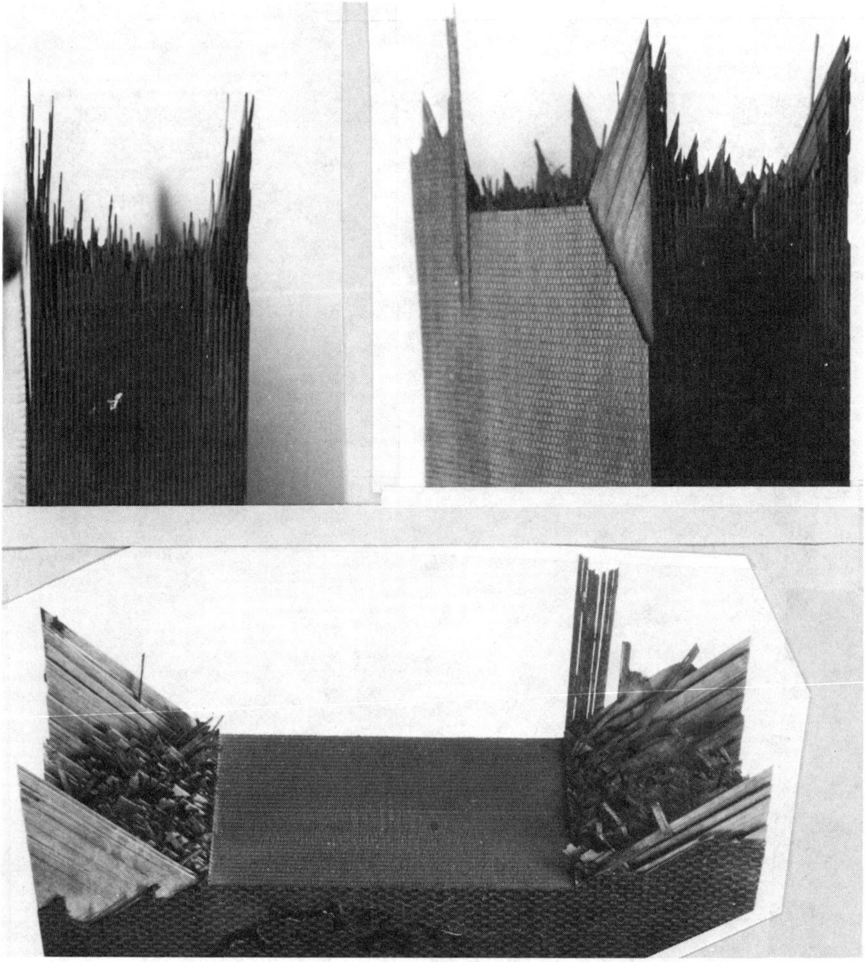

FIG. 16—*Photographs of the fracture surfaces of several* $[0/\pm45]_{20s}$ *center-cracked specimens.*

occurred.) For fracture to occur, fibers must break in all the interior plies. Therefore, a higher fracture energy per unit thickness was required to fracture the thick laminates. This resulted in an increase in fracture toughness with increasing laminate thickness.

Summary and Conclusions

The study of three laminate types of T300/5208 graphite/epoxy has shown the fracture toughness of center-notched laminates to be significantly influenced by laminate thickness. The fracture of the thin $[0/\pm45/90]_s$, $[0/90]_{2s}$, and $[0/\pm45]_s$ laminates was dominated by substantial damage that formed at the crack-tip prior to fracture. The effect of laminate thickness was to provide sufficient constraint in the laminate interior to prevent delaminations from occurring and to minimize the size of the crack-tip damage zone. While the thin $[0/\pm45/90]_s$, $[0/90]_{2s}$, and $[0/\pm45]_s$ laminates exhibited dramatically different types of fracture, their thick laminate counterparts fractured in a similar manner. For each laminate type, the fracture surface of the thick laminates was nearly uniform in the laminate interior and coplanar with the surface of the machined notch. Each laminate type exhibited a surface boundary layer where deviate types of damage occurred that resembled the damage that characterized their thin laminate counterparts.

The fracture toughness of the thin and thick $[0/\pm45/90]_{ns}$ and $[0/\pm45]_{ns}$ center-cracked laminates was independent of crack size. The thin $[0/90]_{2s}$ center-cracked laminate exhibited fracture toughness that increased sharply with increasing crack size. This was because laminates with larger cracks were much more susceptible to axial splitting. The fracture toughness of the $[0/90]_{24s}$ laminate that did not form axial splits was independent of crack size.

Finally, for the $[0/\pm45/90]_{ns}$ and $[0/90]_{ns}$ laminates, the center-cracked tension, compact tension, and three-point bend specimens defined the same lower plateau value of fracture toughness. In the case of the $[0/\pm45]_{ns}$ laminate, the difference between the fracture toughness values of the three specimen configurations was about 15% and the values did not tend to coalesce as with the other two laminate types.

Acknowledgments

The financial support provided for this work by the National Aeronautics and Space Administration (NASA) Grant NAG-1-264 from the Fatigue and Fracture Branch of NASA-Langley Research Center is gratefully acknowledged. Sincere appreciation is extended to the NASA Contract Monitor, C. C. Poe, Jr., for his encouragement and helpful discussion.

V. Sarma Avva,[1] J. Rao Vala,[2] and Muthiah Jeyaseelan[1]

Effect of Impact and Fatigue Loads on the Strength of Graphite/Epoxy Composites

REFERENCE: Avva, V. S., Vala, J. R., and Jeyaseelan, M., "**Effect of Impact and Fatigue Loads on the Strength of Graphite/Epoxy Composites,**" *Composite Materials: Testing and Design (Seventh Conference), ASTM STP 893,* J. M. Whitney, Ed., American Society for Testing and Materials, Philadelphia, 1986, pp. 196–206.

ABSTRACT: An extensive experimental investigation on a graphite/epoxy (T300/934) composite laminate having an orientation and a stacking sequence of $(\pm 45,0,90)_{2s}$ was conducted to study the behavior of the composite material subjected to several combinations of low-velocity projectile impact and tension-tension fatigue loads. The sequence of loading (impact/cycle or cycle/impact) was changed to study its effect on the strength or the life of the composite materials. The experimental data in impact tests were compared with the corresponding results obtained by using a theoretical model.

KEY WORDS: composite materials, impact and fatigue studies, residual strength, sequence of loading, failure threshold, critical velocity

The use of laminated composites as primary structural material has generated the need to characterize the response of these materials under all anticipated loading conditions such as impact loads, cyclical loads, etc. One of the major problems encountered in the structural application of composite materials was their low-impact resistance. The behavior of the laminated composites subjected to fatigue loads is known to be dependent on several factors such as ply-orientation and stacking sequence, fabrication variables, frequency of loading, environmental conditions, etc. [*1*]. In addition, artificially implanted flaws or mechanically damaged laminates would lead to reduced laminate strength [*2*]. The fatigue strength, in general, would vary with the cyclical loading. Further, the fatigue strength of a typical composite laminate with an implanted flaw such as a smooth circular hole through the entire thickness will also degrade as a function of the number of cycles to failure. In general, the fatigue failures would be accompanied

[1]Professor of Mechanical Engineering and graduate research assistant, respectively, North Carolina A & T State University, Greensboro, NC 27411. Mr. Jeyaseelan is presently a graduate student, Department of Mechanical Engineering, University of Maryland, College Park, MD.
[2]Senior engineer, FMC Corporation, San Jose, CA 95052.

by progressive delaminations, fiber splitting, and (in compression mode) fiber buckling [*3*]. Just like the fatigue strength, the residual strength of the laminated composites subjected to low-velocity projectile impact is also one of the important factors to be considered in the design of components with fiber composites. Low-velocity impact damage could take place in composites due to falling objects such as hand tools, runway debris, etc. Such a damage may or may not be visible clearly and, yet, it results in a loss of the laminate strength [*4*]. The impact-damaged laminates under controlled conditions exhibit many characteristics similar to the fatigue-damaged composites. Strength reduction is one of the common features. The damaged laminate under fatigue [*5*] or impact loads [*6*] typically exhibits delaminations, debonding, fiber breakage, interfacial phenomena, etc. There are many experimental techniques presently available to study the characteristics of the damaged laminates. There are also some analytical models to predict the behavior of the laminates subjected to impact [*7–10*] and fatigue loads [*11–14*].

The behavior of the laminated composite materials subjected to a combination of low-velocity projectile impact and axial fatigue loads is to be understood fully. The order in which these two loads (impact and fatigue) are applied to the laminate could have an effect on the residual strength of the composite material. The results reported here are concerned with the study and the analysis of the effect of changing the order of applying the impact and fatigue loads on the load carrying ability of a fiber-reinforced composite material.

Experimental Procedure

The composite specimens were fabricated with Thornel 300/Fiberite 934 material system having an orientation and a stacking sequence of $(\pm 45,0,90)_{2s}$. Each lamina had a nominal cured thickness of 127×10^{-6} m (0.005 in.).[3] The specimens were provided with tapered tabs of glass/epoxy or glass-phenolic with a geometry as shown in Fig. 1. The nominal dimensions of the specimen are also shown in Fig. 1.

All the tests were performed using a closed-loop servohydraulic testing machine with hydraulic grips. The average static strength and strain values of the laminate were determined from tests on eleven specimens. The average static strength value was used as a basis in selecting the other applied stress values in subsequent tests.

The impact tests were performed using an air gun to propel an aluminum sphere 1.27 cm (0.5 in.) in diameter. A detailed description of the operation of the air gun leading to the measurement of impact velocity of the projectile was given in Ref *15*. These tests were conducted to determine the failure threshold level of the composite laminate. The specimens were subjected to different preloads in tension and were then impacted by the projectile at different velocities. Some

[3]Original measurements were in English Units.

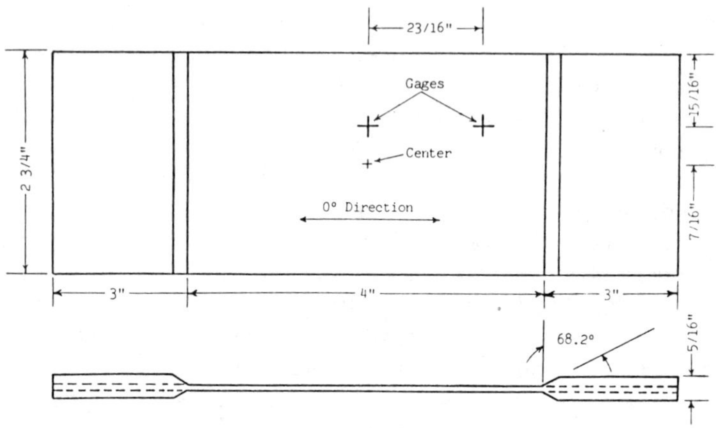

FIG. 1—*Specimen geometry and configuration (1 in. = 2.54 cm).*

of the specimens failed catastrophically upon impact. Those specimens that survived the impact were subjected to continued loading until failure occurred and the residual strength of impact-damaged laminates was thus determined.

The tension-tension fatigue tests were conducted at a frequency of 10 Hz (haversine function) and a stress ratio (ratio of minimum stress to the maximum applied stress on a specimen) of 0.1. The σ-N diagram was generated by using fatigue lives at different stress levels. At each stress level, about five specimens were tested. Those specimens that have completed a million load cycles were considered as runouts and were tested for their residual strength.

The objective of the impact/fatigue tests was to study the behavior of the composite laminate subjected to impact loads followed by cyclic loads and vice-versa. The cyclic frequency of loading and the stress amplitude ratio were the same as before. The specimen was preloaded statically to a (maximum) stress value (with $R = 0.1$) and was then impacted at a predetermined velocity. If it had survived the impact, it was then subjected to cyclic loading until fatigue failure took place. On the other hand, if the specimen failed upon impact, the magnitude of the static tensile stress was reduced and the process was repeated. Sometimes, the velocity of the impacting projectile was reduced while maintaining the cyclic stress amplitude the same as before. In the fatigue/impact tests, the specimen was subjected to a specified number of fatigue cycles and was then impacted to determine the minimum velocity that can cause catastrophic failure (failure on impact) of the specimen. At any one stress amplitude level (with $R = 0.1$) past the predetermined number of fatigue cycles, the velocity of the impacting projectile was either increased or decreased to affect a catastrophic failure. This procedure was repeated at different maximum stress levels ($R = 0.1$) and fatigue lives in establishing the minimum projectile velocity at which the catastrophic failure in the laminate takes place.

Results and Analyses

The average static tensile strength (σ_u) and strain values were found to be 570.63 MPa (82.76 ksi) and 0.011, respectively.

Impact Tests

Several low-velocity projectile impact tests were performed to understand the extent of damage and the resulting behavior of the laminates. The experimental data are plotted as shown in Fig. 2. Typically, the data points shown in Fig. 2 refer to the prestress (the load applied to the specimen prior to impact shown as an open circle), the catastrophic failure stress (the stress at which the specimen failed upon impact—shown as solid circle), and the residual stress (the post-impact strength of the laminate that survived the projectile impact—shown as an open circle with a slash). For the purpose of comparison, the static ultimate strengths of the laminate, with and without hole, are also shown as broken horizontal lines. The solid curve shown is a faired curve drawn through appropriate data points and is designated as a failure threshold curve for the laminate under consideration. In general, the failure threshold curve may be considered to form the boundary of the survival and the failure regions of the laminate subjected to impact. The percentage reduction in the strength of the laminate subjected to the projectile impact is shown in Fig. 2. The data from this graph may be used to assess the residual strength of the laminate at various impact energy levels. The strength of the laminate appears to degrade rapidly up to about

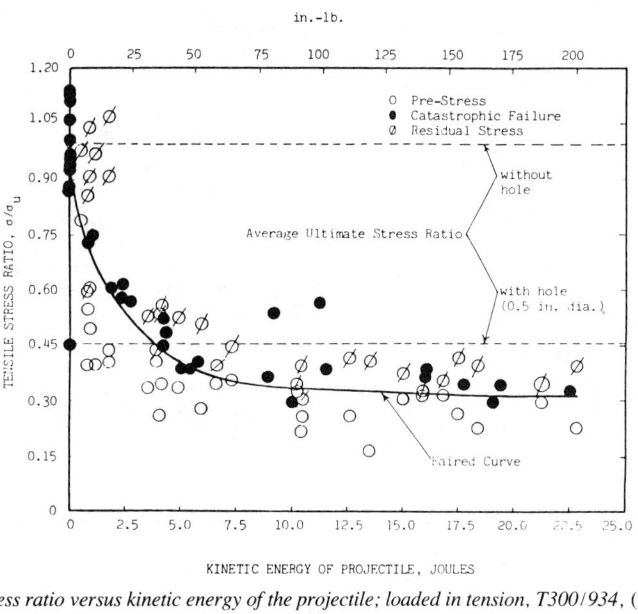

FIG. 2—*Stress ratio versus kinetic energy of the projectile; loaded in tension, T300/934,* ($\pm 45,0,90)_{2s}$.

5 J of impact energy and then tends to remain essentially constant with increasing impact energy of the projectile. In other words, the strength of the laminate degrades rapidly at higher preloads and lower impact energy. As the applied load is decreased while the energy of impact is increased, the failure threshold curve exhibits asymptotic behavior with respect to the energy axis. It may also be noted that the failure threshold curve is below the horizontal line representing the laminate strength with a clean hole. Such an observation should not be surprising since the projectile causes more damage to the laminate during and immediately after the penetration takes place than in a laminate with an implanted clean notch (hole). If the projectile could penetrate the laminate and produce literally a clean hole in it, then the failure threshold curve should coincide with the horizontal line representing the notched strength at higher energy levels.

Recently, an analytical model [*6*] was developed to predict the residual strength of the impact-damaged laminates. The validity of that model was verified experimentally and the comparison was found to be good. In order to apply that model to the present impact data, the rebound velocity was assumed here to be 10% of the forward velocity. For the purpose of calculating the theoretical values using the preceding model, the error introduced through the assumed rebound velocities was considered to be small. With the help of the model, and knowing the ultimate strength of the specimen and the energy imparted per unit thickness of the specimen, one can calculate the residual strength of the impact-damaged specimens. Figure 3 shows the theoretical and the experimental (faired) results plotted with σ/σ_u as a function of impact energy/unit thickness. It may be observed from Fig. 3 that beyond an impact energy of about 15 J/cm (375 in. ·lb/in.), the error involved between the two graphs is less than 10%.

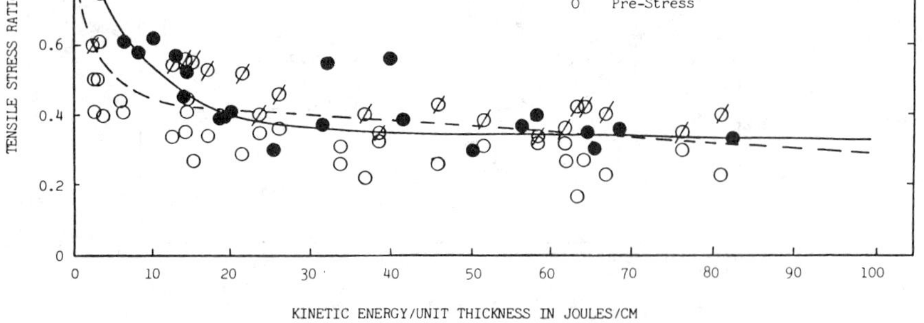

FIG. 3—*Comparison of experimental and theoretical impact data; loaded in tension, T300/934,* $(\pm 45, 0, 90)_{2s}$.

Impact-Fatigue/Fatigue-Impact Tests

The objective of these tests was to assess the effect of impact loads on the fatigue life of the composite material under consideration. Consequently, two σ-N curves were first generated using unnotched (projectile impact velocity $= 0$) and notched (projectile velocity $= \infty$) laminates. The experimental data and additional details pertaining to the two σ-N curves were reported in Ref *15*, and these curves are reproduced here as shown in Fig. 4. The data points for the impact/fatigue-damaged specimens were expected to be bounded by the notched and the unnotched σ-N curves. Hence, the values of the applied stress and the impact velocity were chosen accordingly and the number of cycles the specimen could undergo after the impact was determined. In cases where the specimen survived a million fatigue cycles after impact, the test was stopped and the specimen was tested for its residual strength. The experimental data for the impact/fatigue tests is plotted in Fig. 4. This graph shows that most of the data points from the impact/fatigue tests do indeed lie in the region bounded by the notched and the unnotched σ-N curves. From Fig. 4, it may be observed that when the prestress prior to impact was close to the stress amplitude of the unnotched σ-N curve, the laminate usually fails catastrophically irrespective of the magnitude of the impact velocity. On the other hand, when the pre-impact stress was in the vicinity of the notched-laminate strength, the laminate not only appears to survive a projectile impact velocity of about 30.48 m/s (100 ft/s) but also may be seen to exhibit a residual strength close to the static ultimate strength of the laminate even after seeing a million cycles of fatigue life.

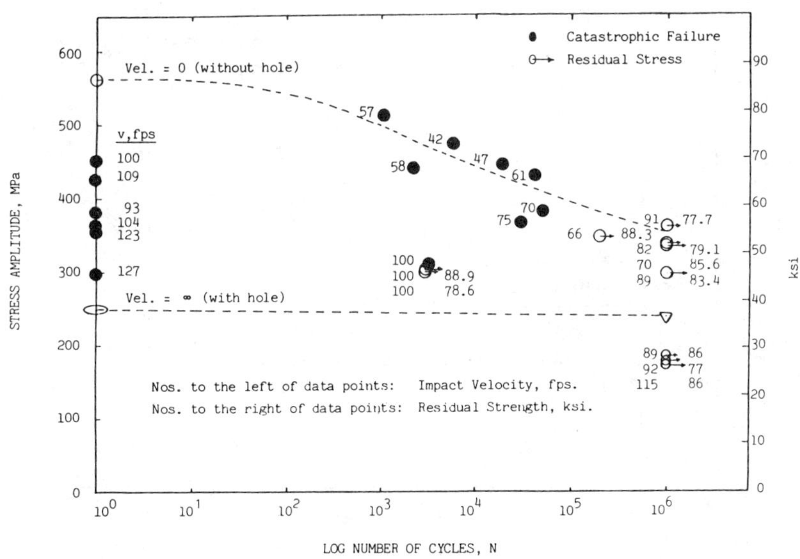

FIG. 4—*Impact/fatigue behavior; tension-tension cyclic loading with stress ratio,* R $= 0.1$ *at 10 Hz; T300/934,* $(\pm 45,0,90)_{2s}$.

As an extension of the previous loading sequence (impact/cycle), it was decided to study the behavior of the laminate with reversed sequence of loading (cycle/impact/and cycle, if necessary). Consequently, three combinations of the applied stress and velocity of impact values were selected. The specimen was first subjected to a predetermined number of fatigue cycles at which another specimen failed in a previous test (see Fig. 4). It was then impacted and the cyclic loading was continued if it did not fail on impact. These results are shown in Fig. 5. Altogether, three tests were conducted. These limited number of results indicate that the fatigue life is longer when the specimen is first subjected to cyclic loading and then impacted as compared with the reversed sequence of loading. This observation may be interpreted to signify that the relative damage in the composite is more severe in impact/cyclic sequence of loading resulting in reduced life of the laminate. Any quantitative assessment of the extent of such relative damage can only be arrived at by testing additional specimens.

Some additional tests were conducted to determine the minimum velocity of impact that can cause catastrophic failure in a specimen at a particular applied stress level and after a particular number of fatigue cycles. Two stress levels, namely, 46 ksi (56% of σ_u) and 61 ksi (74% of σ_u) and three different numbers of cycles, namely, 10^4, 10^5, and 10^6 up to which the specimen will be cyclically loaded were selected. On the average, about seven specimens were tested for each test condition. Higher impact velocity was selected initially and then the velocity was gradually reduced to find the minimum velocity that could cause catastrophic failure of the specimen. The experimental data points are plotted as

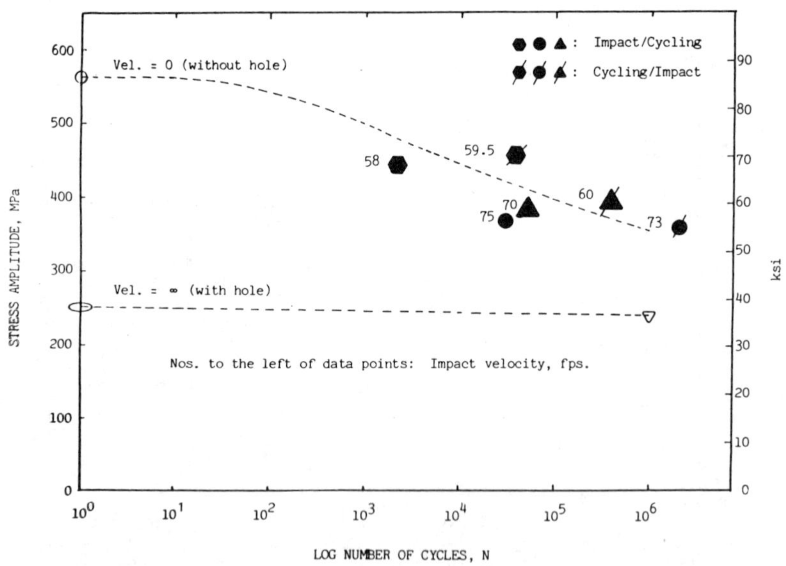

FIG. 5—*Fatigue/impact behavior; tension-tension cyclic loading with stress ratio,* R = *0.1 at 10 Hz; T300/934,* $(\pm 45,0,90)_{2s}$.

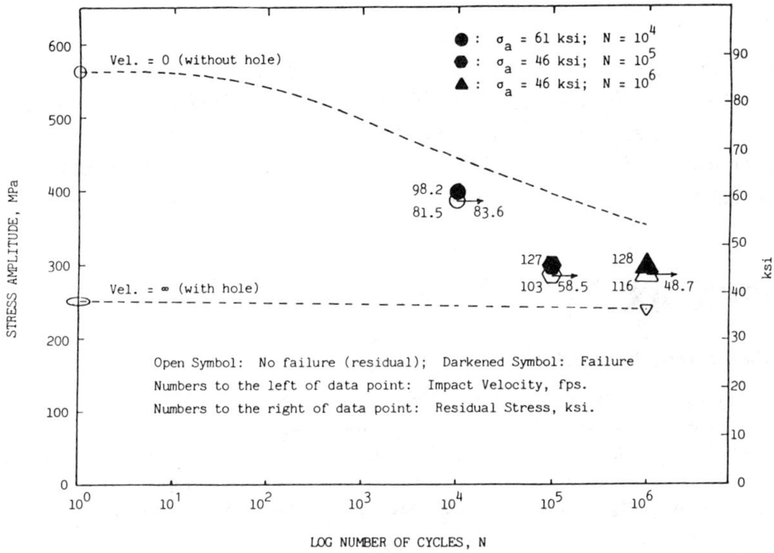

FIG. 6—*Critical impact velocity studies; fatigue/impact sequence of loading with stress ratio, R = 0.1 at 10 Hz; T300/934,* $(\pm 45,0,90)_{2s}$. *Note: even though the points in each of the three pairs of data points overlap, they are shown here slightly apart for clarity.*

shown in Fig. 6. It may be seen from this figure that as the applied stress amplitude is decreased, the impact velocity leading to catastrophic failure increases from 29.93 m/s (98.2 ft/s) to about 38.71 m/s (127 ft/s). At any constant-applied stress amplitude (such as 46 ksi) and impact velocity in the range of 31.39 m/s (103 ft/s) to 35.36 m/s (116 ft/s), the fatigue life of the laminate appears to improve from 10^5 to 10^6 cycles. However, it should be noted that such an improvement appears to take place at the expense of loss of residual strength (58.5 to 48.7 ksi) in the laminate.

Conclusions

Based on the experimental results, the following conclusions may be drawn:

1. The residual strength of the impact-damaged laminates can be predicted using an analytical model. The correlation between the experimental and the analytical results was found to be good. Extensive experimental data is not necessary to use the analytical model.
2. Impact loading followed by cyclic loading was found to be more damaging (in reducing the life of the laminate) than the reversed sequence of loading.
3. The magnitude of the minimum projectile velocity causing catastrophic failure in the laminates tested was found to be a function of the applied stress and the number of fatigue cycles.

Acknowledgments

The financial support provided for this work by the Department of Air Force through Contract No. F33615-80-K-3243 is gratefully acknowledged. The authors express their sincere appreciation to Dr. George P. Sendeckyj, Air Force Program Monitor for the contract, for his assistance and guidance in the course of this investigation.

APPENDIX

Residual Strength Model

At very low-impact energy levels, many laminated composite materials do not show any significant static strength degradation. Since significant damage does not occur at velocities less than the threshold damage velocity, it may be assumed that the damage is proportional to the difference between the kinetic energy of impact and the threshold damage kinetic energy. The preceding statement may be expressed as [*16*].

$$c = k\,(W - W_o) \tag{1}$$

where

c = damage as an effective through-the-thickness crack of length $2c$,

k = constant depending on the material and the laminate,

W = kinetic energy of impact per unit thickness, and

W_o = threshold damage kinetic energy per unit thickness.

The average stress criterion developed by Whitney and Nuismer [*8*] for an isotropic laminate is given by

$$(\sigma_N^\infty/\sigma_o) = [(1 - \zeta_1)/(1 + \zeta_1)]^{1/2} \tag{2}$$

where

σ_N^∞ = failure stress of an infinite plate containing a crack (notch) of length $2c$,

σ_o = failure stress of an unnotched plate,

ζ_1 = $c/(c + a_o)$, and

a_o = characteristic dimension adjacent to discontinuity.

Substituting for ζ_1 in Eq 2

$$(\sigma_N^\infty/\sigma_o) = [(a_o)/(2c + a_o)]^{1/2} \tag{3}$$

Equations 1 and 3 can be combined resulting in the following

$$(\sigma_N^\infty/\sigma_o) = 1/[(2k/a_o)(W - W_o) + 1]^{1/2} \tag{4}$$

Assuming $\sigma_N^\infty = \sigma_r$, residual strength of the specimen and $(k/a_o) = K$, a constant, Eq 4 can be re-written in the following form

$$(\sigma_r/\sigma_o) = [2K(W - W_o) + 1]^{-1/2} \quad (5)$$

Equation 5 can be used to predict the residual strength of a laminate for different values of kinetic energy. In order to determine the values of $2K$ and W_o, Eq 5 is re-written in linear form as

$$y = ax + b$$

where

$$y = (\sigma_o/\sigma_r)^2,$$

$$x = W,$$

$$a = 2K, \text{ and}$$

$$b = (1 - 2KW_o)$$

The ultimate strength, σ_o, and the post-impact residual strength, σ_r, at different velocities are determined by performing tests on a few specimens. Using linear regression analysis, a best fit is obtained for the data and the values for $2K$ and W_o are thus determined. These values are used in Eq 5 to predict the residual strength at any velocity of impact. Additional details on the application of this model with a numerical example can be found in a report by Avva [*15*].

References

[*1*] Reifsnider, K. L., *ASTM Standardization News*, Feb. 1980, p. 15.

[*2*] Ryder, J. T. in *Proceedings*, Fifth Annual Mechanics of Composites Review, AFWAL-TR-80-4020, Jan. 1980, p. 73.

[*3*] Ratwani, M. M., "Influence of Penetrants Used in X-Ray Radiography on Compression Fatigue Life of Graphite/Epoxy Laminates," *Composites Technology Review*, No. 1, Vol. 2, Winter, 1980.

[*4*] Rhodes, M. D., "Impact Fracture of Composite Sandwich Structures," *Proceedings*, ASME/AIAA/SAE, 16th Structural Dynamics and Materials Conference, Denver, CO, American Society of Mechanical Engineers, May 1975.

[*5*] Sendeckyj, G. P. in *Test Methods and Design Allowables for Fibrous Composites, ASTM STP 734*, C. C. Chamis, Ed., American Society for Testing and Materials, Philadelphia, 1981, pp. 245–260.

[*6*] Avva, V. S. and Padmanabha, H. L., "Compressive Residual Strength Prediction in Fiber-Reinforced Laminated Composites Subjected to Impact Loads," *Proceedings*, Sixth International Conference on Fracture, New Delhi, India, 4–10 Dec. 1984. Also shown in Report No. AFOSR-TR-82-1038 (AD-A122166), Air Force Office of Scientific Research, Bolling AFB, DC.

[*7*] Husman, G. E., Whitney, J. M., and Halpin, J. C. in *Foreign Object Impact Damage to Composites, STP 568, ASTM 1974*, American Society for Testing and Materials, Philadelphia, 1975, pp. 92–113.

[*8*] Waddoups, M. E., Eisenmann, J. R., and Kaminski, B. E., *Journal of Composite Materials*, Vol. 5, Oct. 1971, pp. 446–454.

[*9*] Nuismer, R. J. and Whitney, J. M., *Fracture Mechanics of Composites, ASTM STP 593*, American Society for Testing and Materials, Philadelphia, 1976, pp. 117–142.

[10] Yeow, Y. T., Morris, D. H., and Brinson, H. F., *Journal of Testing and Evaluation*, Vol. 7, No. 2, March 1979, pp. 117–125.

[11] Hahn, H. T., "Fatigue Behavior and Life Prediction of Composite Laminates," AFML-TR-78-43, Air Force Materials Laboratory, April 1978.

[12] McLaughlin, P. V., Jr., Kulkarni, S. V., Huang, S. N., and Rosen B. W., "Fatigue of Notched Fiber Composite Laminates," Part I: Analytical Model, NASA CR-132747, National Aeronautics and Space Administration, March 1975.

[13] Ramkumar, R. L., Kulkarni, S. V., and Pipes, R. B., "Evaluation and Expansion of an Analytical Model for Fatigue of Notched Composite Laminates," Materials Sciences Corporation, Blue Bell, PA, March 1978, p. 129.

[14] Kulkarni, S. V., McLaughlin, P. V., Jr., and Pipes, R. B., "Fatigue of Notched Fiber Composite Laminates," Part II: Analytical and Experimental Evaluation, NASA CR-145039, National Aeronautics and Space Administration, April 1976.

[15] Avva, V. S., "Fatigue/Impact Studies in Laminated Composites," Final Report AFWAL-TR-83-3060, Flight Dynamics Laboratory, Air Force Wright Aeronautical Laboratories, Wright-Patterson Air Force Base, Ohio, May 1983.

[16] Whitney, J. M., private communication, Air Force Materials Laboratory, Wright Aeronautical Laboratories, Wright-Patterson Air Force Base, Ohio, 45433.

Delamination

Herzl Chai[1]

Bond Thickness Effect in Adhesive Joints and Its Significance for Mode I Interlaminar Fracture of Composites

REFERENCE: Chai, H., "**Bond Thickness Effect in Adhesive Joints and Its Significance for Mode I Interlaminar Fracture of Composites,**" *Composite Materials: Testing and Design (Seventh Conference), ASTM STP 893*, J. M. Whitney, Ed., American Society for Testing and Materials, Philadelphia, 1986, pp. 209–231.

ABSTRACT: The effect of constraining the crack-tip damage zone on the Mode I fracture of a toughened epoxy resin (BP-907) was studied systematically using the double-cantilever-beam (DCB) adhesive joint specimen and scanning electron microscopy (SEM). While a broad bond thickness range was employed ($5 < t < 875$ μm), particular attention was given to the lower bond spectrum and to the applicability of the results to interlaminar fracture of toughened composites.

Depending on t, interesting fracture characteristics were found including continuous, unstable, and a combination of both types of growth. For $t > 33$ μm the fracture energy exhibited dramatic variations with t that were shown to follow from the variations in size of a certain dimple-like morphological feature. In constrast, the arrest energy was fixed, in consistency with the unchanging morphology. The fracture behavior for $t < 33$ μm was equally interesting, now though because G_{IC} was fixed while $G_{IC}{}^a$ varied with t. These changes in failure conditions were accompanied by morphological changes; as t decreased from 33 μm, the failure locus shifted from the bond center to the metal/matrix interface, and the failure itself occurred by shear yielding.

Similarities were found in the failure conditions of butt joint and DCB adhesive specimens, on the one hand, and in that of the latter and composite interlaminar fracture, on the other hand, which suggest an interrelationship among these three failure phenomena.

KEY WORDS: composite materials, adhesive joints, Mode I, fracture energy, toughened matrix, bond thickness, interlaminar fracture

The poor resistance to delamination growth of conventional fiber-reinforced epoxy laminates is a major obstacle to the widespread use of such materials in advanced aerospace structures. The realization that this type of failure is matrix dominated has led to various interlaminar strengthening concepts including interlayering [*1*], modifications or toughening of existing brittle epoxy systems [*2–5*], and use of thermoplastics [*6*]. There appears to be no simplified testing methodology to assess the improvement to delamination damage tolerance offered

[1]Visiting scientist, Materials Laboratories, AFWAL/MLBM, Wright-Patterson Air Force Base, Dayton, OH 45433.

by such materials, however. In Ref *5*, for instance, a rather comprehensive test program including impact and shear tests were employed to aid in the matrix screening.

Fracture mechanics is a widely accepted mean for material toughness ranking. In recent years, such an approach has become also useful in composite damage tolerance evaluations. Unfortunately, however, in the case of ductile or toughened resins, the significance of test results obtained from the neat resin constituent to the interlaminar fracture behavior of laminated composites is far from understood. In fact, comparative Mode I fracture tests show the bulk resin toughness to largely exceed the composite toughness, sometimes by an order of magnitude [*2–4*]. It has been argued [*2,3*] that this toughness loss is merely due to the constraint on the crack-tip plastic deformation zone imposed by the adjoining composite layers. Although the generality of this argument will be challenged in this work, it does appeal to the common sense. This point was demonstrated by Scott and Phillips [*2*] who evaluated the Mode I fracture energy in bulk, adhesive joint (see Fig. 1), and unidirectional composite forms over a broad range of matrix ductility; the latter was controlled by the weight fraction of the elastomer particles added to the unmodified epoxy. The results, summarized in Table 1, show that for the highly modified resins the adhesive joint specimen has indeed produced much lower fracture energies relative to the bulk, the differences which increased with the rubber content. Nevertheless, these values are still larger than their composite counterparts. In retrospect, this discrepancy may be attributed to the large bond thickness (0.2 mm) used. It was the purpose of this work to study systematically the effect of bond thickness on the fracture behavior in adhesive joints so that the prospect of a matrix-composite toughness correlation can be studied unbiasedly.

The effect of bond thickness on the Mode I fracture work of adhesive joint specimens has been considered by several authors treating brittle [*7*] or rubber-toughened [*8–11*] polymer systems. These works were largely concerned with adhesive bonding applications that call for a relatively large bond thickness. In the present investigation, the emphasis was placed on extremely thin bonds that physically simulate the interlaminar matrix layer of laminated composites, though larger bonds were also respectfully treated.

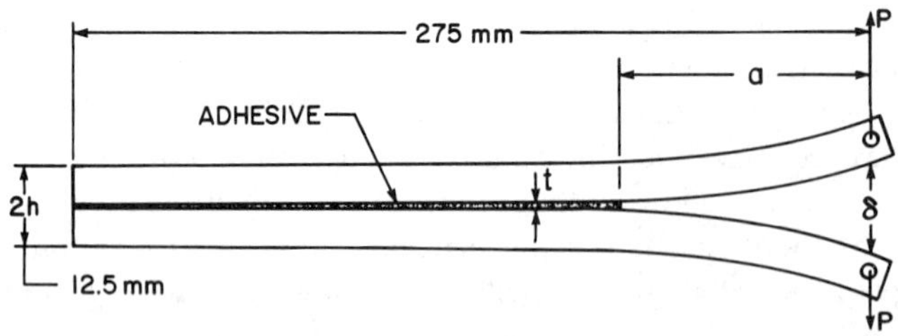

FIG. 1—*Mode I adhesive joint specimen. Note the bond thickness may be uniform or vary linearly with crack length. Beam width = 10 mm.*

TABLE 1—*Comparison of Mode I Fracture energy in bulk, adhesive film, and composite configurations.*[a]

Resin System	Fracture Energy, J/m^2		
	Bulk Resin	Adhesive Film (thickness = 0.2 mm)	Unidirectional Composite
My750: piperidine	330 ± 20	280 ± 10%	280 ± 30
My750 + 3.2% CTBN	1400 ± 100	330	370 ± 30
My750 + 6.2%	2200 ± 200	1350	360 ± 40
My750 + 9%	3200 ± 300	1500	490 ± 70

[a]Data taken from Ref *2*.

Materials and Fabrication

The adhesive was a Cycom 907 (BP-907), a modified test matrix (Bis-Phenol A Latent Aliphatic Amine Vinyl Resin) manufactured by American Cyanamid. As shown in Fig. 2 this material exhibits a nonlinear stress-strain behavior, with a 4.8% uniaxial failure strain.

Mode I DCB adhesive joint specimens were prepared from two aluminum (2024-T3) halves as shown in Fig. 1. The bond surfaces were mill-finished and then lightly sanded using a 400 grit Carbimet polishing paper, which provided a random texture to the surface. Following degreasing with acetone, the bond surfaces were chemically etched in sulfuric acid-sodium dichromate solution (FPL standard). The resin was degassed under vacuum at 105°C (220°F) and then

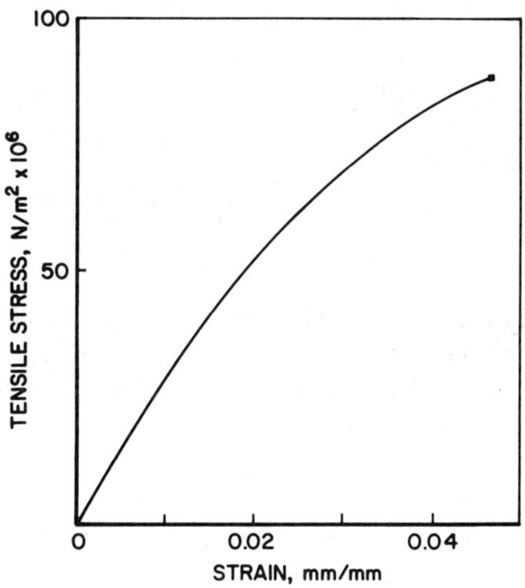

FIG. 2—*Tensile stress-strain curve for BP-907. Graph was reproduced from Ref 5.*

applied to the bond surface. The bond thickness, controlled by two spacers placed at each end of the bond, was chosen to be either uniform or to vary linearly in the specimen length. The motivation for the latter choice will become clear later. For thin bonds (<0.05 mm) no spacers were used. Instead, the bond thickness was controlled by the amount of pressure applied to the adherends via clips of various stiffnesses that were distributed along the bondline.

The assembled specimens were cured in an oven as follows: dwell at 121°C (250°F) for 1 h; cure at 176°C (350°F) for 1 h; cool slowly. After curing, the specimen's edges were flattened and polished, and the bond thickness measured from both edges, along the entire bondline, using an optical microscope. Only edge-to-edge variations not exceeding 10% were accepted.

Testing Procedure

Testing and data interpretation were as those of the interlaminar fracture tests [*12*]. Cracks a few centimetres long were introduced into the specimen using a razor blade. The specimens were split in a screw-driven Instron machine at a rate of 0.13 cm/min while the load, p, and the deflection, δ, were continuously recorded. When δ became sufficiently large, the crack grew. This growth was either continuous or unstable, depending on the bond thickness. Crack length, a, was recorded each time a noticeable growth occurred.

Fracture energy/unit area, G_{IC}, was calculated as a function of crack length using the compliance method together with Berry's empirical relationship

$$\delta/p = \alpha a^n, \ \alpha, n = \text{constants} \tag{1}$$

which gives (see account in Ref *12*)

$$G_{IC} = \frac{n\,\delta P}{2b\,a} \tag{2}$$

where b is the beam width and n is the slope of a straight line fit to the compliance versus crack length data plotted on a log-log paper. When crack jump occurred, two values of G_{IC} were calculated corresponding to the conditions at crack initiation (i) and at crack arrest (a):

$$G_{IC}{}^i = \frac{n\,\delta_i P_i}{2b\,a_i} \qquad G_{IC}{}^a = \frac{n\,\delta_a P_a}{2b\,a_a} \tag{3}$$

Test Results

Figure 3 gives the fracture energy as a function of bond thickness. The failure was always fully cohesive. This rather interesting fracture behavior seems to be characterized by the following pivotal bond thicknesses

$$
\begin{aligned}
t_b &\approx 0.5 \text{ mm} \\
t_m &\approx 0.22 \text{ mm} \\
t_c &\approx 0.15 \text{ mm} \\
t_t &\approx 0.033 \text{ mm}
\end{aligned}
\tag{4}
$$

The crack grew unstably (that is, in jumps a few centimetres long) for $t > t_c$, continuously for $t_t < t < t_c$, and in either of these modes (typically switching modes at random during the growth) for $t < t_t$. In the latter case, the initiation and the continuous fracture energies were quite indistinguishable. The energy seems to stabilize at large t ($>t_b$) or at small t ($<t_t$) while varying impressively intermediately, incurring a maximum at $t = t_m$. The arrest energy was also affected by the bond thickness. As shown, $G_{IC}{}^a$ is fixed for $t > t_t$ (except for mild variations in the transition range $t_t < t < 0.13$ mm) but varies sharply for $t < t_t$, virtually approaching G_{IC} in the limit for small bonds. [Note that normally no crack jump occurred for $t_t < t < t_c$. Fast cracking in that range was affected, however, by using the linearly-varying bond thickness specimen, with the crack initiating from the thinner bond ($<t_t$) but arresting at the thicker one ($>t_t$).]

It is now noted that the arrest energy was calculated based on quasistatic crack growth conditions (Eq 3), whereas the actual crack jump event is a dynamic process involving initiation, growth, and arrest phases. Although this simplification has been applied to a variety of beam-type test pieces, its validity has not been demonstrated. In a somewhat related work, Kalthoff et al [*13,14*] have studied the process of crack arrest in a homogeneous, relatively brittle material (Arldite B) tailored to either compact tension (CT, $a \approx h$) or rectangular DCB (RDCB, $a \approx 3h$) configurations. The crack tip was intentionally blunted so as to produce a crack jump. High-speed photography coupled with the shadow optical technique was used to record the instantaneous crack-tip position and stress intensity factor (SIF). It was found that the dynamic effects were largely dependent on the specimen geometry and the initiation SIF. For the CT specimen, these effects were small, but for the RDCB specimen, the recorded SIF at crack arrest differed significantly from that calculated from the static analysis, the difference that increased with increasing the initiation SIF.

Although these findings do not bear directly on the present results due to the differences in material (that is, rate sensitive versus rate insensitive) and geometry (that is, $a/h \gg 1$ versus $a/h \approx 1$ to 3), they are a cause for concern. As was also done in Ref *15*, the arrest energy from Fig. 3 was surveyed for a possible

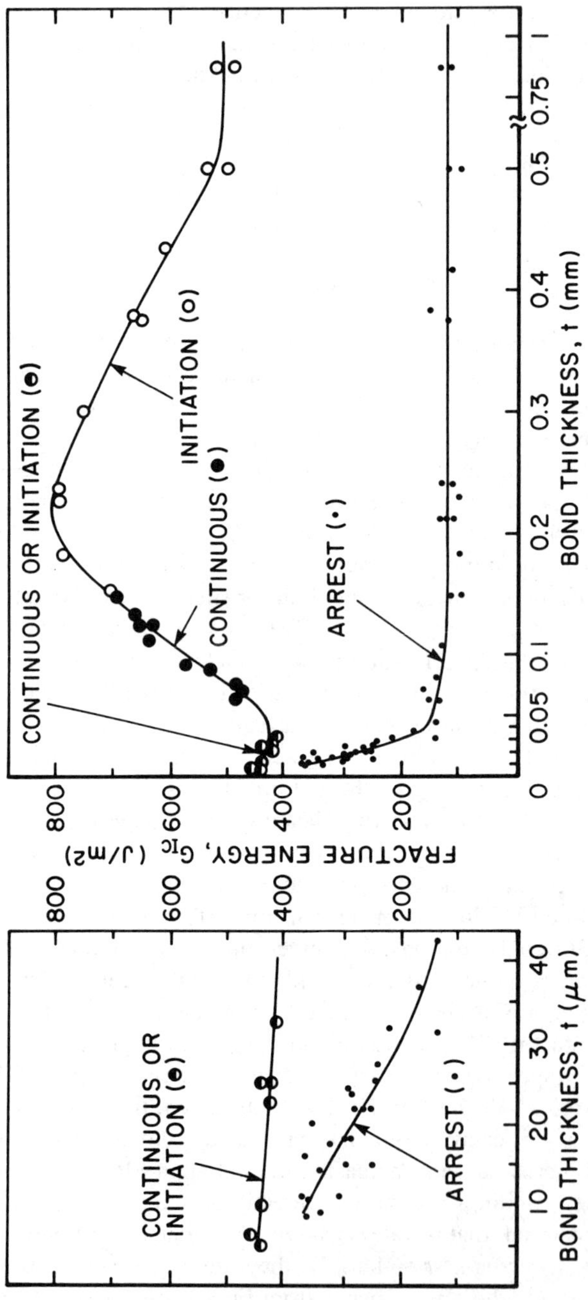

FIG. 3—*Adhesive joint fracture energy as a function of bond thickness; figure to the left is an enlargement of the thin-bond region from the figure to the right. Note the arrest energy was calculated from Eq 3 assuming quasistatic crack growth conditions.*

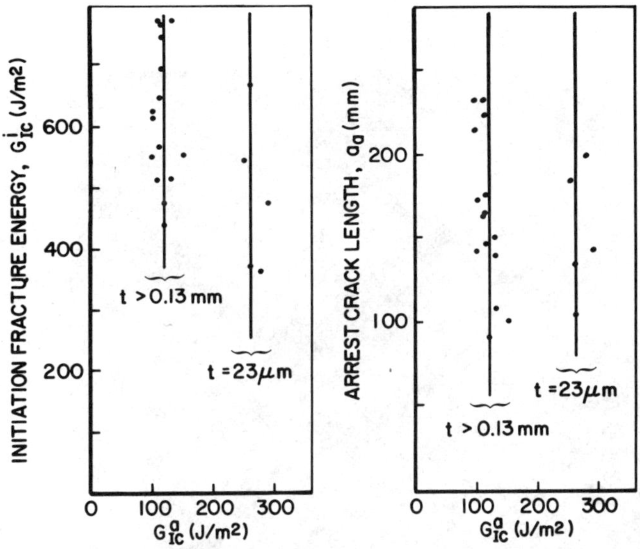

FIG. 4—*The dependence of arrest energy on initiation energy and crack length at arrest. Data collected from Fig. 3. The vertical solid lines are the corresponding arrest energies from Fig. 3.*

dependence on the specimen geometry and the initiation energy. Figure 4 plots G_{IC}^{a} as a function of G_{IC}^{i} and a_a, the crack length at arrest, for two representative bond configurations, that is, $t > 0.13$ mm where G_{IC}^{a} is fixed and $t = 23$ μm where G_{IC}^{a} strongly varies with t. To broaden the range of G_{IC}^{i} values normally available, several improvisions were made. First, for $t > 0.13$ mm, initiation values as little as 430 J/m^2 were achieved by using the linearly-varying bond thickness specimen in which the crack was initiated from the thin bond ($t < t_t$) but arrested at the thick bond (> 0.13 mm). Second, for $t = 23$ μm, initiation values exceeding 430 J/m^2 were affected by reinforcing the bondline from both the specimen edges with a droplet of fast-curing epoxy, the amount of which controlled G_{IC}^{i}. Figure 4 reveals no particular dependence of G_{IC}^{a} on either G_{IC}^{i} or a_a; the data seem to scatter evenly about the solid lines representing the corresponding arrest energies from Fig. 3. This finding suggests the so-calculated arrest energy to be a material property. However, to firmly establish this trend would necessitate additional favorable arrest data pertaining to very small cracks.

Fractography

Scanning electron microscopy was used to study the fracture surfaces. Each characteristic bond thickness regime will now be treated separately.

Crack Initiation, $t > t_c$

Figure 5 shows the crack arrest-reinitiation site for two bond thicknesses. Both the sites exhibit a narrow, parabolically-shaped band (ductile band) of a rough

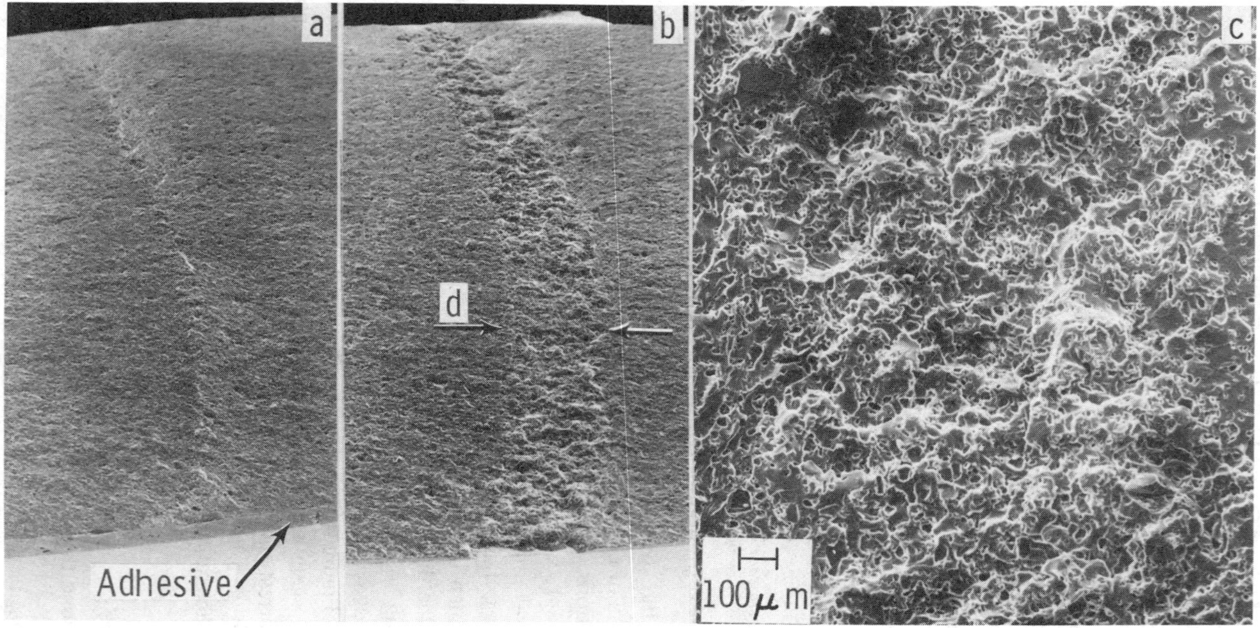

FIG. 5—*Overviews of crack arrest—reinitiation sites for two different bonds;* (a) t = *0.875 mm;* (b) t = *0.18 mm;* (c) *magnified view of the central part of the ductile band in* (b) *showing dimple-like topography. Note the ductile band is well-contained within the adhesive layer in* (a) *but it is extended lengthwisely in* (b) *due to damage zone-adherend interaction. This extension is denoted as* d. *Crack growth is from left to right.*

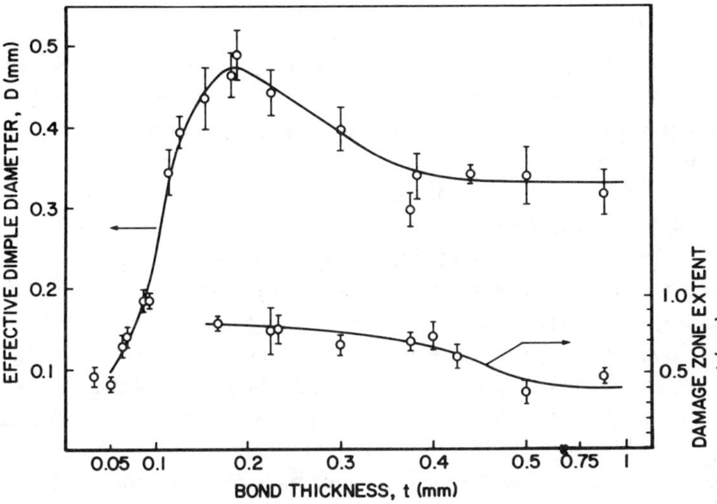

FIG. 6—*The dependence of the morphological Parameters* D *and* d *(see definitions in Figs. 5 and 7) on the bond thickness. Each data point represents the mean from ten independent measurements taken from a given sample. Bars indicate standard deviations.*

surface that is preceded and followed by flat surfaces. This topographical contrast reflects the time scale difference of the failure; whereas the ductile band was generated during the loading period (typically a few minutes), the flat surfaces were formed by a fast-running crack. At a larger magnification, Fig. 5*c*, the ductile band displays a dimple-like macrostructure indicative of the significant material drawing that preceded the final rupture. (Note that the pits or voids in these and subsequent prints are due to the low curing pressure used. No such voids were found on the fracture surface of a T300/BP-907 composite cured while under pressure of 0.69 MPa (100 psi).

The characteristic damage zone in Fig. 5*a* is well contained within the adhesive layer. In contrast, the thinner bond, Fig. 5*b*, exhibits a pronounced lengthwise damage spread attributed to the interaction of the damage zone with the adherends' surfaces. The extent of this spread, d (see Fig. 5*b*), is a valuable morphological landmark that could be easily measured using an optical microscope. The results as a function of t are given in Fig. 6.[2] As shown, d is invariable for large t but it increases as t decreases from 0.5 mm, first significantly but later only mildly. The onset of variation at t = 0.5 mm, which seems to be shared also by the energy trace from Fig. 3, signifies the transition from bulk to adhesive joint configuration. Interestingly, at this transition, d and t are of the same size, which suggests the unconstrained damage zone to have comparable axial and lateral extensions.

[2]Note that on occasions repeated tests at a given bond thickness produced differing values of d. This scatter, which increased as t approached t_c, was clearly due to a continuous crack growth. In such a case, the smallest measure was taken as the critical value.

Continuous Growth, $t_t < t < t_c$

Figures 7 to 9 show a series of photos pertaining to this bond thickness segment in which G_{IC} drastically increases with t. All surfaces display the same dimple topography typifying the crack initiation site. A closer examination of a dimple surface, Fig. 7*c*, reveals numerous river markings or plastic flow lines of which extensions and orientations vary at random.

Comparison of Figs. 7 to 9 shows both the dimple size and the deformation level to gradually diminish with decreasing t. In fact, for $t < t_t$ the dimple topography vanished. Association of these morphological variations with the corresponding variations in fracture energy is a foregone conclusion. To express the former quantitatively, the effective dimple diameter, D (see Fig. 7*b*), was considered. The results as a function of t, including the unstable growth bond segment, are shown in Fig. 6. As shown, the points data produced a well-defined curve of which variations with t are quite similar to those of the energy trace from Fig. 3; for small t, the dimple diameter rises sharply with t but for large t it is unchanged. Moreover, in the intermediate range, D exhibits a max at $t \approx t_m$.

Continuous or Unstable Growth, $t < t_t$

As for the fracture conditions, the morphology for this bond segment, typified in Figs. 10 and 11, also differed from that for larger bonds. The fracture surface in Fig. 10*a* consists of dark, randomly-shaped islands that are scattered unevenly on an otherwise gray background. A closer examination of a typical island, Fig. 10*b*, clearly reveals that the gray surface is adjacent to a one metal/matrix interface whereas the dark surface to the other. Therefore, excluding the islands' boundaries, the failure as a whole occurred near the metal/matrix interface. On a casual observation, this failure may be mistaken for adhesive failure. However, the magnified view in Fig. 10*c* clearly shows the adherend surface to be well covered by a matrix layer that, although thin compared to the bond thickness, is evidently thick enough so as to eliminate all adherend surface texture details from the fracture surface. This, therefore, constitutes a fully cohesive failure. As to the failure itself, Fig. 10*b* shows a consistent pattern of plastic flow lines that suggests a shear type failure. Over a given area, the river markings seem to follow a well-defined pattern but that pattern may change from one area to another. In some locations, as specified by the arrows in Fig. 10*c*, local damage growth seems to have originated from a void ahead of the crack tip.

This type of failure always occurred for $t < t_t$, see Fig. 11. This figure differs though from Fig. 10 in that the characteristic islands are smaller in size and their boundaries more rounded.

Crack Arrest

Figures 12 and 13 typify the morphology for the two bond segments where $G_{IC}{}^a$ is fixed ($t > t_t$) or descends with t ($t < t_t$), respectively. As shown in Fig.

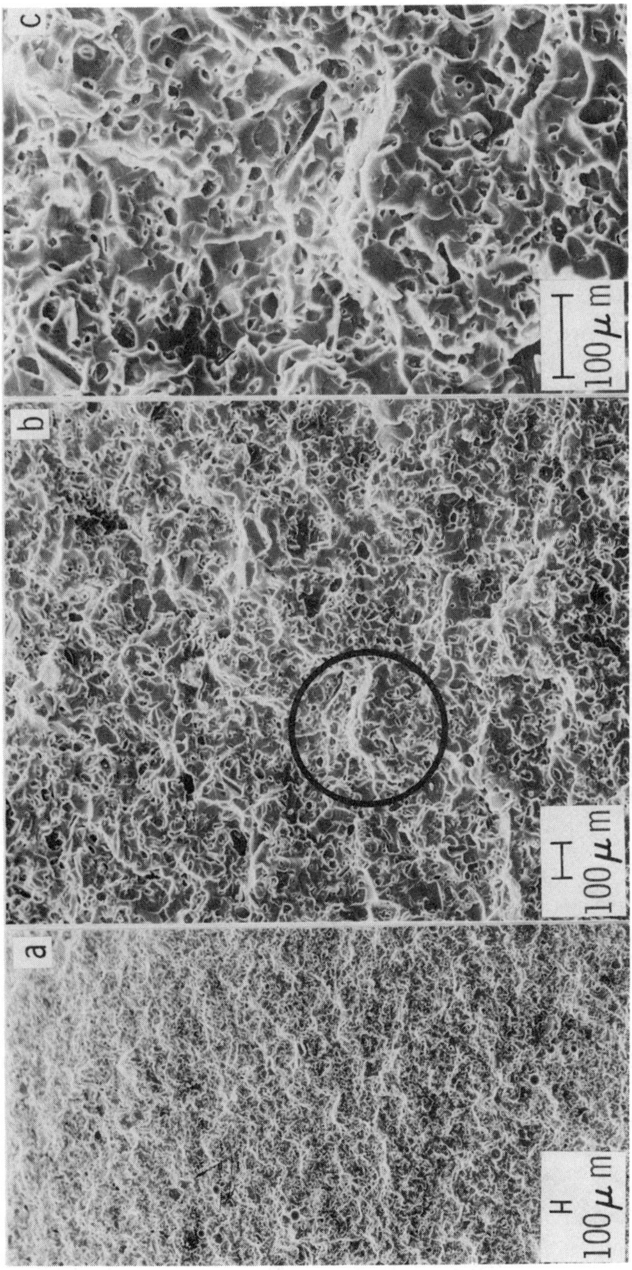

FIG. 7—*Fracture surface for a continuous type growth at three different magnifications,* t = 0.15 *mm. The surface exhibits the same dimple topography as for the crack initiation site, Fig. 5c. The diameter of the circled area in* (b) *is the effective dimple diameter,* D. *Crack growth is from left to right. In these and all subsequent prints, the tilt angle is 20°.*

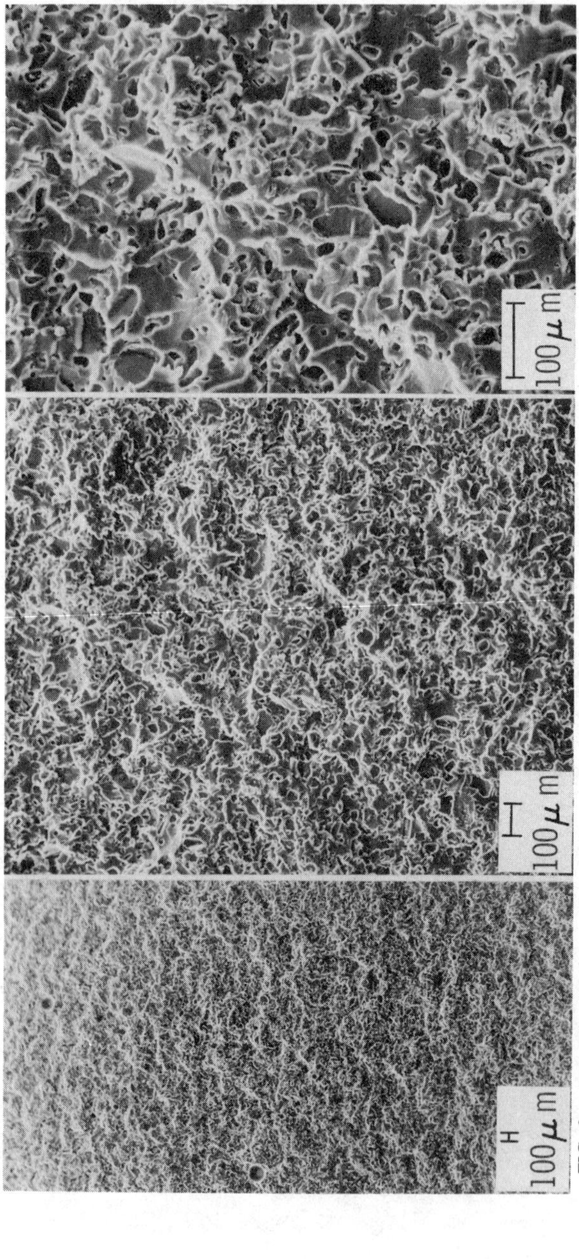

FIG. 8—*As in Fig. 7,* t = *0.125 mm. Note the reduced dimple size as compared to Fig. 7. In these and all subsequent prints, the crack growth is from right to left.*

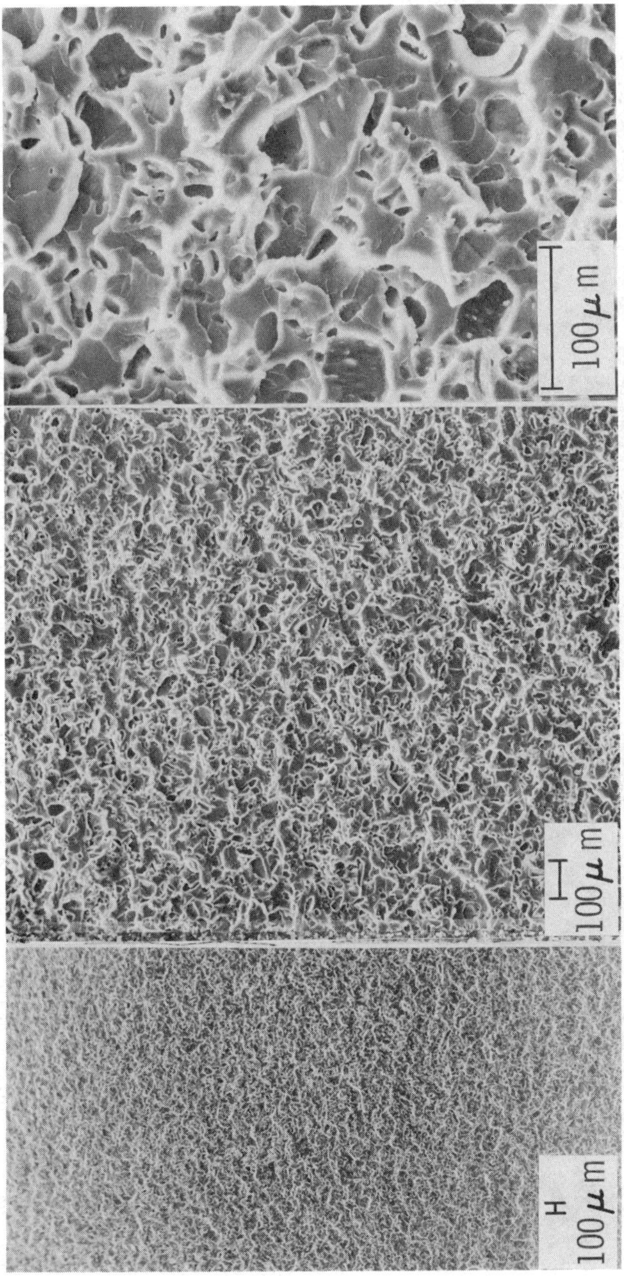

FIG. 9—As in Fig. 8, t = 0.068 mm. Note the reduced dimple size as compared to Fig. 8.

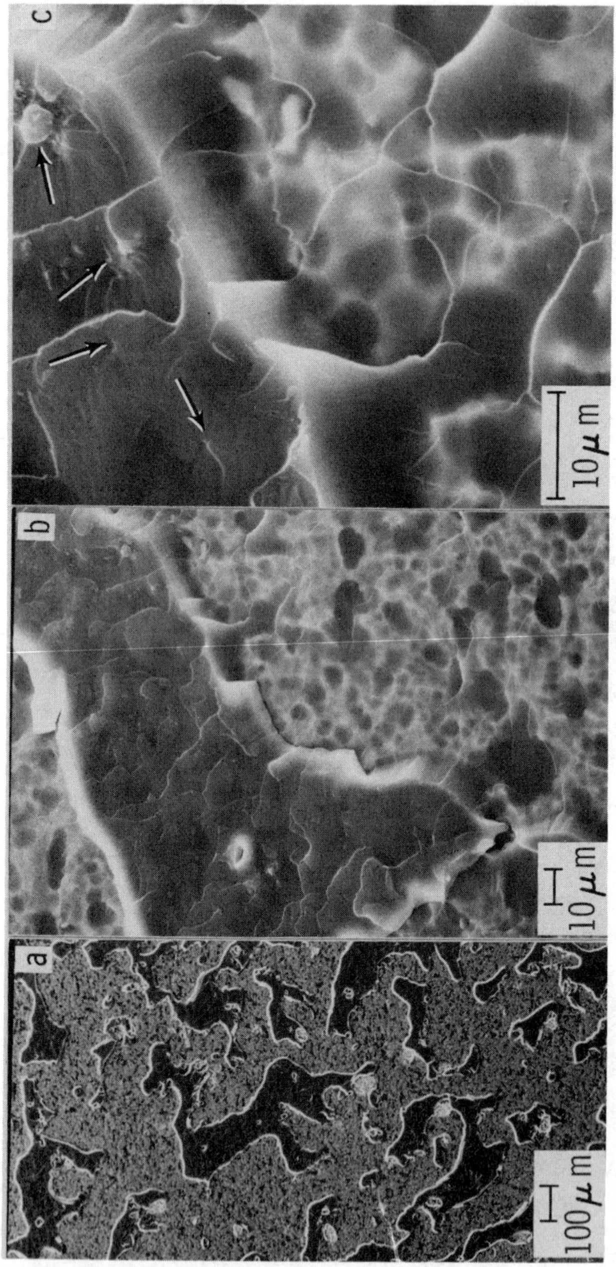

FIG. 10—*Continuous crack growth,* t = *18* μm. (a) *The surface exhibits dark islands that are randomly distributed on a gray background.* (b) *A magnified view of a typical island showing that the gray and dark regions correspond to failure near the metal/matrix interface.* (c) *A magnified view of* (b) *showing the fracture surface to be covered by a thin matrix layer that eliminated imprints or texture details of the adherend. The failure itself seems to be characterized by shear yielding. Arrows in* (c) *indicate possible damage nucleation sites.*

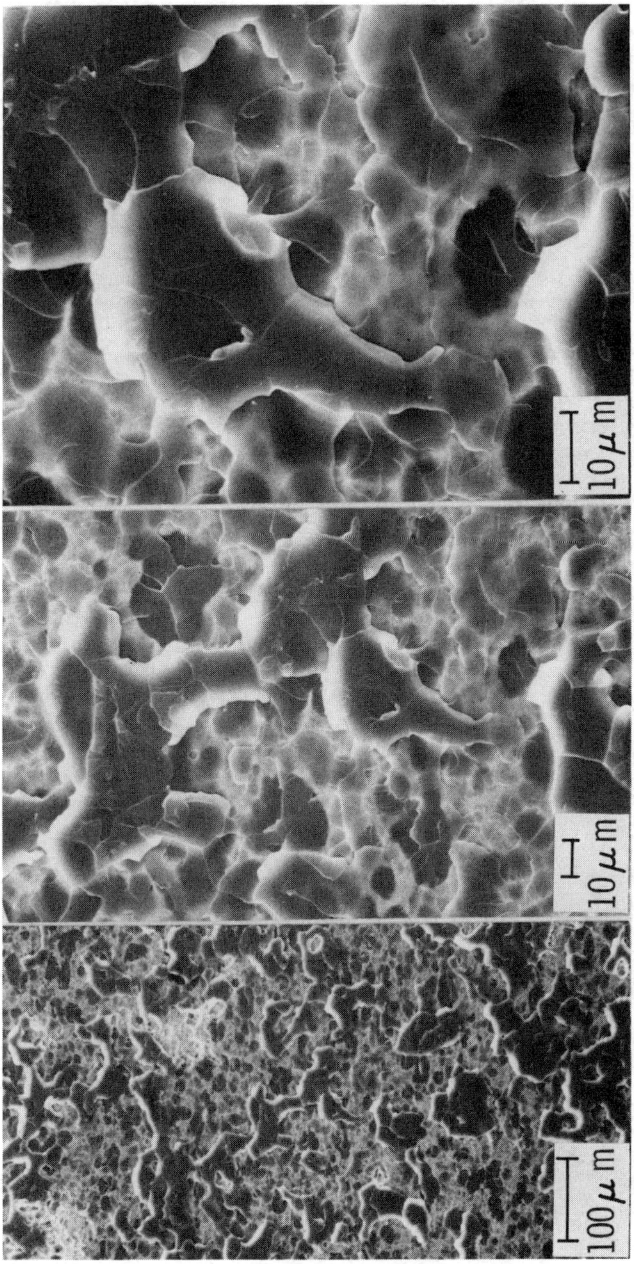

FIG. 11—*As in Fig. 10*, t = 5 *μm.*

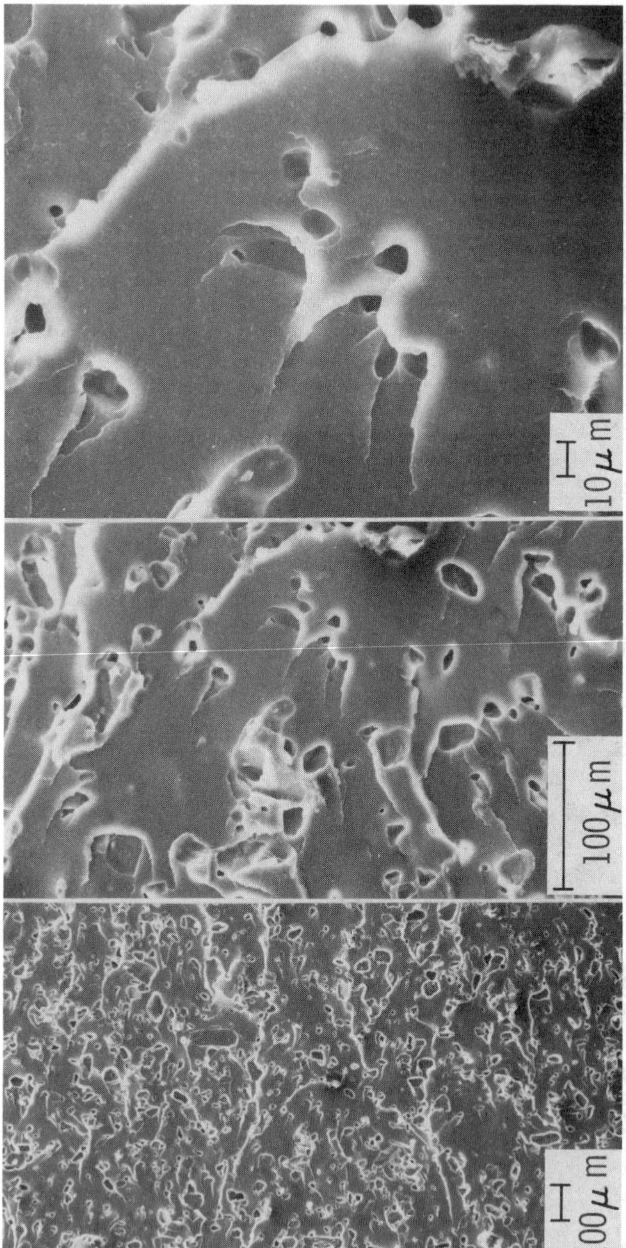

FIG. 12—*Fracture surface of a fast-running crack at three different magnifications,* t = *0.18 mm. Pictures taken 2 mm behind the arrest curve. The morphology is relatively flat and featureless (except for voids), which is consistent with the small value of* G_{IC}^{a} *and with the invariability of* G_{IC}^{a} *at large* t.

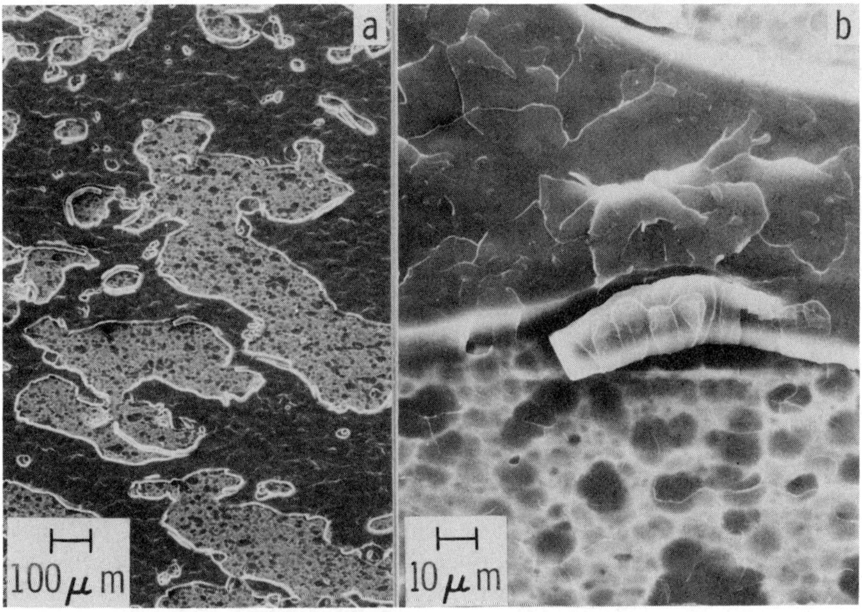

FIG. 13—*Fracture surface of a fast-running crack,* t = *8.3* μ*m, where* (b) *is a magnification of* (a). *The morphology is essentially similar to that in Fig. 10.*

12, except for voids, the failure is mostly smooth and featureless, similarly to cleavage failure of brittle epoxy. Evidently, the short time scale of the cracking has precluded the evolution of the energy-consuming dimple formation. The fracture surface reveals river markings that are again oriented randomly relative to the crack path. The elevation difference among these markings is negligibly small compared to the bond thickness, which may explain the invariability of G_{IC}^{a} at large t.

The fracture surface for the thin bond, Fig. 13, is essentially as that in Fig. 10 corresponding to the continuous crack growth. The minute details reveal that the surface is flatter, and the river markings less frequent than in Fig. 10. This difference, again attributed to the faster cracking rate, is probably responsible for the lower value of G_{IC}^{a} as compared to G_{IC}.

Discussion

Related Work

It is interesting to note that some of the bond thickness effects shown in Fig. 3 were also observed in a similar test utilizing a rubber-modified epoxy as the adhesive [*8,9*]. The similarities include a maximum in G_{IC} (at $t_m \approx 0.6$ mm), a plateau in G_{IC} at large t, and an invariability in G_{IC}^{a} for $t > t_m$. There are differences, however, in the behavior for $t < t_m$; whereas in Ref *8* G_{IC} decreases monotonically as t decreases from t_m, approaching G_{IC}^{a} at $t \approx 0.05$ mm, in Fig.

3 G_{IC} stabilizes for small t at a value 3.5 times that of G_{IC}^a. This noticeable difference for thin bonds is somewhat surprising in virtue of the similarities for large bonds. However, it is possible that this behavior from Ref *8* is an artifact of adhesion failure. It was pointed out [*9*] that the failure for $t < t_m$ occurred near the metal/matrix interface. Unlike in Fig. 10, the morphology of such a failure (given in Fig. 10 of Ref *9*), reveals the texture or imprints of the adherend surface. Although it was argued [*9*] that a thin polymeric film was deposited on the fracture surface, it is doubtful whether such a failure can be classified as cohesive.

It was proposed [*9*] that the fracture work is maximized when the plastic zone extent, r_{yc}, coincides with the bond thickness, and that the rapid decline in G_{IC} as t decreases from t_m is merely due to the restraint on the resin plasticity.[3] This notion was furthered [*11*] by stipulating a direct relationship between the damage zone volume and the fracture energy. These propositions are not supported by the present results. As to the maximum in G_{IC}, Fig. 3 shows that a plastic zone-adherend surface interaction occurs at $t = 0.5$ mm whereas G_{IC} is maximized at $t = 0.22$ mm. As to the energy-zone volume correspondence, this hypothesis is clearly at odds with the invariability of G_{IC} for $t < 33$ μm. (It is unlikely for d to vary like $1/t$ in that range so as to make the product $t \times d$ fixed.) Moreover, Fig. 6 shows that over the bond thickness range 0.15 to 0.3 mm d is quite fixed, which implies that the zone volume should vary proportionally to t. However, as shown in Fig. 3, this type of variation is not shared by G_{IC}.

As discussed earlier, the key to understanding energy dissipation rests on the morphology; for $t > t_t$, the fracture energy was shown to vary according to D, and, to a lesser extent, d whereas the invariability of G_{IC} for $t < t_t$ was associated with the shift of failure from the bond center to the metal/matrix vicinity. Some factors affecting D, d, and that shift are discussed in the following.

The Effect of Bond Thickness on Damage Extent and Dimple Size

An insight into the variation of the damage extent, d, with t can be gained by considering the crack tip stress field in the adhesive joint. Linear elastic solutions for this have been reported by several authors [*16–18*]. The results show that in the immediate crack-tip vicinity, the normal stress decay monotonically at a rate greatly affected by the bond thickness. For thick bonds, this decay is rapid, following the inverse-square-root singularity pattern, but for thin bonds the stress was found to decay only mildly. (A graphical comparison of a 0.127-mm-thick bond and a 2.54-mm-thick bond, Figure 6–15 of Ref *16*, shows that at a distance 0.25 mm ahead of the crack tip the normal stress for the thinner bond is more than three times that of the thicker bond.) This result suggests d to increase from its "bulk" or large t value when t becomes sufficiently small, which is indeed the case from Fig. 6.

[3]In Refs *9* and *11*, r_{yc} was estimated using a fracture mechanics approach and was claimed to compare reasonably well with t_m. However, in those calculations, some far-reaching simplifications were adopted including the idealization of the material as elastic-plastic and the neglection of the adherend boundary effect.

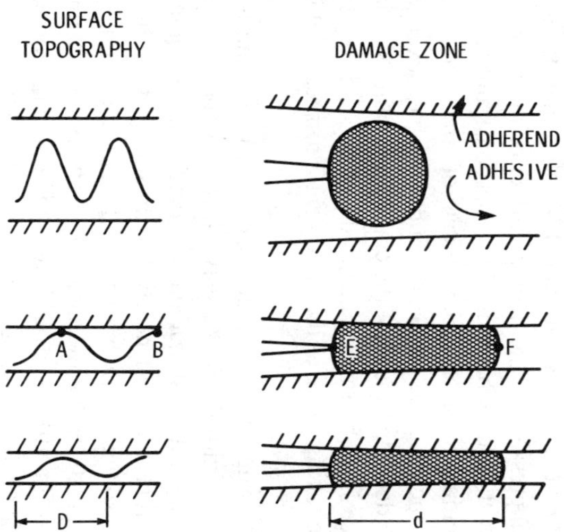

FRACTOGRAPHIC DETAILS					
t(mm)	d (mm)	D(mm)	G_{IC}(J/m^2)	G^a_{IC}(J/m^2)	COMMENTS
0.5± 0.05	0.45± 0.07	0.33± 0.03	510± 25	123± 20	At transition from "bulk" to "joint"
0.22± 0.01	0.8± 0.1	0.46± 0.03	810± 20	123± 20	At maximum fracture energy
0.15	0.82± 0.1	0.45± 0.04	700± 20	123± 20	At transition from unstable to continuous crack growth

FIG. 14—*Crack-tip damage zone characteristics for three pivotal bond thicknesses. Note the dimple profiles* (left column) *and the damage zone shapes were idealized.*

The analysis of D is much more complex owing to the three dimensionality of the fracture surface. With the lack of any appropriate analytical treatment, the discussion is bound to be highly speculative. Suppose a typical dimple evolves from a certain point ahead of the crack tip, possibly as a result of void coalescence. This point is denoted as A in Fig. 14. Symmetry suggests the ensuing growth to take the form of a surface of revolution, the particular shape that is controlled by the triaxial stress field at Point A. Now the formation of this free surface relieves the stresses around Point A so that the next dimple source, say Point B, must be sufficiently removed from A. The distance, $\overline{AB}$, which allows for a full stress recovery, is therefore the dimple diameter, D. For $t \gg d$, the only geometrical factor affecting the stresses ahead of the crack tip is d. Since the latter is fixed for $t > 0.5$ mm, one expects D to also be fixed there, as is indeed the case. When t is comparable to D, it affects the stresses around Point A directly, which may explain the monotonic rise in D as t increases from t_t. Finally, the unique behavior in the intermediate range may be attributed to the interaction of the diverging trends from small and large t.

Analogies from Butt Joint Specimen

The slight increase in toughness with decreasing t experienced at extremely thin bonds is an interesting phenomenon, particularly alien to the energy-zone volume hypothesis. Interestingly, similar effects have long been observed in butt joint tension specimens (that is, a thin adhesive layer joining two stiff cylinders) employing either metallic [*19,20*] or polymeric fillers [*21*]. Among the explanations given for this increase in strength with decreasing joint thickness were the reduced probability of incurring internal flaws that may trigger premature failure [*20*] and the suppression of the necking effect or the lateral deformation in the adhesive layer by the stiff adherends [*19*]. This constraint that is manifested following adhesive yielding leads to a triaxial state of stress within the adhesive layer that is known to increase the material strength (see, for example, Ref *19*). Such a stress field also develops ahead of the crack tip in thin DCB joints, which may explain the change in failure mode incurred as t is decreased from t_t, from a combined-stress type taking place at the bond center to shear yielding near the metal/matrix interface.

This reasoning is also supported from another result pertaining to butt joints. The problem of a thin, rigid-plastic adhesive layer compressed between two rough and rigid adherends was first studied by Prandtl (see account in Ref *20*). This analysis suggests that a state of nearly "hydrostatic" tension exists in the majority of bond region so that failure by shear will commence from the adhesive/adherend interface as it is there that the shear stress attains its maximum. It must be stated, however, that Prandtl's analysis bears only qualitatively on the present test due to the difference in material behavior.

Factors Affecting the Crack Growth Manner

Crack instability (or catastrophic failure) is a major liability of toughened resins. This phenomenon will now be assessed qualitatively from Fig. 14, which shows grossly the fully-developed crack-tip damage zone for a few pivotal bonds. Consider a cracked specimen loaded to failure. When the critical load is reached, a strand of material at the crack tip, for example, Point E, ruptures. This causes a stress redistribution in the sample and consequently a damage growth at Point F. Clearly, that growth will be stable only if the stress redistribution rate (SRR) at Point F is sufficiently small so as to allow for the formation of the dimple structure at that point. Otherwise, the growth will be unstable. Two parameters that affect the SRR are d and $G_{IC}{}^{i}$; increasing the former or decreasing the latter clearly reduces the SRR at Point F, which tends to stabilize the growth.

With this in mind, consider the effect of bond thickness on the cracking manner. Starting with large t, crack jump in the bulk specimen is easily understood as the emergence of a sharp crack consuming relatively little energy from the blunted tip. As t decreases from 0.5 mm, both d and $G_{IC}{}^{i}$ increase, which have a mixed effect on the cracking manner. However, upon decreasing t from t_m, only $G_{IC}{}^{i}$ significantly varies, in fact it decreases sharply. Therefore, as t is sufficiently reduced from t_m, one expects the growth to change from unstable to continuous. This evidently occurs at $t = t_c$.

The transition from continuous to unstable growth at $t = t_t$ is an interesting phenomenon that seems to be associated with the shift of failure locus from the bond center to the metal/matrix interface. The concentration of failure over an extremely thin layer near the interface increases the SRR. (In the analogous butt joint specimen, for instance, the straining rate of the adhesive layer is proportional to $1/t$). This, together with the carrying over of the rate sensitivity of the fracture work into thin bonds may explain the crack jump. To test this argument, several thin-bond specimens were loaded at rates larger than normally used. It was found that the tendency for fast cracking has indeed increased with the loading rate. Whereas at the normal rate of 1.3 mm/min the growth changed from continuous to unstable or vice versa rather randomly as the crack grew, at the largest rate attempted (127 mm/min), the crack advanced in a sequence of jumps.

Correlation with Interlaminar Fracture

Interlaminar fracture testing of composites is a tedious process that becomes highly time and cost ineffective when a parametric study is of concern. Moreover, interpretation of test results may be complicated by such effects as fiber bridging, multiple crackings, and time-temperature dependent deformations away from the crack tip. The prospect of applying a simplified testing methodology such as the adhesive joint to predict the interlaminar fracture behavior is therefore challenging.

In Ref *12*, the interlaminar fracture of a T300/BP-907 multidirectional laminate was studied in detail. By defining such a failure as a process involving no fiber breakage, multiple crackings, etc., the fracture energy was found to be a material property independent of the orientations of fibers above and below the delaminating interface. Moreover, the interlaminar crack was found to grow either continuously or unstably, with $G_{IC} = 368 \pm 35$ J/m^2 and $G_{IC}{}^{a} \approx 300$ J/m^2. As shown in Fig. 10 of Ref *12*, the thickness of the interlaminar matrix layer typically varied between 0 and 30 μm. Over this range, Fig. 3 shows a rather similar fracture behavior, the agreement is both in the crack growth manner and in the energy levels. One obvious difference though between the joint and composite test results is in the morphology. As was shown [*12*], the interlaminar failure was rich in microstructural details that varied with the type of interface. Nevertheless, such details were shown not to affect the fracture work.

Several conclusions relevant to interlaminar fracture of toughened composites can be made from the thin-bond test results in Fig. 3. First, in such composites, both continuous and unstable types of growth are to be expected; moreover, the intensity of the crack jump may vary considerably from sample to sample due to unavoidable thickness variations of the interlaminar matrix layer. Second, resin rich areas may promote catastrophic failure. Third, in using interlayers to toughen brittle-matrix based composites, as far as the Mode I fraction is of concern a thin layer is preferable over a thick one. This is because, in addition to the obvious gain in overall laminate shear performance, such a choice may greatly reduce crack jump yet at no cost to "static" toughness.

Summary and Conclusions

The effect of bond thickness on the fracture behavior of a ductile adhesive was characterized in details. The adhesive joint performance can be characterized by "bulk," "joint," and "thin-bond" regimes corresponding to intact, constrained, and largely suppressed crack-tip damage zone, respectively. Each of these regimes exhibited interesting yet differing fracture characteristics and energy dissipation mechanisms that were not directly related to the size of the crack tip plastic zone but instead to the fracture surface morphology.

Over the "joint" and "bulk" regimes, the failure has originated from the bond center, and the variations in G_{IC} and $G_{IC}{}^{i}$ with t followed chiefly from variations in the size of a dimple-like macrostructure characterizing the fracture surface. Over the "thin-bond" regime ($t < t_t = 33$ μm), failure occurred by shear yielding near the metal/matrix interface, with G_{IC} being quite fixed while $G_{IC}{}^{a}$ decreased sharply with t. With the aid of an analogous result from butt joints, the change in fracture behavior incurred at small bonds ($t < t_t$) was attributed to the increase in the triaxiality of the stress field. This eliminated a combined stress type failure at the bond center, thereby leading to shear yielding at the metal/matrix vicinity that occurred at higher loads.

The "thin-bond" test produced quite accurately the interlaminar fracture behavior of its composite counterpart, which makes it a viable alternative to *in situ* composite testing. The preservation of toughness in extremely thin bonds is advantageous to toughened resins. Whether the mechanisms responsible for this also operate in a great variety of other toughened resins is obviously an issue of great concern. An apparent exception is a rubber-toughened epoxy [*8,9*] but, as previously discussed, that toughness loss may be an artifact of adhesion failure.

If that interpretation is correct, then it highlights the need for a better matrix-adherend (or fiber) adhesion. As opposed to the static toughness advantage, the carrying over of the fracture work rate sensitivity into the thin-bond regime is a drawback to toughened resins. One source of comfort is that this effect may be reduced by designing for extremely thin matrix interlayers.

References

[1] Krieger, R. B., Jr., "The Relation Between Graphite Composite Toughness and Matrix Shear-Strain Properties," 29th National SAMPE Symposium, Society for the Advancement of Materials and Process Engineering, 3–5 April 1984, pp. 1570–1584.

[2] Scott, J. M. and Phillips, D. C., *Journal of Materials Science*, Vol. 10, 1975, pp. 551–562.

[3] Bascom, W. D., Bitner, J. L., Moulton, R. J., and Siebert, A. R., *Composites*, Vol. 11, No. 1, Jan. 1980, pp. 9–18.

[4] Moulton, R. J. and Ting, R. Y., *Composite Structures*, I. H. Marshall, Ed., 1981, pp. 674–689.

[5] Palmer, R. J., "Investigation of the Effect of Resin Material on Impact Damage to Graphite/Epoxy Composites," NASA CR 165677, National Aeronautics and Space Administration, March 1981.

[6] Ting, R. Y. and Cottington, R. L., *Journal of Adhesion*, Vol. 12, 1981, pp. 243–255.

[7] Mostovoy, S., Ripling, E. J., and Bersch, C. F., *Journal of Adhesion*, Vol. 3, 1971, pp. 125–144.

[8] Bascom, W. D. and Cottington, R. L., *Journal of Adhesion*, Vol. 7, 1976, pp. 333–346.

[9] Bascom, W. D., Cottington, R. L., Jones, R. L., and Peyser, P., *Journal of Applied Polymer Science*, Vol. 19, 1975, pp. 2545–2562.

[10] Hunston, D. L., Bitner, J. L., Rushford, J. L., Oroshnik, J., and Rose, W. S., *Journal of Elastomers and Plastics*, Vol. 12, July 1980, pp. 133–149.

[11] Kinloch, A. J. and Shaw, S. J., *Journal of Adhesion*, Vol. 12, 1981, pp. 59–77.

[12] Chai, H., *Composites*, Vol. 15, No. 4, Oct. 1984, pp. 277–290.

[13] Kalthoff, J. F., Beinert, J., and Winkler, S., in *Fast Fracture and Crack Arrest, ASTM STP 627*, Hahn and Kanninen, Eds., American Society for Testing and Materials, Philadelphia, 1977, pp. 161–176.

[14] Kalthoff, J. F., Beinert, J., Winkler, S., and Klemm, W., in *Crack Arrest Methodology and Applications, ASTM STP 711*, Hahn and Kanninen, Eds., American Society for Testing and Materials, Philadelphia, 1980, pp. 109–127.

[15] Crosley, P. B. and Ripling, E. J., in *Crack Arrest Methodology and Applications, ASTM STP 711*, Hahn and Kanninen, Eds., American Society for Testing and Materials, Philadelphia, 1980, pp. 321–337.

[16] Anderson, G. P., Bennett, S. J., and DeVries, K. L., *Analysis and Testing of Adhesive Bonds*, Academic Press, New York, 1977.

[17] Farhad, F., Muki, R., and Westmann, R. A., *International Journal of Solids and Structures*, Vol. 13, 1977, pp. 561–570.

[18] Wang, S. S., Mandell, J. F., and McGarry, F. J., *International Journal of Fracture*, Vol. 14, No. 1, Feb. 1978, pp. 39–58.

[19] Bredzs, N., *Welding Journal*, Nov. 1954, p. 545–S.

[20] Bikerman, J. J., *The Science of Adhesive Joints*, Academic Press, New York, 1968.

[21] Bryant, R. W. and Dukes, W. A., "The Effect of Joint Design and Dimensions on Adhesive Strength," Aeronautics and Space Engineering and Manufacturing Meeting, Los Angeles, Paper No. 670855, 2–6 Oct. 1967.

Erian A. Armanios,[1] *Lawrence W. Rehfield,*[1] *and Ambur D. Reddy*[1]

Design Analysis and Testing for Mixed-Mode and Mode II Interlaminar Fracture of Composites

REFERENCE: Armanios, E. A., Rehfield, L. W., and Reddy, A. D., **"Design Analysis and Testing for Mixed-Mode and Mode II Interlaminar Fracture of Composites,"** *Composite Materials: Testing and Design (Seventh Conference), ASTM STP 893,* J. M. Whitney, Ed., American Society for Testing and Materials, Philadelphia, 1986, pp. 232–255.

ABSTRACT: This paper presents the design analysis and results of an experimental program that was conducted to evaluate mixed-mode and Mode II interlaminar fracture behavior of a resin matrix composite material system. A double cracked-lap-shear specimen was designed utilizing a simple, new analysis method. The specimen was made of AS4/3502 graphite/epoxy material with a $[\pm45, 0, 90]_{6s}$ quasi-isotropic balanced symmetric layup. The lap interface studied was at $\pm45°$ orientations to the loading direction. A fundamental feature of the designed specimen is its ability to be tested under net tensile and compressive loadings. The specimen exhibits mixed-mode or Mode II behavior depending on the loading direction. This paper summarizes the tension testing only. The crack growth during testing was monitored by observing the isochromatic fringes in photoelastic coatings mounted on the two lap surfaces. Preliminary results indicate that the fracture behavior follows a resistance curve. In order to assess the double cracked-lap-shear specimen data and test, a comparison was made with results obtained with single cracked-lap-shear specimens. Less scatter in the data from the double cracked-lap-shear specimens was found.

KEY WORDS: composite materials, fiber-reinforced composites, composite structures, graphite-epoxy, crack propagation, fracture (materials), delamination, scanning electron microscopy, test methods

Laminated composite structures exhibit a number of different failure modes. These include fiber-matrix debonding within individual layers, delamination or separation of the layers, transverse cracks through one or more layers, and fiber fracture. A complex state of stress with steep gradients exists in regions near free edges, ply terminations, cutouts, holes, and voids. These sites often initiate interlaminar delamination cracks due to high local interlaminar shear and normal stresses. Delamination is observed to be one of the most common damage modes

[1]Research engineer, professor, and senior research engineer, respectively, School of Aerospace Engineering, Georgia Institute of Technology, Atlanta, GA 30332.

in composites. A material characteristic that could be used to determine resistance to delamination growth is the strain-energy release rate, *G*. Investigations on the strain-energy release rate associated with Mode I fracture have been reported by several authors [*1–7*].

Mixed-mode interlaminar tests include the single cracked-lap-shear specimen [*7,8*], the edge delamination specimen [*9*], and the mixed-mode flexural specimen [*7*]. The most popular test to determine Mode II interlaminar fracture toughness is based on the use of the end-notched flexural specimen [*7*]. A complete survey of mixed-mode and Mode II analysis and test methods is provided in Refs *10* and *11*. It is observed that these specimens require usually extensive computational effort to analyze the data and, most of all, are limited to one type of loading. The objective of the present work is to develop a simpler approach where mixed-mode and Mode II contributions to fracture can be evaluated using the same specimen and to be able to analyze and interpret the result by elementary methods. A specimen that can accept reverse loadings is required to determine stress range effects under cyclic alternating stresses and other fatigue environments.

The heart of the approach is a simple, new analysis method that permits the rapidly varying interlaminar stresses to be determined by elementary means. Extensive numerical computations are thus avoided and the results are obtained in closed form. The parameters governing the behavior are easily identified. The specimen selected for testing is a symmetric double cracked-lap-shear type specimen with end tabs as shown in Fig. 1. The layup is selected to produce balanced symmetric specimens with differently oriented interface plies at the lap/strap junction. This also minimizes ply nesting effects that would occur if similar plies were used at the interface. This specimen is designed to be tested in both tension and compression. The geometry of the lap, strap, and tabs is designed to precipitate crack growth in the gage section. Overall dimensions of the specimen are determined by the boundary layer effects from tab ends and from the lap/strap junction. Interaction among the local stresses in the preceding regions is avoided. The lengths of the lap and strap regions are selected so as to provide an adequate crack length to be monitored and to prevent overall buckling of the specimen. Symmetric crack growth is enhanced by choosing a large specimen thickness and by strict control of manufacturing quality.

The overall objective of this work is to systematically examine this specimen behavior in static tension and compression. This paper, however, summarizes

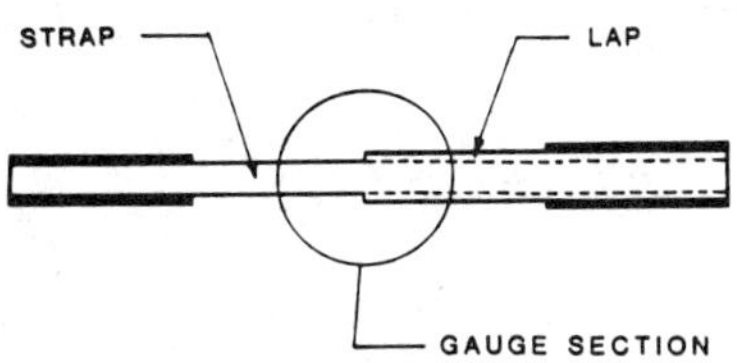

FIG. 1—*The double cracked-lap-shear specimen.*

the design analysis, manufacturing, and tension testing only. The resistance curve developed from the test data is also presented.

Overview of Analysis Approach

The present analysis models are based on the theory developed in Ref *12*. This theory includes transverse shear strain, transverse normal strain, and section warping effects. These effects, which are ignored by classical theories, can be significant in structures constructed of modern composite materials. For laminated composite structures, the theory will be applied by a ply-by-ply basis or to a group of plies (sublaminates) as if they are homogeneous bodies in equilibrium. Interfacial stresses are initially unknown. Enforcement of continuity conditions at interfaces and boundary conditions in an overall sense leads to a solution for these stresses. A complete development of the analytical model is provided in Ref *11*. A summary of the governing equations is given in the Appendix for completeness.

A simplification can be made by neglecting overall bending of the sublaminates. This model will be referred to as the sublaminate membrane (SM) model. The basic equations for this model are derived from the Appendix equations by neglecting the moment terms. For a sublaminate in a plane strain situation shown in Fig. 2, these equations encompass the following main categories:

Overall Equilibrium

$$N_{,x}^{k} + T_2^{k} - T_1^{k} = 0 \qquad (1)$$

$$Q_{,x}^{k} + p_2^{k} - p_1^{k} = 0 \qquad (2)$$

$$\frac{h^k}{2}(T_2^{k} + T_1^{k}) - Q^{k} = 0 \qquad (3)$$

The normal and transverse shear stress resultants of the kth element are N^k and Q^k, respectively. The interfacial shear stresses on the top and bottom surfaces of the kth element are T_2 and T_1. The interfacial transverse normal stresses are p_2 and p_1. The thickness of the kth element is h^k. Superscript k, which identifies the element, will be dropped in the subsequent equations for convenience.

Constitutive Relationship

$$N = h\, E_{11}\, \tilde{u}_{,x} + h\nu_{13} r - \frac{h^2}{12}\nu_{13} N_{,xx} \qquad (4)$$

where

$$r = \frac{1}{2}(p_2 + p_1) \qquad (5)$$

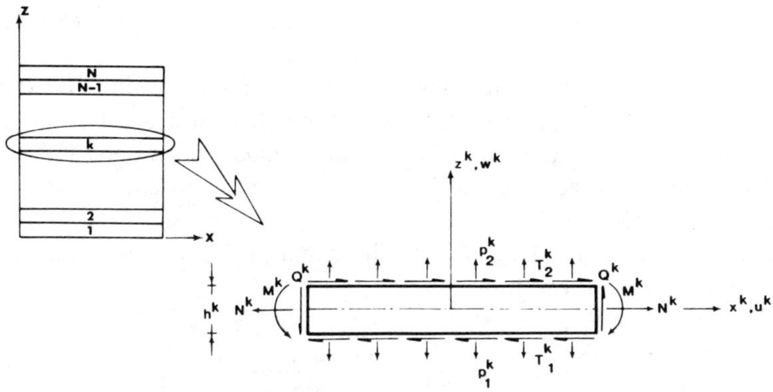

FIG. 2—*Notation and sign convention for the* k*th element.*

The elastic modulus associated with the x direction is E_{11}. Poisson's ratio is ν_{13}. The average axial displacement is $\tilde{u}$.

Stress Distribution

$$\sigma_{xx}(x,z) = \frac{N}{h} + \frac{h}{24}\left(\frac{E_{11}}{G_{13}} - 2\nu_{13}\right)\left(1 - \frac{12z^2}{h_2}\right)N_{,xx} \quad (6)$$

$$\sigma_{xz}(x,z) = -\frac{z}{h}N_{,x} + \frac{Q}{h} \quad (7)$$

$$\sigma_{zz}(x,z) = r - \frac{h}{8}\left(1 - \frac{4z^2}{h^2}\right)N_{,xx} - \frac{z}{h}Q_{,x} \quad (8)$$

The stress components follow the usual elasticity notations. The transverse shear modulus is G_{13}.

Displacement Distribution

$$w(x,z) = \tilde{w} - \frac{\nu_{13}}{E_{11}}\frac{z}{h}N \quad (9)$$

$$u(x,z) = \tilde{u} - z\tilde{w}_{,x} + \frac{1}{2hE_{11}}$$

$$\times \left[\frac{h^2}{12}\left(\frac{E_{11}}{G_{13}} - \nu_{13}\right)\left(1 - \frac{12z^2}{h^2}\right)N_{,x} + 2\frac{E_{11}}{G_{13}}zQ\right] \quad (10)$$

$$\tilde{\phi} = -\tilde{w}_{,x} + \frac{1}{hG_{13}}Q \quad (11)$$

The average transverse displacement is $\tilde{w}$, and the average section rotation is $\tilde{\phi}$.

The boundary conditions appropriate to this model are limited to extensional behavior. The N or $\tilde{u}$ must be prescribed at the element ends. The interlaminar continuity conditions (Eq 37) and the traction boundary conditions at the extreme upper and lower surfaces of the laminate must be imposed, and they remain unchanged for this model.

For an applied tensile load, P, the shear and peel stress distributions at the lap/strap interface are given by

$$T = -\frac{P}{2A_r}\left(\frac{m}{t}\right)\frac{\sinh m\,x/t}{\cosh m\,L/t} \tag{12}$$

$$p = \frac{h}{2}T_{,x} \tag{13}$$

where

$$A_r = 1 + \frac{t}{h}\frac{\overline{E_{11}}}{E_{11}} \tag{14}$$

$$m = \sqrt{\frac{3A_r}{z_1}} \tag{15}$$

$$z_1 = \frac{h}{t}[2\alpha\,\overline{E_{11}}/E_{11} + 3(\nu_{13}\,\overline{E_{11}}/E_{11} - \overline{\nu_{13}})] + 2\overline{\alpha} \tag{16}$$

$$\alpha = \frac{1}{2}(E_{11}/G_{13} - 2\,\nu_{13}) \tag{17}$$

The lap thickness is h, and the strap half-thickness is t. The length of the uncracked portion of the lap is L. The engineering constants for the strap are denoted with a bar, A_r is the ratio of the axial stiffnesses of the lap, and the portion of the strap under the lap to the lap axial stiffness. Parameter α is unity for an isotropic material and can be as high as 50 for a laminated composite material.

For a specimen where the lap and strap layups are balanced and symmetric, Eq 15 takes the form

$$m = \sqrt{\frac{3}{2}\left(\frac{t}{h}\right)\frac{1}{\alpha}} \tag{18}$$

from a design point of view, Eq 15 or 18 can be a base for determining the geometry and selecting the material system needed for a desirable response. A detailed derivation of the stress expressions in Eqs 12 through 17 is provided in Ref *11*.

Bending effects at the sublaminate level are considered in the sublaminate bending (SB) model. For this model, the equations used are those given in the Appendix. The boundary conditions appropriate to this model are limited to extensional and shear behavior. The N or $\tilde{u}$ and Q or $\tilde{w}$ must be prescribed at the sublaminate ends. Bending moment and section rotation conditions cannot be independently prescribed at this level of approximation. The interlaminar continuity conditions (Eq 37) and the traction boundary conditions at the extreme upper and lower surfaces of the laminate remain unchanged. Interfacial stress expressions similar to Eqs 12 through 17 based on the SB model are provided in Ref *11*.

Delamination crack growth is usually characterized by the strain-energy release rate. For linear, elastic material behavior [*13*], and the proposed specimen geometries

$$G = P^2(dC/da)/2b \tag{18}$$

where P is the applied load, C is the specimen compliance, a is the crack length, and b is the specimen width. The critical strain-energy release rate, G_c, is obtained by substituting the critical load, P_c, for P in Eq 18. In this context, the load required to cause a crack extension is defined as critical load. An expression for the compliance as a function of crack length can be determined by using Eqs 1 to 5, 9, and 10, along with the appropriate boundary and loading conditions.

A simplification can be made in calculating the energy release rate by neglecting the stresses in the cracked portion of the lap with crack growth. For this situation, the result is

$$G = P^2(1 - m^2F \operatorname{sech}^2 m\, L/t)/8A_r t\, \overline{E_{11}} \tag{19}$$

where

$$\begin{aligned} F = {} & \frac{h}{6t}(2\overline{E_{11}}/G_{13} + 2\nu_{13}\,\overline{E_{11}}/E_{11} - 3\overline{\nu_{13}}) \\ & + \frac{1}{3}(\overline{E_{11}}/\overline{G_{13}} - \overline{\nu_{13}}) - (A_r - 1)/m^2 \end{aligned} \tag{20}$$

A similar expression for the energy release rate based on the SB model is given in Ref *11*.

The three components of the strain-energy release rate, G_{I}, G_{II}, and G_{III}, associated with the opening mode, the forward shearing mode, and the tearing mode can be calculated by using the stress field predicted from the proposed models and the crack-closure method [*14*]. Since the present specimen is assumed to be under plane strain condition, Mode III effects are not present.

In the crack-closure method, the elastic strain energy released during an incremental crack extension, δ, can be equated to the work done in closing the incremental crack. This work is computed based on the near stress and displacement fields. Since the shear stresses are obtained more reliably than peel stresses regardless of the level of modeling used, G_{II} will be determined using the crack-closure method and G_I deduced as

$$G_I = G - G_{II} \tag{21}$$

For an applied tensile load, P, the expression for G_{II} based on the SM model is

$$G_{II} = \frac{P^2}{8A_r t\, E_{11}} \left\{ 1 - \frac{t}{\delta m} \left[\tanh m \frac{L}{t} - \frac{\sinh m\,(L - \delta)/t}{\cosh m\, L/t} \right] \right\} \tag{22}$$

Energy release rates are strongly influenced by conditions near the crack tip. For this reason, the crack step size, δ, is chosen as the shortest decay length of the stress solution that is sufficient to capture the essential behavioral features. In the finite-element representation of the crack-closure method [*15*], recommended *ad hoc* values for the crack step size, δ, as a percentage of the initial crack length have been proposed. Since the energy release rates are controlled by the near field stress distribution, a crack extension size that approaches the near-field stress boundary layer decay length rather than a percentage of the original crack length is more appropriate.

An appropriate estimate of the decay length, or the dimension of the boundary zone, based on the SM model can be obtained by equating $e^{-m\delta}$ to e^{-3}, which is approximately 0.05 in value.

Under tensile loading, the crack in the double lap specimen opens and the interfacial peel stress at the crack tip is tensile. This opening (Mode I) action is coupled with a sliding (Mode II) one and the crack extends in a mixed-mode manner. Under compressive loading, the situation is different. The crack closes as the interfacial peel stress at the crack tip reverses its sign and becomes compressive. This situation suppresses Mode I opening action and the crack extends in pure Mode II behavior. (Since the crack extends in a plane strain situation, Mode III effects are not present.) Therefore, in the double cracked-lap-shear specimen, Mode II contribution to fracture can be isolated by simply reversing the applied loading.

Specimen Design

Specimen design will be based on the predictions of the SB model. In order to determine the overall dimensions of the specimen, a parametric study has been conducted in which the material properties of the specimen are those of AS4/3502 graphite/epoxy. Woven fiber glass end tabs are utilized. The number of plies in the lap and strap regions is varied. Three interfaces at the lap/strap junction are considered, namely, (+45/−45), (0/90), and (0/45). The level of model used does not permit a distinction for different stacking sequences or interface plies. The optimum number of plies and specimen dimensions were determined based on the following:

1. preventing overall buckling,
2. initiating a crack at lap/strap juncture first,
3. providing an adequate length to monitor crack growth,
4. ensuring no interaction between stress boundary layers at the tab/lap, tab/strap, and the lap/strap regions, and
5. minimizing ply nesting effects.

Ply nesting effects are minimized when differently oriented interface plies at the lap/strap junction are used. A (+45/−45) interface is selected, among other possibilities, as it provides balanced contributions along the principal material directions. A crack will start first at the lap/strap juncture if the dimensions of the lap, strap, and tabs are determined so as to produce the higher stress intensity at the lap/strap juncture. This is illustrated in Fig. 3 by comparing the absolute values of the interlaminar shear stress distribution at the lap/strap interface with the tab/lap and tab/strap distributions. The interface length is L. Moreover, interaction of stress boundary layers at the tab/lap and the lap/strap regions is eliminated if the distance between these regions is adequate for the stress boundary layers to decay to an allowable value. This situation is illustrated in Fig. 4. The axial stress distribution through the depth of the specimen strap is plotted. The strap length is d, and the strap half-thickness is t. The depth coordinate, z, is

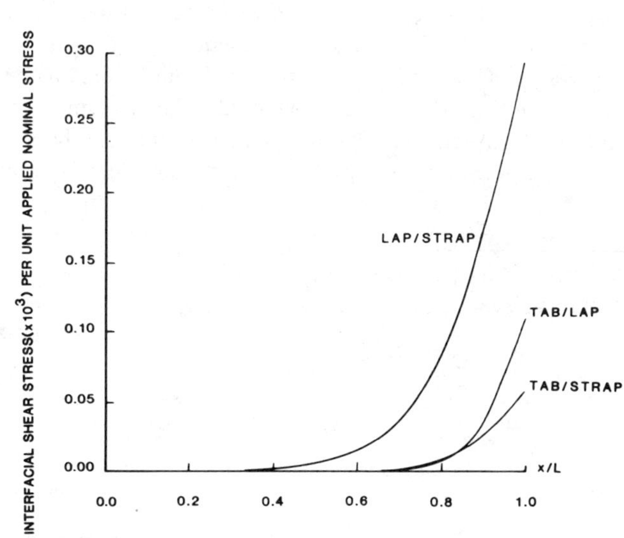

FIG. 3—*Shear stress distribution at interfaces with fiber glass tabs.*

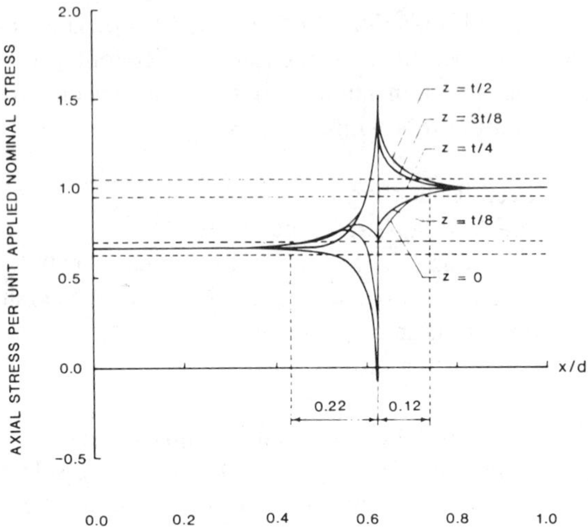

FIG. 4—*Through-depth axial stress distribution and decay length in the step/body region.*

measured from the strap middle-plane. For the stress to decay to within 5% of the uniform stress value in the lap/strap region and the gage section, the decay length is 0.22*d* and 0.12*d* from the lap end.

The final design is shown in Fig. 5. The number of plies in the lap and strap region is 8 and 32, respectively. The lap/strap body interface is [+45/−45]. The lap layup is $[\pm 45, 0, 90]_s$, and the strap layup is $[\pm 45, 0, 90]_{4s}$. A single panel was fabricated by the Lockheed-Georgia Company and inspected for quality using standard aerospace industry practice before sectioning into 13 test specimens. During layup, a folded Kapton film was placed at the end of the lap/strap interface to initiate a delamination under subsequent loading.

In order to verify the accuracy of the present formulations, a comparison with a finite-element solution is presented. The double-lap specimen shown in Fig. 1 is subjected to uniform tension. Due to symmetry, only one half of the problem is analyzed. The finite-element simulation of the problem is shown in Fig. 6.

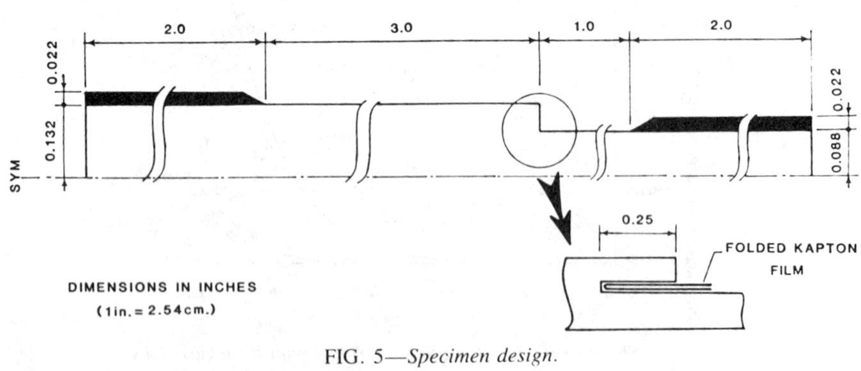

FIG. 5—*Specimen design.*

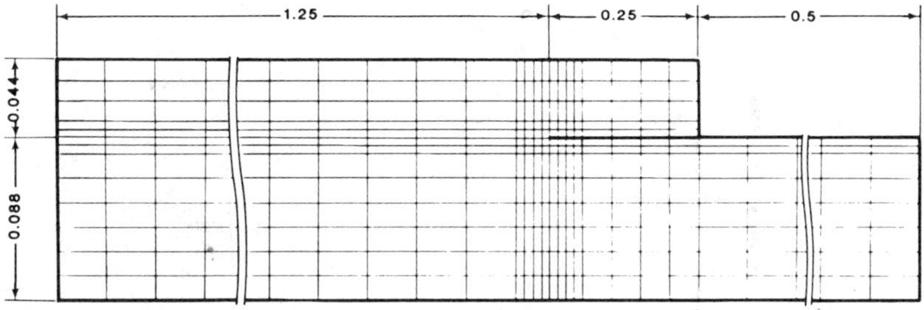

FIG. 6—*Finite-element discretization of the double cracked-lap-shear specimen.*

The element used is a constant strain four-node quadrilateral element. The total number of elements used is 784, and the total number of degrees of freedom is 1752. The finite-element code used is the Engineering Analysis Language (EAL). A comparison of the stress distribution at the interface between the lap and the strap is shown in Figs. 7*a* and *b* for a specimen made of T300/5208 graphite/epoxy with [+45/−45] interface at the lap/strap junction. The number of plies in the lap region is 8 and in the strap is 32. The shear stress distribution predicted by the present model shows good agreement with thc finite-element solution. The agreement in the peel stress distributions is less satisfactory since they are derivatives of the shear stresses. The interfacial peel stress distribution of the SM model given in Eq 13 is not in equilibrium due to the violation of the transverse shear boundary condition at the lap end. This anomaly can be corrected by superimposing a boundary function that will restore transverse force equilibrium. This approach is presented in detail in Ref *11*.

The available strain-energy release rate for the designed specimen is given in Eqs 19 and 20 based on the SM model, and a similar expression can be derived using the SB model [*11*]. It is found that the specimen compliance consists of two parts. The first is a linear function of the crack length. This contribution is similar to the strength of materials extensional model reported in Ref *8*. The second, an exponential function of the crack length, can be regarded as a refinement added to the first term. For the specimen geometry and material used, the refined term is found to be negligible for practical range of crack lengths. This situation is illustrated in Fig. 8. The energy release rate is plotted against crack growth from 0 to 10 times the strap half-thickness, t. For that range of crack growth, the SM and the SB predictions are within 1% of the strength of materials prediction.

A comparison of the strain-energy release rate components appears in Table 1. In the finite-element crack-closure simulation, the forces required to close the virtual crack extension, δ, are computed using two approaches. The first, by using very stiff (or zero length) bar elements. The second, by applying a unit load and computing an effective local compliance from which the forces that are required to hold the extended crack closed are determined. Two crack extension sizes are selected. For the zero length element method, the first equals to 1.57%

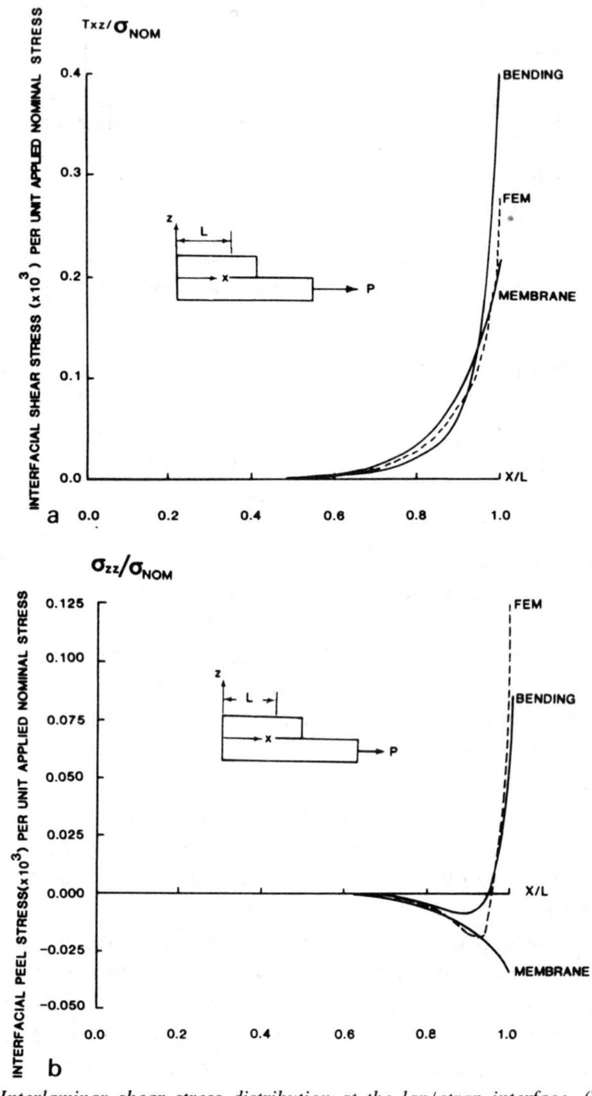

FIG. 7—(a) *Interlaminar shear stress distribution at the lap/strap interface.* (b) *Interlaminar transverse normal stress distribution at the lap/strap interface.*

of the original crack length, a_0, while the second is twice that value. The finite-element unit load method is based on a crack extension length equal to 1.57% of the original crack length. In the SM and SB models, the crack step size is selected to be the shortest stress boundary decay length. The values appearing in Table 1 are based on a nominal applied load of 44 482 N (10 000 lb).

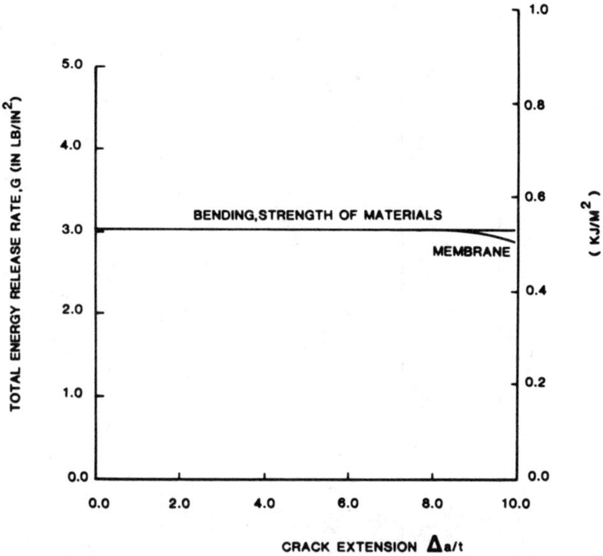

FIG. 8—*Strain-energy release rate comparison.*

Experimental Procedure and Preliminary Results

The static tension tests on the double cracked-lap-shear specimens were performed on a displacement controlled Baldwin screw-type testing machine. The specimens were carefully positioned and the tab ends were tightly held between serrated grips. Two techniques were tried to measure the crack growth in the specimen. The first method involved visual observation of the cracks on the white painted (typewriter correction fluid) specimen edge [*8*]. In the second method, a sheet of photoelastic material was bonded on each of the surfaces of the specimen lap portion [*16*]. Isochromatic fringes develop at the crack front as a result of the high strain gradient in that vicinity when the specimen is loaded. These fringes were tracked through the analyzer of a reflection polariscope to locate the crack front. By comparison, the edge cracks observed from the first method did not correspond to the averaged crack length of the curved crack front observed by the second method. This was because the crack fronts were irregular regardless of the care exercised in initially aligning the specimen. Hence, the first method was considered unreliable and was abandoned. The second method was used on all the specimens due to its simple and direct nature in monitoring the crack front. Also, the isochromatic fringes served to check initial uniformity of the applied load on the specimen. The specimen was also instrumented with a custom-built, linear variable differential transformer (LVDT) displacement transducer-based extensometer to measure its compliance.

The specimen was loaded continuously and the front and back cracks were

TABLE 1—*Comparison of strain-energy release rate for a nominal applied tensile load of 44 482 N (10 000 lb).*

Method	G_I		G_{II}		G		G_I/G	Crack Step Size
	J/m^2	(in.·lb/in.^{2})	J/m^2	(in.·lb/in.^{2})	J/m^2	(in.·lb/in.^{2})		
			FEM[a]					
								(δ/a_0)
Zero length element	196.8	(1.124)	334.1	(1.908)	530.9	(3.032)	0.37	0.0157
	142.6	(0.815)	306.1	(1.748)	448.7	(2.563)	0.32	0.0314
Unit load	193.0	(1.102)	325.5	(1.859)	518.5	(2.961)	0.37	0.0157
			PROPOSED MODELS[b]					
								(δ/L)[c]
SB	184.1	(1.052)	350.0	(1.999)	534.1	(3.050)	0.34	0.09
SM	169.4	(0.968)	364.7	(2.083)	534.1	(3.050)	0.32	0.33

[a]Based on a numerical simulation with 1752 degrees of freedom.
[b]Based on closed form expressions.
[c]Based on the shortest stress decay length.

followed through separate polariscopes. At a certain critical load, the first threshold of the crack growth occurred at the lap/strap juncture and it grew beyond the Kapton film. With crack growth, the specimen stiffness reduces and hence the load drops in a controlled displacement machine. Loading was stopped whenever crack growth was observed and the extent of propagation was recorded together with the load. The crack growth was initially stable but intermittent on the two fronts with one crack trying to maintain parity with the other. During this phase, crack growth was possible only under increasing load. The crack front was not planar and it curved forward at the specimen edges. Some Poisson effects are inevitable due to the difference in constraint in the unloaded lap and the loaded strap and the free edges composed of dissimilar plies. This explains the delamination front curving forward at the specimen edges. Near the edge, G_I, G_{II}, and G_{III} are all present. The specimen failure occurred finally with both the cracks growing in an unstable manner. The load value corresponding to this event was recorded. In general, there was no crack wandering through the plies. Pictures of the isochromatic fringes before the initial crack propagation and during the stable crack growth phase are presented in Figs. 9 and 10. The crack growth data, taken at five locations along the width of the specimen and from both the front and back faces, was averaged to obtain a resultant crack length corresponding to each load. To eliminate the edge effects, crack length values at the edges

FIG. 9—*Isochromatic fringes before initial crack propagation.*

FIG. 10—*Isochromatic fringes during stable crack propagation.*

were discarded in the averaging process. Altogether, three specimens were tested and the data is presented in Fig. 11. Numbers 1, 2, and 3 in Fig. 11 correspond to data points from Specimens 1, 2, and 3, respectively. The data were plotted together as the specimens were generated from the same parent panel.

The failed specimens were sectioned and the fracture surfaces were observed under a scanning electron microscope. Typical photomicrographs of a fracture surface immediately after the Kapton film in the stable region and in the unstable region of crack growth are presented in Figs. 12*a* and *b*, respectively.

Discussion

The load versus crack growth data appears in Fig. 11. The data shows that there is a consistent increase in the critical load with crack length. A linear least squares fit through this data is also plotted in Fig. 11. Although the fit corresponding to the least standard deviation is parabolic, its improvement over the linear fit is only 2.5%. The simplicity and advantages of the linear fit is preferred over the parabolic fit. The multiple correlation coefficient, r^2, corresponding to the linear fit is 0.926, and the standard error percent relative to the mean is 3.63.

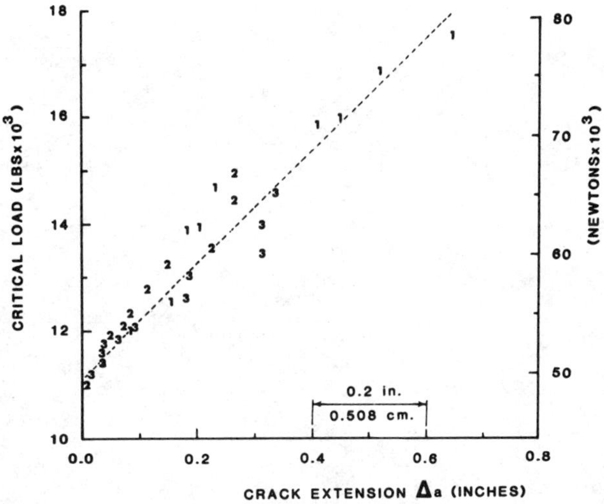

FIG. 11—*Double cracked-lap-shear specimen test data (graphite/epoxy AS4/3502 material) linear fit.*

The linear fit equation is

$$\begin{aligned} P_c &= 50\,590 + 18\,359\ \Delta a \text{ (N, } \Delta a \text{ in centimetres)} \\ P_c &= 11\,373 + 10\,483\ \Delta a \text{ (lb, } \Delta a \text{ in inches)} \end{aligned} \quad (23)$$

where P_c is the critical load corresponding to a crack extension, Δa.

As outlined in the previous section and illustrated in Fig. 8, the dC/da values can be considered independent of crack length and will be computed using the strength of materials approach (Ref *8*). This also checks with the dC/da values computed from the experimental specimen compliance data. From this and Eq 18, it follows that the critical strain-energy release rate, G_c, is proportional to the square of the critical load only. Figure 11 can also be regarded as a plot of scaled $\sqrt{G_c}$ versus crack length that is similar to a crack growth resistance curve.

The substitution of Eq 23 into Eq 18 gives the following fit between $\sqrt{G_c}$ and crack extension, Δa

$$\begin{aligned} \sqrt{G_c} &= 19.19 + 692.94\ \Delta a \text{ (}\sqrt{\text{J/m}}\text{, } \Delta a \text{ in metres)} \\ \sqrt{G_c} &= 1.45 + 1.33\ \Delta a \quad \text{(}\sqrt{\text{in. lb/in.}}\text{, } \Delta a \text{ in inches)} \end{aligned} \quad (24)$$

There are three phases in the resistance curve. A starting region corresponding to the onset of crack growth, an intermediate region corresponding to stable crack growth, and a final or failure region of unstable crack growth. The fit shown in Figure 11 and represented in Eqs 23 and 24 describes the stable crack growth region.

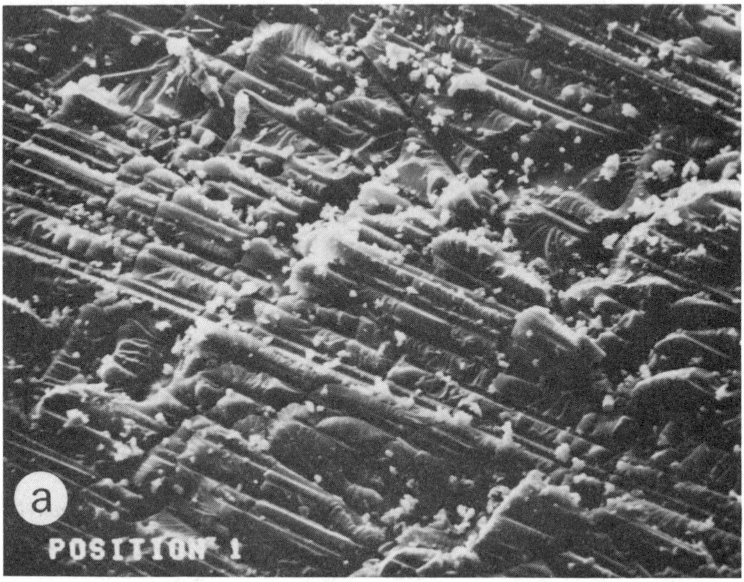

FIG. 12—(a) *Photomicrograph of fracture surface (original magnification ×300) stable crack growth region, and* (b) *photomicrograph of fracture surface (original magnification ×300) unstable crack growth region.*

The results on the three specimens are summarized in Table 2. The critical loads that initiated the crack growth beyond the Kapton film and the corresponding energy release rates are listed. The total critical strain-energy release rate is G_c. The small amount of scatter in the observed critical load data may be attributable to the design and testing approach chosen. The specimen geometry and symmetry virtually eliminates geometrically nonlinear effects and the ±45° interface minimized nesting and the resulting crack wandering. Observation of the crack front by means of the photoelastic coating provided complete characterization across the width of the specimen and helped accurately determine the initial and subsequent crack growths with loading. This situation is different compared to the problems encountered in single cracked-lap-shear specimen testing.

Preliminary tests on 48-ply (±45, 0, 90) quasi-isotropic layup (8 plies in the lap) single cracked-lap-shear specimens made of AS4/3502 graphite/epoxy material exhibited the following problems:

1. the first crack did not always initiate at the lap/strap juncture;
2. multiple, isolated cracks occurred and sometimes grew in opposite directions; and
3. the crack at lap/strap juncture wandered.

Figure 13 presents a typical failed single cracked-lap-shear specimen depicting the preceding problems. These factors might contribute to the data scatter appreciably. Test results performed on three single cracked-lap-shear specimens also indicated that the fracture behavior follows a resistance curve. In Fig. 14, the load versus crack extension data from two single cracked-lap-shear specimens are plotted. Numbers 1 and 2 in this figure correspond to data points from Specimens 1 and 2 of Table 3. The crack extension data in Fig. 14 are based on visual observation of the cracks on the white painted specimen edge. This crack tracking method (based on edge crack information only) may contribute to the data scatter. In the third tension test, the photoelastic coating method was tried for the first time. The crack front was tracked through the analyzer of a reflection polariscope, and the loads corresponding to initial crack growth and final unstable growth only were recorded. In the subsequent tests, graduated tapes were glued

TABLE 2—*Summary of experimental data on double cracked-lap-shear specimens.*

	P_c		$(dC/da) \times 10^8$		G_c	
Specimen	N	(lb)	m/J	(in./lb·in.)	J/m²	(in.·lb/in.²)
1	53 379	(12 000)	1.411	(6.277)	396.7	(2.265)
2	48 753	(10 960)	1.470	(6.540)	343.8	(1.963)
3	49 731	(11 180)	1.481	(6.590)	360.6	(2.059)

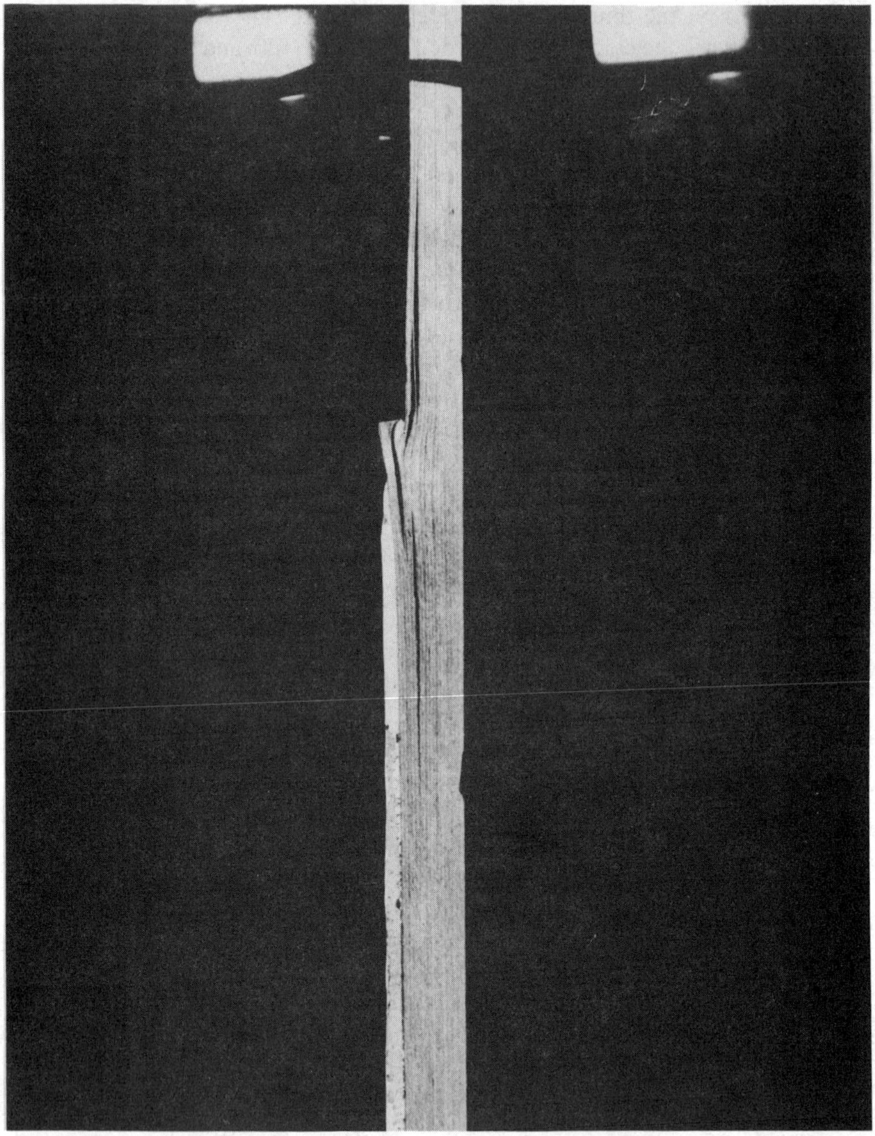

FIG. 13—*Failed single cracked-lap-shear specimen.*

to the photoelastic coating in order to help in measuring crack front location with loading during the stable phase of crack growth.

The data in Fig. 14 show that there is a consistent increase in the critical load with crack length. Among polynomial curve fits through this data, the linear fit corresponds to the least standard deviation. The data provides a multiple correlation coefficient, r^2, equals to 0.366 for the linear fit, and the associated

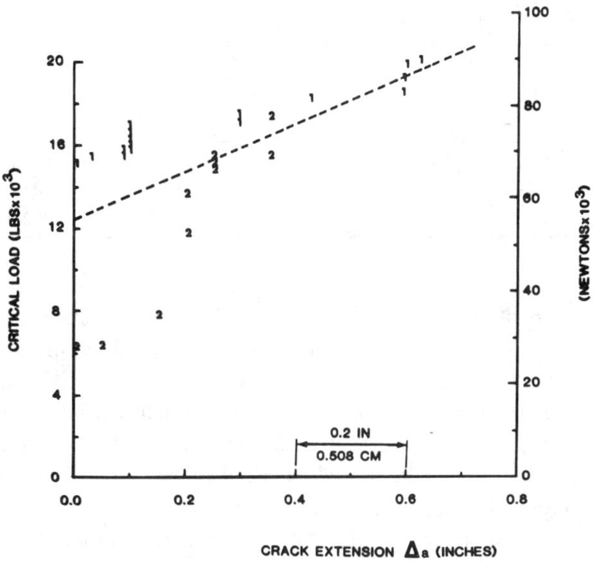

FIG. 14—*Single cracked-lap-shear specimen test data (graphite/epoxy AS4/3502 material) linear fit.*

standard error percent relative to the mean is 19.22. The linear fit equation is

$$\begin{aligned} P_c &= 56\,288 + 19\,865\ \Delta a\ (\text{N}, \Delta a \text{ in centimetres}) \\ P_c &= 12\,654 + 11\,343\ \Delta a\ (\text{lb}, \Delta a \text{ in inches}) \end{aligned} \quad (25)$$

The corresponding fit between $\sqrt{G_c}$ and crack extension Δa gives

$$\begin{aligned} \sqrt{G_c} &= 25.94 + 916.97\ \Delta a\ (\sqrt{\text{J}}/\text{m}, \Delta a \text{ in metres}) \\ \sqrt{G_c} &= 1.96 + 1.76\ \Delta a\ (\sqrt{\text{in. lb}}/\text{in.}, \Delta a \text{ in inches}) \end{aligned} \quad (26)$$

The linear fit shown in Fig. 14 and represented in Eqs 25 and 26 describes the stable crack growth region.

The results on the three single cracked-lap-shear specimens tested in tension are summarized in Table 3. The critical loads that initiated the crack growth beyond the Kapton film and the corresponding energy release rates are listed, along with the rate of change of compliance with crack extension. The total critical strain-energy release rate is G_c. The standard deviation in the critical load values in Table 3 is 16 601 N (3732 lb) for a mean value of 50 652 N (11 387 lb). The standard deviation in the critical load values for the double lap specimens in Table 2 is 1993 N (448 lb) for a mean value of 50 621 N (11 380 lb). The amount of scatter in the observed critical load data may be attributable to the single cracked-lap-shear specimen and the crack tracking method. Similar scatter were found in single lap specimen testing by other investigators [*7,8*].

TABLE 3—*Summary of experimental data on single cracked-lap-shear specimens.*

	P_c		$(dC/da) \times 10^8$		G_c	
Specimen	N	(lb)	m/J	(in./lb·in.)	J/m^2	(in.·lb/in.^{2})
1	67 613	(15 200)	1.240	(5.515)	558.0	(3.186)
2	28 113	(6 320)	1.125	(5.003)	87.6	(0.500)
3	56 226	(12 640)	1.230	(5.472)	383.4	(2.189)

A comparison of the general features of the photomicrographs in Figs. 12*a* and *b* indicates that crack growth in the unstable region occurs at a much higher speed than the corresponding growth in the stable region. The fracture surface in the stable growth region appears to have predominantly high matrix deformation with lacerated resin bands and matrix cracks. These are believed to correspond to the stop-start regions of the load dependent crack fronts observed. The surface reflects "hackles" that are characteristic of matrix cleavage failure. The unstable crack growth region shown in Fig. 12*b* is characterized by a relatively smooth surface with a small amount of fiber pull out and breakage. Striations are observed throughout the fiber pull-out region. This is characteristic of a shear failure.

Conclusions

A double cracked-lap-shear specimen has been designed using simple analytical models. These models provided good estimates for stresses compared to the finite-element model used. This has been tested in static tension to determine the mixed-mode interlaminar fracture behavior. The preliminary data generated are consistent and provide similar behavioral trends on the specimens tested.

The specimen behaved as designed and the crack started at the lap/strap interface and grew in an intermittent but stable fashion with increasing applied load. This stable crack growth phase was followed by an unstable one that resulted in the disbonding of the laps from the strap to the vicinity of the grips. All these events were followed by observing the isochromatic fringes in photoelastic coatings bonded to the two-lap portions. There was no crack wandering at the lap/strap juncture. The scatter in the critical load data at initial cracking was small. This demonstrates the advantages of using a double cracked-lap-shear specimen.

The preliminary results provided delamination growth resistance curves for the AS4/3502 graphite/epoxy material system with a ±45° lap/strap interface.

Acknowledgments

This work was supported by the U. S. Air Force Office of Scientific Research under Grant AFOSR-83-0056. This support is gratefully acknowledged. Specimens used in this test program were manufactured and supplied by the Lockheed-Georgia Company, Marietta, GA. We acknowledge Dr. Kris Kathiresan for his cooperation. The finite-element code, EAL, was provided by Engineering Information Systems, Inc., Saratoga, CA; this assistance is also gratefully acknowledged.

APPENDIX

Sublaminate Bending Model

Consider a laminate made of N perfectly bonded layers or plies, each ply having a plane of material symmetry parallel to the plane of the laminate. Let k denote a particular ply that is singled out for study. The ply thickness is h^k. Let x^k and y^k represent the coordinates in the mid-plane of the ply. The interlaminar stresses, σ_{zz} and σ_{xz}, at the top of the ply are denoted by p_2 and t_2, respectively, while the corresponding stresses at the bottom of the ply are designated as p_1 and t_1. Notation and sign convention appear in Fig. 2.

The basic equations for the sublaminate bending model encompass the following main categories.

Overall Equilibrium

$$N^k_{,x} + T^k_2 - T^k_1 = 0$$

$$Q^k_{,x} + p^k_2 - p^k_1 = 0$$

$$M^k_{,x} - Q^k + \frac{h^k}{2}(T^k_2 + T^k_1) = 0 \qquad (27)$$

The force, moment, and transverse shear stress resultants of the kth ply are N^k, M^k, and Q^k, respectively. These bear the classical definitions.

$$(N^k, M^k, Q^k) = \int_{-h^k/2}^{h^k/2} (\sigma_{xx}, z\sigma_{xx}, \sigma_{xz})^k dz \qquad (28)$$

Superscript k, which identifies the ply, will be dropped in the subsequent equations for convenience.

Constitutive Relationships

$$N = h\,E_{11}\,\bar{u}_{,x} + \frac{h}{2}\nu_{13}(p_2 + p_1) - \frac{h^2}{12}\nu_{13}N_{,xx} \qquad (29)$$

$$M = -\frac{h^3}{12}E_{11}\,\bar{w}_{,xx} + \frac{h^2}{60}\left(\frac{E_{11}}{G_{13}} - \nu_{13}\right)(5Q_{,x} + M_{,xx}) - \frac{h^2}{60}\nu_{13}M_{,xx} \qquad (30)$$

$$\bar{w} = \int_{-h/2}^{h/2} \frac{w}{h}dz \qquad (31)$$

The elastic modulus associated with the x direction is E_{11}. The Poisson's ratio is ν_{13}, and the transverse shear modulus is G_{13}. The average axial displacement is $\bar{u}$, and the transverse displacement is w.

Stress Distribution

$$\sigma_{xx}(x,z) = \frac{N}{h} + \frac{12z}{h^3}M + \frac{1}{2}\left(\frac{E_{11}}{G_{13}} - 2\nu_{13}\right) \times \left[\frac{h}{12}\left(1 - \frac{12z^2}{h^2}\right)N_{,xx} + \frac{3z}{5h}\left(1 - \frac{20z^2}{3h^2}\right)M_{,xx}\right] \tag{32}$$

$$\sigma_{xz}(x,z) = -\frac{z}{h}N_{,x} + \frac{Q}{h} + \frac{1}{2h}\left(1 - \frac{12z^2}{h^2}\right)M_{,x} \tag{33}$$

$$\sigma_{zz}(x,z) = \frac{1}{2}(p_2 + p_1) - \frac{h}{8}\left(1 - \frac{4z^2}{h^2}\right)N_{,xx} - \frac{z}{h}Q_{,x} - \frac{z}{2h}\left(1 - \frac{4z^2}{h^2}\right)M_{,xx} \tag{34}$$

Displacement Distribution

$$w(x,z) = \tilde{w} - \frac{\nu_{13}}{E_{11}}\left[\frac{z}{h}N - \frac{M}{2h}\left(1 - \frac{12z^2}{h^2}\right)\right] \tag{35}$$

$$u(x,z) = \tilde{u} - z\tilde{w}_{,x} + \frac{z}{hG_{13}}Q + \frac{1}{2}\left(\frac{1}{G_{13}} - \frac{\nu_{13}}{E_{11}}\right) \times \left[\frac{h}{12}\left(1 - \frac{12z^2}{h^2}\right)N_{,x} + \frac{z}{h}\left(1 - \frac{4z^2}{h^2}\right)M_{,x}\right] \tag{36}$$

The preceding equations are supplemented by interlaminar conditions that ensure reciprocity of tractions and continuity of displacements at the interfaces between plies. At the interface between the kth and $(k - 1)$th plies, they are

$$\begin{aligned} T_1^k &= T_2^{k-1} \\ p_1^k &= p_2^{k-1} \\ u^k\left(x^k, -\frac{h^k}{2}\right) &= u^{k-1}\left(x^{k-1}, \frac{h^{k-1}}{2}\right) \\ w^k\left(x^k, -\frac{h^k}{2}\right) &= w^{k-1}\left(x^{k-1}, \frac{h^{k-1}}{2}\right) \end{aligned} \tag{37}$$

The kth ply axial component of the displacement is u^k, and the lateral component is w^k. In addition, traction boundary conditions are normally specified at the extreme upper and lower surfaces of the laminate. At the boundaries corresponding to constant values of x, the laminate sections, boundary conditions will be enforced in an overall sense as is commonly done in engineering theories of bending and stretching.

References

[1] Wilkins et al in *Damage in Composite Materials, ASTM STP 775*, K. Reifsnider, Ed., American Society for Testing and Materials, Philadelphia, 1982, pp. 168–183.

[2] Whitney, J. M., Browning, L. E., and Hougsteden, W., *Journal of Reinforced Plastics and Composites*, Oct. 1982, pp. 297–313.

[3] Nicholls, D. J. and Gallagher, J. P., *Journal of Reinforced Plastics and Composites*, Jan. 1983, pp. 2–17.
[4] Whitney, J. M. and Browning, C. G. in *Effects of Defects in Composite Materials, ASTM STP 836*, American Society for Testing and Materials, Philadelphia, 1984, pp. 104–124.
[5] Wang, S. S. in *Composite Materials: Testing and Design (Fifth Conference), ASTM STP 674*, S. Tsai, Ed., American Society for Testing and Materials, Philadelphia, 1980, pp. 642–663.
[6] O'Brien, T. K., in *Damage in Composite Materials, ASTM STP 775*, K. Reifsnider, Ed., American Society for Testing and Materials, Philadelphia, 1982, pp. 140–167.
[7] Russell, A. J. and Street, K. N., "Moisture and Temperature Effects on the Mixed-Mode Delamination Fracture of Unidirectional Graphite/Epoxy," ASTM Symposium on Delamination and Debonding of Materials, 8–10 Nov 1983, Pittsburgh, American Society for Testing and Materials.
[8] Wilkins, D. J., "A Comparison of the Delamination and Environmental Resistance of a Graphite-Epoxy and a Graphite-Bismaleimide," Technical Report, NAV-GD-0037, Sept. 1981.
[9] O'Brien, T. K. in *Effects of Defects in Composite Materials, ASTM STP 836*, American Society for Testing and Materials, Philadelphia, 1984, pp. 125–142.
[10] O'Brien, T. K., "Interlaminar Fracture of Composites," NASA Technical Memorandum 85768, USAAVSCOM TR-84-B-2, National Aeronautics and Space Administration, June 1984.
[11] Armanios, E. A., "New Methods of Sublaminate Analysis for Composite Structures and Applications to Fracture Processes," Ph.D. thesis, Georgia Institute of Technology, March 1985.
[12] Rehfield, L. W. and Murthy, P. L. N., *AIAA Journal*, American Institute of Aeronautics and Astronautics, Vol. 20, No. 5, May 1982, pp. 693–699.
[13] Paris, P. C. and Sih, G. C. in *Fracture Toughness Testing and Its Applications, ASTM STP 381*, American Society for Testing and Materials, Philadelphia, 1970.
[14] Eshelby, J. C., *Stress Analysis of Cracks*, ISI Publication 121, International Steel Institute, 1968.
[15] Rybicki, E. F. and Kanninen, M. F., *Engineering Fracture Mechanics*, Vol. 9, 1977, pp. 931–938.
[16] Roderick, G. L., Everett, R. A., and Crews, J. H., Jr., in *Fatigue of Composite Materials, ASTM STP 569*, American Society for Testing and Materials, Philadelphia, 1975, pp. 295–306.

Charles E. Browning[1] *and Herbert S. Schwartz*[1]

Delamination Resistant Composite Concepts

REFERENCE: Browning, C. E. and Schwartz, H. S., "**Delamination Resistant Composite Concepts,**" *Composite Materials: Testing and Design (Seventh Conference), ASTM STP 893*, J. M. Whitney, Ed., American Society for Testing and Materials, Philadelphia, 1986, pp. 256–265.

ABSTRACT: Delamination is the most prevalent life-limiting failure mode in advanced composite structures and is a fundamental issue in their evaluation for both durability and damage tolerance. Consequently, the attainment of improved delamination resistance and, in turn, more damage tolerant composite structures has been a major goal of a significant number of materials development activities. One approach to more delamination resistant composites is a laminate construction concept in which materials are utilized in a composite construction that is tailored for maximum resistance to interlaminar forces. The double cantilever beam (DCB) test was used as a convenient method for evaluating and ranking the Mode I delamination resistance of these construction concepts since the interlaminar crack growth region of the test coupon could easily be altered and characterized as to its influence on Mode I delamination behavior. Hercules' AS1/3502 graphite/epoxy was chosen as a baseline material and the construction of the interlaminar crack growth region in the mid-plane of the laminate was varied using inner-layer materials that have previously demonstrated the potential for resisting matrix crack growth—a rubber toughened epoxy adhesive, organic fiber mats, and glass fiber mat and scrim. Resultant laminates were characterized as to their Mode I behavior using the DCB test and were also subjected to compression testing. Fractured surfaces were examined with a scanning electron microscope (SEM). Results show that significant improvements in delamination resistance can be achieved by specifically tailoring the construction of the laminate.

KEY WORDS: composite materials, delamination resistance, double cantilever beam test, toughness, graphite/epoxy, scanning electron microscopy

Delamination is the failure mechanism that is the most prevalent life-limiting growth mode in advanced composite structures. It is a fundamental issue in the evaluation of laminated composite structures for both durability and damage tolerance. Many of the common design features in today's laminated composite structures can result in interlaminar forces that can induce the growth of delaminations from either pre-existent manufacturing defects or in-service damage. Consequently, the attainment of improved delamination resistance (or toughness) and, in turn, more damage tolerant composite structures has been a major goal of a significant number of materials development activities.

One approach to attaining more delamination resistant composites is a laminate

[1]Senior materials research engineers, Materials Laboratory, Air Force Wright Aeronautical Laboratories, Wright-Patterson Air Force Base, Dayton, OH 45433.

construction concept in which today's commercially available materials are utilized in a composite construction that is specified in such a manner so as to optimally tailor the properties of the laminate for maximum resistance to interlaminar forces. This is considered to be a lower risk approach than developing totally new materials.

The double cantilever beam (DCB) test has been found to be a convenient method for evaluating the Mode I delamination resistance of several construction concepts. During the development of this test method,[2] it became apparent that the nature of the interlaminar crack growth region could easily be altered and characterized as to its influence on Mode I delamination behavior (that is, it could be used as a test bed for screening a variety of materials approaches to delamination resistance). Accordingly, this test was used to assess the relative merits of a variety of construction approaches for improving delamination resistance. Hercules' AS1/3502 graphite epoxy was chosen as the baseline material and the construction of the interlaminar crack growth region in the mid-plane of the laminate was varied using inner-layer materials that have previously demonstrated the potential for resisting matrix crack growth. These inner-layer materials included a rubber-toughened epoxy adhesive, organic fiber mats, and glass fiber mat and fabric scrim. The materials were used individually as well as in a variety of combinations. In addition, postcured AS1/3502 and AS4/3502 laminates were utilized for reference comparisons.

All of the composite laminates were characterized as to their delamination resistance using the DCB Mode I test. They were also examined microscopically and were measured for effects on composite per ply thickness as a measure of fiber volume. In addition, selected specimens were tested in compression to determine if the laminate constructions produced any deleterious effects on matrix-sensitive in-plane properties. All of the test data were compared to AS1/3502 as the reference baseline.

The results show that significant improvements in the delamination resistance of today's materials can be achieved by specifically tailoring the construction of the laminate.

Experimental Procedures

All composite data were generated on test specimens machined from large, unidirectional composite panels.

AS1/3502 and AS4/3502 graphite/epoxy panels were vacuum bagged/autoclave cured using 153-mm (6 in.)-wide graphite/epoxy prepreg tape from Hercules. Standard lay-up and bagging techniques were used with a final autoclave cure temperature of 177°C (350°F) for 2 h. One AS1/3502 panel was postcured 6 h at 177°C (350°F) in an air circulating oven.

Materials used as inner-layers are shown in Table 1. These materials were

[2]Whitney, J. M., Browning, C. E., and Hoogsteden, W., *Journal of Reinforced Plastics and Composites*, Oct. 1982, pp. 297–313.

TABLE 1—*Materials used as delamination resistant in-lays for graphite/epoxy composites.*

Material	Description	Manufacturer
AF-163 adhesive	toughened epoxy unsupported film adhesive	3M
Glass mat	OCF surfacing mat TRT 236	Owens Corning Fiberglass
Glass scrim	Style 104 glass scrim cloth	Owens Corning Fiberglass
Kevlar mat	100% Kevlar 29 random short fibers	International Paper Company
Polyester mat	100% Polyester random short fibers	International Paper Company

placed in the mid-plane of AS1/3502 graphite/epoxy laminates during lay-up. The Kevlar and polyester (PE) mats were incorporated into the graphite/epoxy lay-up in two different forms: (1) dry, as received; or (2) preimpregnated with the epoxy resin. With the exception of the toughened AF-163 adhesive, all laminates containing the in-lays were cured identically to the AS1/3502 baseline. The AS1/3502/AF-163 composite was cured similarly to the other laminates with the exception that a 2-h hold at 121°C (250°F) was used to accommodate the adhesive's cure requirements.

Test specimens were machined from the panels using a diamond-impregnated cut-off wheel. Photomicrographs and per ply thickness measurements were taken on several specimens from each panel.

The DCB test method and data reduction method used in this investigation are thoroughly described in Ref *1* and will not be discussed in detail here. The test geometry is shown in Fig. 1. All of the test specimens contained 24 plies of unidirectional graphite/epoxy tape. As shown in Fig. 1, each coupon contains a starter crack generated by the inclusion of a Teflon film in the mid-plane of the

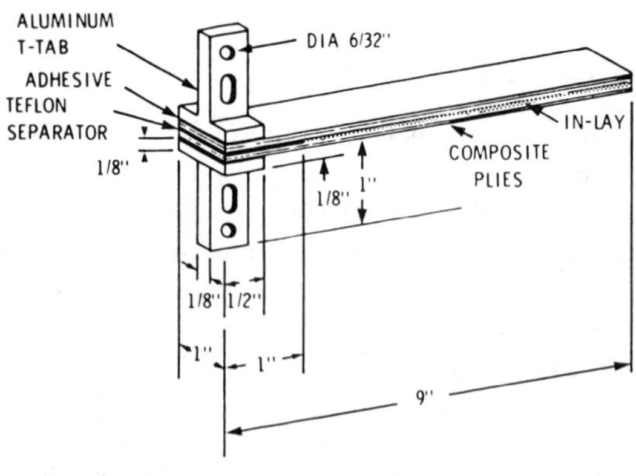

FIG. 1—*Double cantilever beam (DCB) specimen geometry.*

composite during fabrication. The material in-lays were placed in this mid-plane, adjacent to the Teflon starter strip during lay-up.

Compression tests performed in this study were conducted on end-tabbed, side-supported coupons.

Delamination Resistant Concepts

Since the DCB test measures the energy required to move a delamination, any method of increasing this required energy (without adversely affecting other properties) would result in increased delamination resistance. Accordingly, three approaches capable of increasing the graphite/epoxy strain-energy release rate were investigated—(1) the use of a fibrous mat or cloth in the crack growth plane, (2) the use of a tough adhesive in the crack growth plane, and (3) a combination of these two.

The fibrous mats and cloth potentially could increase the energy required to move the delamination as a result of their fibrillar, multidirectional nature. Energy could be expended in creating new surfaces around these fibers, in multiple directions, instead of going solely into propagating a crack through a pure matrix resin region. The use of a tough adhesive in the crack growth plane had the potential for increasing the energy release rate in a manner analogous to that by which it increases the peel strength of an adhesive bonded joint. The combination of both approaches had the potential for a synergistic effect, increasing the strain-energy release rate to a value greater than either one alone.

The graphite/epoxy and graphite/epoxy/in-lay materials evaluated in this study are shown in Table 2. As discussed, in addition to a monolayer in-lay of one component, multi-layer in-lays of multiple components were investigated to determine if any synergism could be generated.

Figure 2 shows cross-sectional photomicrographs of the control and selected other in-lay containing specimens. As can be seen, void-free composites were obtained. The photomicrographs serve to illustrate the relative thickness of the in-lays as compared to the nominal cured per ply thickness for graphite/epoxy of 0.0143 cm (0.0057 in.).

Discussion of Results

Test data (critical strain-energy release rate, G_{Ic}; compression strength; and cured per ply thickness) for all materials systems are summarized in Table 3. The G_{Ic} values are averages of five data points per coupon for at least three DCB coupons, with less than 10% scatter in the data. Specific failure modes are discussed later for each concept; however, in all cases, crack growth appeared to occur in the plane of the inner-layer and was considered to be colinear. Compression values shown in the table are averages of two specimens. Per ply thickness values are averages of at least five measurements per specimen.

The baseline control, AS1/3502, had a G_{Ic} of 148.7 J/m^2 (0.85 in.·lb/in.^{2}). A scanning electron micrograph (SEM) of the control's failure surface generated

TABLE 2—*Graphite/epoxy and graphite/epoxy/in-lay combinations evaluated.*

In-Lay Approach	Material Designation	Description
Baseline graphite/epoxy systems	AS1/3502	control
	AS1/3502 (PC)	postcured control
	AS4/3502	reference graphite/epoxy
Adhesive in-lay	AS1/3502/163	AF-163 adhesive in-lay (one layer)
Mats and scrim in-lays	AS1/3502/Kev (P)	preimpregnated Kevlar mat in-lay (one-layer)
	AS1/3502/Kev	dry Kevlar mat in-lay (one layer)
	AS1/3502/G1 (m)	glass mat in-lay (one layer)
	AS1/3502/G1 (s)	in-lay of glass scrim cloth (one layer)
	AS1/3502/PE (P)	preimpregnated polyester mat in-lay (one layer)
Combinations of adhesive and mats	AS1/3502/Kev-163	in-lay of one layer of Kevlar mat and one layer of AF-163
	AS1/3502/163-Kev-163	in-lay of one layer of AF-163, one layer of Kevlar mat, and one layer of AF-163

during the propagation of the delamination is shown in Fig. 3. The SEMs were taken of both the top and bottom surfaces generated in the DCB coupon for all specimens tested. No major differences in the two surfaces were observed for any of the materials.

Postcuring of the control material [AS1/3502/(PC) in Table 3] resulted in a reduction of the G_{Ic} value to 112 J/m^2 (0.64 in.·lb/in.^{2}). This is anticipated behavior since an extended postcure produces a more brittle matrix resin.

The effect of an alternative fiber was investigated by characterizing AS4/3502 composites. This graphite/epoxy system provides a higher strain graphite fiber for comparison while keeping the matrix constant. As can be seen in Table 3, a slight increase in the G_{Ic} was found.

The incorporation of in-lays in the delamination/crack growth plane of the DCB coupon in all cases resulted in an increased strain-energy release rate compared to the baseline control. Increases ranged from a modest 245 J/m^2 (1.4 in.·lb/in.^{2}) for the glass scrim and mat in-lays to a significant 1855 J/m^2 (10.6 in.·lb/in^2) for the 163-Kevlar-163 combination in-lay.

The effect of a rubber-toughened adhesive in-lay (AS1/3502/163 in Table 3) was dramatic. A G_{Ic} value of 1268.8 J/m^2 (7.25 in.·lb/in.^{2}) represented nearly an order of magnitude increase in delamination resistance versus the control. An SEM of the failure surface (Fig. 3) is very revealing in that it shows a substantial amount of adhesive residue on the failure surface. This would indicate that crack

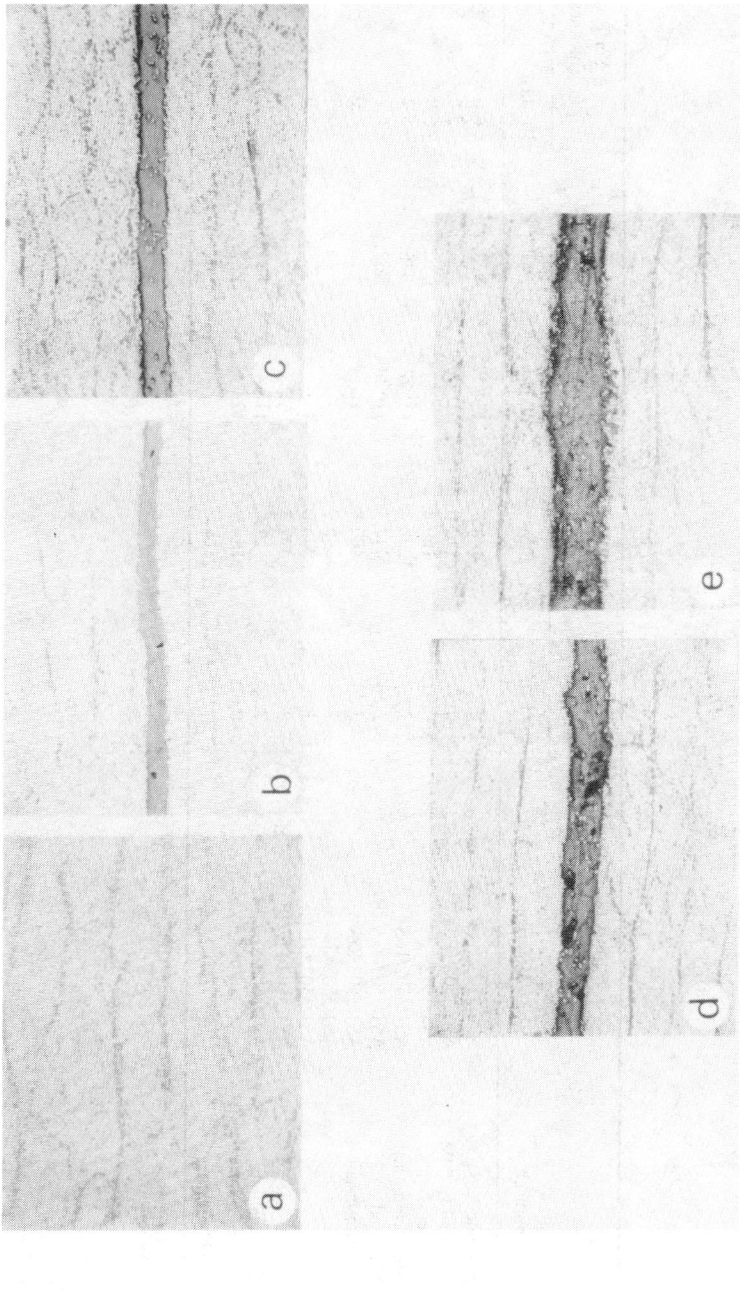

FIG. 2—*Cross-sectional photomicrographs (all* $\times 75$*) of (a) AS1/3502, (b) AS1/3502/163, (c) AS1/3502/Kev, (d) AS1/3502/Kev-163, and (e) AS1/3502/163-Kev-163 composites.*

TABLE 3—*Test data for graphite/epoxy and graphite/epoxy/in-lay materials.*

In-Lay Approach	Material	G_{Ic}		Compression Strength		Per Ply Thickness	
		in. · lb/in.^{2}	J/M^2	Ksi	MPa	10^{-3}in.	10^{-2}cm
Baseline	AS1/3502	0.85	148.7	134.1	928.6	6.0	1.52
graphite/epoxy	AS1/3502 (PC)	0.64	112.0	...	...	5.6	1.42
systems	AS4/3502	0.97	169.8	...	...	5.9	1.50
Adhesive in-lay	AS1/3502/163	7.25	1268.8	135.5	936.2	5.7	1.45
Mats and	AS1/3502/Kev (P)	2.54	444.5	110.5	763.5	6.2	1.57
scrim in-lays	AS1/3502/Kev	2.29	400.8	...	...	6.0	1.52
	AS1/3502/G1 (m)	1.40	245.0	...	...	5.9	1.50
	AS1/3502/G1 (s)	1.40	245.0	...	...	5.9	1.50
	AS1/3502/PE (P)	2.50	437.5	...	...	6.0	1.52
Combinations	AS1/3502/Kev-163	7.40	1295	116.4	804.2	6.0	1.52
of adhesive	AS1/3502/163-Kev-163	10.60	1855	128.4	887.1	6.1	1.55
and mats							

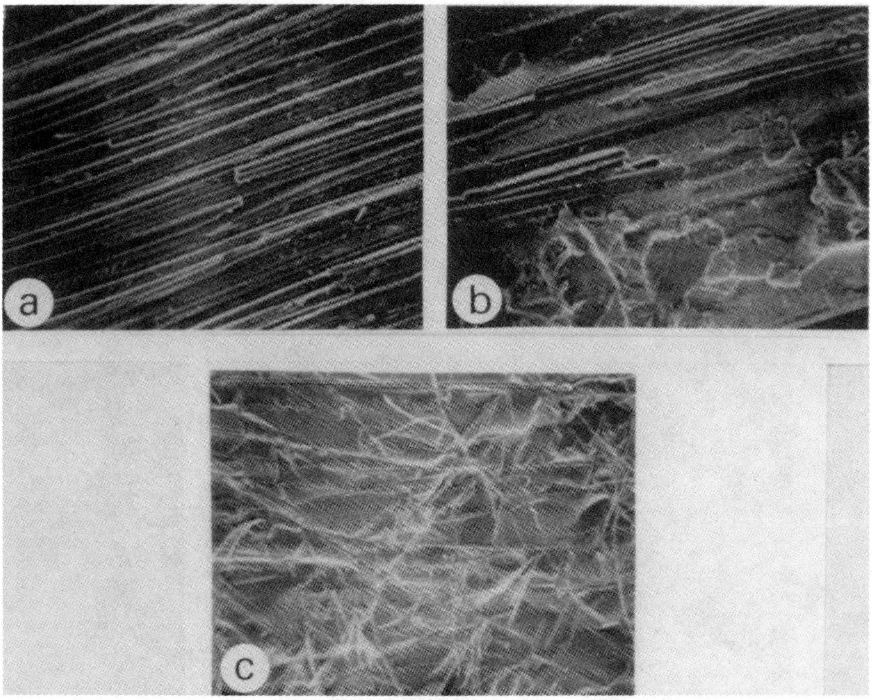

FIG. 3—*SEM fractographs of DCB failure surfaces: (a) AS1/3502 (×284), (b) AS1/3502/163, and (c) AS1/3502/Kev (×67).*

growth had taken place in the adhesive as would be desired to achieve high G_{Ic} values.

The inclusion of fibrous mats and cloth in-lays provided modest increases in G_{Ic} values compared to the adhesive increase. The largest increase in this class was produced by an in-lay of preimpregnated Kevlar paper [AS1/3502/Kev (P) in Table 3]. A G_{Ic} value of 444.5 J/m^2 (2.54 in.·lb/in.^{2}) represented about a three-fold increase over the control. An SEM fractograph of this material's failure surface (Fig. 3) showed that new surfaces were created around the randomly oriented, multi-directional fibers of the in-lay as was expected. In addition, the SEM shows that the mat provided a degree of three-dimensional reinforcement as evidenced by the out-of-plane orientations of some of its fibers. Fiber oriented in out-of-plane directions would be expected to produce an increase in the energy required to grow a delamination. The SEM also shows a substantial amount of the Kevlar fibers were pulled out of, or debonded from, the resin's surface. Increased adhesion of the Kevlar to the resin would be expected to produce an increase in the G_{Ic}.

The effect of combining the aforementioned approaches was investigated with the panels designated AS1/3502/Kev-163 and AS1/3502/163-Kev-163. The AS1/3502/Kev-163 contained an in-lay of one layer of AF-163 adhesive and one layer of Kevlar fibrous mat. This combination produced a dramatic increase in G_{Ic} versus the control. A G_{Ic} value of 1295 J/m^2 (7.4 in.·lb/in.^{2}) is substantially higher than the control and only slightly higher than the adhesive in-lay by itself. An SEM of this combination's failure surface (Fig. 4) showed a substantially

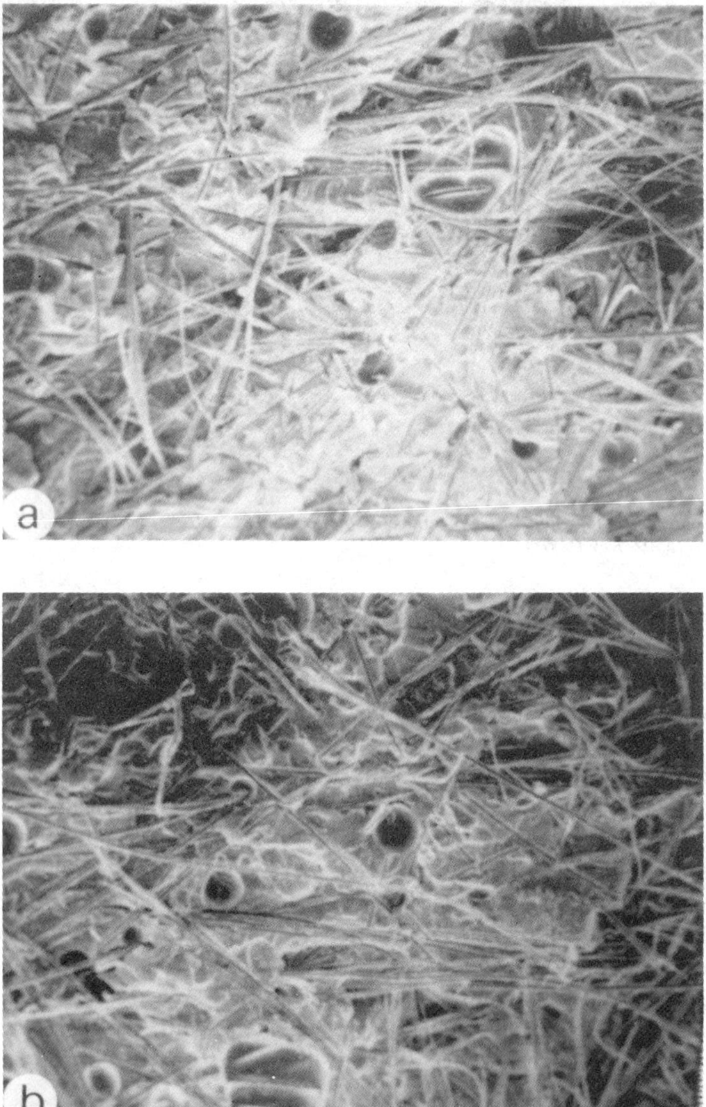

FIG. 4—*SEM fractographs of DCB failure surfaces: (a) AS1/3502/Kev-163 (×68); and (b) AS1/3502/163-Kev-163 (×66).*

lesser amount of fiber pull-out and the presence of the adhesive around the Kevlar fibers.

The AS1/3502/163-Kev-163 contained an in-lay of one layer of AF-163 adhesive, one layer of Kevlar mat, and one layer of AF-163. This combination produced the highest G_{Ic} of any material investigated. The value of 1855 J/m^2 (10.6 in.·lb/in.^{2}) was greater than an order of magnitude increase over the control. An SEM of this material's failure surface (Fig. 4) was very similar to that of the 163-Kevlar combination. A substantial amount of adhesive residue around the Kevlar fibers, with reduced fiber pull-out, is observable.

It was not unanticipated that the in-lays could produce deleterious effects on certain composite properties. Two of these that were likely to be sensitive to the incorportion of in-lays were compression strength and composite thickness.

Compression strength is sensitive to matrix properties and, accordingly, was utilized as the test to screen for deleterious effects on in-plane composite properties. As can be seen in Table 3, relative low values were obtained for the control specimens. However, using this as a comparative, reference baseline, very little degradation in compression strength was detected, with the lowest compression strength value measured (AS1/3502/Kev in Table 3) being 82% of the control value.

Composite per ply thickness was used as a measure of in-lay effects on laminate thickness and, in turn, fiber volume. As can be seen in Table 3, negligible effects were measured.

Conclusions

The DCB test has been found to be an extremely useful tool for screening approaches to improving composite delamination resistance. The results discussed herein show that significant improvements in the delamination resistance of today's materials can be achieved by specifically tailoring the construction of the laminate. In particular, the incorporation of an adhesive in-lay, either by itself or in combination with a Kevlar mat, was found to be extremely effective in increasing delamination resistance.

Wen S. Chan,[1] *Charles Rogers,*[1] *and Sam Aker*[1]

Improvement of Edge Delamination Strength of Composite Laminates Using Adhesive Layers

REFERENCE: Chan, W. S., Rogers, C., and Aker, S., "**Improvement of Edge Delamination Strength of Composite Laminates Using Adhesive Layers,**" *Composite Materials: Testing and Design (Seventh Conference), ASTM STP 893*, J. M. Whitney, Ed., American Society for Testing and Materials, Philadelphia, 1986, pp. 266–285.

ABSTRACT: A thin but tough adhesive inner layer placed in laminates at the proper interface has been found to give improved delamination strength. Static tension tests of laminates with $(\pm 30_2/\bar{90}_3/\mp 30_2)_T$ and $(\pm 35/0/90)_S$ layups of graphite/epoxy show that the added adhesive inner layer effectively suppresses delamination until final failure. The effects of adhesive layers on interlaminar stresses and strain-energy release rate have been studied analytically using a finite-element method. It was found that the introduction of adhesive layers reduced the interlaminar stresses. Although there was no significant change in total strain-energy release rate, there was a significant decrease in the G_I (tension mode) percentage of the total strain-energy release rate.

KEY WORDS: edge delamination, interlaminar stress, strain-energy release rate, adhesive layer, graphite/epoxy, composite materials, damage tolerance, interlaminar fracture toughness

Free-edge delamination has been found to be one of the most important failure modes of helicopter composite rotor hubs during both testing and service. The presence of delamination initiated by interlaminar stresses causes redistribution of the stresses among plies in a laminate that may result in a reduction of tensile strength. Considerable research has been devoted to suppression of delamination, including changing the ply stacking sequence [*1*], hybridizing a softening strip [*2*], stitching along the edges [*3*], and wrapping an outer layer reinforcement around the edge of the laminate [*4*]. Changing the ply stacking sequence in the laminate may change the interlaminar tension normal stress to compressive stress so that delamination can be suppressed or delayed; however, if the applied load is reversed, the advantage of changing the stacking sequence is lost since the stress is also reversed. Hybridizing a softening strip in a laminate, though it reduces interlaminar stresses, may cause a stiffness loss. Stitching or wrapping an outer layer along the edges gives reinforcement in the thickness direction that suppresses delamination crack opening; however, in the practical application,

[1]Senior engineering specialist, Research Group; chief of LHX Airframe and Stress; and former staff member, Research Group; respectively, Bell Helicopter Textron Inc., Fort Worth, TX 76101.

these two approaches may not be easy to accomplish. Present work is directed towards the development of toughened matrix resins for composites that will improve delamination strength and thus fracture toughness. Toughened matrix resins give more transverse or normal-to-the-plane strain capability to the laminate to achieve increased toughness. However, toughened resin development has shown relatively small improvement in fracture toughness to date [*5,6*].

In the past, experience with metal bonded structures showed that a layer of brittle, low-peel adhesive laminated with a ductile, high-peel adhesive behaved in combination like a high-peel adhesive. This principle should work in a laminate; that is, a ductile adhesive placed in the interface between plies where critical interlaminar stresses exist should increase the overall fracture toughness. A toughened laminating resin alone is not enough to provide sufficient elongation in the very thin region between plies. With this in mind, the addition of a thin and discrete layer of tough but unreinforced resin (adhesive film) placed at the proper interface should improve the laminate fracture toughness.

The purpose of the work described in this paper is to investigate an adhesive layer placed in laminates at critical locations for improved delamination strength. Specimens of graphite/epoxy laminates were tested to evaluate the effects of the adhesive layer on delamination strength. Finally, a finite-element analysis was employed to calculate interlaminar stresses and strain-energy release rate.

Specimen Description and Test Procedure

Two laminates suggested by NASA Publication 1092 [*7*] for edge delamination tension testing were investigated: $(\pm 30_2/90_3/\mp 30_2)_T$, a matrix-dominated laminate, and $(\pm 35/0/90)_S$, a fiber-dominated laminate. Hercules AS4/3501-6 graphite/epoxy prepreg tapes were used to construct the laminates. The adhesive layer was American Cyanamid FM1000 adhesive film. This film has a weight of 0.15 kg/m^2 (0.03 lb/ft^2) and a thickness of 0.13 mm (0.005 in.). The adhesive film was placed only at the $-30/90$ interfaces in the $(\pm 30_2/90_3/\mp 30_2)_T$ laminates and at the 0/90 interfaces in the $(\pm 35/0/90)_S$ laminates. These locations were selected after observing tests of the laminate without an adhesive layer.

The laminates were fabricated into square panels measuring 254 mm (10 in.) on the side. Some panels contained no adhesive layer while others contained a complete layer of adhesive film. In addition, one of the $(\pm 35/0/90)_S$ panels had 12.7-mm (1/2-in.) wide strips of the adhesive placed 25.4 mm (1 in.) apart in the direction of 0° ply at the 0/90 interface, starting 6.35 mm (0.25 in.) from the panel edges. The panels with adhesive layers were cured in the same batch, and all panels were cured by the method described in the Appendix. Woven glass-laminate loading tabs were adhesively bonded at room temperature for 72 h. After curing, the panels were cut into straight-edge specimens approximately 38 mm (1.5 in.) wide (see Fig. 1).

Before running the tests, the specimens were checked by X-ray for any delamination that might have occurred during specimen preparation. The specimens

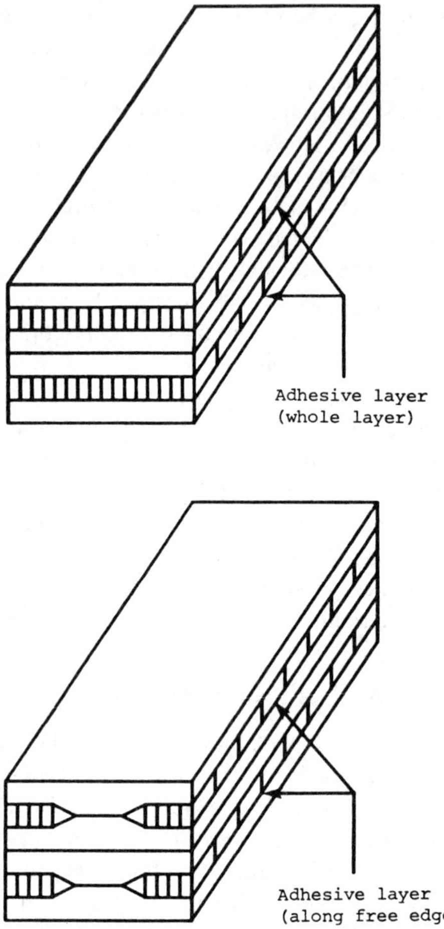

FIG. 1—*Configuration of laminates with adhesive layer.*

were then loaded in tension at a rate of 1.27 mm/min (0.05 in./min) in a Tinius Olsen servohydraulic test machine and monitored by a 10-cm (4-in.) extensometer on the loading axis. An *x-y* plotter was attached to record stress-strain behavior during loading. The free edges of the test specimens were carefully inspected by the unaided eye during the test and monitored for audible sounds. When audible sounds were first heard, loading was halted and the specimen was removed from the machine and examined under a microscope and by X-ray. The onset of delamination required confirmations by both microscope and X-ray examinations. The delamination interface was identified through the microscope. Several loading increments were set for delamination growth investigation to failure. Dye penetrant, a zinc iodide solution opaque to X-rays, was used for enhancement to monitor delamination. All of the test specimens were exposed to X-rays for

1.5 min. at 15 kV. Ultimate strength was then determined by testing to complete fracture.

Material Characterization of Adhesive Layer (FM1000)

A set of 12.7-mm (0.5-in.) wide dogbone specimens of the adhesive film was used for stress-strain characterization. The specimens consisted of 16 plies of the adhesive film. Each ply was 0.27 mm (0.0105 in.) thick. Five coupons were fabricated from the same panel.

Static tension tests were then conducted. A high-elongation axial strain gage, EP08-10CBE-120 of Micro-Measurements, was used to measure the axial strain. A strain gage was also placed to measure the transverse strain.

The typical stress-strain curve for the FM1000 adhesive film, shown in Fig. 2, exhibits a nonlinear behavior throughout the entire loading. Because of this

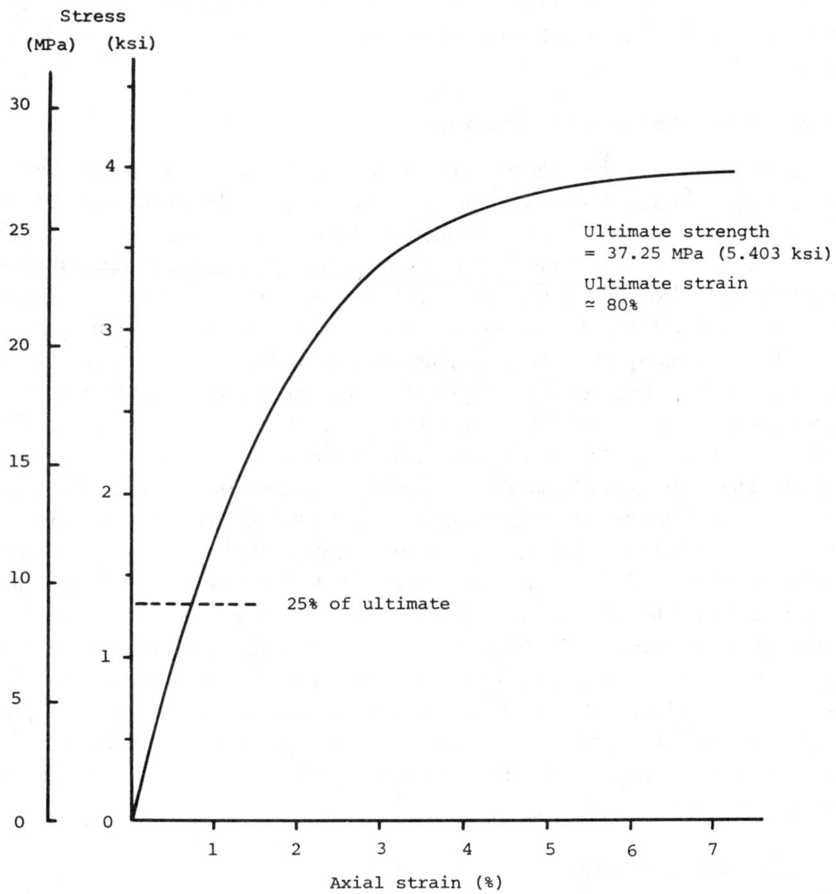

FIG. 2—*Typical stress-strain curve of FM1000 adhesive film.*

TABLE 1—*Modulus and Poisson's ratio of FM1000 adhesive film.*

Specimen Number	Ultimate Strength		Secant Modulus at 25% of Ultimate Strength		Poisson's Ratio at 25% of Ultimate Strength
	MPa	ksi	GPa	Msi	
1	37.25	5.403	1.39	0.202	0.463
2	37.03	5.371	1.45	0.210	0.438
3	34.72	5.035	1.50	0.217	0.379
4	35.08	5.088	1.51	0.219	0.448
5	37.62	5.456	1.39	0.201	0.379
Average	36.34	5.271	1.45	0.210	0.425

nonlinearity, the secant elastic modulus of the material was measured at 25% of ultimate load, where the stress-strain curve is approximately linear. Poisson's ratio was calculated as the ratio of transverse strain to longitudinal strain at 25% of ultimate load. The results of modulus and Poisson's ratio are listed in Table 1.

Experimental Results and Observation

Laminates without the adhesive layer exhibited 90° ply cracking followed by delamination. Figure 3 shows transverse cracking (90° ply crack) and delamination of the $(\pm30_2/90_3/\mp30_2)_T$ laminate without the adhesive layer at a stress of 281 MPa (40.7 ksi). The onset of transverse cracking occurred at approximately 161 MPa (23.3 ksi). Delaminations formed and grew in the $-30/90$ interfaces, typically shifting from one interface, through the 90° ply cracks, to its symmetric $-30/90$ counterpart. This same phenomenon of delamination crack patterns was observed by O'Brien [*6*]. Figure 4 shows transverse cracking and onset delamination in a $(\pm35/0/90)_S$ laminate without an adhesive layer at a load of 313 MPa (45.4 ksi). Delamination occurred at the 0/90 interfaces.

Laminates with a whole adhesive layer or with strips of adhesive film along the free edge exhibited no delamination before final failure. This is shown in Figs. 5, 6, and 7. Figure 5 shows only transverse cracking of a $(\pm30_2/90_3/\mp30_2)_T$ laminate with the adhesive layers at a failing load of 521 MPa (75.6 ksi), which is very near the ultimate load of 542 MPa (78.6 ksi). Figures 6 and 7 also show only transverse cracking of a $(\pm35/0/90)_S$ laminate with a whole layer and with edge strips at the load of 542 MPa (78.6 ksi). These laminates were failed at 544 MPa (78.8 ksi) and 711 MPa (103.2 ksi), respectively. In Fig. 8, a photomicrograph of the same specimen as shown in Fig. 5 shows that the adhesive inner layer effectively isolates the transverse crack at its boundary and the delamination is suppressed.

Transverse Crack Density

For the laminates tested, the density of the transverse cracks was found to be affected by the addition of the adhesive layer. As indicated in Figs. 3 and 4, the

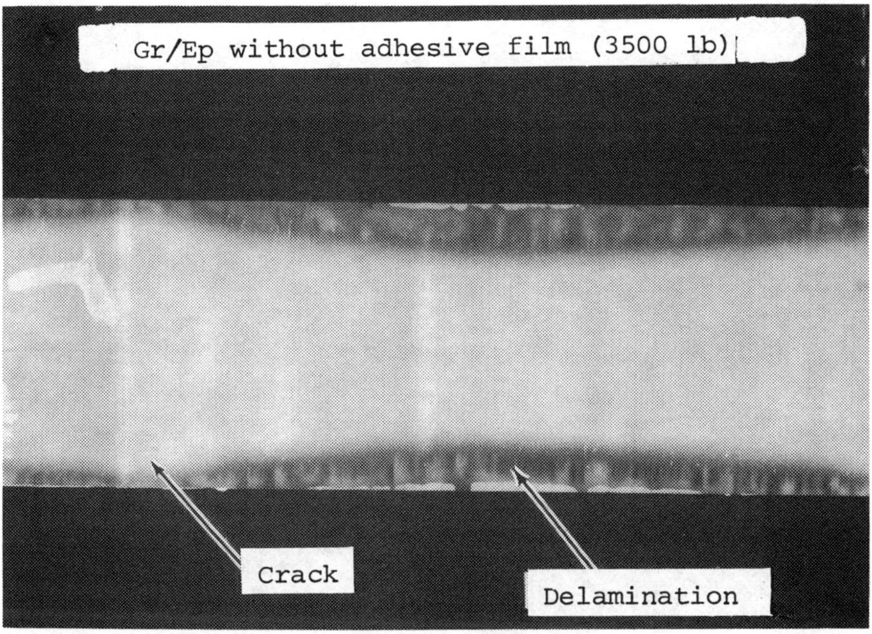

FIG. 3—*X-ray radiography of a* $(\pm 30_2/90_3/\mp 30_2)_T$ *graphite/epoxy laminate at the load of 281 MPa (40.7 ksi).*

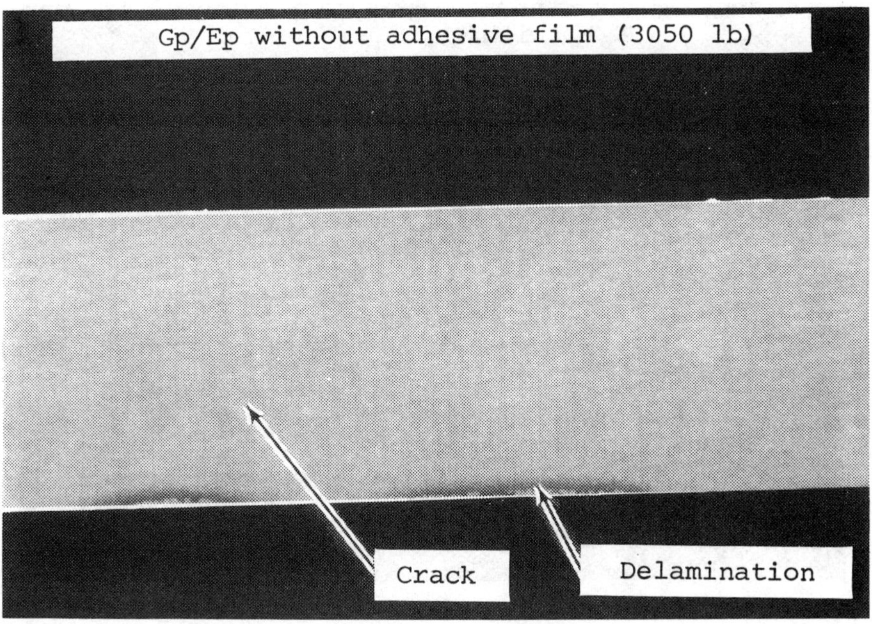

FIG. 4—*X-ray radiography at the onset of delamination in a* $(\pm 35/0/90)_S$ *graphite/epoxy laminate at the load of 313 MPa (45.4 ksi).*

FIG. 5—*X-ray radiography of a* $(\pm 30_2/90_3/\mp 30_2)_T$ *graphite/epoxy laminate with adhesive layers at the load of 521 MPa (75.6 ksi).*

FIG. 6—*X-ray radiography of a* $(\pm 35/0/90)_S$ *graphite/epoxy laminate with adhesive layers at the load of 542 MPa (78.6 ksi).*

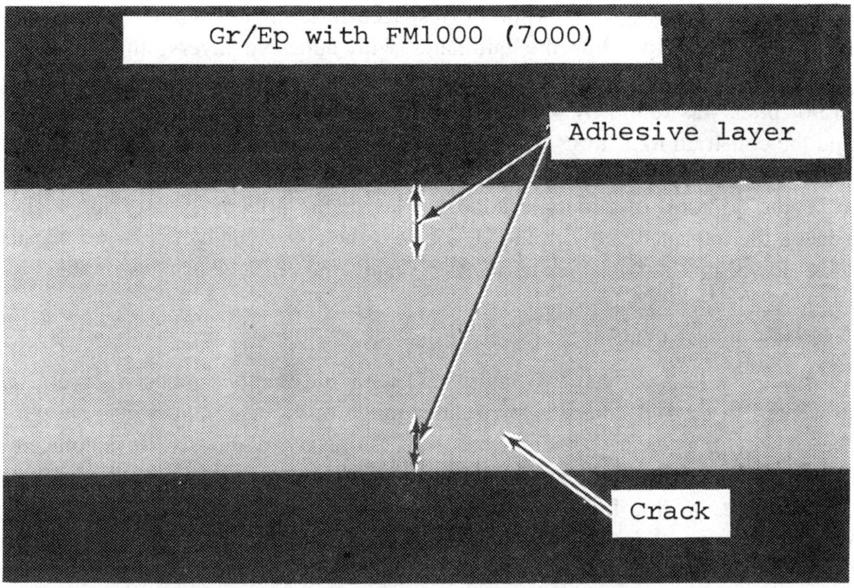

FIG. 7—*X-ray radiography of a* $(\pm 35/0/90)_S$ *laminate with adhesive layers along the edges at the load of 626 MPa (90.8 ksi).*

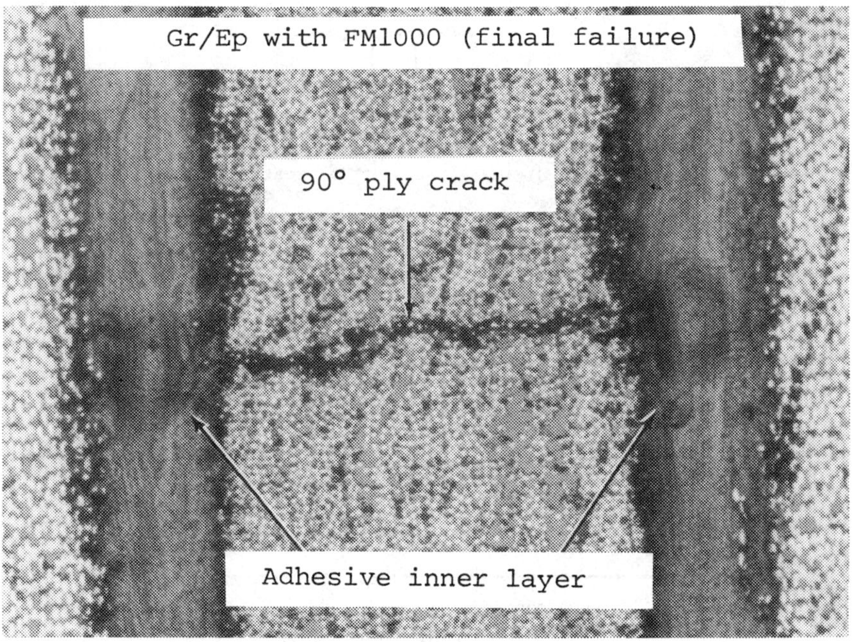

FIG. 8—*Photomicrograph of a* $(\pm 30_2/90_3/\mp 30_2)_T$ *graphite/epoxy laminate with adhesive layers at the final failure load of 542 MPa (78.6 ksi).*

transverse cracks are fine and densely spaced in laminates without an adhesive layer or edge strips. For the laminates with adhesive layers, the density of transverse cracks is reduced, as shown in Figs. 5, 6, and 7. The crack density in 90° plies was found by Wang [*8*] to be dependent on the 90° layer thickness and the constraining actions of the adjacent ply. The spacing of transverse cracks in the 90° layer appears to be approximately equal to the shear-lag zone distance. With the addition of a ductile adhesive layer, the effective interlaminar shear zone at the interface between the 90° ply and the adhesive layer is larger because of its lower modulus of elasticity. As a result, the crack density is reduced.

Delamination Strength

The test results for the two laminates, with and without adhesive layers, are presented in Figs. 9 and 10. The scatter range of the test data is also shown in the figures. It is clearly evident that the addition of an adhesive layer placed at the critical interface in the laminate significantly improves delamination strength. Furthermore, in this limited number of tests, the $(\pm30_2/90_3/\mp30_2)_T$ laminates exhibit an increase in ultimate strength of approximately 21% with the addition of an adhesive layer. However, the fiber-dominant $(\pm35/0/90)_S$ laminates have only a 12 to 16% increase in ultimate strength. Furthermore, the whole adhesive layer was found to effectively eliminate delamination prior to final failure. In

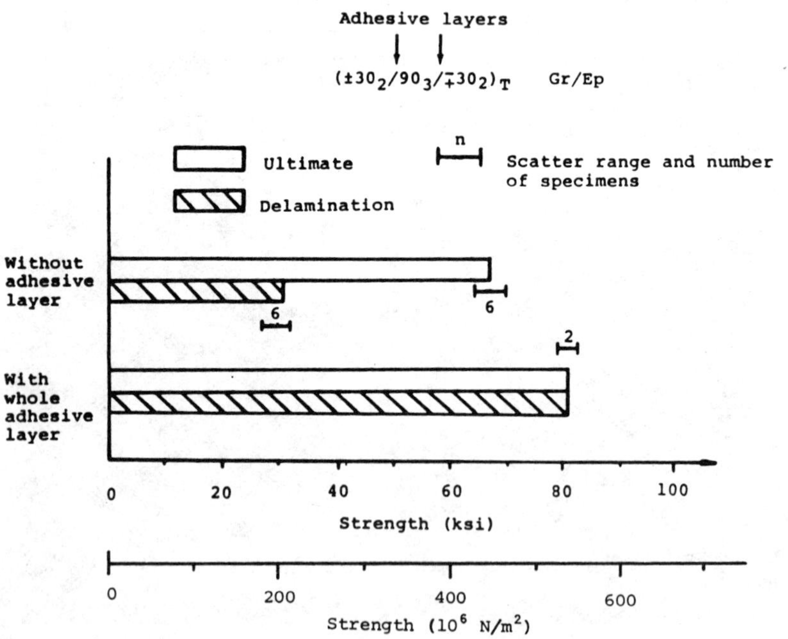

FIG. 9—*Delamination and ultimate strength of* $(\pm30_2/90_3/\mp30_2)_T$ *AS4/3501 graphite laminates with and without adhesive layer.*

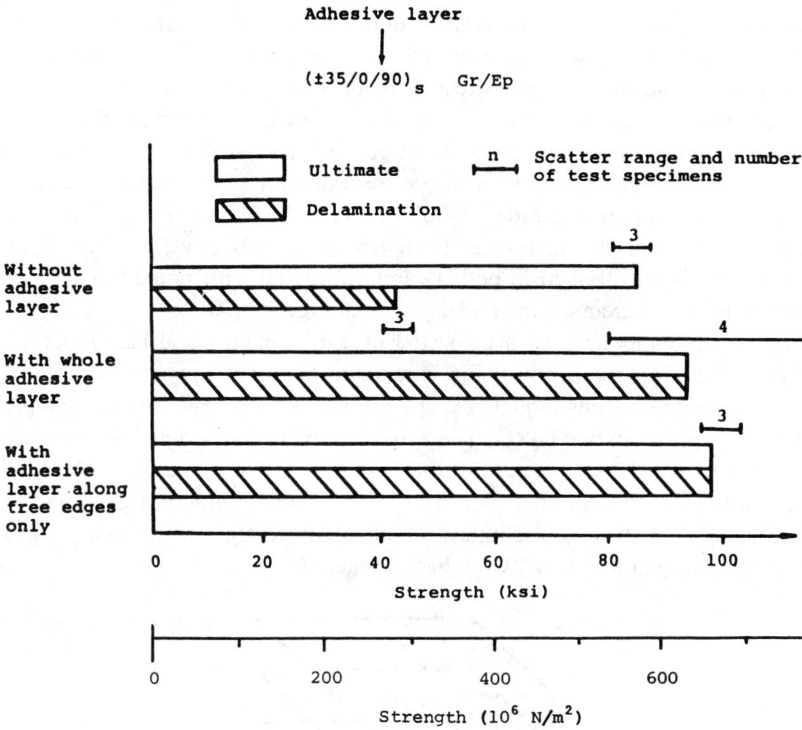

FIG. 10—*Delamination and ultimate strength of* $(\pm 35/0/90)_S$ *AS4/3501 graphite/epoxy laminates with and without adhesive layer.*

addition, the edge strips of adhesive film have the same effect as a whole adhesive layer. This can be seen in Fig. 10.

Analysis

Since delamination is initiated by high interlaminar stress and is dependent on the strain-energy release rate, a finite-element analysis was conducted to investigate the effect of the adhesive layer on these two parameters.

Finite Element and Modeling

A quasi three-dimensional finite-element program by Arrenburg [9] was modified and used for this study. An eight-node, quadrilateral, isoparametric element with three degrees of freedom per node was used. A displacement field of the form

$$u = \epsilon_0 x + U(y,z)$$
$$v = V(y,z)$$
$$w = W(y,z)$$

was assumed, where ϵ_0 was a prescribed uniform axial strain. Because of symmetries inherent in the layup considered, only a quarter of the laminate was modeled. Since the interlaminar stresses at the free edge are local to the free edge, only a portion of the entire width of the laminate is needed for the model. The width modeled was equal to about three thicknesses of the laminate. This width was always greater than the distance from the free edge at which the stress converged to the classical lamination solution (see the sketch in Fig. 11). Only the region 5.08 mm (0.2 in.) away from free edge was modeled. The finite-element model is constrained in both x- and y-directions for nodal points along $y = 0$ and in the z-direction for nodal points along $Z^1 = 0$ (see Fig. 11).

A typical finite-element mesh consisted of 120 elements and had about 410 nodes (see Fig. 11). Each ply of the composite laminate was modeled as one layer in the finite-element model, except for the 90° ply and its adjacent ply, which were modeled in two layers. The adhesive layer was modeled in two layers of finite elements. The aspect ratio of mesh sizes near the free edge or the crack tip was about one. Fine meshes were used at the neighborhood of delamination and free edge. The stress at the interface was obtained from linear interpolation of stress at Gaussian points of two adjacent elements.

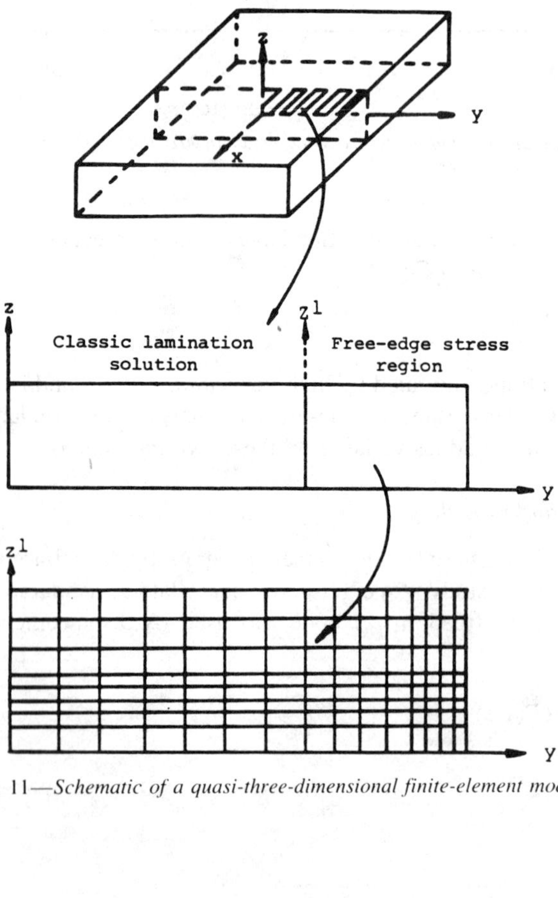

FIG. 11—*Schematic of a quasi-three-dimensional finite-element modeling.*

Verification of the Computer Program

To gain confidence in using this program, an example problem considered by O'Brien [*10*] was repeated. A $(\pm 45/0/90)_S$ graphite/epoxy laminate, in which delamination occurs at the interface between the 0°/90° plies, was loaded at a uniaxial strain of 4000 $\mu\epsilon$. The material constants used were as follows:

$$E_{11} = 134 \text{ GPa}$$
$$E_{22} = 10.2 \text{ GPa}$$
$$G_{12} = 5.5 \text{ GPa}$$
$$\nu_{12} = 0.30$$

The strain-energy release rate, which is relatively insensitive to mesh size, was used for comparison. The strain-energy release rate was computed by a virtual crack closure technique described in Ref *6*. The results of the analysis compare well with the results published in Ref *10*, as shown in Fig. 12.

Free-Edge Stress

Laminates considered for investigating the effect of interlaminar stress were the $(\pm 30_2/90_3/\mp 30_2)_T$ laminate, with and without adhesive layers. The material

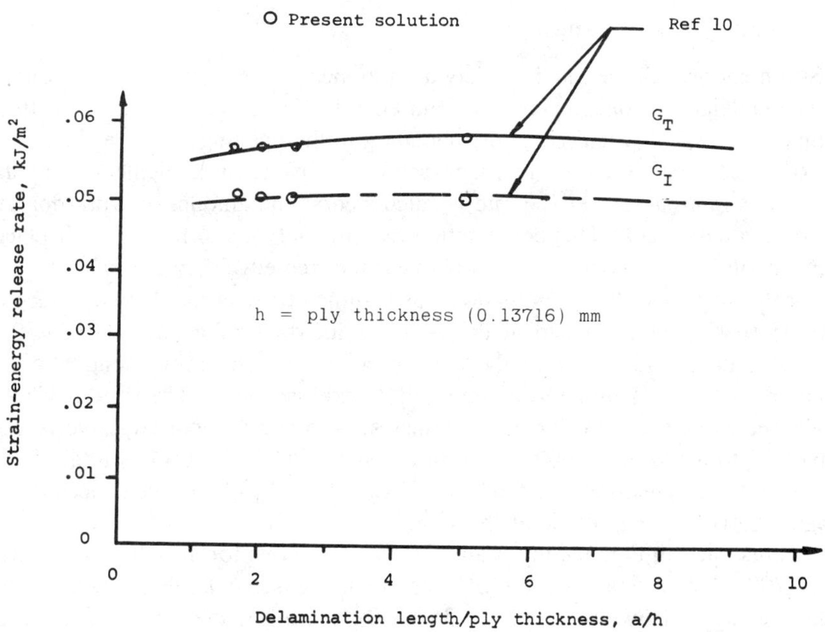

FIG. 12—*Comparison of strain-energy release rates.*

constants were as follows:

$$E_{11} = 133 \text{ GPa } (19.3 \text{ Msi})$$
$$E_{22} = 11.2 \text{ GPa } (1.62 \text{ Msi})$$
$$G_{12} = 7.03 \text{ GPa } (1.02 \text{ Msi})$$
$$\nu_{12} = 0.288$$

The adhesive layer properties are listed in Table 1. The shear modulus of this adhesive layer was approximated by using Young's modulus and Poisson's ratio based on an isotropy assumption.

The calculated interlaminar stresses normalized by applied nominal strain are plotted in Fig. 13. The interlaminar normal stress, σ_z, across the laminate thickness near the free edge reaches a high tensile value at the $-30/90$ interface and throughout the 90° plies. The interlaminar stress distributions, σ_z and τ_{yz} along that interface, are shown in Figs. 14 and 15. It is seen that the added adhesive layer in the laminates reduces both the interlaminar normal and shear stresses.

It has been found that understanding of the interlaminar stress state alone in laminates is not sufficient to predict the location of delamination and its strength [*8*]. This is because the singular behavior at both the free edge [*11*] and the ply interface [*12*] preclude using a failure criterion based on maximum interlaminar stress. Therefore, an analysis was done to evaluate the effect of the adhesive layer on strain-energy release rate.

Strain-Energy Release Rate

Strain-energy release rate is widely used to measure the strain energy required to extend delamination. In general, strain-energy release rate consists of the three components, G_{I}, G_{II}, and G_{III}, corresponding to the opening, inplane shear, and out-of-plane shear fracture modes, respectively. In the finite-element computation, the strain-energy release rate is calculated as the amount of work done to close up a new crack. This computation requires only nodal force and displacement involving no singular stress field near the free edge.

For this analysis, the delamination was assumed to be at the 0/90 interface of a $(\pm35/0/90)_S$ laminate and at the $-30/90$ interface of a $(\pm30_2/90_3/\mp30_2)_T$ laminate, in accordance with the test observations. The initial length of the delamination was assumed to be three ply thicknesses long. The results of total strain-energy release rate for both laminates, with and without adhesive layers, subjected to a uniform 1000-$\mu\epsilon$ strain are shown in Fig. 16. The total strain-energy release rate is the sum of G_{I} and G_{II}. The G_{III} is neglected because its magnitude is less than 0.5% of the sum.

As illustrated in Fig. 16, the strain-energy release rate for the adhesive-layered $(\pm30_2/90_3/\mp30_2)_T$ laminate is 13% less than the same laminate without the adhesive layer. However, for a $(\pm35/0/90)_S$ laminate, the total strain-energy release rate is about the same for laminates with and without adhesive layers.

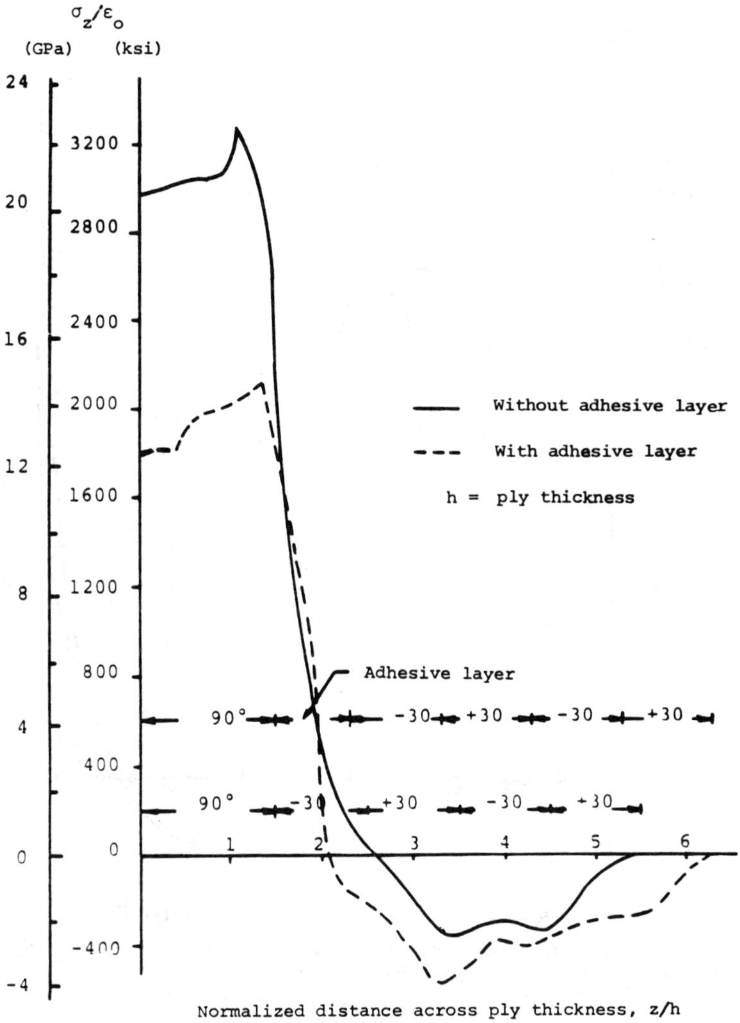

FIG. 13—*Normalized interlaminar normal stress at the free edge across the thickness of a* $(\pm 30_2/90_3/\mp 30_2)_T$ *graphite/epoxy laminate.*

The percentage of G_I in total strain-energy release rate is shown in Fig. 17. As seen in the figure, G_I is significantly reduced by the adhesive layers. The low modulus and high elongation of the adhesive layer can significantly reduce the interlaminar forces but results in more deformation. The increased deformation normal to the interface due to the adhesive layer is relatively small because the adhesive is confined to a very thin region between plies. As a result, G_I is reduced. On the other hand, the shear deformation of the adhesive at the interface is significantly large since there is no constraint in the width direction. This con-

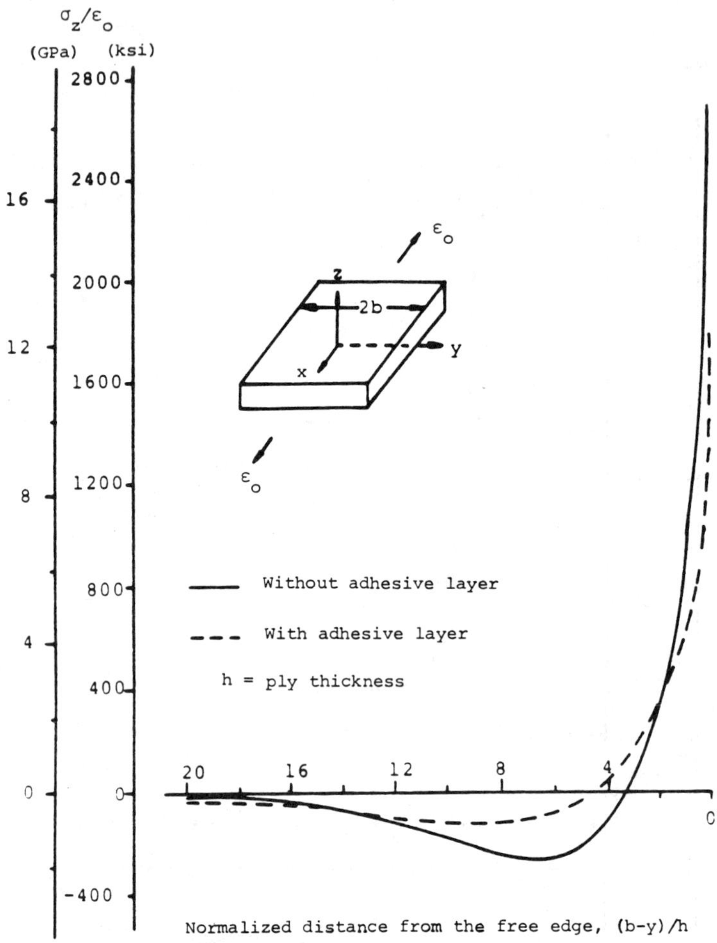

FIG. 14—*Normalized interlaminar normal stress along the interface between the 90° ply and its adjacent ply of a* $(\pm 30_2/90_3/\mp 30_2)_T$ *graphite/epoxy laminate.*

tributes to a large increase of G_{II}. It has been suggested that G_I, and not the total G, appears to control the onset of edge delamination for the static case [*10,13*]. Therefore, the significant reduction of G_I found in the adhesive-layered laminates should explain the improved delamination strength.

Comparison of Measured and Calculated Interlaminar Fracture Toughness

The interlaminar fracture toughness, G_c, of a composite is the critical value of the strain-energy release rate. An accurate measurement of this quantity is very difficult to obtain because it depends on the morphology of the fracture surface [*14*]. A simple equation derived by O'Brien [*6*] was used for estimating

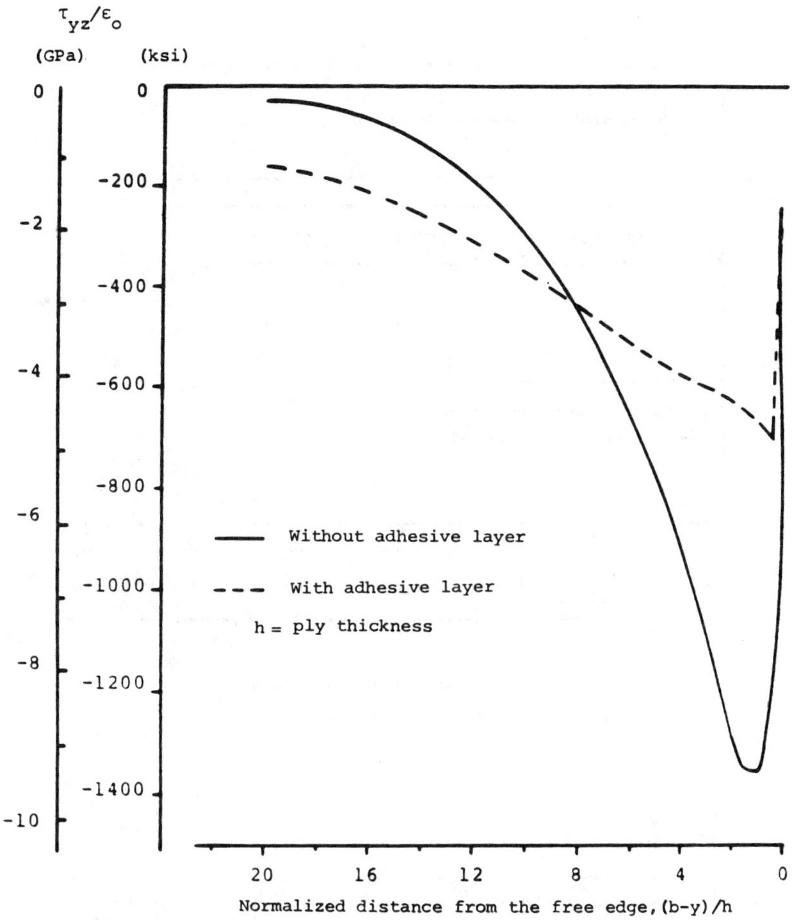

FIG. 15—*Normalized interlaminar shear stress along the interface between the 90° ply and its adjacent ply of a* ($\pm30_2/90_3/\mp30_2)_T$ *graphite/epoxy laminate.*

the measured interlaminar fracture toughness. This equation is

$$G_c = \frac{\epsilon_c^2 t}{2} (E_{LAM} - E^*)$$

where E_{LAM} is a laminate stiffness that is dependent on layup; E^*, which is dependent on delamination location, is the stiffness of a laminate at a delaminated interface; t is the laminate thickness; ϵ_c is the applied strain at the onset of delamination; and E_{LAM} and E^* can be computed by lamination theory and the rule of mixtures.

A comparison of interlaminar fracture toughness from the test and finite-element calculation is presented in Table 2. Agreement within 10% is shown for

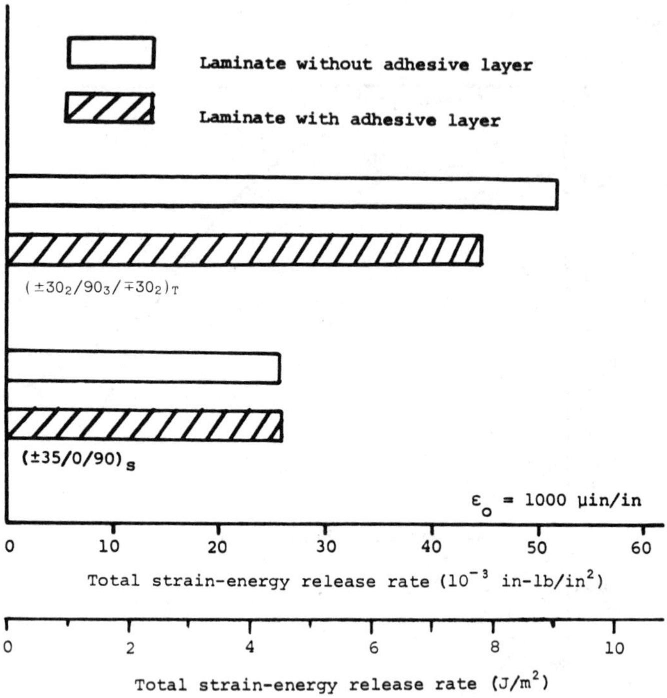

FIG. 16—*Effect of total strain-energy release rate.*

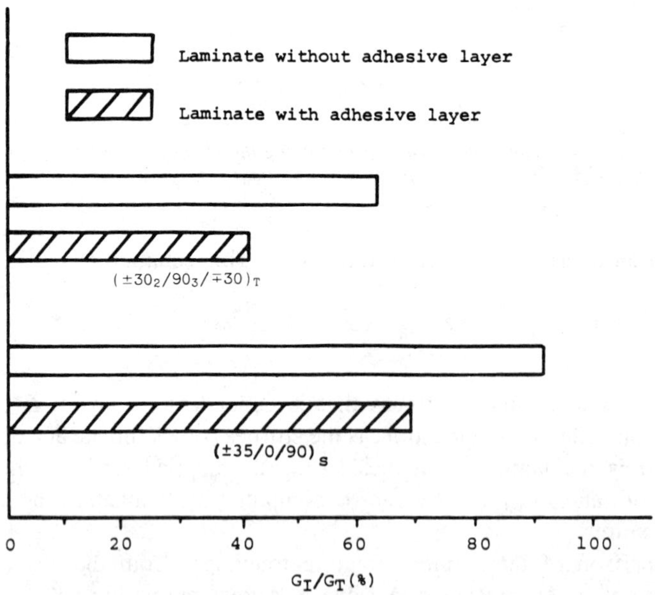

FIG. 17—*Percentage of G_I in total strain-energy release rate.*

TABLE 2—*Comparison of measured and calculated interlaminar fracture toughness.*

Specimen	Test kJ/m^2	Test in · lb/in.^{2}	Finite-Element Analysis kJ/m^2	Finite-Element Analysis in · lb/in.^{2}
			$(\pm30_2/90_3/\mp30_2)_T$	
Without adhesive layer	0.144	0.82	0.131	0.75
With adhesive layer	1.350	7.71	1.189	6.79
			$(\pm35/0/90)_S$	
Without adhesive layer	0.100	0.57	0.103	0.59
With adhesive layer	0.690	3.94	0.751	4.29

all laminates except the matrix-dominated $(\pm30_2/90_3/\mp30_2)_T$ laminate with adhesive layer, which is 12% lower than the test values. A fine mesh to compute the high-stress gradient between the composite layer and the adhesive layer should improve the correlation.

Conclusions

An adhesive layer placed at critical locations in laminates has successfully been demonstrated to improve the laminate strength. For AS4/3501 graphite laminates with FM1000 adhesive film, edge delamination is eliminated until final failure. Ultimate strength is also increased. The suppression of delamination is attributed to reduction of the G_I percentage of total strain-energy release rate in the adhesive-layered laminates. Moreover, the adhesive layer placed adjacent to the 90° ply provides a larger shear-lag zone and consequently reduces the 90° ply crack density. The interlaminar normal and shear stresses of the laminate with adhesive layers are reduced with respect to the laminates without adhesive layers. It is concluded that such a strip of adhesive layer, if placed in laminates away from the edge, can be used as a delamination arrester under static tension loading, and it is expected that the adhesive layer will also be effective under fatigue loading [*15*].

Acknowledgments

This work was supported under the 1983 Bell Helicopter Textron IR&D program. Useful discussions with Professor C. T. Sun of Purdue University are appreciated. The finite-element program for interlaminar stress computation was written by Mr. R. T. Arenburg of Boeing at Seattle and permission to use it is appreciated.

APPENDIX

Laminate Curing Procedure

The panel of prepreg laminate was first vacuumed in a bag with two 6.35-mm (9.25-in.) thick aluminum caul plates at 660.4 mm (26 in.) of mercury and then placed in an

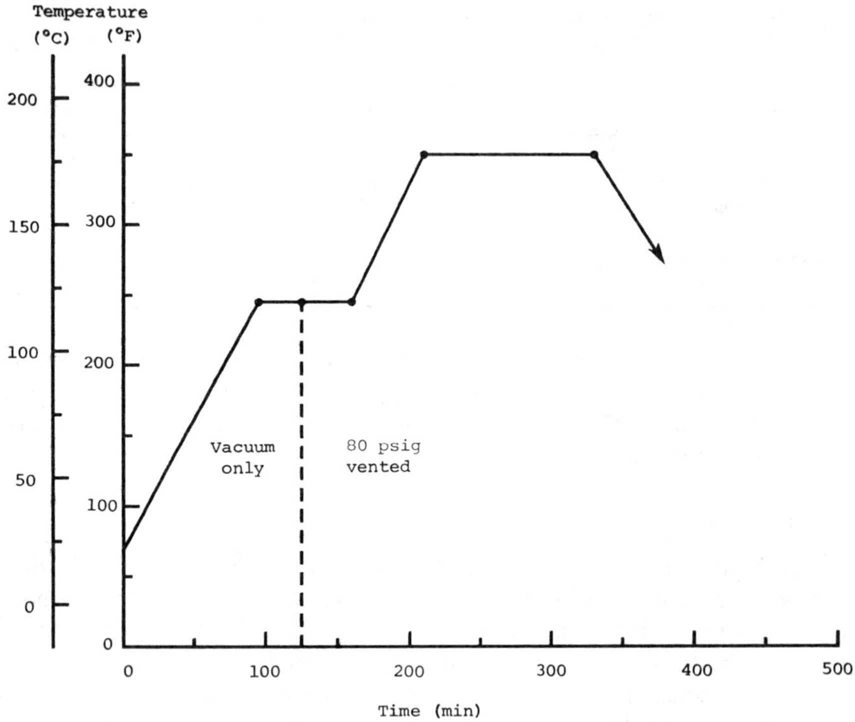

FIG. 18—*Panel curing procedure.*

autoclave. The autoclave was heated to 116 ± 3°C (240 ± 5°F) at a rate of 0.56 to 1.67°C (1 to 3°F) per min. After holding the panel at this temperature and this pressure for 30 ± 5 min, the autoclave gage pressure was raised to 551 ± 35 kPa (80 ± 5 psig). When the autoclave gage pressure reached 172 kPa (25 psig), the panel was vented to the atmosphere. The panel was then held at 116 ± 3°C (245 ± 5°F) and 551 ± 35 kPa (80 ± 5 psig) autoclave pressure for another 30 ± 5 min. The autoclave temperature was raised to 177 ± 3°C (350 ± 5°F) at the same rate as before. The panel was then kept in the autoclave for 2 h at 177 ± 3°C (350 ± 5°F) and 551 ± 35 kPa (80 ± 5 psig). Upon completion of the foregoing stages, the heat was shut off.

The autoclave pressure was removed when the temperature cooled to 66°C (150°F) or less. The panel was then debagged and checked by C-scan for any imperfections. The curing cycle is depicted in Fig. 18.

References

[1] Crossman, F. W., "Analysis of Free Edge Induced Failure of Composite Laminates," *Fracture of Composite Materials*, G. C. Sih and V. P. Tamuzs, Eds., Proceedings of First USA-USSR Symposium, 1979.

[2] Veretee, R. M. and, Labor, J. D., "Structural Criteria for Advanced Composites," Vol. 1, AFFDL-TR-76-142, Air Force Flight Dynamics Laboratory, March 1977.

[3] Mignery, L. A., Tan, T. M., and Sun, C. T., "The Use of Stitching to Suppress Delamination in Laminated Composites," presented at the ASTM Symposium on Delamination and Debonding of Materials, Pittsburgh, American Society for Testing and Materials, Nov. 1983.

[*4*] Kim, R. Y., "Prevention of Free-Edge Delamination," 28th National SAMPE Symposium, Society for the Advancement of Materials and Process Engineering, April 1983.

[*5*] Johnson, N. J., O'Brien, T. K., Morris, D. H., and Simonds, R. A., "Interlaminar Fracture Toughness of Composites—Refinement of the Edge Delamination Test and Application to Thermoplastics," 28th National SAMPE Symposium, Society for the Advancement of Materials and Process Engineering, April 1983.

[*6*] O'Brien, T. K. in *Damage in Composite Materials, ASTM STP 775*, E. Reifsnider, Ed., American Society for Testing and Materials, Philadelphia, 1982, pp. 140–167.

[*7*] *Standard Test for Toughened Composites*, NASA Reference Publication 1092, National Aeronautics and Space Administration, April 1982.

[*8*] Wang, A. S. D., "Fracture Mechanics of Sublaminate Cracks in Composites," *Characterization, Analysis and Significance of Defects in Composite Materials*, AGARD Conference Proceedings No. 355, France, 1983.

[*9*] Arrenburg, R. T., "Analysis of the Effect of Matrix Degradation on Fatigue Behavior of a Graphite/Epoxy Laminate," Master's thesis, Texas A&M University, 1983.

[*10*] O'Brien, T. K., "Mixed-Mode Strain Energy Release Rate Effects on Edge Delamination of Composites," NASA Report TM84592, National Aeronautics and Space Administration, Jan. 1983.

[*11*] Wang, S. S and Choi, I., "Boundary-Layer Effects in Composite Laminates: Free Edge Singularities," NASA CR165440, National Aeronautics and Space Administration, April 1981.

[*12*] Wang, S. S. and Choi, I., "The Mechanics of Delamination in Fiber-Reinforced Composite Materials, Part I—Stress Singularities and Solution Structure," NASA Report CR172269, National Aeronautics and Space Administration, Nov. 1983.

[*13*] Wilkins, D. J., Eisenmann, J. R., Camin, R. A., Margolis, W. S., and Benson, R. A. in *Damage in Composite Materials, ASTM STP 775*, E. Reifsnider, Ed., American Society for Testing and Materials, Philadelphia, 1982, pp. 168–183.

[*14*] Wang, A. S. D., Slomiana, M., and Bucinell, R. B., "Delamination Crack Growth in Composite Laminates," presented at the ASTM Symposium on Delamination and Debonding of Materials, Pittsburgh, American Society for Testing and Materials, Nov. 1983.

[*15*] Chan, W. S., "Delamination Arrester—An Adhesive Inner Layer in Laminated Composites," presented at the ASTM Symposium on Composite Materials: Fracture and Fatigue, American Society for Testing and Materials, Oct. 1984.

Som R. Soni[1] *and Ran Y. Kim*[1]

Delamination of Composite Laminates Stimulated by Interlaminar Shear

REFERENCE: Soni, S. R. and Kim, R. Y., "**Delamination of Composite Laminates Stimulated by Interlaminar Shear,**" *Composite Materials: Testing and Design (Seventh Conference), ASTM STP 893,* J. M. Whitney, Ed., American Society for Testing and Materials, Philadelphia, 1986, pp. 286–307.

ABSTRACT: Although interlaminar shear stresses become very large in composite laminates, in the literature there is no conclusive demonstration of the onset of delamination in composites due to interlaminar shear. The present work addresses the title problem by accurate stress analysis, application of an average stress failure criterion over a predominant stress component, and testing chosen laminates made of T300/1034-C graphite epoxy to the onset of delamination. The stress analysis has been used to define the laminate stacking sequence and the required loading direction to eliminate modes of failure other than that stimulated by interlaminar shear stress. The predicted strains at the onset of delamination are close to experimental values in some of the laminates, and the predicted interfaces of delamination are the same as experimentally observed delaminated interfaces. This work leads to an understanding of validity of average stress failure criterion and probable causes of difference between the predicted and experimental values.

KEY WORDS: delamination, composite materials, interlaminar shear, failure, edge effect, stress analysis, Global-local model, photomicrographs, X-rays, static tension, static compression, fatigue (materials), first-ply failure, lamina, laminate

The work described in this paper is closely related to recent publications on the elastic response of layered composites [*1–8*]. There are numerous other publications addressing the onset of delamination and free-edge effects in composite laminates; some of them are given in Ref *2*. We will discuss here, only those articles that are not included in Ref *2* and are directly related to the present work. The first two papers [*1,2*] deal with the development of an analytical approach, "Global-Local Laminate Variational Model," to conduct an accurate three-dimensional stress analysis of composite laminates. The rest of the papers [*3–7*] deal with the application of this model to predict the onset of delamination and its experimental verification. In most of the laminates considered in Refs *3* to *7*, the interlaminar normal stress, σ_z, was responsible for onset of delamination.

In Ref *3*, the authors have determined the interlaminar normal stress, σ_z, near the midsurface of the free edge of the laminate for a number of laminates by use of miniature strain gages and constitutive equations. A good agreement exists

[1]Senior research scientist and senior research engineer, respectively, University of Dayton Research Institute, Dayton, OH 45469. Dr. Soni is now at AdTech Systems Research Inc., Fairborn, OH 45324.

between the predicted and experimental values of σ_z. Reference *4* showed the advantage of the use of hybrid laminates over the pure laminates with regard to the reduced magnitude of interlaminar out-of-plane stress components, σ_z, τ_{xz}, and τ_{yz}, developed due to an applied inplane load. Further, the difference in the first-ply failure strength for two types of laminate systems was investigated.

References *5* and *6* dealt with the onset of σ_z delamination of composite laminates. The prediction of onset of delamination load was done on the basis of stress analysis by Global-local model and average stress failure criterion applied to the stress component, σ_z. This failure criterion is analogous to the one suggested by Whitney and Nuismer [*9*]. Reference *7* gives a comparative look on commonly-used strength theories for layered composites. With the help of studies similar to those in Ref *6*, the significantly lower experimental strength of $(15/0/-15)_s$ laminate than its predicted inplane first-ply failure strength [*7*] was attributed to the interlaminar shear stress.

Pagano and Pipes, in their excellent work [*10*], have emphasized the importance of interlaminar shear stresses in the failure of composite laminates. The present work is made possible because of the Global-local laminate variational model [*1*].

Considerable effort has gone into the investigation of delamination of layered composites using the fracture mechanics approach. In this work, we refrain from going into the controversy of predicted singular stress at the free edge and the continuity of the resin in cured laminates. This issue was discussed in detail in Refs *12* and *13*. Instead, average stress components are used in the delamination failure criterion for the present work.

A large number of laminates made of T300/1034-C graphite epoxy were analyzed using the Global-local laminate model [*1*]. The laminates exhibiting the predominant interlaminar shear stress component, τ_{xz}, were tested to delamination. The laminates were chosen so that the delamination occurred prior to transverse cracking, except for the $(0/\pm45/90)_s$ laminate under tensile loading. The onset of delamination was first observed via visual inspection during loading and later by inspecting the laminate through the microscope. Predictions of onset of delamination strain are made on the basis of average stress failure criterion [*6*] on τ_{xz} calculated using Ref *1*. Significant out-of-plane stress components, σ_z, τ_{xz}, and τ_{yz}, for some of the laminates are given in graphical form. Photomicrographs for representative laminates showing the onset of delamination at appropriate interfaces are also given. A table summarizing the significant results of this paper is presented. Curing stresses are not included in the analysis, although their effect on the failure of laminates may be significant.

Problem Description

Figure 1 shows the laminate geometry, coordinate axis, and the direction of applied displacement. The magnitude of the stress components is dependent upon the stacking sequence and the applied load. Any one of these stress components or a combination thereof may be responsible for failure of the laminate.

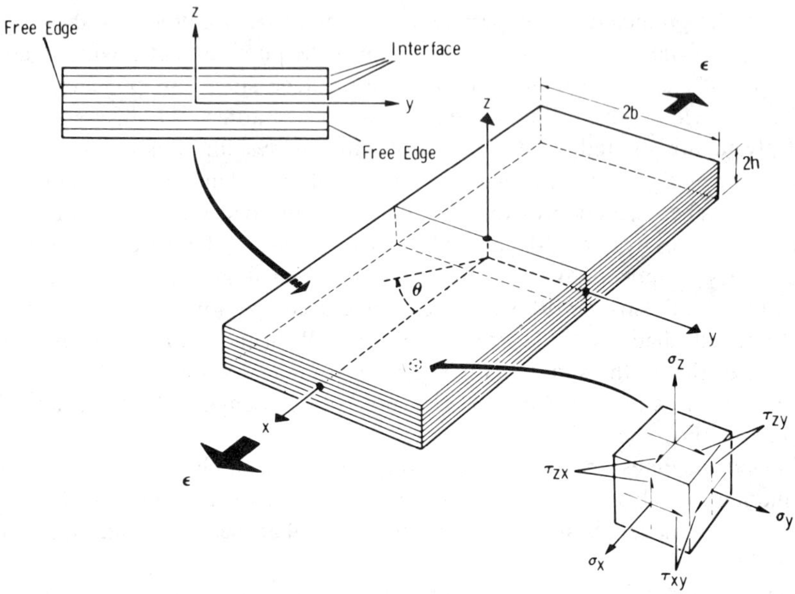

LAMINATE GEOMETRY

FIG. 1—*Boundary-value problem.*

The distribution of interlaminar stress components, σ_z, τ_{xz}, and τ_{yz}, for a number of laminates made of T300/1034-C graphite/epoxy material have been determined by the Global-local model [*1*]. The following material properties were determined for the stress analysis:

$$E_1 = 137.9 \text{ GPa (20 Msi)}$$

$$E_2 = E_3 = 9.653 \text{ GPa (1.4 Msi)}$$

$$G_{12} = G_{13} = 5.516 \text{ GPa (0.8 Msi)}$$

$$G_{32} = 4.137 \text{ GPa (0.6 Msi)}$$

$$\nu_{12} = \nu_{13} = 0.3$$

$$\nu_{23} = 0.6$$

where E, G, ν, and Subscripts 1, 2, and 3 have their usual meaning. It will be convenient to introduce a normalized parameter, along the width

$$Y = \frac{y - b + h}{h} \qquad b - h < y < b$$

Thus, the stress distribution is given in the free-edge region.

Onset of Delamination—Predictions and Experiments

A large number of composites were analyzed [1] to make a choice of laminates in which τ_{xz} predominates. On the basis of this analysis, the following laminates were chosen for further investigation of onset of delamination:

$$(0_n/\phi_n/-\phi_n)_s \qquad n = 2,4; \quad \phi = 15°,30°$$

$$(\phi_2/0_2/-\phi_2)_s$$

$$(\phi_m/-\phi_m)_s \qquad m = 3,4$$

$$(0_3/45_3/-45_3/90_3)_s$$

The stress distribution computed on the basis of the Global-local laminate model [1] at different interfaces of the laminate for the aforementioned laminates is given in Figs. 2 through 8. The interface for which the results are given in the figures

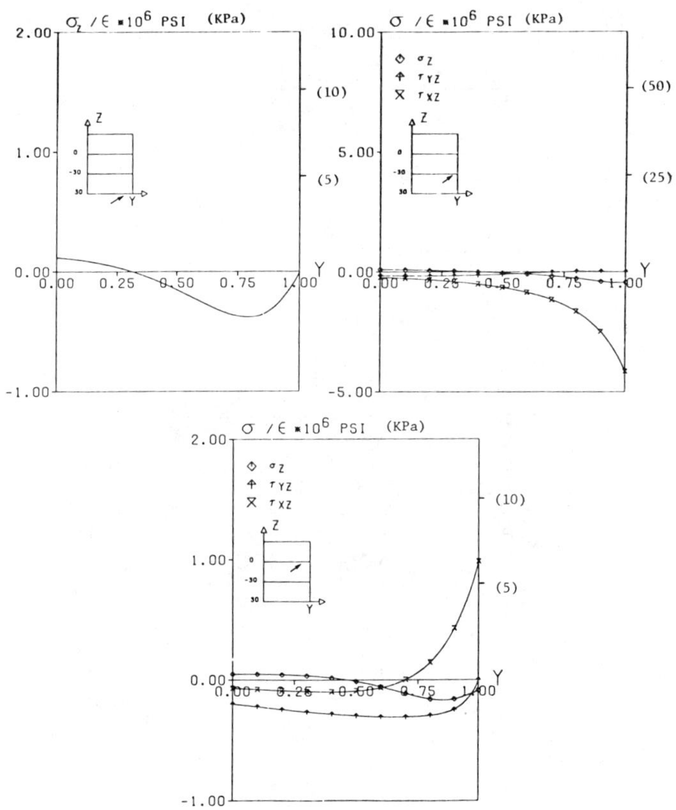

FIG. 2—*Variation of stress components,* σ_z, τ_{xz}, *and* τ_{yz}, *for a* $(0/\pm30)_s$ *laminate along the width at different interfaces.*

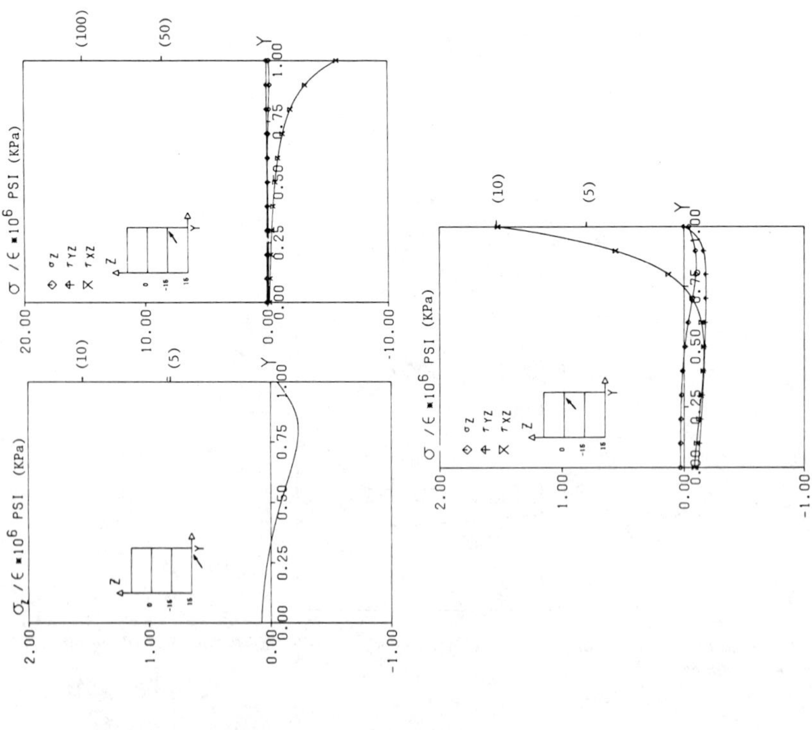

FIG. 3—*Variation of stress components,* σ_z, τ_{xz}, *and* τ_{yz}, *for a* $(0/\pm 15)_s$ *laminate along the width at different interfaces.*

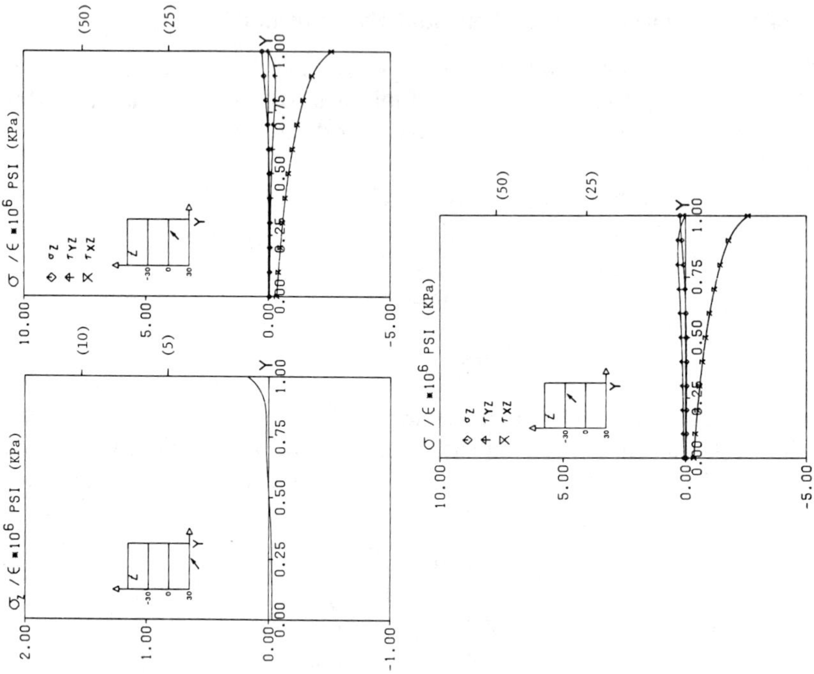

FIG. 4—*Variation of stress components,* σ_z, τ_{xz}, *and* τ_{yz}, *for a* $(30/0/-30)_s$ *laminate along the width at different interfaces.*

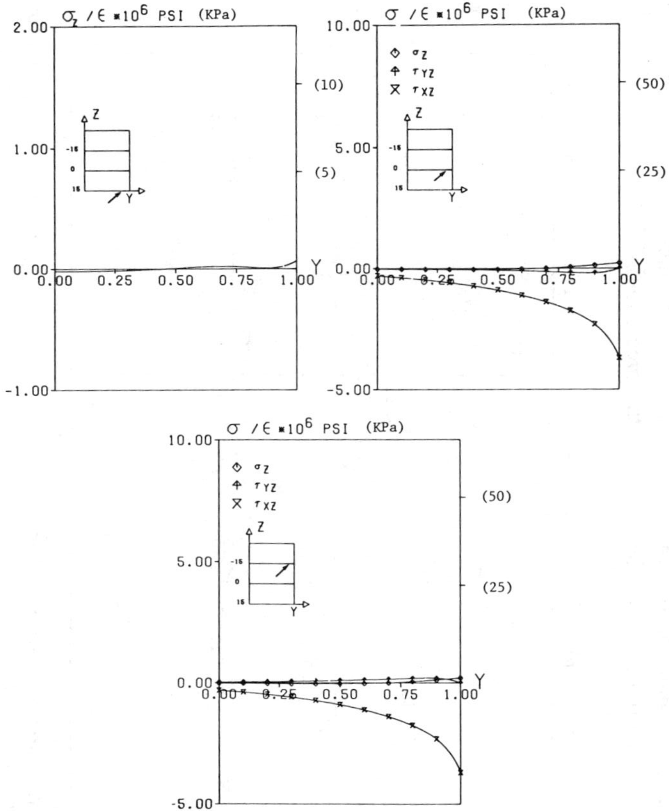

FIG. 5—*Variation of stress components,* σ_z, τ_{xz}, *and* τ_{yz}, *for a* $(15/0/-15)_s$ *laminate along the width at different interfaces.*

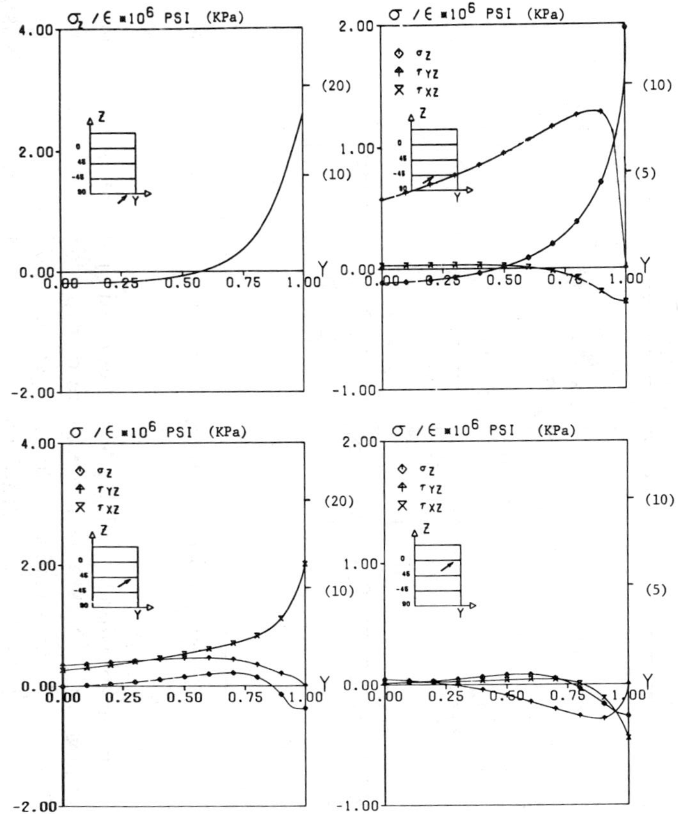

FIG. 6—*Variation of stress components,* σ_z, τ_{xz}, *and* τ_{yz}, *for a* $(0/45/-45/90)_s$ *laminate along the width at different interfaces.*

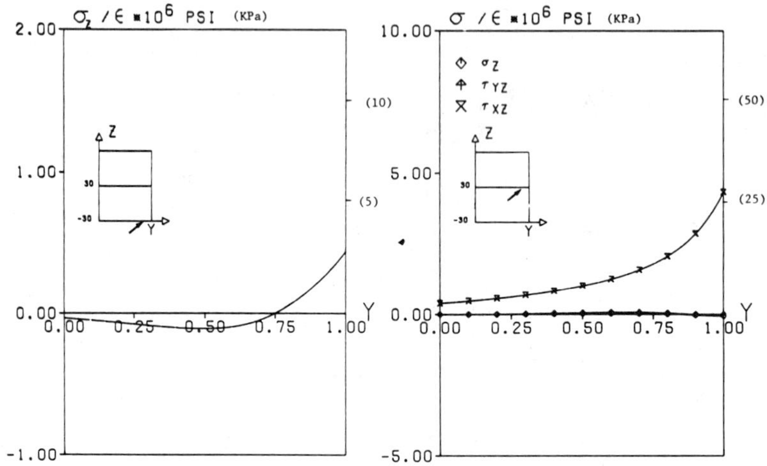

FIG. 7—*Variation of stress components,* σ_z, τ_{xz}, *and* τ_{yz}, *for a* $(\pm 30)_s$ *laminate at different interfaces.*

is marked by an arrowhead. In all the laminates considered here, τ_{xz} is predominant, except in $(0/\pm 45/90)_s$ laminate where σ_z is predominant; however, the $(0/\pm 45/90)_s$ laminate delaminates due to τ_{xz} at the $45/-45$ interface under compressive load. In the stress distribution, Figs. 2–8, the horizontal axis is normalized so that the whole scale corresponds to the half laminate thickness. Thus, Fig. 2 is valid for $(0_2/\pm 30_2)_s$ as well as $(0_4/\pm 30_4)_s$ laminates. The difference lies in the interpretation of the results in the prediction of failure.

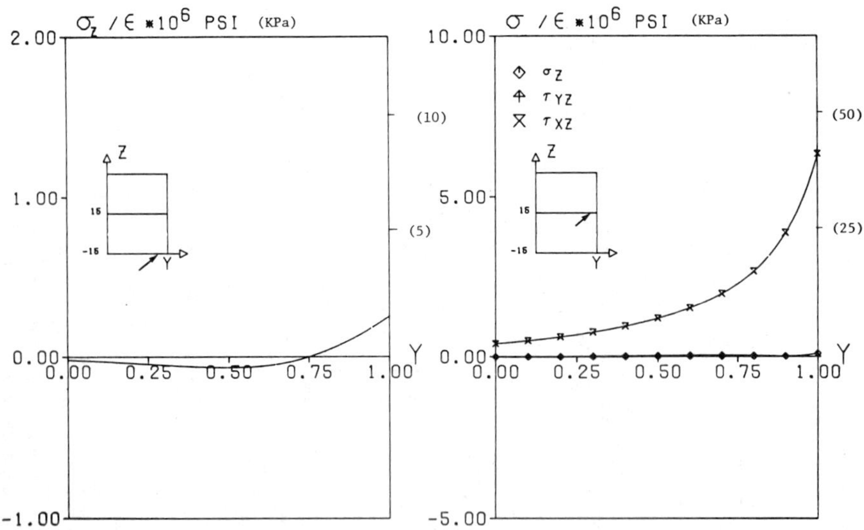

FIG. 8—*Variation of stress components,* σ_z, τ_{xz}, *and* τ_{yz}, *for a* $(\pm 15)_s$ *laminate along the width at different interfaces.*

For the prediction of the interface at which the onset of delamination occurs, we inspect the stress levels at the interfaces. The interface having the lowest relevant directional strength to stress component (in this case, average stress component) ratio should develop damage. At delamination, the average value of τ_{xz} on the appropriate interface over a distance of h_o from the free edge, where h_o is the single-ply thickness of the material, reaches the shear strength of the material. In computing the delamination threshold strain in this study, we have assumed that the interlaminar shear strength of the material is equal to its inplane shear strength (13.5 ksi or 93.1 MPa). Further, we have assumed that the interlaminar shear strength is not dependent upon the orientations of the layers at the delaminated interface. Table 1 shows a comparison between the predicted and observed incipient delamination strains. This table also shows the corresponding interface at which the delamination was predicted. The observed delaminated interfaces were the same as those predicted. The average stress component, $\bar{\sigma}_z$, given in Table 1 helps to explain the difference in experimental onset of delamination strain between tensile and compressive applied strains as discussed later.

Figures 9 through 19 show the photomicrographs and X-rays of laminates at the onset of delamination. Figure 9 shows the onset of delamination of the $(0_4/30_4/-30_4)_s$ laminate under static tension. In the upper-right photomicrograph, the delamination is, as expected, at the $30/-30$ interface, whereas at another location of the sample shown in the upper-left photomicrograph, the delamination is accompanied by transverse cracks. It appears that the development of cracks is due to redistribution of loads in different plies after delamination in the ply having highest transverse stress. Under tensile loading of the laminate, the first-ply failure strain, predicted on the basis of Tsai-Wu [7] failure criterion, is about 0.9%, which is close to the onset of delamination strain of about 0.85% given in Table 1. Since the transverse compressive strength of the material is more than five times higher than its tensile strength, the compression test can eliminate the transverse cracking of $-30°$ layers. This fact has been demonstrated by tests under compressive loads. Figure 10 shows the photomicrographs and X-rays of onset of delamination of a $(0_4/\pm30_4)_s$ laminate under static compression. Table 1 shows the predicted and observed onset of delamination strains and predicted average stress components $\bar{\sigma}_z$ and $\bar{\tau}_{xz}$ for this laminate. The difference between the compressive and the tensile strains at onset of delamination is due to the fact that under compressive loads the effective stresses ($\bar{\sigma}_z$) developed near the free edge is tensile, whereas under tensile loads this stress is compressive. The interaction of σ_z and τ_{xz} in composites is studied by Pagano and Kim [*11*], and the present observation is consistent with their results. A similar behavior was observed for $(0_4/\pm15_4)_s$ laminates, Figs. 11 and 12. Figure 13 shows the onset of delamination under compressive loads in $(0_2/\pm15_2)_s$ and $(0_2/\pm30_2)_s$ laminates. In these cases, too, the onset of delamination was at the predicted interfaces and at predicted and experimental strains shown in Table 1.

Figures 4 and 5 show the interlaminar stress distribution for $(30/0/-30)_s$ and $(15/0/-15)_s$ laminates. On the basis of this analysis, the onset of delamination

TABLE 1—*Facts and figures for onset of delamination.*

Laminate	Observed ϵ (%) at Delamination	Load Type	Average Stress Components for ϵ = 0.1%			Predicted ϵ (%) at Delamination
			$\bar{\sigma}_z$, MPa	$\bar{\tau}_{xz}$, MPa	Interface	
$(0_4/30_4/-30_4)_s$	0.347	static compression	3.03	25.5	±30	0.365
	0.849	static tension	−3.03	−25.5	...	0.365
$(0_2/30_2/-30_2)$	0.537	static compression	2.83	20.0	±30	0.468
$(0_4/15_4/-15_4)_s$	0.349	static compression	1.45	34.5	±15	0.269
	0.570	static tension	−1.45	−34.5	...	0.269
$(0_2/15_2/-15_2)_s$	0.514	static compression	1.59	26.3	±15	0.354
$(30_2/0_2/-30_2)_s$	...	fatigue	1.22	−13.7	0/−30	0.678
$(15_2/0_2/-15_2)_s$	...	fatigue	1.24	−22.8	0/−15	0.409
$(0_3/45_3/-45_3/90_3)_s$	0.363 FPF[a]	static tension	14.5	0	midplane	0.362 DEL[b]
	0.516 DEL[b]	static tension				
	0.706	static compression	2.55	−11.9	±45	0.784
		C-C[c] fatigue 10 K cycle			±45	
$(15_3/-15_3)_s$	0.390	static compression	−0.2	−29.6	±15	0.314
$(15_4/-15_4)_s$	0.350	static compression	−0.28	−34.1	±15	0.270
$(30_3/-30_3)_s$	0.46	static compression	0.2	−22.5	±30	0.431
$(30_4/-30_4)_s$	0.41	static compression	0.3	−24.2	±30	0.385

[a]FPF = first-ply failure.
[b]DEL = delamination.
[c]C-C = compression-compression.

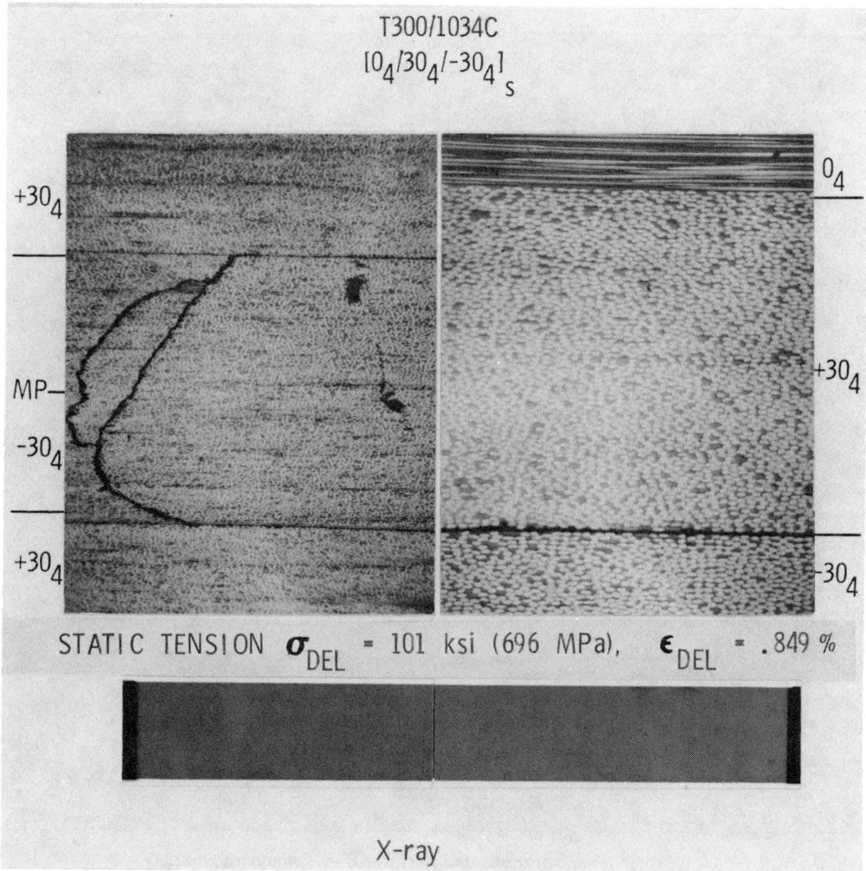

FIG. 9—*Photomicrographs and X-ray of a delaminated laminate.*

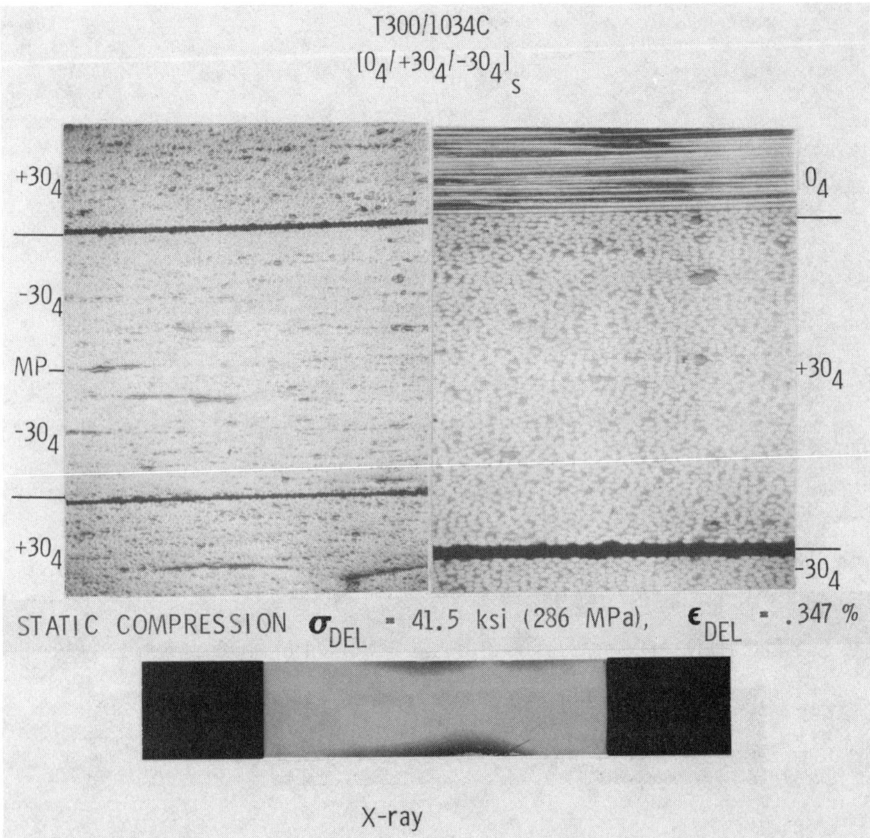

FIG. 10—*Photomicrographs and X-ray of a delaminated laminate.*

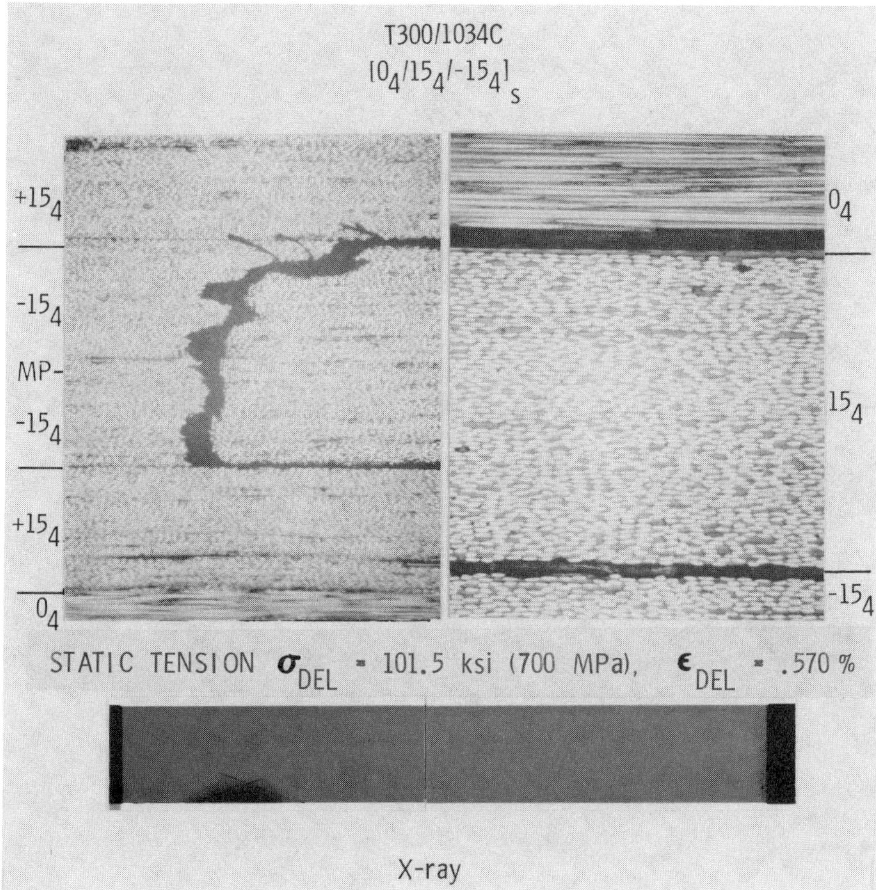

FIG. 11—*Photomicrographs and X-ray of a delaminated laminate.*

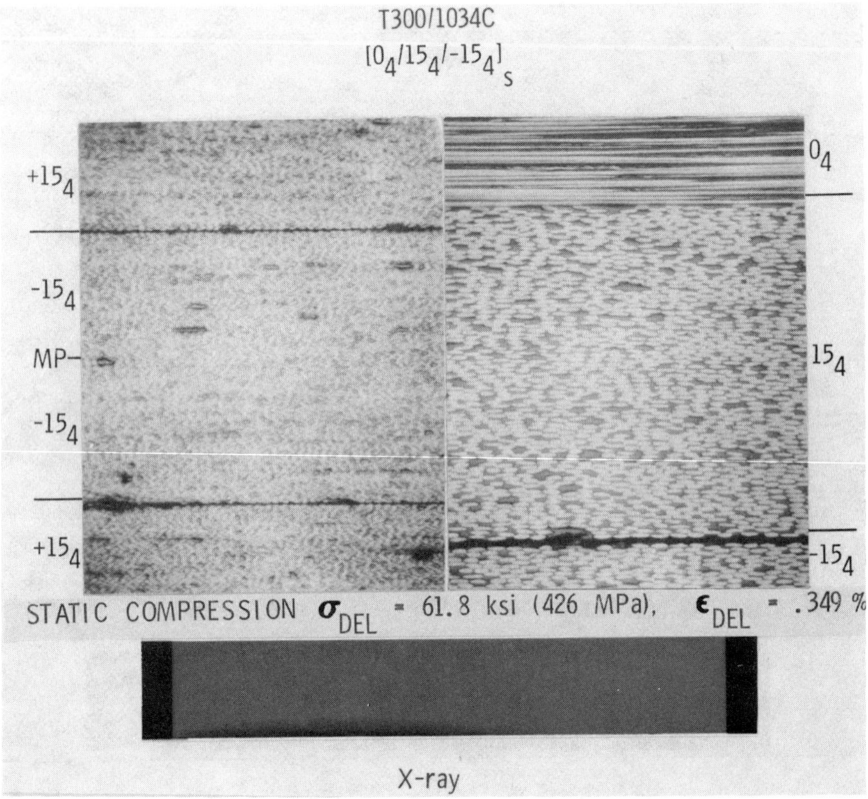

FIG. 12—*Photomicrographs and X-ray of a delaminated laminate.*

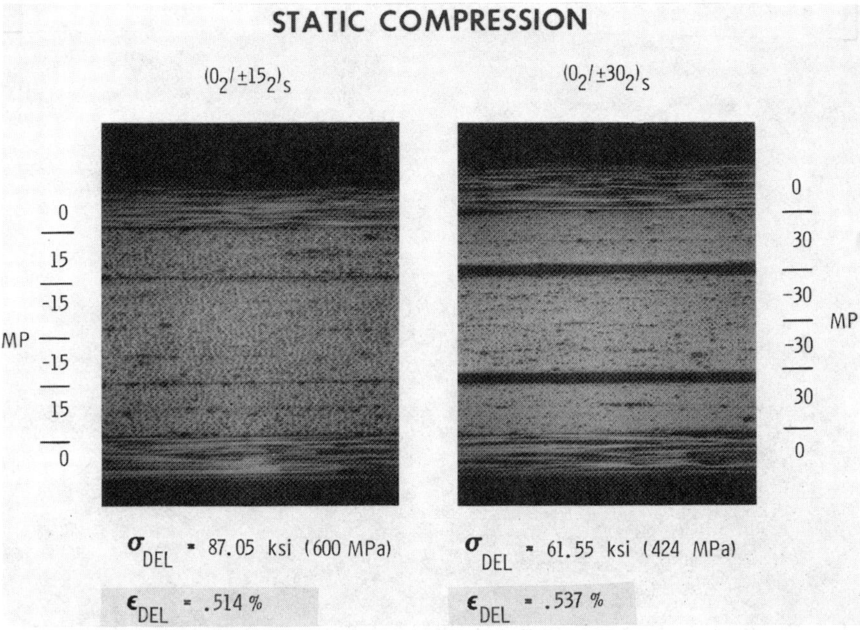

FIG. 13—*Photomicrographs of a delaminated laminate.*

should take place either at the 30/0 or at the 0/ −30 interface. On static tension testing of these laminates, it was found that the laminates failed instantaneously without any prior visual sign of delamination. We expect that the delamination and the first-ply failure occurred at almost the same time and that caused catastrophic failure in the laminate. To confirm this understanding, we fatigued the samples under tension-tension fatigue and compression-compression fatigue loads. Figures 14 through 16 show the photomicrographs and X-rays of delaminated laminates, and it has been shown that the delamination takes place at the predicted interfaces.

Figure 6 shows the stress analysis of a $(0/\pm45/90)_s$ laminate. According to this analysis, the delamination should take place at the midsurface due to σ_z, under tensile loadings. For the compressive loading condition, the delamination should take place at the 45/ −45 interface due to interlaminar shear stress component, τ_{xz}. In Fig. 17, the upper-left photomicrograph shows the development of transverse cracks before the onset of delamination (that is, first-ply failure), the upper-right photomicrograph shows the onset of delamination, and lower photomicrograph shows a region having transverse cracks and the onset of delamination at the midsurface and −45/90 interface. Because of the first-ply failure prior to delamination of the laminate, our analysis may not be valid for the prediction of onset of delamination under tensile loads. The considerably higher transverse compressive strength of the lamina enables us to exhibit the

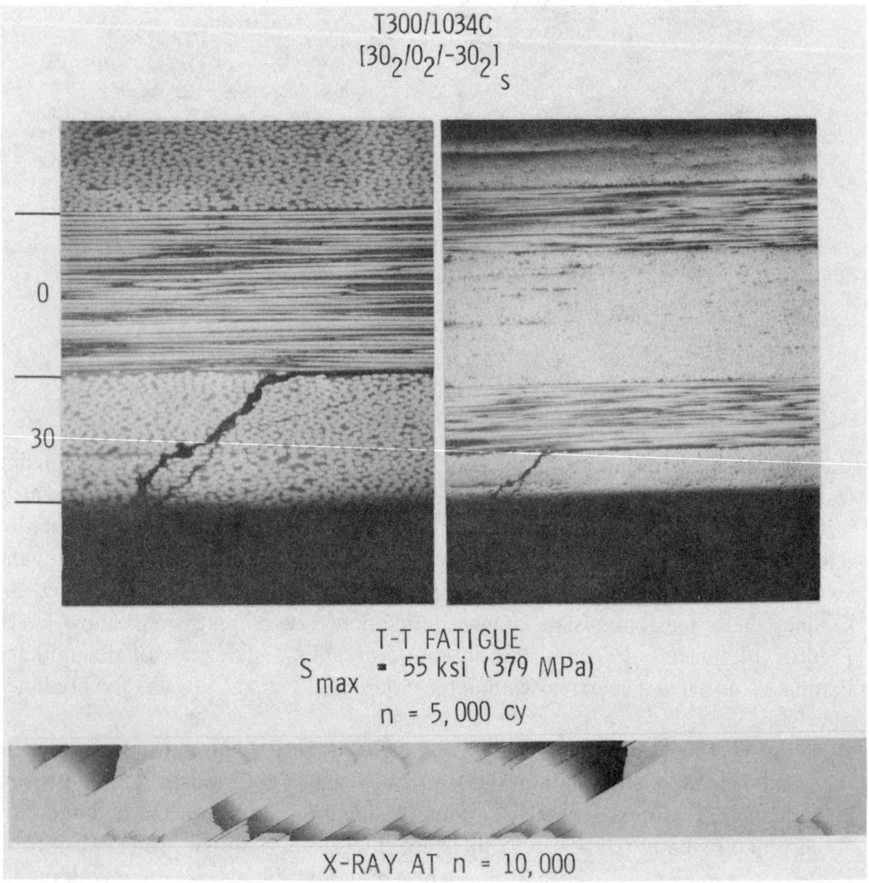

FIG. 14—*Photomicrographs and X-ray of a delaminated laminate.*

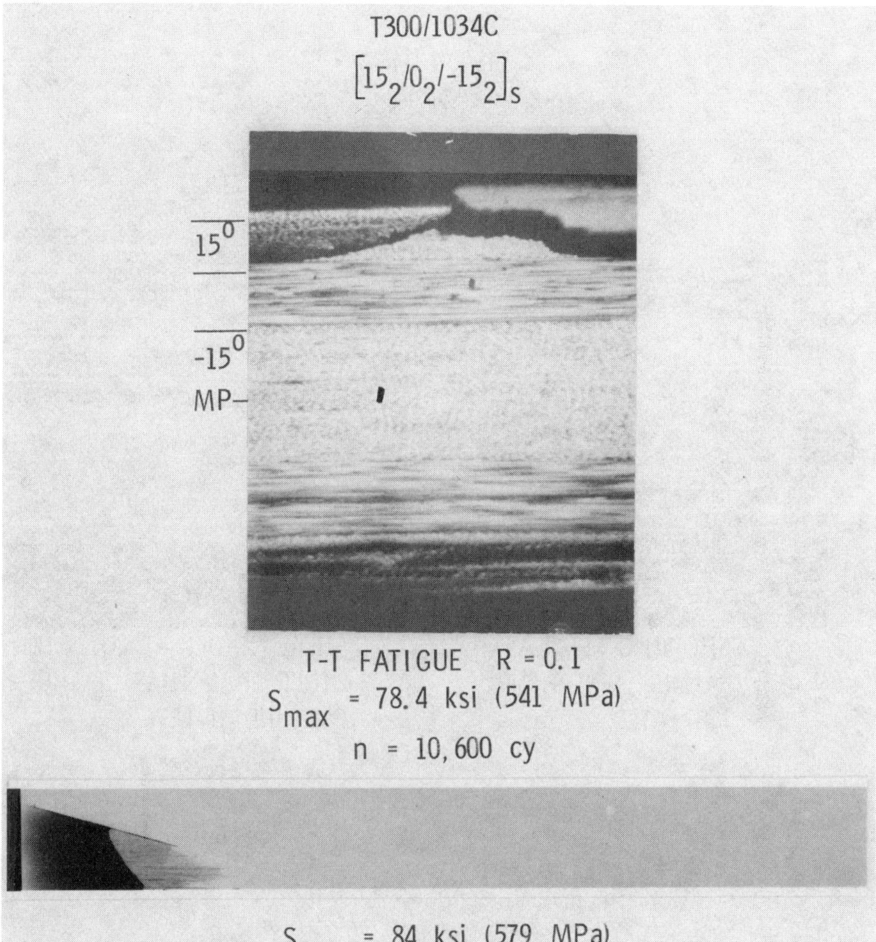

FIG. 15—*Photomicrographs and X-ray of a delaminated laminate.*

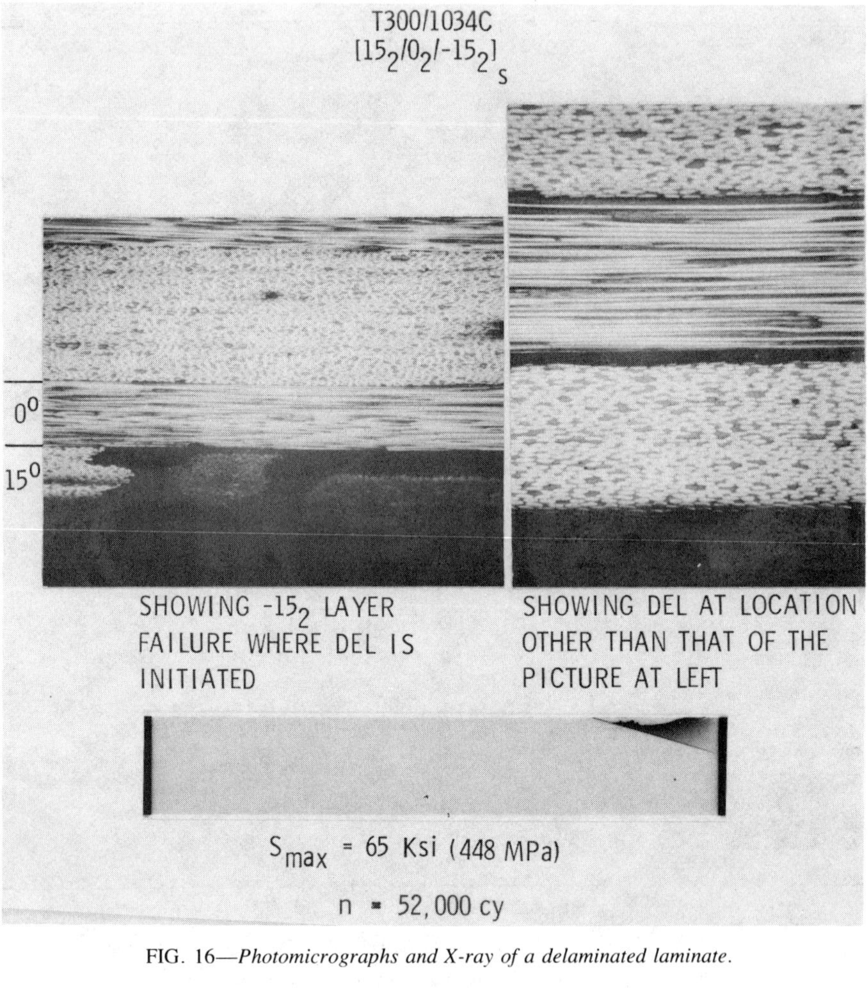

FIG. 16—*Photomicrographs and X-ray of a delaminated laminate.*

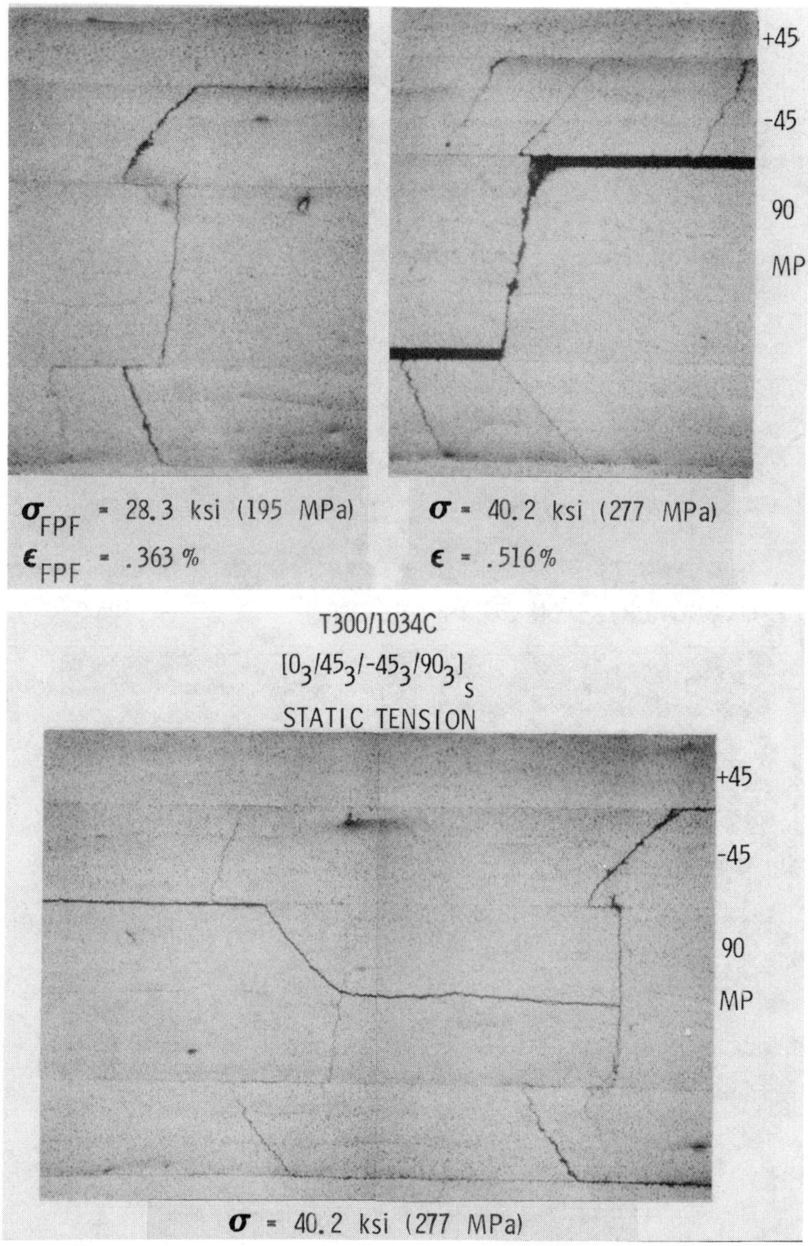

FIG. 17—*Photomicrographs of a delaminated laminate.*

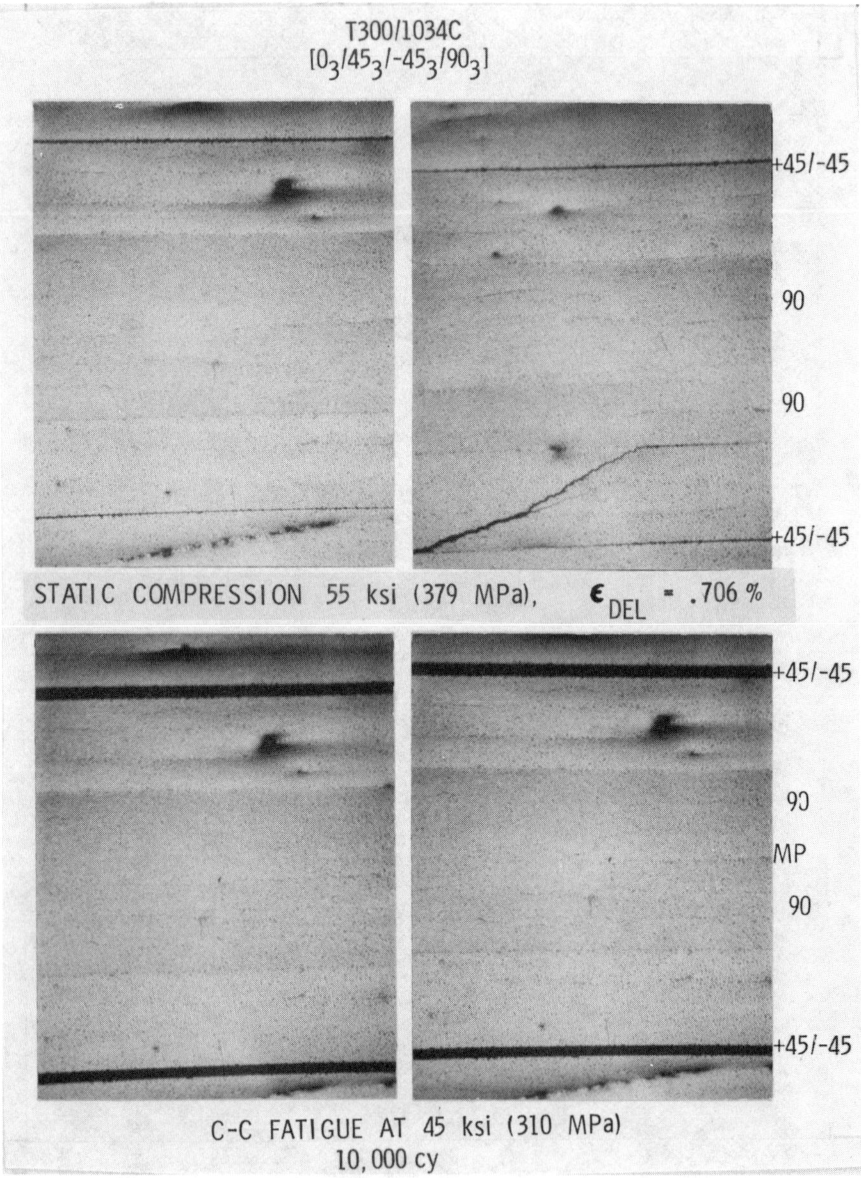

FIG. 18—*Photomicrographs of a delaminated laminate.*

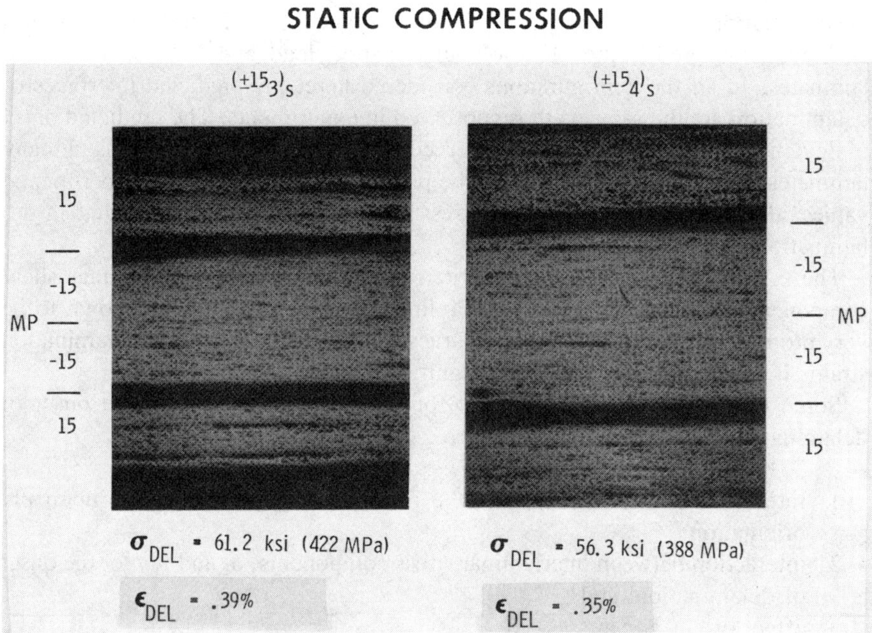

FIG. 19—*Photomicrographs of a delaminated laminate.*

onset of delamination without any other prior failure in this laminate. Figure 18 contains the photomicrographs of onset of delamination, top at two locations under static compressive load and bottom at two locations under compressive-compressive fatigue; in both cases the delamination took place at the predicted locations. The predicted and observed onset of delamination strains for static tension and static compression tests are given in Table 1. In this table, the value (0.516% strain) corresponds to a load level at which the photomicrographs of Fig. 17 were taken and delamination had taken place prior to this load.

Figure 19 shows the photomicrographs of laminates $(\pm 15_3)_s$ and $(\pm 15_4)_s$ at the onset of delamination. The onset of delamination interface is $15/-15$ interface in both the cases and is the same as expected from the stress analysis given in Fig. 8. The observed and predicted onset of delamination strains for both of these laminates are given in Table 1. Since the behavior of the onset of delamination in $(\pm 30_3)_s$ and $(\pm 30_4)_s$ laminates was similar to that for $(\pm 15_3)_s$ and $(\pm 15_4)_s$ laminates, the photomicrographs for these laminates are not given. The predicted and observed onset of delamination strains in these cases as given in the table are close.

Conclusions

Onset of delamination, in layered composites at the free edge, stimulated by interlaminar shear has been studied. The stress analysis, on the basis of Global-local variational model employed in the average stress failure criterion, has enabled us to predict the onset of delamination load and location in all the laminates. In all the configurations considered here, the predicted interfaces of delamination are the same as those observed in experiments. The predicted onset of delamination strains is close to observed onset of delamination strains in many laminates. In those laminates where the predictions are away from experimental values, the interaction of different stress components may be important in delamination.

The effectiveness of the failure criterion based upon the interlaminar shear component has been demonstrated. Preliminary results have been shown to be very promising, although a loss of accuracy in the predicted onset of delamination strains is high in cases where σ_z is significant.

Some of the necessary ingredients for the better prediction of the onset of delamination load are:

1. interlaminar shear strength of the material and its dependence upon ply orientation;
2. interaction between interlaminar stress components, σ_z and τ_{xz}, for the onset of delamination; and
3. effect of curing stresses.

These studies will be the subject of future investigations.

Acknowledgments

The authors wish to acknowledge R. Esterline of the University of Dayton Research Institute for preparation and testing of specimens, and Dr. N. J. Pagano of the Air Force Materials Laboratory for his helpful suggestions on this study. This work was sponsored by the Nonmetallic Materials Division, Air Force Materials Laboratory, under Contract No. F33615-81-C-5056.

References

[1] Pagano, N. J. and Soni, S. R., *International Journal of Solids and Structures*, Vol. 19, No. 3, 1983, pp. 207–288.

[2] Soni, S. R. and Pagano, N. J. in *Proceedings*, IUTAM Symposium on Mechanics of Composite Materials, 16–19 Aug. 1982, Virginia Polytechnic Institute and State University, Blacksburg, VA, pp. 227–241.

[3] Kim, R. Y. and Soni, S. R., "Initiation of Delamination of Composite Laminates," *Proceedings*, 1982 Joint SESA and JSME Conference on Experimental Mechanics, Society for Experimental Stress Analysis, Oahu-Maui, Hawaii, May 1982.

[4] Soni, S. R. in *Progress in Science and Engineering of Composites*, Vol. I, Japan Society for Composite Materials, T. Hayashi, K. Kawata and S. Umekawa, Eds., 1982, pp. 251–260; also, *Proceedings* ICCM-IV, Tokyo, Japan.

[5] Soni, S. R. in *Proceedings*, 28th National SAMPE Symposium and Exhibition, Anaheim, CA, Society for the Advancement of Materials and Process Engineering, 12–14 April 1983, pp. 189–199.

[6] Kim, R. Y. and Soni, S. R., *Journal of Composite Materials*, Vol. 18, Jan. 1984, pp. 70–80.

[7] Soni, S. R. in *Proceedings,* 24th Structures, Structural Dynamics and Materials Conference, AIAA/ASME/ASCE/AHS, Lake Tahoe, CA, May 1983, pp. 171–179.
[8] Kim, R. Y. in *Proceedings,* 28th National SAMPE Symposium and Exhibition, Anaheim, CA, 12–14 April 1983, pp. 200–209.
[9] Whitney, J. M. and Nuismer, R. J., *Journal of Composite Materials,* Vol. 8, July 1974, pp. 253–265.
[10] Pagano, N. J. and Pipes, R. B., *International Journal of Mechanical Sciences,* Vol. 17, 1973, pp. 679–688.
[11] Pagano, N. J. and Kim, R. Y., "Torsion Test of Carbon-Carbon Composites," private communication.
[12] Pagano, N. J., *International Journal of Solids and Structures,* Vol. 14, 1978, pp. 385–400.
[13] Pagano, N. J. and Rybicki, E. F., *Journal of Composite Materials,* Vol. 8, July 1974, pp. 214–228.

Mohan M. Ratwani[1] *and Han-Pin Kan*[1]

Analysis of Delamination Propagation in the Composite Layer of Metal-to-Composite Stepped-Lap Joints

REFERENCE: Ratwani, M. M. and Kan, H.-P., "**Analysis of Delamination Propagation in the Composite Layer of Metal-to-Composite Stepped-Lap Joints,**" *Composite Materials: Testing and Design (Seventh Conference)*, J. M. Whitney, Ed., American Society for Testing and Materials, Philadelphia, 1986, pp. 308–321.

ABSTRACT: The initiation and propagation of delaminations in the composite layers of typical step-lap joints, with and without bondline flaws, has been investigated under typical spectrum loading. The typical bondline flaws considered are end gaps and debonds at various steps. A mathematical analysis of the stress in the joints is first carried out with the composite taken as a continuum, and the critical locations of delamination initiation are determined. The interlaminar stress analysis of critical areas in the composite is then carried out with the NASTRAN finite-element computer code. The location and magnitude of the maximum interlaminar stresses is determined. Using these interlaminar stresses in a delamination propagation model, the possibility of initiation and propagation of delamination in the composite layers is investigated. It is shown that the delaminations do not grow to failure under spectrum load in typical joints, even after four lifetimes of spectrum loading.

KEY WORDS: composite materials, delamination, fatigue (materials), interlaminar stresses, residual strength, spectrum loading

Adhesive bonded joints have features that make them attractive for advanced composite-to-metal joining and are currently being used in several important primary aircraft structural applications. The most significant concerns regarding the structural reliability of such types of bonded joints are possible delaminations occuring in the composite layer and debonding taking place in the adhesive during fatigue loads. In order to utilize adhesive bonding of metal-to-composite joints to their fullest potential, improved analysis methods must be developed particularly in the area of fatigue life and residual strength predictions. This can be achieved more effectively if the fatigue behavior of composite-to-metal joints is properly understood and various factors influencing the delamination of composites in the joint are thoroughly investigated.

[1] Manager and engineering specialist, respectively, Strength and Life Assurance Research, Northrop Corporation, Hawthorne, CA 90250.

The fatigue failure of a bonded joint, with or without initial flaws, may be due to the failure of either the adhesive or adherend. A flaw may develop (if not preexisting) in the adhesive, thus causing local debonding that may grow under cyclic loading. The propagation of these local debonds in the adhesive joint will depend on the stress level and stress concentration around the debond. As a result of the debonding and consequent local load redistributions, a flaw may also initiate in the adherend and propagate. Thus, the flaw may initiate in the adhesive or adherend at any of the steps of a step-lap joint and propagate. The resulting load transfer to the composite may initiate failure in the composite. Failure in composites generally initiates as a delamination within the laminate and propagates under subsequent flight loads.

The stress analysis methods of bonded joints, using mathematical and finite-element techniques, have been discussed by various investigators; some of these methods are discussed in Refs *1* to *14*. The majority of the analytical work reported in the literature deals with adhesive joints without bondline flaws. The analysis of adhesive joints with flaws has been carried out in Refs *15* and *16*. Mathematical and finite-element methods of analyses have been used in these references. In all the analytical work on composite-to-metal bonded joints reported in the literature, the composite is treated as an anisotropic or orthotropic material, and the plies are generally lumped together, with the result that the state of stress between the plies (interlaminar stresses) is not known. Hence, these analyses cannot be used to study delamination initiation or propagation in the composite layer of composite-to-metal bonded joints.

The objective of this study is to develop a method capable of predicting delamination initiation and propagation in the composite layer of composite-to-metal bonded joints. The delamination propagation model (Ref *17*), based on the interlaminar stresses in a composite laminate, has been used to predict delamination initiation and propagation in the composite layer of composite-to-metal bonded joints. A four-step joint configuration has been investigated. Loading conditions under which delaminations grow to cause failure have also been investigated.

Analysis Procedure

The analysis method for predicting delamination initiation and propagation in the composite layer of a step-lap joint consists of the following steps.

1. Perform a mathematical analysis of the step-lap joint, assuming bondline flaws likely to occur in service or fabrication. In this analysis, the composite is considered to be an orthotropic continuum. A method of analysis is presented in the next section of this paper.
2. Determine the critical areas of the joint where the shear stress in the adhesive is large and the stress in the composite is maximum. These critical areas are generally near the first and the last steps.
3. Perform three-dimensional finite-element analyses of the critical areas of

the joint, modeling each ply in the composite as a separate layer. The forces, obtained from the mathematical analysis, are applied as nodal forces in the finite-element model.

4. Determine the location and the magnitude of the maximum interlaminar stresses in the joint from the finite-element analysis.
5. Determine the fatigue load spectrum and the maximum load in the spectrum to which the joint is subjected.
6. Determine the maximum interlaminar stress, corresponding to the maximum load in the spectrum as well as any thermal mismatch in the joint.
7. If the maximum interlaminar shear stress exceeds the fatigue threshold of the resin in the composite material, the delamination will propagate, otherwise it will not.
8. Using the interlaminar stresses in the fatigue spectrum, predict the fatigue life of the joint under compression-dominated spectrum loading.

Analysis of Stepped-Lap Joint with Bondline Flaws

The elastic analyses of bonded joints with flaws in the form of end gaps and debonding at various steps (Fig. 1) under external applied loads have been discussed in Ref *16*. The analysis of the joint under applied loading and thermal mismatch between metallic and composite layers is developed here. Consider the stepped-lap joint shown in Fig. 1. The axial strains in Material 1, ϵ_1 (metallic), and Material 2, ϵ_2 (composite) at a point, distance x, in the ith step are given as

$$\begin{aligned} \epsilon_1 &= \frac{du_1}{dx} = \frac{1 - \nu_1^2}{E_1 h_{1i}} [p_0 - p_{2i}(x)] + \alpha_1 \Delta T \\ \epsilon_2 &= \frac{du_2}{dx} = \frac{1 - \nu_{2x}{}^i \nu_{2z}{}^i}{E_{2x}{}^i h_{2i}} p_{2i}(x) + \alpha_{2i} \Delta T \end{aligned} \qquad (1)$$

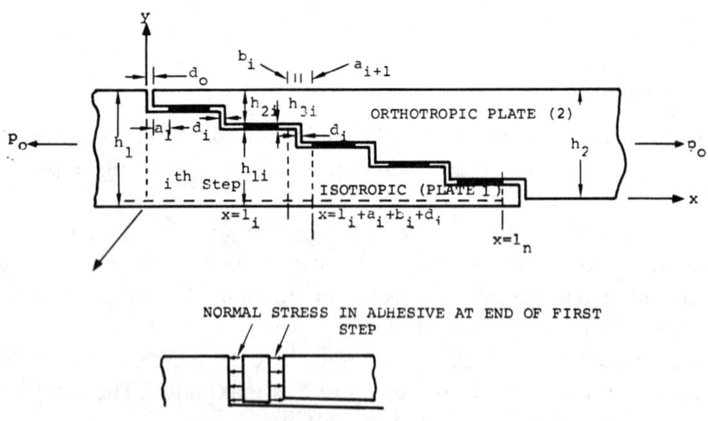

FIG. 1—*Stepped-lap joint with bondline flaws.*

where

$p_{2i}^{(x)}$ = the force in composite material at ith step,
p_0 = the remote applied load,
h_{1i} = the metallic layer thickness at ith step,
h_{2i} = the composite layer thickness at ith step,
E_1 = Young's modulus of metallic material,
ν_1 = Poisson's ratio of metallic material,
$E_{2x}^{\ i}$ = Young's modulus of the composite in x direction at ith step,
$\nu_{2x}^{\ i}$, $\nu_{2z}^{\ i}$ = x and z direction Poisson's ratios, respectively, of composite layer at ith step,
α_1 = the coefficient of thermal expansion of metallic Material 1,
α_{2i} = the coefficient of thermal expansion of Material 2 (composite) in the ith step, and
ΔT = the change in temperature.

Following the procedure outlined in Refs *1* and *16*, it can be shown that the governing differential equation for the force in the ith step of the composite layer can be written as

$$\frac{d^2 p_{2i}}{dx^2} - \phi_i^2\, p_{2i} = \beta_i p_0 + \gamma_i \Delta T \tag{2}$$

where

$$\gamma_i = \frac{G_3}{h_3}\,(\alpha_{2i} - \alpha_1), \tag{3}$$

$$\phi_i^2 = \frac{G_3}{h_{3i}}\left[\frac{1 - \nu_1^2}{E_1 h_{1i}} + \frac{1 - \nu_{2x}^{\ i}\nu_{2z}^{\ i}}{E_{2x}^{\ i} h_{2i}}\right]; \tag{4}$$

$$\beta_i = \frac{G_3}{h_{3i}}\,\frac{1 - \nu_1^2}{E_1 h_{1i}} \text{ (for plane strain)},$$

G_3 = the adhesive shear modulus, and

h_{3i} = the adhesive layer thickness at the ith step.

The solution of differential Eq 2 is given by

$$p_{2i}\,(x) = A_i e^{-\phi_i x} + B_i e^{\phi_i x} - \frac{\beta_i}{\phi_i^2}\, p_0 - \frac{\gamma_i}{\phi_i^2}\,\Delta T \tag{5}$$

The boundary conditions are given by

$$p_{21}(a_1) = 0$$
$$p_{2n}(\ell_n) = p_0 \tag{6}$$

$$p_{2i}(\ell_i) = p_{2,i+1}(\ell_i + a_{i+1} + b_i) \qquad (i = 1,2 \ldots n - 1) \tag{7}$$

$$\frac{d}{dx} p_{2,i+1}(\ell_i + a_{i+1} + b_i) = \frac{h_{3i}}{h_{3,i+1}} \frac{d}{dx} p_{2i}(\ell_i) + \Delta\tau_i \; (i = 1,2,\ldots n - 1) \tag{8}$$

where

$$\Delta\tau_i = \frac{G_3}{h_{3,i+1}}(\Delta u_2 - \Delta u_1),$$

$$\Delta u_2 = \epsilon_2^1 b_i + \epsilon_2^2 a_{i+1} + (\alpha_{2i} b_i + \alpha_{2,i+1} a_{i+1}) \Delta T,$$

$$\Delta u_1 = \epsilon_1^1 (b_i - d_i) + \epsilon_1^2 (a_{i+1} + d_i) + \alpha_1 (a_{i+1} + b_i) \Delta T,$$

$$\epsilon_2^1 = \frac{1 - \nu_{2x}^i \nu_{2z}^i}{E_{2x}^i h_{2i}} p_{2i}(\ell_i), \quad \epsilon_2^2 = \frac{1 - \nu_{2x}^{i+1} \nu_{2z}^{i+1}}{E_{2x}^{i+1} h_{2,i+1}} p_{2i}(\ell_i),$$

$$\epsilon_1^1 = \frac{1 - \nu_1^2}{E_1 h_{1i}} [p_0 - p_{2i}(\ell_i)], \text{ and}$$

$$\epsilon_1^2 = \frac{1 - \nu_1^2}{E_1 h_{1,i+1}} [p_0 - p_{2i}(\ell_i)].$$

A generalized computer program has been written to solve numerically the differential Eq 2. The titanium to composite four-step joint, shown in Fig. 2, was analyzed using this computer program. The influence of thermal mismatch on the force in the composite is shown in Fig. 3. It is seen that the thermal mismatch has very little influence on the force in the composite under an applied load of 1226 N/mm (7000 lb/in.). Figure 4 shows the influence of a thermal mismatch on the shear stresses in the adhesive layer of the joint. The influence of a thermal mismatch on the adhesive shear stresses is seen to be very small. It may be noted that the influence of a thermal mismatch on the force in the composite and on the adhesive shear stresses is small due to the small difference in the coefficient of linear expansion of titanium and the composite layer in the joint.

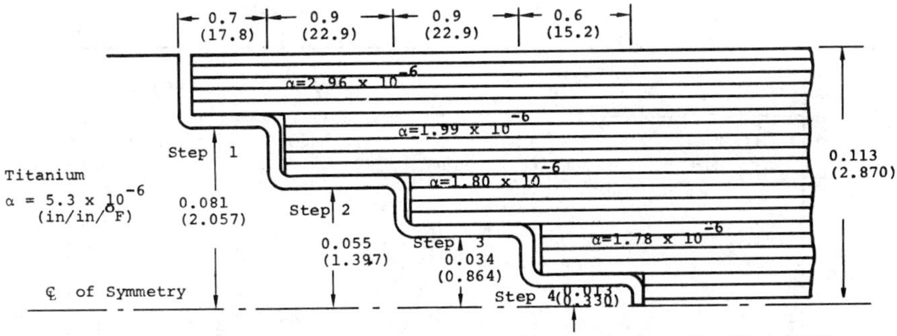

Isotropic Young's Modulus E1 = 0.160000E+8 Poisson Ratio NUI = 0.310
Adhesive Shear Modulus G3 = 0.570000E+5
Composite Property

Step	E2X	E2Z	NUX	NUZ
1	0.9749E+7	0.6584E+7	0.3145	0.2124
2	0.1122E+8	0.4954E+7	0.4140	0.1838
3	0.1231E+8	0.5349E+7	0.3131	0.1361
4	0.1207E+8	0.4862E+7	0.3700	0.1521

All properties in psi. Joint dimensions in inch with mm in parenthesis.

FIG. 2—*Geometry of four-step-lap joint.*

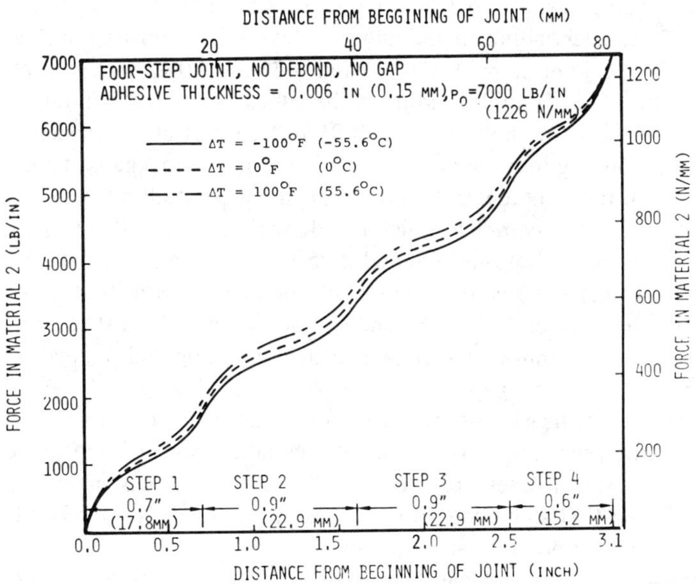

FIG. 3—*Influence of thermal mismatch on the force in Material 2 (composite).*

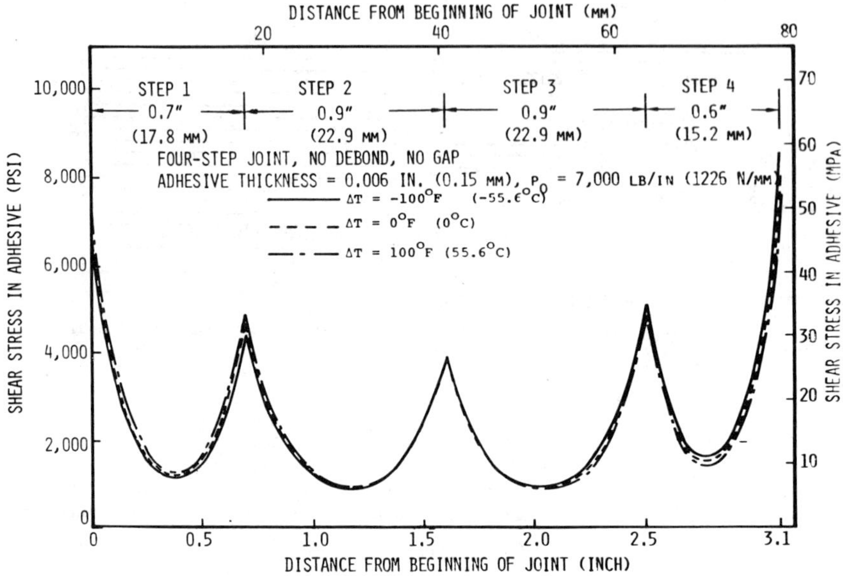

FIG. 4—*Influence of thermal mismatch on the shear stress in the adhesive.*

Types of Defects Investigated in the Example Case

The analysis of the four-step-lap joint shown in Fig. 2 was carried out for the following seven cases: (1) no debonding in the adhesive and no end gaps present, Fig. 5*a*; (2) no debonding in the adhesive, with 2.5 mm (0.1 in.) gap at the beginning of the joint ($x = 0$, Fig. 1) and 0.25 mm (0.01 in.) gaps at all other step ends, Fig. 5*b*; (3) no debonding in the adhesive, with 2.5 mm (0.1 in.) gap at the end of the joint and 0.25 mm (0.01 in.) gaps at all other step ends, Fig. 5*c*; (4) no debonding in the adhesive, with 2.5 mm (0.1 in.) gaps at the beginning and the end of the joint and 0.25 mm (0.01 in.) gaps at all other step ends, Fig. 5*d*; (5) the first step completely debonded, with 2.5 mm (0.1 in.) gaps at the beginning and the end of the joint and 0.25 mm (0.01 in.) gaps at all other step ends, Fig. 5*e*; (6) the last step completely debonded, with 0.25 mm (0.1 in.) gaps at all other step ends, Fig. 5*f*; and (7) the first and the last steps completely debonded, with 2.5 mm (0.1 in.) gaps at the beginning and the end of the joint and 0.25 mm (0.01 in.) gaps at all other step ends, Fig. 5*g*.

The stresses at the end of various steps in the four-step joint are shown in Table 1 for the preceding seven cases of debonding and end gaps. The end gap forces for the seven cases are shown in Table 2. The maximum stress in the adhesive in the first step occurs for the case when there is no debonding in the adhesive layer and no end gaps are present.

The absolute maximum shear stress in the joint occurs for the case of debonding in the adhesive layer and the end gaps present. The maximum stress in the composite near the first step occurs when the end gaps are present and there is

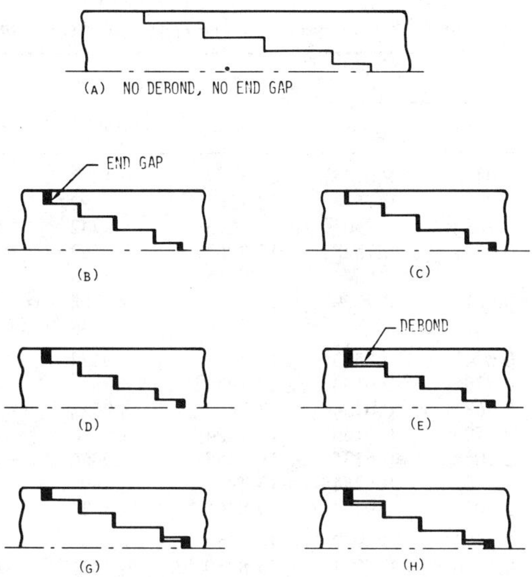

FIG. 5—*Type of defects investigated.*

no debonding in the adhesive layer. The absolute maximum stress in the composite occurs at Step 3 for the case of the fourth step being debonded and the end gaps present.

Three-Dimensional Finite-Element Modeling of the Composite in the Four-Step-Lap Joint

All four steps of the joint were modeled for the finite-element analysis. All plies in the composite are modeled as individual layers, except coupled 0° plies. The coupled 0° plies are modeled as one layer as shown in Fig. 5. The loading to be applied to the joint is obtained from the mathematical analysis described earlier in this paper. The normal loads are distributed at the nodal points in each ply, using laminate theory. The shear loads are applied along the metal-to-composite interface.

Interlaminar Stresses in the Composite Layer of the Four-Step-Lap Joint

An examination of the forces in the composite layers and the shear stresses in the adhesive indicated the critical cases for interlaminar stress analyses to be: (1) no debonding in the adhesive layer and no end gaps present, Fig. 5*a*; (2) first and last steps completely debonded and end gaps present (2.5 mm (0.1 in.) gaps at the beginning of the first and the end of the last steps and 0.25 mm (0.01 in.) gaps at all other step ends), Fig. 5*g*; (3) no debonding in the adhesive layer and end gaps present (2.5 mm (0.1 in.) gaps at the beginning of the first and the end of the last steps and 0.25 mm (0.01 in.) gaps at all other step ends), Fig. 5*d*.

TABLE 1—*Stresses at the beginning and the end of various steps in the four-step joint.*

		$\tau_a{}^a/p_0$		$\sigma_1{}^b/p_0$		$\sigma_2{}^c/p_0$	
Case	Step	Beginning	End	Beginning	End	Beginning	End
1	1	1.0418	0.6747	12.3457	9.1313	0.0000	10.0141
2	1	0.9811	0.7528	12.2263	8.9792	0.3720	10.4878
3	1	0.7382	0.7501	11.7613	8.9742	1.8207	10.5035
4	1	0.9811	0.7528	12.2263	8.9792	0.3720	10.4878
5	1	...	...	11.8748	11.8748	1.4670	1.4670
6	1	0.9811	0.7530	12.2263	8.9788	0.3720	10.4890
7	1	...	...	11.8748	11.8748	1.4671	1.4671
1	2	0.6747	0.5583	13.4479	9.0474	5.0071	9.6614
2	2	0.7706	0.6134	13.7934	8.8541	4.6416	9.8659
3	2	0.7681	0.6134	13.7842	8.8538	4.6513	9.8661
4	2	0.7706	0.6135	13.7934	8.8540	4.6416	9.8660
5	2	2.2068	0.6135	19.0358	8.9066	−0.9033	9.8103
6	2	0.7708	0.6284	13.7929	8.7994	4.6421	9.9237
7	2	2.2072	0.6464	19.0360	8.8521	−0.9035	9.8679
1	3	0.5583	0.6992	14.6352	7.5160	6.8822	10.1980
2	3	0.6036	0.7974	14.8850	6.9536	6.7659	10.4600
3	3	0.6036	0.8000	14.8847	6.9385	6.7660	10.4670
4	3	0.6036	0.8000	14.8849	6.9385	6.7659	10.4670
5	3	0.6209	0.8002	14.9845	6.9392	6.7196	10.4667
6	3	0.6192	2.2011	14.8092	−1.1576	6.8012	14.2378
7	3	0.6365	2.2015	14.9088	−1.1579	6.7548	14.2379
1	4	0.6992	1.1470	19.6562	0.0000	7.9199	10.6382
2	4	0.7278	0.9062	20.0485	2.6515	7.8656	10.2716
3	4	0.7305	1.1015	20.0147	0.5096	7.8703	10.5678
4	4	0.7305	1.1015	20.0147	0.5094	7.8703	10.5679
5	4	0.7308	1.1015	20.0170	0.5094	7.8700	10.5679
6	4	...	...	1.8824	1.8824	10.3780	10.3780
7	4	...	...	1.8825	1.8825	10.3780	10.3780

[a] τ_a = shear stress in the adhesive.
[b] σ_1 = stress in titanium.
[c] σ_2 = stress in composite.

TABLE 2—*End-gap forces in the four-step joint (normalized to applied load).*

	Step				
Case	0	1	2	3	4
1	...	...	...	...	...
2	0.00967	−0.03132	−0.01913	−0.02422	0.03446
3	0.04734	−0.03122	−0.01913	−0.02430	0.00662
4	0.00967	−0.03132	−0.01913	−0.02430	0.00662
5	0.03814	−0.08511	−0.01962	−0.02430	0.00662
6	0.00967	−0.03133	−0.01955	−0.06387	0.02447
7	0.03814	−0.08513	−0.02004	−0.06388	0.02447

The interlaminar stresses have been obtained at Interfaces A through O (Fig. 6) for these three cases. The absolute maximum interlaminar stresses at various interfaces are shown in Table 3. The table also shows the location and the cases under which these maximum stresses occur. The maximum interlaminar shear stress is 0.291 of the applied stress and occurs at Location *K*. The interlaminar normal stresses are small compared to the shear stresses.

Delamination Propagation Prediction

A macromechanics model, based on the delamination between plies, has been developed for fatigue life prediction in Ref *17*. The model is based on the assumption that initial defects of macroscopic levels exist in the composite between plies. During fatigue loads, the defects along the free edge or around a stress raiser are opened by induced interlaminar shear or normal stresses or both. During fatigue cycling, these defects propagate causing progressive delamination of the composite. When the delamination reaches a critical size, buckling of a composite lamina, or group of lamina separated from the total laminate, will take place, causing the collapse of the structure. The model assumes that there is a threshold stress below which the delamination will not initiate or propagate.

The initiation and propagation of delamination in the four-step joint (Fig. 2) was investigated under a typical vertical tail spectrum loading shown in Fig. 7 where the 100% load corresponding to the maximum compression load. In the spectrum, one lifetime (LT) is 6000 flight hours. An initial flaw of 0.25 mm (0.01 in.) is assumed in the delamination propagation model. The maximum interlaminar shear stress in the joint is 0.291 of the applied stress. Corresponding to the threshold interlaminar shear stress of 48.3 MPa (7.0 ksi) (Ref *17*), if the

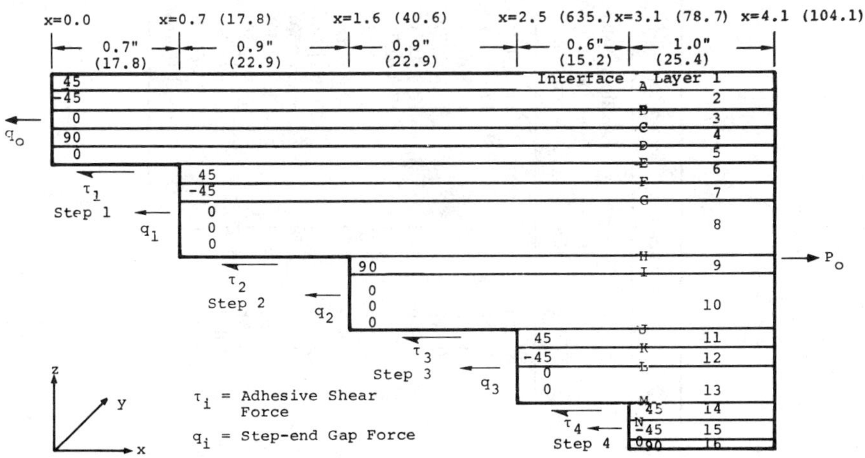

FIG. 6—*Definition of layers and interfaces and the applied load in the finite-element model of the four-step joint.*

TABLE 3—*Location and magnitude of maximum interlaminar stresses in the four-step-lap joint.*

	Maximum Shear, τ_z/σ_0			Maximum Positive, σ_z/σ_0			Maximum Negative, σ_z/σ_0		
Interface	Value	Case	Location (distance from end of joint), mm (in.)	Value	Case	Location (distance from end of joint), mm (in.)	Value	Case	Location (distance from end of joint), mm (in.)
A	0.054	2	17.8 (0.7)	0.007	2	0.0 (0.0)	0.009	2	46.4 (1.825)
B	0.080	2	17.8 (0.7)	0.011	2	17.8 (0.7)	0.012	2	46.4 (1.825)
C	0.088	2	17.8 (0.7)	0.015	3	29.2 (1.15)	0.013	2	67.3 (2.65)
D	0.079	2	17.8 (0.7)	0.018	3	29.2 (1.15)	0.019	2	67.3 (2.65)
E	0.096	2	17.8 (0.7)	0.018	2	63.5 (2.50)	0.025	2	67.3 (2.65)
F	0.083	3	17.8 (0.7)	0.022	2	63.5 (2.50)	0.033	2	67.3 (2.65)
G	0.076	2	67.3 (2.65)	0.029	2	63.5 (2.50)	0.047	2	67.3 (2.65)
H	0.069	2	40.6 (1.60)	0.059	2	63.5 (2.50)	0.050	2	67.3 (2.65)
I	0.099	2	40.6 (1.60)	0.040	2	71.1 (2.80)	0.061	2	67.3 (2.65)
J	0.155	2	63.5 (2.50)	0.084	2	63.5 (2.50)	0.066	2	67.3 (2.65)
K	0.291	2	63.5 (2.50)	0.076	2	63.5 (2.50)	0.063	2	67.3 (2.65)
L	0.198	2	63.5 (2.50)	0.050	2	63.5 (2.50)	0.037	2	78.7 (3.10)
M	0.014	3	104.1 (4.10)	0.014	3	78.7 (3.10)	0.051	2	78.7 (3.10)
N	0.031	1	78.7 (3.10)	0.009	3	78.7 (3.10)	0.064	2	78.7 (3.10)
O	0.020	1	104.1 (4.10)	0.016	3	78.7 (3.10)	0.060	2	78.7 (3.10)

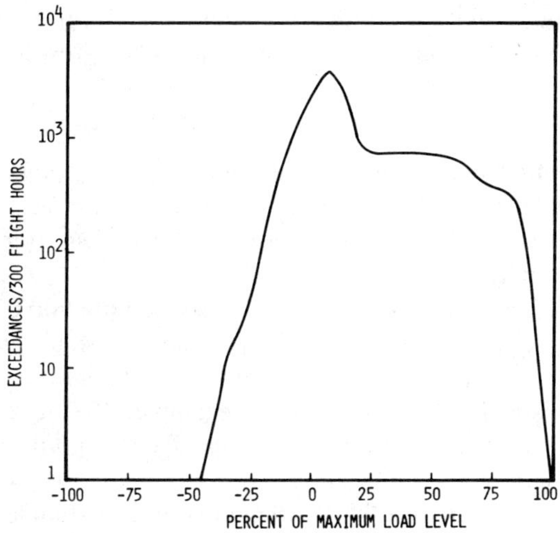

FIG. 7—*Vertical tail spectrum loading.*

stress level in the spectrum exceeds 165.9 MPa (24.06 ksi), the initial flaw will propagate. A typical spectrum has only a few high loads; with the majority of the loads being small enough such that the delamination was found to not propagate. The design load in a typical joint, such as shown in Fig. 2 is about 1226 KN/m (7000 lb/in.). The fatigue life prediction model, based on the delamination propagation, showed no failure after 24 000 flight hours (4 LT) of spectrum loading. Assuming the static strength of the composite as 413.7 MPa (60 ksi), very little reduction in the residual strength is observed, as shown in Table 4. The table also shows the residual strength of the four-step joint after 24 000 flight hours (4 LT) for various values of the maximum stress in the spectrum. It is seen that only at load levels higher than 1489 KN/m (8500 lb/in.) does the failure occur within four lifetimes of the spectrum loading. This value of the maximum load is very high and is not likely to occur in the spectrum loading.

TABLE 4—*Predicted life and residual strength of four-step-lap joint.*

Maximum Load in Spectrum, KN/m (lb/in.)	Maximum Stress, MPa (ksi)	Flight Hours	Residual Strength	
			Stress, MPa (ksi)	Load, KN/m (lb/in.)
−1 226 (−7 000)	−226.2 (−32.81)	24 000	−413.7 (−59.996)	−2 240 (−12 791)
−1 313 (−7 500)	−242.7 (−35.20)	24 000	−413.6 (−59.987)	−2 240 (−12 789)
−1 401 (−8 000)	−258.9 (−37.55)	24 000	−413.4 (−59.963)	−2 239 (−12 784)
−1 489 (−8 500)	−275.1 (−39.90)	24 000	−412.8 (−59.874)	−2 235 (−12 765)
−1 576 (−9 000)	−291.2 (−42.24)	20 903	failure	failure
−1 671 (−9 543)	−308.6 (−44.76)	10 954	failure	failure
−1 751 (−10 000)	−323.7 (−46.94)	6 544	failure	failure

Thus, even in the extreme case of debonding, that is, first and fourth steps debonded in the four-step joint, the delaminations do not grow to be catastrophic.

Conclusions

1. Delaminations do not grow to be catastrophic in the composite layer of the four-step joint under vertical tail spectrum loading, even in four lifetimes, if the fatigue threshold of the resin is about 48.3 MPa (7 ksi) (threshold in room temperature dry environments).

2. A few large loads in a spectrum may cause the initiation of delaminations. However, the majority of the loads in a spectrum are small and do not cause delamination propagation provided the fatigue threshold of the resin is about 48.3 MPa (7 ksi). For small threshold stress, for example, 20.7 to 27.6 MPa (3 to 4 ksi), a significant number of loads in a typical spectrum will contribute to the delamination initiation and propagation.

3. Negligible reduction in residual strength is predicted under a typical spectrum with maximum stress equal to commonly used design stress.

4. The presence of both debonding and step-end gaps in the adhesive layer represents a critical condition from the point of view of delamination initiation and propagation.

5. The stresses produced by thermal mismatch between titanium and graphite/epoxy are small and may be neglected for investigating delamination propagation.

Acknowledgment

The work reported here was performed under a contract from the Naval Air Systems Command (NAVAIR). Special thanks to Dr. Dan Mulville of NAVAIR for useful suggestions during the course of investigations.

References

[*1*] Erdogan, F. and Ratwani, M. M., "Stress Distribution in Bonded Joints," *Journal of Composite Materials*, Vol. 5, July 1971.
[*2*] Hart-Smith, L. J., "Adhesively Bonded Double Lap Joints," NASA CR 112235, National Aeronautics and Space Administration, Jan. 1973.
[*3*] Hart-Smith, L. J., "Adhesive-Bonded Single Lap Joints," NASA CR 112236, National Aeronautics and Space Administration, Jan. 1973.
[*4*] Grimes, G. C., et al, "The Development of Nonlinear Analysis Methods for Bonded Joints in Advanced Filamentary Composite Structures," AFFDL-TR-72-97, Air Force Flight Dynamics Laboratory, Sept. 1972.
[*5*] Segerline, L. J., "On the Shear Stress in Bonded Joints," *Journal of Applied Mechanics*, March 1968.
[*6*] Cornell, R. W., "Determination of Stresses in Cemented Lap Joints," *Journal of Applied Mechanics*, Sept. 1953.
[*7*] Lunford, L. R., "Stress Analysis of Bonded Joints," Applied Polymer Symposium No. 3, Applied Science Publishers, New York, 1966, pp. 57–73.
[*8*] Lubkin, J. L., "A Theory of Adhesive Scarf Joints," *Journal of Applied Mechanics*, 1957.
[*9*] Harrison, N. L. and Harrison, W. J., "The Stresses in an Adhesive Layer," *Journal of Adhesion*, Vol. 3, 1972.

[*10*] Anderson, G. P., DeVries, K. L., and Williams, M. L., "Finite Element in Adhesion Analyses," *International Journal of Fracture*, Vol. 9, No. 4, Dec. 1973.
[*11*] Goland, W. and Reissner, G., "The Stresses in Cemented Joints," *Journal of Applied Mechanics*, March 1944.
[*12*] Renton, J. W. and Vinson, J. R., "On the Behavior of Bonded Joints in Composite Material Structure," *Engineering Fracture Mechanics*, Vol. 7, 1975.
[*13*] Renton, J. W. and Vinson, J. R., "Analysis of Adhesively Bonded Joints Between Panels of Composite Materials," *Journal of Applied Mechanics*, Vol. 44, March 1977.
[*14*] Yuceoglu, U. and Updike, D. P., "Stress Analysis of Bonded Joints," *Journal*, Engineering Mechanics Division, American Society of Civil Engineers, Feb. 1980.
[*15*] Hart-Smith, L. J., "Adhesive Joints with Flaws," presented at ASTM Meeting in Minneapolis, American Society for Testing and Materials, 1980.
[*16*] Kan, H. P. and Ratwani, M. M., "Stress Analysis of Step-Lap Joint with Bondline Flaws," *Proceedings*, AIAA/ASME/ASCE/AHS 23rd Structures, Structural Dynamics & Materials Conference, 10–12 May 1982, New Orleans.
[*17*] Ratwani, M. M. and Kan, H. P., *Journal of Aircraft*, Vol. 18, No. 6, pp. 458–462.

S. Mall[1] *and W. S. Johnson*[2]

Characterization of Mode I and Mixed-Mode Failure of Adhesive Bonds Between Composite Adherends

REFERENCE: Mall, S. and Johnson, W. S., "**Characterization of Mode I and Mixed-Mode Failure of Adhesive Bonds Between Composite Adherends,**" *Composite Materials: Testing and Design (Seventh Conference), ASTM STP 893*, J. M. Whitney, Ed., American Society for Testing and Materials, Philadelphia, 1986, pp. 322–334.

ABSTRACT: A combined experimental and analytical investigation of an adhesively bonded composite joint was conducted to characterize both the static and fatigue debond growth mechanism under Mode I and Mixed-Mode I and II loadings. Two bonded systems were studied: graphite/epoxy adherends bonded with EC 3445 and FM-300 adhesives. For each bonded system, two specimen types were tested: (1) a double-cantilever-beam specimen for Mode I loading and (2) a cracked-lap-shear specimen for Mixed-Mode I and II loading. In all specimens tested, failure occurred in the form of debond growth. Debonding always occurred in a cohesive manner with EC 3445 adhesive. The FM-300 adhesive debonded in a cohesive manner under Mixed-Mode I and II loading, but in a cohesive, adhesive, or combined cohesive and adhesive manner under Mode I loading. Total strain-energy release rate appeared to be the driving parameter for debond growth under static and fatigue loadings.

KEY WORDS: composite materials, adhesive bonding, debond propagation, strain-energy release rates, fracture mechanics, fatigue (materials)

Nomenclature

a	Length of debond, mm
A_1,A_2	Constants from least square fit
$\frac{da}{dN}$	Debond growth rate, mm/cycle.
C	Compliance, mm/N
c	Curve-fit parameters for power-law equation
G_I	Mode I strain-energy release rate, J/m^2

[1]Associate professor, Department of Engineering Mechanics, University of Missouri, Rolla, MO 65401.

[2]Research engineer, National Aeronautics and Space Administration, Langley Research Center, Hampton, VA 23665.

G_{Ic} Fracture toughness, J/m^2
G_{II} Mode II strain-energy release rate, J/m^2
G_T Total strain-energy release rate ($G_I + G_{II}$), J/m^2
G_{Tc} Critical total strain-energy release rate, J/m^2
N Number of cycles
n Curve-fit parameter for power-law equation
P_{cr} Critical load, N
w Width of specimen, mm

To achieve the maximum saving in weight without sacrificing strength, engineers are faced with the problem of developing methods of joining structural composite components without weakening or damaging them. It is impossible to use conventional fastening techniques without drastically affecting the strength of fiber-reinforced composites. Hence, adhesive bonding is a desirable alternative to mechanical fastening in composite structures. Even with all the potential advantages and encouraging experience with adhesive bonding, manufacturers still hesitate to use this technology in primary structural components. This reluctance is due, in part, to the lack of understanding of failure mechanism and durability. Several studies have been reported on the static strength of adhesively bonded composite joints (for example, see Refs *1–3*); however, very little information is available on their fatigue behavior.

The possible fatigue failure modes for bonded composites are: cyclic debonding (namely, progressive separation of the adhesive bond under cyclic load), cyclic delamination, adherend fatigue, or a combination of these. In a previous study [*4*], cyclic debonding of adhesively bonded composites was investigated under a mixed-mode loading, which introduced a combination of opening (Mode I) and sliding (Mode II) at the debond front. Graphite/epoxy (T300/5208) cracked-lap-shear (CLS) specimens were tested under constant amplitude cyclic loading in an ambient laboratory environment. Two adhesives were used: EC 3445 and FM-300. A fracture mechanics approach, employed previously in fatigue studies of composite-to-metal joints [*5,6*] and metal-to-metal joints [*7,8*], was used to model the cyclic debonding. The strain-energy release rate associated with cyclic failure of the adhesive bond was correlated with the measured cyclic debond growth rate, da/dN. Two different geometries of the CLS specimens were tested. These two specimen geometries provided the different ratio of G_I/G_{II}, where G_I and G_{II} are the strain-energy release rates for opening Mode I and for sliding Mode II, respectively. Data from these two specimens were used to determine fracture mode dependence of the adhesive. The cyclic debond growth rate correlated better with the total strain-energy release rate, G_T, than it did with either G_I or G_{II} independently [*4*].

Since the previous study [*4*] was based on a rather narrow range of G_I/G_{II} ratios (that is, 0.25 to 0.38), further investigations were required to substantiate that G_T is the cyclic debond driver for tough structural adhesives. The previously tested CLS specimens produced mostly shear stresses at the debond tip, therefore, this study will address a specimen configuration that is entirely loaded in peel (namely, G_I), the double-cantilever-beam specimen (DCB). Also, the previous study focused solely on cyclic debonding. The present study included static toughness results as well as cyclic results from the CLS and DCB specimens to evaluate the mixed-mode loading effects. The data obtained from both CLS and DCB specimens gave a very wide range of mixed-mode loading conditions for evaluation.

Specimen Preparation and Configuration

Two bonded systems were studied: graphite/epoxy (T300/5208) adherends bonded with either EC 3445 adhesive or with FM-300 adhesive [*9*]. The EC 3445 adhesive is a thermosetting paste with a cure temperature of 121°C; specimens were fabricated by conventional secondary bonding procedures. The FM-300 is a modified epoxy adhesive supported with a carrier cloth with a cure temperature of 177°C. The FM-300 specimens were fabricated by cocure, whereby adherends were cured and bonded simultaneously. The bonding processes followed the manufacturers' recommended procedures for each adhesive. The nominal adhesive thickness was 0.10 mm and 0.25 mm for the EC 3445 and FM-300, respectively.

Two specimen types were fabricated: DCB specimens and CLS specimens. The DCB and CLS specimens (shown in Figs. 1 and 2) were used to characterize debond growth under opening Mode I loading and the mixed-mode loading, respectively. The DCB specimen consisted of two bonded adherends, each having 14 unidirectional plies with an initial debond of 38-mm length. This debond was introduced by a Teflon film of thickness equal to the adhesive bondline. Two 0.5-mm-thick aluminum end tabs were bonded to the DCB specimen, along with two 1.3-mm-thick aluminum reinforcing plates. The peeling load was applied through these tabs.

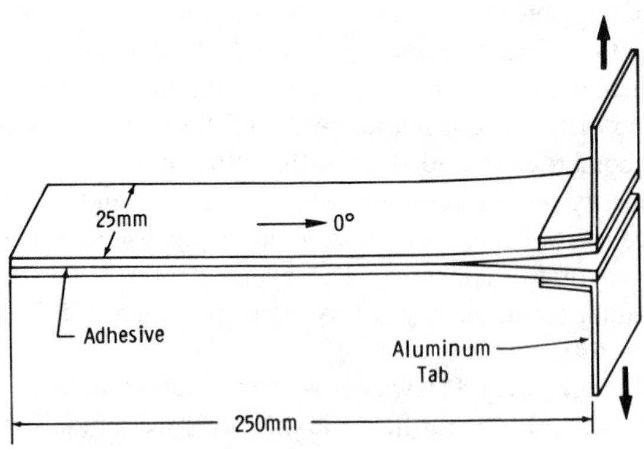

FIG. 1—*Double-cantilever-beam (DCB) specimen.*

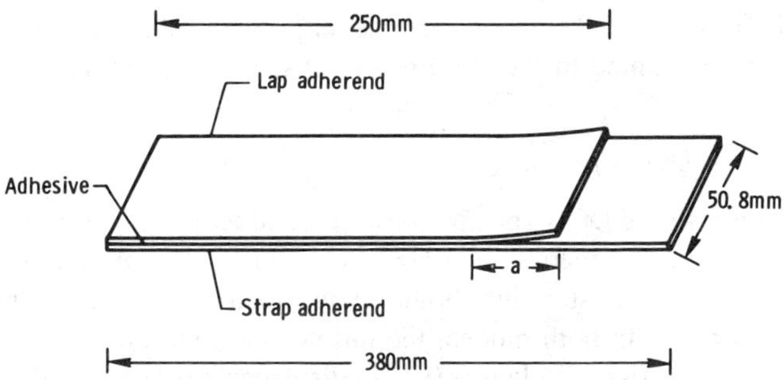

FIG. 2—*Cracked-lap-shear (CLS) specimen.*

The adherends of the cracked-lap-shear specimens consisted of quasi-isotropic lay-ups, $[0/45/-45/90]_s$ and $[0/45/-45/90]_{2s}$. Two configurations of CLS specimens were tested: 8-ply strap to 16-ply lap and 16-ply strap to 8-ply lap. The two adhesive systems with these two geometries resulted in four sets of specimens. The CLS specimen did not have an initial debond like the DCB specimens.

Testing Procedure

The test program included the static and fatigue tests for both types of specimens. The objective of the test program was to measure the critical strain-energy release rate under the static loading, and to measure the debond growth rate under the cyclic loading. These are described separately for each specimen in the following paragraphs.

Static Tests of DCB Specimen

All static tests of DCB specimens were performed in a displacement-controlled test machine. Both edges of the specimen were coated with a white brittle fluid, to aid in visually locating the debond tip. Fine visible marks were put on these edges, at 1-mm intervals, to aid in measuring the debond length. The debond length was measured visually on both sides with two microscopes having a magnification factor of 20. Prior to testing, either for static or fatigue loading, these specimens were fatigued to create a debond of at least 6 mm beyond the end of Teflon film. The static test involved the application of displacement at a slow crosshead speed (approximately 1.0 mm/min). The load corresponding to the applied displacement was also recorded. When the load reached the critical value, the debond grew. The onset of growth results in a deviation from linearity in the load versus crosshead displacement record. The applied displacement was then decreased until a zero load reading was observed. After each static test, the specimen was fatigued until the debond grew at least 6 mm further, thus forming a sharp crack for the next static test. A series of static tests was performed on each specimen, which provided compliance and critical load measurements at several debond lengths. These measurements provided the critical strain-energy release rate as explained in the Results and Discussions section of this paper.

Fatigue Tests of DCB Specimen

The fatigue tests of DCB specimens were conducted in a servohydraulic test machine at a cyclic frequency of 3 Hz. Two constant-amplitude testing modes were employed: (1) constant-amplitude cyclic load and (2) constant-amplitude cyclic displacement. In both modes, the ratio of minimum to maximum load (or displacement) in a fatigue cycle was 0.1. In displacement-controlled tests, debond growth rates reduced as the debond propagated, while in the case of load-controlled tests, debond growth rates increased as the debond propagated. Debond lengths, fatigue cycles, applied loads, and displacements were monitored continuously throughout each test. The measured relationship between the debond length and fatigue cycle provided the debond growth rate, da/dN. The strain-energy release rate, G_I, was computed from the measured compliance and applied load, as explained in the Results and Discussions section. Thus, a relationship between G_I and da/dN was established for the cyclic debonding under Mode I loading.

Static Tests of CLS Specimen

Static tension tests on CLS specimens were conducted in a displacement-controlled mode. Prior to static testing, this specimen was fatigued, and thus it had an initial sharp debond. During the test, the axial load and displacement were recorded. The displacement was measured with two displacement transducers attached on the opposite sides of the specimen. The applied load was increased slowly until the debond propagated. The critical load corresponding to unstable debond growth was measured and verified by the deviation from linearity in the recorded load-displacement curve. Only one such measurement could be obtained from each specimen, since debonds grew into the composite strap adherend. Static tests were conducted on all four sets of specimens (that is, two geometries and two adhesives).

Fatigue Tests of CLS Specimen

A detailed investigation of cyclic debonding under mixed-mode loading was conducted in a previous study [*4*]. In that study, the CLS specimen was tested under constant-amplitude cyclic load at 10 Hz frequency and stress ratio, $R = 0.1$. In the present study, fatigue tests of the CLS specimen were conducted at 3 Hz in order to compare mixed-mode results with Mode I results from DCB specimens that were also obtained at 3 Hz frequency. Only the 8-ply strap bonded to the 16-ply lap with EC 3445 adhesive system was tested at 3 Hz.

Analysis

Static tests on DCB specimens, conducted as described earlier, provided the critical load, P_{cr}, and the compliance, C, for each debonded length. The measured values of P_{cr} and C were used with linear beam theory to compute the fracture toughness, G_{Ic}. The details of this procedure are elaborated by Wilkins et al [*10*]. A brief description of the Wilkins et al technique is given subsequently. Figure 3 shows the variation of compliance with the debond length in a typical DCB specimen with EC 3445 adhesive. A compliance relationship of

$$C = A_1 a^3 \tag{1}$$

was fitted through the experimental data points by the method of least squares that is shown in Fig. 3 as a solid line. This relationship, based on linear beam theory, fits very well with the experimental data. The constant, A_1, in Eq 1 is $2/3EI$, where E is the extensional stiffness and I is the moment of inertia of each adherend of the DCB specimen. The experimental values of A_1 are ±7% of the linear beam theory value of 3.77×10^{-7}.

Finite-element analysis [*11*] was also used to analyze the DCB results. The adhesive was modeled with eight layers of elements. The analysis was conducted assuming plane-strain conditions. The experimental values of compliance were within ±5% of those given by a geometric linear finite-element analysis. The computer compliances at several debond lengths were within 5% of the experimental values. These computed values are also shown in Fig. 3. Further, the geometric nonlinear analysis of this specimen did not show any significant change from the linear analysis. The maximum difference in the computed compliance from nonlinear and linear analyses was 5% for the maximum debond length

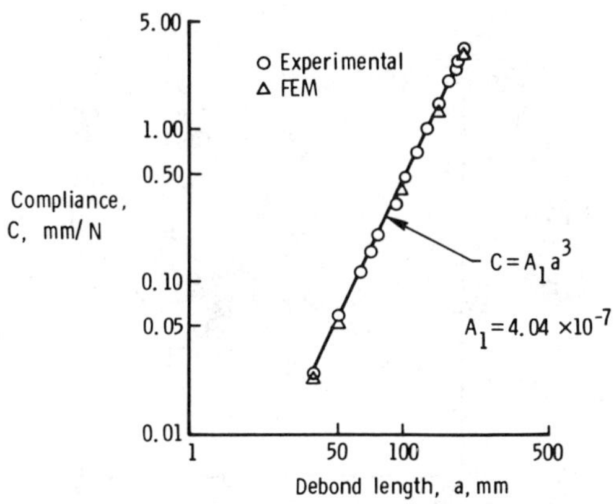

FIG. 3—*Relationship between compliance and debond length for a DCB specimen with EC 3445 adhesive.*

employed in the investigation (that is, 200 mm) at its maximum or critical load. Thus, the compliance-debond length relationship, expressed by Eq 1, represents the appropriate behavior of the presently employed DCB specimen. All results from the DCB specimen in this study are calculated using the linear beam theory.

Figure 4 shows the measured critical load as a function of debond length for a typical specimen with EC 3445 adhesive. Based on the linear beam theory [*10*], a relationship between the critical load, P_{cr}, and the debond length, a, is

$$P_{cr} = A_2/a \tag{2}$$

A solid line shown in Fig. 4 with a slope of -1 was fitted to the experimental data with the method of least squares. Then, the averaged value of G_{Ic} for each specimen was computed from the relationship

$$G_{Ic} = \frac{P_{cr}^2}{2w}\frac{\partial C}{\partial a} = 3A_1A_2^2/(2w) \tag{3}$$

where w is the specimen width. A similar procedure was used to compute the strain-energy release rate, G_I, associated with the cyclic debonding where the critical load was replaced by the maximum load of the fatigue cycle. The details of the analysis for the CLS specimens are given in Ref *4*.

Results and Discussions

Debond Locations

All DCB and CLS specimens with both adhesives EC 3445 and FM-300 failed by debond propagation during both static and fatigue tests. However, the debond grew in a different manner in each case. In the case of DCB specimens with EC

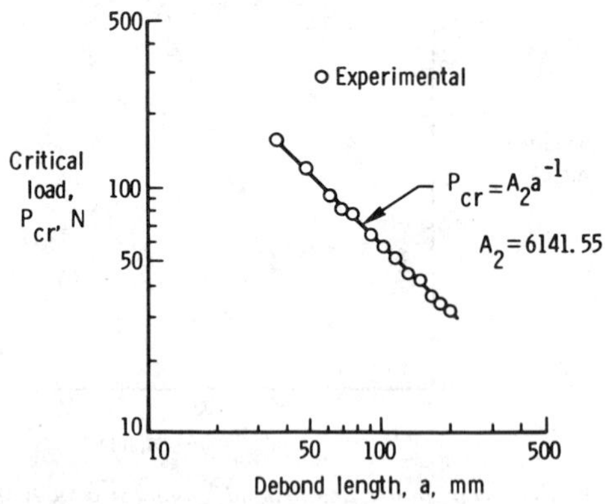

FIG. 4—*Relationship between critical load and debond length for a DCB specimen with EC 3445 adhesive.*

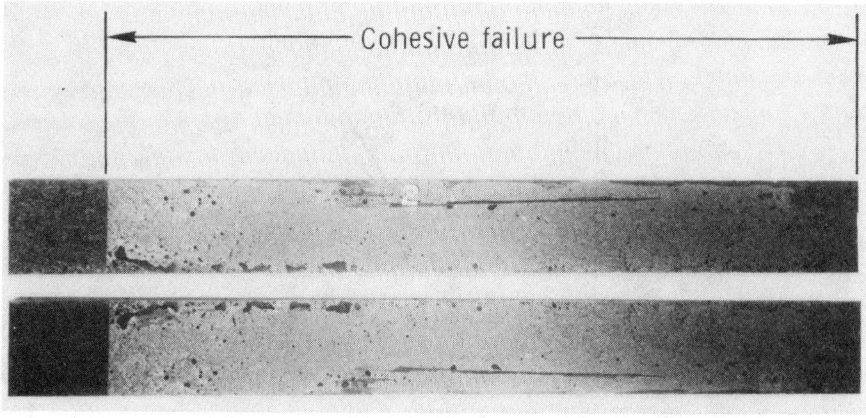

FIG. 5—*Debonded surfaces of DCB specimen with EC 3445 adhesive.*

3445 adhesives, the debond grew in a cohesive manner during both static and fatigue tests. Here the debond grew consistently in the middle portion of the adhesive layer. In DCB specimens with FM-300, the debond propagated in an irregular manner during both static and fatigue tests, involving cohesive, adhesive, or mixed cohesive-adhesive debonding. Typical debonded surfaces with these failure details are shown in Figs. 5 and 6 for both adhesives.

The CLS specimens debonded in a cohesive manner during fatigue tests for both adhesive systems. The debond grew in the vicinity of strap-adhesive interface. A possible explanation of this phenomenon is given in the previous study [*4*].

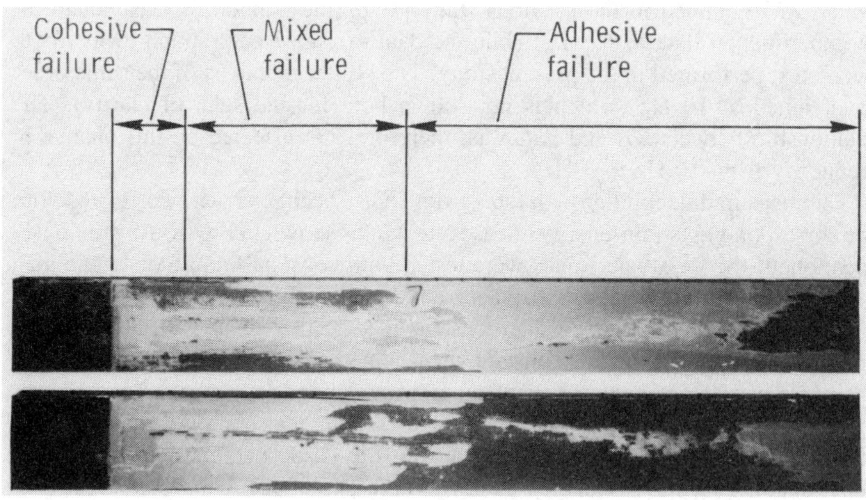

FIG. 6—*Debonded surfaces of DCB specimen with FM-300 adhesive.*

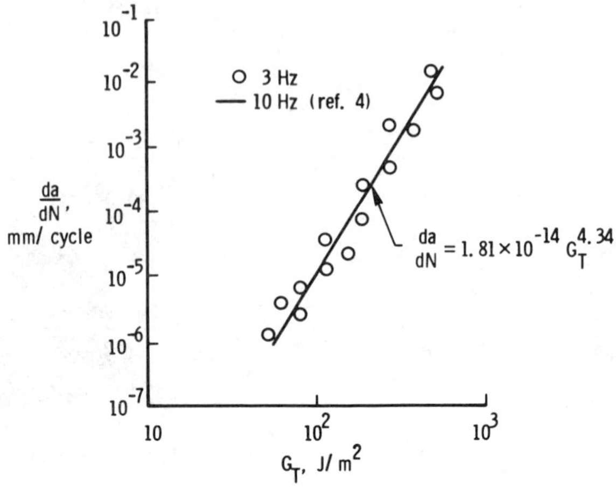

FIG. 7—*Relationship between total strain-energy release rate and debond growth rate of EC 3445 adhesive at two cyclic frequencies using CLS specimens.*

Cyclic Debonding Behavior

In the present study, all fatigue tests with both DCB and CLS specimens were conducted at 3 Hz. Figure 7 shows the comparison between the G_T versus da/dN relationship for two cyclic frequencies, 10 Hz and 3 Hz, obtained from CLS specimens with EC 3445 adhesive. The solid line shown is a power-law relationship

$$da/dN = cG_T^n. \tag{4}$$

which was obtained in the previous study [*4*] by the method of least square fit to experimental data at 10 Hz, while the data in Fig. 7 correspond to the 3 Hz cyclic test performed in the present study. The scatter in data is of the same order as obtained at 10 Hz (which is not shown here for the sake of clarity). The relationship between G_T and da/dN is, therefore, not affected by this change in frequency from 10 Hz to 3 Hz.

The measured debond growth rates from DCB specimens were correlated with the corresponding strain-energy release rate, G_I, as shown in Fig. 8. As previously mentioned, the DCB specimens were tested with constant-amplitude cyclic load and constant-amplitude cyclic displacement. Data obtained from these two testing modes are shown in Fig. 8. The constant load testing results in G increasing with debond length while the constant displacement results in G decreasing. Since the constant displacement tests resulted in faster debond rates, the debond process appears to be influenced by the G gradient. Figure 8 also shows G_T versus da/dN and G_I versus da/dN relationships from the CLS specimens under mixed-mode loading [*4*]. The scatter in data from the DCB specimens was larger than from the CLS specimens. The CLS data points [*4*] are not shown herein for the

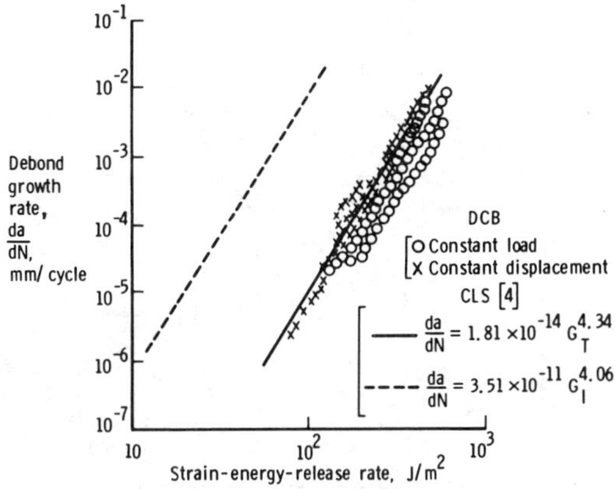

FIG. 8—*Relationship between strain-energy release rates and debond growth rate of EC 3445 adhesive for DCB and CLS specimens.*

sake of clarity. The G_I versus da/dN data from the DCB specimen are in good agreement with the G_T versus da/dN relationship from the CLS specimen, represented by the solid line. On the other hand, the G_I versus da/dN relationship from the CLS specimen represented by the dashed line, did not agree with the DCB specimen. This indicates that the cyclic debond growth is a function of total strain-energy release rate.

A similar phenomenon was also observed in the case of FM-300 adhesive. Figure 9 shows the comparison of the G_T versus da/dN relationship from the DCB specimen. The data shown from the DCB specimen were obtained under the constant-amplitude cyclic displacement. As previously mentioned, cyclic debonding occurred in cohesive manner, adhesive manner, or a combination of both in the DCB specimens. In Fig. 9, data on the right-hand side correspond to the cohesive failure, and data in between these correspond to the mixed failure.

Static Debonding Behavior

Figure 10 shows the critical strain-energy release rates, G_{Tc} and G_{Ic}, obtained from static tests of CLS and DCB specimens, respectively. The total critical strain-energy release rate, G_{Tc}, from the CLS specimen is in agreement with fracture toughness, G_{Ic}, from DCB specimen in each case. This shows that the total critical strain-energy release is also the driving parameter for debond growth during static loading. The only exceptions are those adhesive failures in the FM-300 DCB specimens. These adhesive failure strengths are 40% lower than the cohesive failure strengths.

General

The cyclic debond growth rate data show, in each case, that the debond propagates at G_T values as much as an order of magnitude below the critical

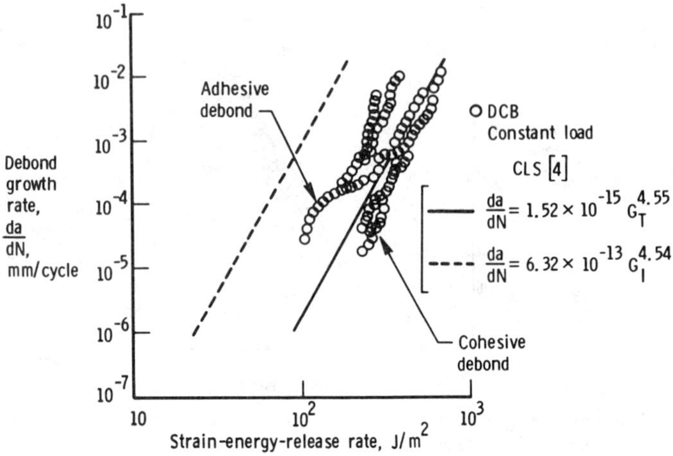

FIG. 9—*Relationship between strain-energy release rates and debond growth rate of FM-300 adhesive for DCB and CLS specimens.*

static value. So static data alone are insufficient for safe joint design. Instead, the G_T associated with cyclic debonding at very slow growth rates is more appropriate as a design value and as a criterion for adhesive selection. The threshold value of G_T has been demonstrated as a viable fracture mechanics approach for designing the adhesive joint [*12*]. If the total strain-energy release rate is a governing parameter for the cyclic debond initiation and propagation, as results of the present and previous studies [*4,12*] have shown, then it would require the characterization of cyclic debonding under Mode I loading only. It would be simpler and easier than testing under mixed-mode loading. This should be, however, verified further for several structural adhesives and various loading conditions. It is suspected that only relatively tough adhesives would demonstrate a G_T governed debond behavior.

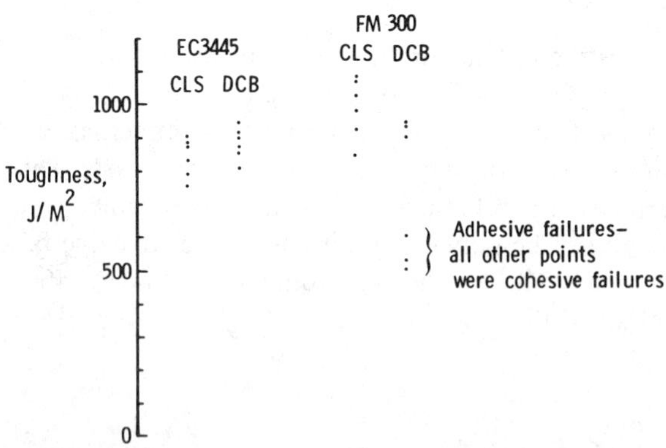

FIG. 10—*Static toughness of FM-300 and EC 3445 adhesives derived from DCB and CLS specimens.*

The G_T is much easier to calculate than are the individual G_I and G_{II} components of a mixed-mode specimen. The G_T can be measured directly from most laboratory specimens. Therefore, even if some error is involved with using G_T as a design parameter, as in Ref [*12*], the ease of calculation may compensate.

Concluding Remarks

A combined experimental and analytical investigation of composite-to-composite bonded joints was undertaken to characterize the debond growth mechanism under Mode I and Mixed-Mode I and II static and fatigue loadings. Two bonded systems were studied: graphite/epoxy adherends bonded with EC 3445 adhesive and with FM-300 adhesives. With each bonded system, two specimen types were tested: (1) a double-cantilever-beam specimen for Mode I loading and (2) a cracked-lap-shear specimen for Mixed-Mode I and II loadings. The following conclusions were obtained.

1. The total strain-energy release rate, G_T, appears to be the governing parameter for cohesive debond growth under static and fatigue loadings. This is indeed significant since in most cases G_T is much easier to calculate or determine experimentally than are G_I and G_{II} components of a mixed-mode specimen.
2. Debond growth was measured at G_T values that were an order of magnitude below the static toughness, G_{Tc}. Therefore, one needs to consider both debond growth threshold values as well as static strength in design and material development and selection.

References

[*1*] Matthews, F. L., Kilty, P. F., and Godwin, E. W., *Composites*, Vol. 13, No. 1, 1982, pp. 29–37.

[*2*] Hart-Smith, L. J., "Analysis and Design of Advanced Composite Bonded Joints," NASA CR-2218, National Aeronautics and Space Administration, 1974.

[*3*] Renton, W. J. and Vinson, J. R., "The Analysis and Design of Composite Material Bonded Joints Under Static and Fatigue Loadings," AFOSR-TR-72-1627, U. S. Air Force, 1973.

[*4*] Mall, S., Johnson, W. S., and Everett, R. A., Jr., "Cyclic Debonding of Adhesively Bonded Composites," NASA TM-84577, National Aeronautics and Space Administration, Nov. 1982; also in *Adhesive Joints: Their Formation, Characteristics, and Testing*, K. L. Mittal, Ed., Plenum Press, New York, 1984.

[*5*] Roderick, G. L., Everett, R. A., Jr., and Crews, J. H., Jr., in *Fatigue of Composite Materials, ASTM STP 569,* American Society for Testing and Materials, Philadelphia, 1975, pp. 295–306.

[*6*] Everett, R. A., Jr., "The Role of Peel Stresses in Cyclic Debonding," NASA TM-84504, National Aeronautics and Space Administration, 1982.

[*7*] Brussat, T. R., Chiu, S. T., and Mostovoy, S., "Fracture Mechanics for Structural Adhesive Bonds," AFML-TR-1963, Air Force Materials Laboratory, 1977.

[*8*] Romanko, J. and Knauss, W. G., "Fatigue Behavior of Adhesively Bonded Joints," Vol. I, AFWAL-TR-80-4037, U. S. Air Force, 1980.

[*9*] Alsmiller, G. R., Jr., and Anderson, W. P., "Advanced Composites Airframe Program—Preliminary Design," USAAVRADCOM-TR-80-D-37A, U. S. Army, 1982.

[*10*] Wilkins, D. J., Eisenmann, J. R., Camin, R. A., Margolis, W. S., and Benson, R. A. in *Damage in Composite Materials, ASTM STP 775,* K. Reifsnider, Ed., American Society for Testing and Materials, Philadelphia, 1981, pp. 168–183.

[*11*] Dattaguru, B., Everett, R. A., Jr., Whitcomb, J. D., and Johnson, W. S., "Geometrically-Nonlinear Analysis of Adhesively Bonded Joints," NASA TM-84562, National Aeronautics and Space Administration, 1982.

[*12*] Johnson, W. S. and Mall, S., "A Fracture Mechanics Approach for Designing Adhesively Bonded Joints," NASA TM-85694, National Aeronautics and Space Administration, Sept. 1983.

Analysis and Characterization

Madhu Madhukar[1] *and Jonathan Awerbuch*[1]

Monitoring Damage Progression in Center-Notched Boron/Aluminum Laminates Through Acoustic Emission

REFERENCE: Madhukar, M. and Awerbuch, J., "**Monitoring Damage Progression in Center-Notched Boron/Aluminum Laminates Through Acoustic Emission,**" *Composite Materials: Testing and Design (Seventh Conference), ASTM STP 893*, J. M. Whitney, Ed., American Society for Testing and Materials, Philadelphia, 1986, pp. 337–367.

ABSTRACT: This study concentrated on monitoring damage initiation and progression through acoustic emission in center-notched boron/aluminum laminates subjected to quasi-static tensile loading. Laminates studied include $[0]_8$, $[90]_8$, $[\pm 45]_{2s}$, $[0/90]_{2s}$, $[0/\pm 45/0]_s$, and $[0/\pm 45/90]_s$ 5.6-mil boron/aluminum 6061-F. The objective of this study was to investigate the potential of the acoustic emission technique for locating existing damage, detecting and tracking damage initiation and progression, identifying potential fracture sites, and distinguishing among the major failure mechanisms of the different laminates. Emphasis was placed on developing the proper test methodology for characterizing a new composite material system through acoustic emission. For base-line data, mechanical properties, fracture behavior, and deformation characteristics were studied and acoustic emission was recorded for the composite constituents (namely, fiber and matrix) as well. A good degree of reproducibility has been established. Acoustic emission results could be correlated with the deformation characteristics of the subject material, and failure modes and processes for the various laminates could be identified.

KEY WORDS: composite materials, nondestructive testing, acoustic emission, metal matrix composites, boron/aluminum

A wide range of nondestructive inspection techniques are used to detect and locate internal damage and damage progression in composites, to identify failure modes, and to evaluate material quality. In resin matrix composites, the delamination and matrix cracking that usually precede fiber fracture are of major concern. The ultrasonic C-scan and X-ray radiography techniques have demonstrated their viability for identifying these types of failure in numerous investigations. In metal matrix composites, however, delamination and matrix cracking are seldom of concern. Fiber fracture, the associated matrix plastic deformation, and interfacial failure are the types of failure that must be detected and identified. For these failure mechanisms, the ultrasonic C-scan and X-ray radiography ex-

[1]Graduate student and associate professor, respectively, Department of Mechanical Engineering and Mechanics, Drexel University, Philadelphia, PA 19104.

amination techniques are not always satisfactory, primarily when multidirectional laminates are being studied.

The acoustic emission (AE) technique appears to offer a very practical procedure for monitoring failure processes in filamentary metal matrix composites. Moreover, this technique is more readily used under actual service conditions. This is important because the composite systems are, in many instances, fabricated during the manufacturing of the complete structure. A nondestructive inspection of the complete component is frequently required to locate non-visual damage and indicate its criticality.

An important aspect of the AE technique is that the information is obtained in real time. Also, once the instrumentation settings have been properly selected for a given problem, the technique is very simple to utilize. On the other hand, AE by its nature is applicable only during proof loading. This may limit its usage in certain aeronautical and aerospace applications, although it is already widely applied to assess the integrity of composite structures in the chemical and automotive industries. For small-scale laboratory testing on well-defined specimens, however, the AE technique can serve as a supportive test procedure for research into the failure process of new composite systems subjected to external stimulation.

A significant amount of research during the past decade has addressed the viability of AE as a nondestructive test technique for composite materials. Most of the work has been directed toward resin matrix systems, primarily glass/epoxy and graphite/epoxy, with some scattered efforts directed toward composite structures. Very few works have been directed toward metal matrix composites, for example, Refs *1–5*. They could be considered as preliminary studies, however, all have demonstrated the viability of the AE technique for metal matrix systems as well.

In this study, the validity of the AE technique for detecting and locating existing damage, determining damage criticality, and identifying the major failure mechanisms in center-notched boron/aluminum laminates has been investigated. Damage initiation and accumulation have been monitored during quasi-static uniaxial tensile loading. Six different laminates were studied: $[0]_8$, $[90]_8$, $[\pm45]_{2s}$, $[0/90]_{2s}$, $[0/\pm45/0]_s$, and $[0/\pm45/90]_s$ 5.6-mil boron/aluminum 6061-F having six different notch length-to-width ratios ranging from 0.05 to 0.5. Constituents, that is, individual fibers and matrix, were tested as well. Results indicated that the failure process in the different laminates can potentially be identified through amplitude analysis of the AE events. Other issues addressed in this paper include: the effect of notch length and laminate configuration on damage initiation and accumulation; the applicability of AE for locating damage and determining damage criticality; correlation of AE results with the actual deformation characteristics; the reproducibility of AE results; and a comparison between failure mechanisms identified through AE and those observed visually and via the scanning electron microscope (SEM).

Experimental Procedure

Specimen Material

Six different 5.6-mil boron/aluminum 6061-F laminates were tested in this program, namely; $[0]_8$, $[90]_8$, $[\pm 45]_{2s}$, $[0/90]_{2s}$, $[0/\pm 45/0]_s$, and $[0/\pm 45/90]_s$. The laminates were fabricated by D. W. A. Composite Specialties, Inc., with approximately 45 to 50% fiber volume fraction. The plates, approximately 200 by 200 by 1.3 mm (8.0 by 8.0 by 0.052 in.), were machined by Electro Discharge Machining (EDM) into 200 by 25 mm (8.0 by 1.0 in.) tensile coupon specimens. About 12.7 mm (0.5 in.) was left at both sides of the plates. Center notches of 1.27, 2.54, 5.08, 7.62, 10.16, and 12.7 mm (0.05, 0.1, 0.2, 0.3, 0.4, and 0.5 in.) long and 0.25 mm (0.01 in.) wide were introduced by EDM. All notches were perpendicular to the loading direction.

Each specimen was carefully measured at three locations along the length to ensure a minimum acceptable uniformity in specimen dimensions. Maximum variation was measured to be 0.05 mm (0.002 in.) in width and less than 0.025 mm (0.001 in.) in thickness. The location of each specimen on each of the plates was recorded so that designation of specimens according to test procedure, notch length, etc., could be made randomly to ensure reliable test results. Plates were cataloged numerically according to supplier code number and each of the specimens within each plate was numbered according to its location and lay-up.

Aluminum end tabs 25 by 25 by 1.3 mm (1.0 by 1.0 by 0.05 in.) were applied to all specimens and glued with epoxy onto the specimen ends. The surface preparation of the aluminum tabs and specimens and the curing procedures were similar to those applied in bonding strain gages. This procedure of specimen tabbing and the choice of tabbing material was found to be more advantageous in reducing unwanted emission emanating from the gripping area than the glass/epoxy tabbing material often used in testing composite coupons. All failures occurred within the gage length of the specimen.

Mechanical Testing Procedure

All specimens were subjected to quasi-static uniaxial tensile loading carried out on a closed-loop servo-hydraulic Instron testing machine (Model 1331). All tests were carried out under stroke control mode at a rate of 0.05 mm/min (0.002 in./min). The grips used were standard 25 mm (1.0 in.) wide wedge-action friction grips. Each specimen was instrumented with a compliance gage (25 mm, 1.0 in. gage length) to obtain "global" stress-strain and load-displacement curves for the unnotched and notched specimens, respectively. A selected number of unnotched specimens were strain gaged to obtain the laminae and laminate stiffnesses. Actual crack opening displacements were also obtained for all laminates, resulting in detailed load-COD curves. The COD was obtained using the laser interferometric technique described in detail in Ref *6*. Strength and AE data on individual boron fibers (203 mm, 8 in. long) were obtained as well.

Failure Modes and Damage Progression

For a selected number of specimens of each laminate configuration, fracture surfaces were examined through SEM. Three-dimensional, or stereo, views of the fracture surfaces were prepared from which the different micro-failure modes could be easily identified. In order to verify the AE results, several specimens were loaded to a predetermined load level, unloaded, and the matrix was dissolved. Notch-tip damage extension and the number of broken fibers were inspected through the SEM. Damage progression was also monitored in real time via a closed-circuit television (CCTV) system that provided a clear view of the notch-tip region at the specimen surface at magnifications of up to $\times 125$. The observed damage progression was also correlated with the AE results.

Acoustic Emission

For all specimens tested in this program (approximately 10 specimens per laminate configuration), AE was monitored using Dunegan/Endevco 3000 series AE instrumentation. The most pertinent operating parameters are: Transducer Type S9204, system threshold level of 1 V, fixed gain of 40-dB preamplifer, post-amplifier gain control of 40 dB, dead time = 3 ms. Transducers were placed at both ends of the specimen with an interim distance of approximately 127 mm (5.0 in.). The length of the specimen between the transducers is divided into 100 segments for the purpose of monitoring line location, and these segments correspond to Locations 0 through 100 in the location distribution histograms.

Spatial filtering (window) has been employed to eliminate unwanted noise, for example, from the grip regions and end tabs. In order to record the emission (that is, counts, events, and events amplitude) generated from the notch-tip regions only, the pre-selected window has been set at 40 to 60, that is, only the emission generated in Locations 40 through 60 was counted and processed. Thus, the experimental results of counts and events versus load curves and the amplitude distribution histograms of events are for the emission generated from within the window. The location distribution histograms of events, however, display all the events occurring throughout the specimen gage length between the pair of transducers. Note that the length of the pre-selected window corresponds, approximately, to the length of the compliance gage of 25 mm.

Data recorded include accumulative events and counts, count-rate, and number of counts per event as a function of applied load, line location, and amplitude distribution histograms of events and Cumulative Event Amplitude Distributions (CEAD) at various load levels, all for emission occurring within a pre-set area of interest (or window) along the specimen. From these data, the potential of the AE technique is assessed for determining material quality, locating existing damage, detecting and tracking damage initiation and progression, identifying potential fracture sites, and distinguishing among the various failure mechanisms. The AE results were correlated with the deformation characteristics, mechanical properties, and failure modes.

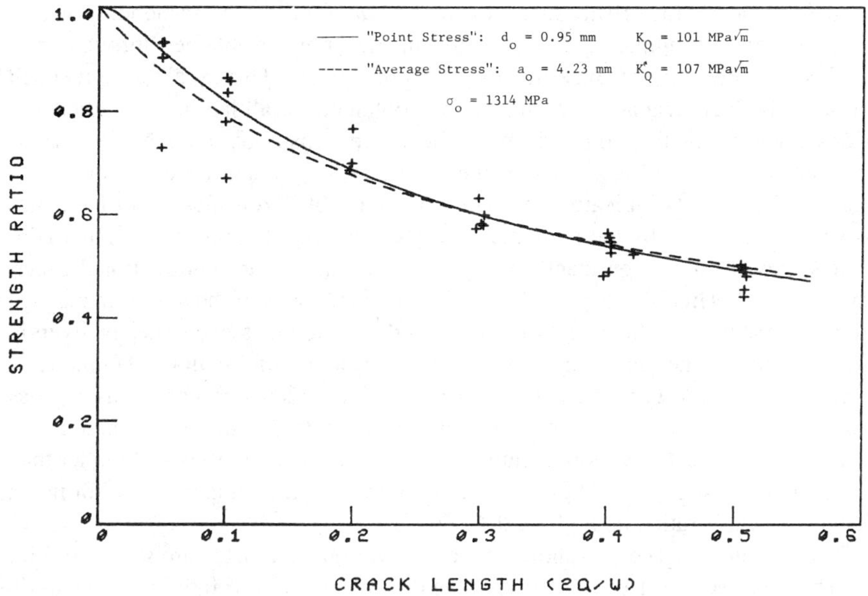

FIG. 1—*Notched strength ratio* (σ_f/σ_0) *versus notch length predictions for unidirectional specimens.*

Results and Discussion

Introduction

The features and characteristics of the AE events and the interpretation of AE results depend upon the type of composite systems studied, the laminate configuration, quality of fabrication, mechanical properties, fracture behavior, and failure modes. Therefore, it is imperative that such information be available before the AE data can be properly assessed. Accordingly, a brief summary of the fracture behavior of center-notched boron/aluminum laminates is given here[2], followed by a detailed discussion of the corresponding AE results.

Fracture Behavior

All boron/aluminum laminates studied exhibited significant notch sensitivity, comparable to that of graphite/epoxy laminates reported by numerous investigators. Figure 1 shows notched strength to unnotched strength ratio (σ_f/σ_0) data for unidirectional specimens together with the prediction according to the "point-stress" and "average-stress" criteria proposed by Whitney and Nuismer [*7*]. The

[2]The acoustic emission results discussed here were obtained during a comprehensive study on the fracture behavior of boron/aluminum laminates at room and elevated temperatures sponsored by the Air Force Office of Scientific Research, results of which will be published elsewhere. This section summarizes a few issues directly related to this article.

values of the characteristic distances, d_0 and a_0, (and the corresponding values of the fracture toughness, K_Q) indicated in Fig. 1 were obtained through error analyses to best fit the experiments with predictions. The excellent agreement between both criteria and experiments was obtained for all laminates studied [*8*]. Consequently, in the comparison of the notch sensitivity among the various laminates, shown in Figs. 2 and 3, the notched strength data were eliminated for clarity. The results indicate that notch sensitivity of boron/aluminum laminates is strongly affected by the laminate configuration, and that multidirectional laminates may become signficantly more notch sensitive than unidirectional specimens, Fig. 2. This is as expected when the characteristics of the notch-tip damage zone are taken into consideration. Detailed discussion of these issues is given in Refs *8* and *9*. The significant notch sensitivity of multidirectional laminates is manifested by the unnotched strength (σ_0) reduction, which can be as high as 30% for notch length-to-width ratios of 0.1. Despite the significant notch sensitivity of unidirectional boron/aluminum, its notched strength is still higher than the unnotched strength of either of the multidirectional laminates, even for notch length-to-width ratios as high as 0.4, Fig. 3.

The fracture surface morphologies for the various laminates are shown in Fig. 4. The detailed SEM examinations revealed various micro-failure modes associated with fracture of the various laminates. In all laminates containing 0° fibers, those fibers are shattered into wedge-shaped fragments. Exposing the fibers

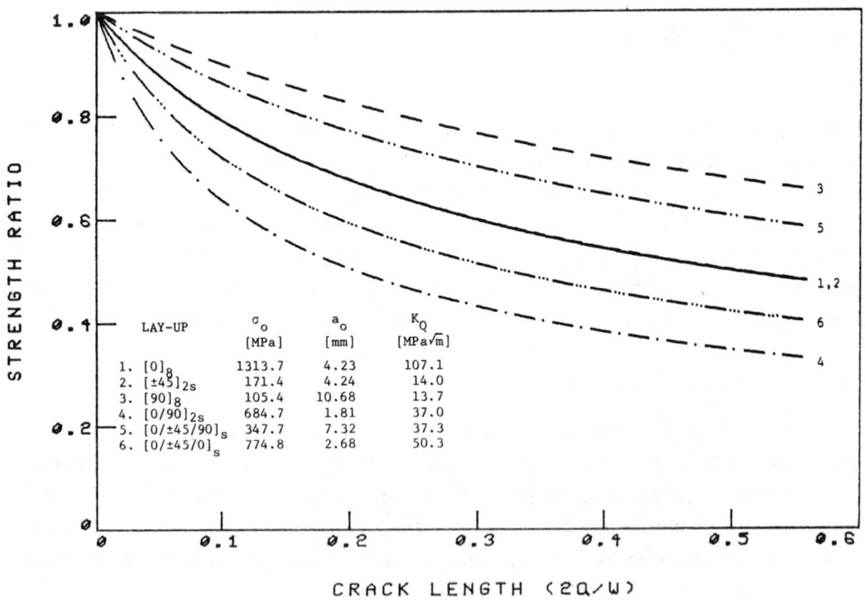

FIG. 2—*Notched strength ratio* (σ_f/σ_0) *versus notch length predictions for six different laminates. Prediction is based on the "average stress" criterion.*

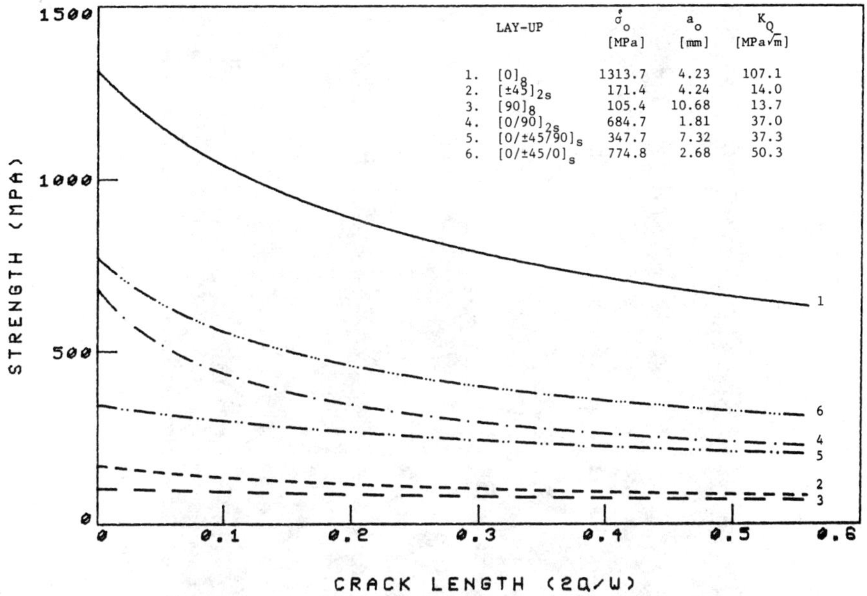

FIG. 3—*Notched strength versus notch length predictions for six different laminates. Prediction is based on the "average stress" criterion.*

through matrix dissolution revealed that this fiber shattering does not extend to a depth greater than two fiber diameters from the fracture surface. The extent of the fiber fragmentation was also confirmed by observations through the stereo viewer. Extensive fiber splitting in the 90° and 45° plies occurred in all multidirectional laminates, while very little fiber splitting has been detected in the $[90]_8$ specimens, except in the vicinity of the notch-tip. Significant matrix yielding has been noticed in all laminates, manifested by the large amount of micro-void coalescence. Matrix plastic shear deformation along the split surfaces of the $[0]_8$ specimens could also be identified. There was very little if any evidence of shear deformation along the 0° and 45° fibers in the multidirectional laminates; however, fiber pull-out occurred randomly throughout the fracture surfaces. In most cases, pull-out length extended to approximately two to three fiber diameters. Interfacial failure was noticeable in all 90° layers, and to a lesser degree along the 45° layers. Very little, if any, interfacial failure could be detected in the $[0]_8$ specimens. Generally, good matrix-matrix consolidation has been determined; however, in the $[0]_8$ specimens, and more so in the $[0/90]_{2s}$ laminates, matrix-matrix disbonding was quite noticeable. The discussions presented in this paper explore the possible correlations between the various failure modes and the AE results.

The deformation characteristics of center-notched unidirectional boron/aluminum can be seen from the load-COD curves shown in Figs. 5 and 6. The COD was measured at the center of the notch surfaces using a laser interferometric

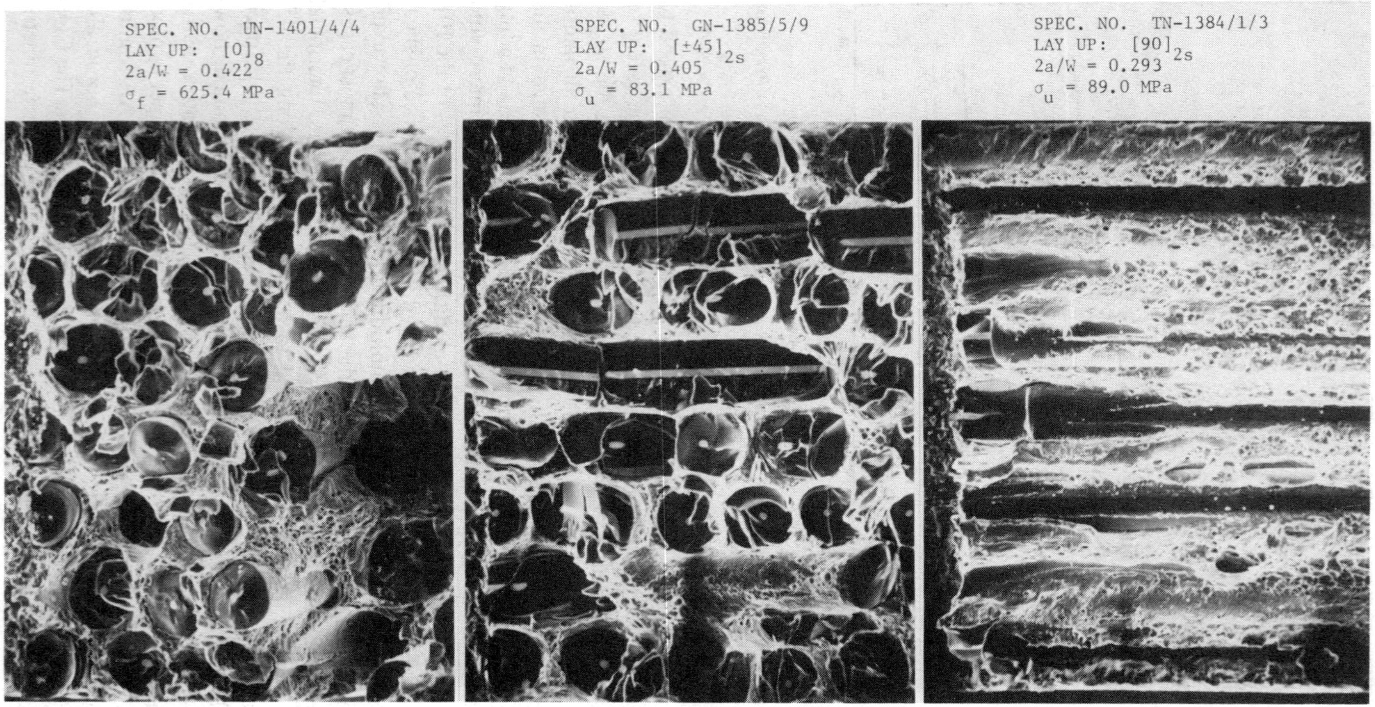

FIG. 4*a*—*Scanning electron microscope micrographs of fracture surfaces of six laminates showing the various micro-failure modes.*

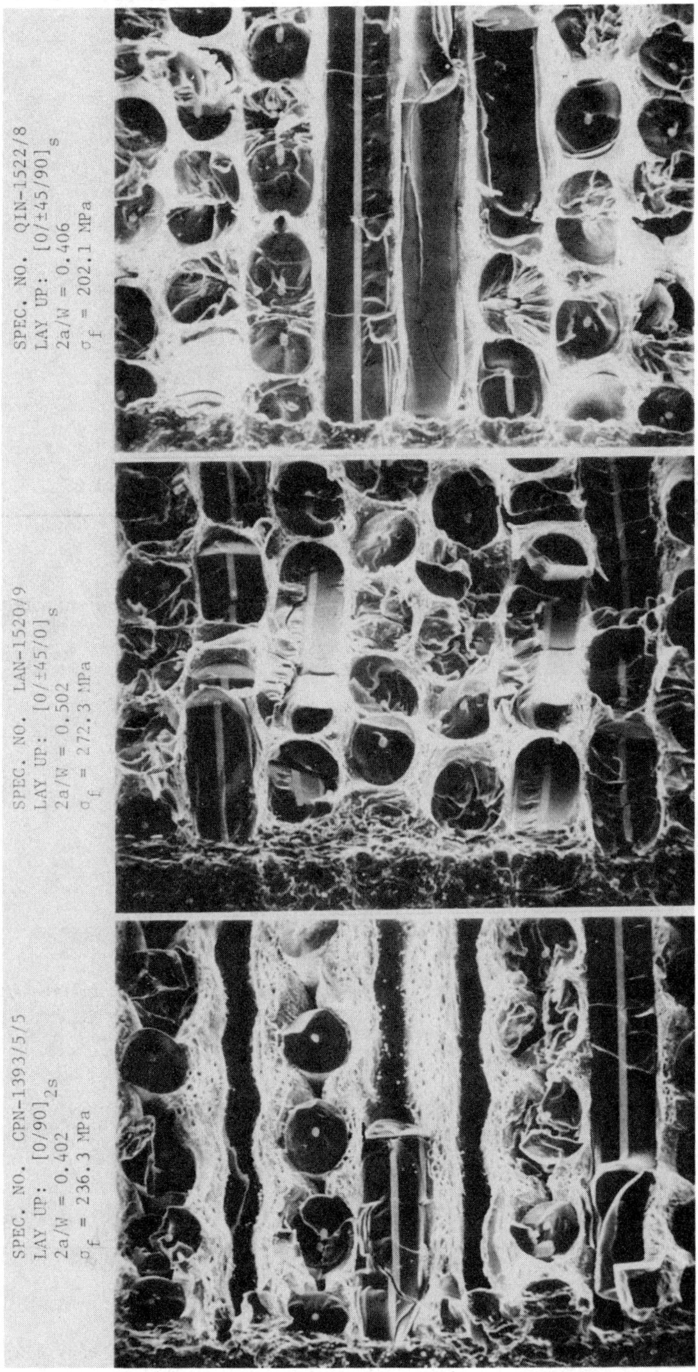

FIG. 4*b*—*Continued.*

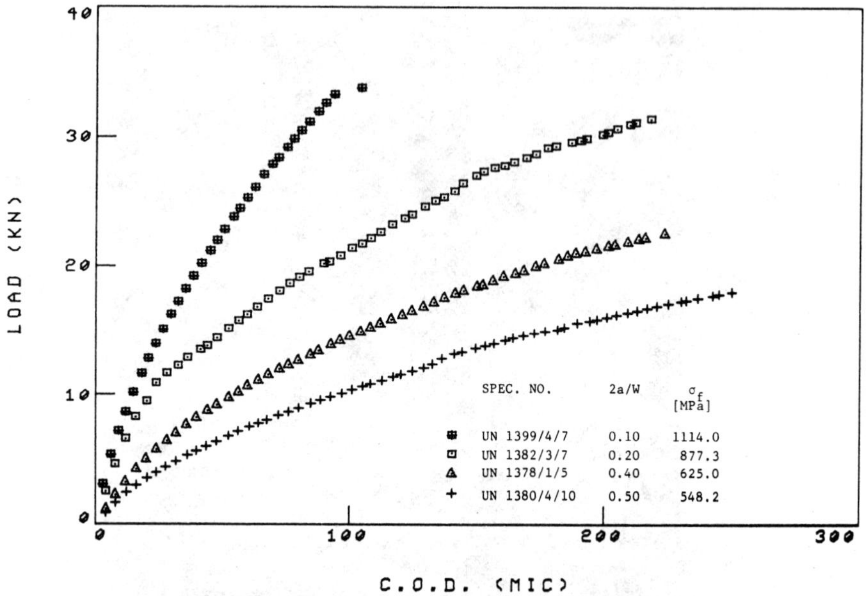

FIG. 5—*Load-COD curves for* $[0]_8$ *specimens for various notch lengths obtained with the IDG. Specimens show significant nonlinearity.*

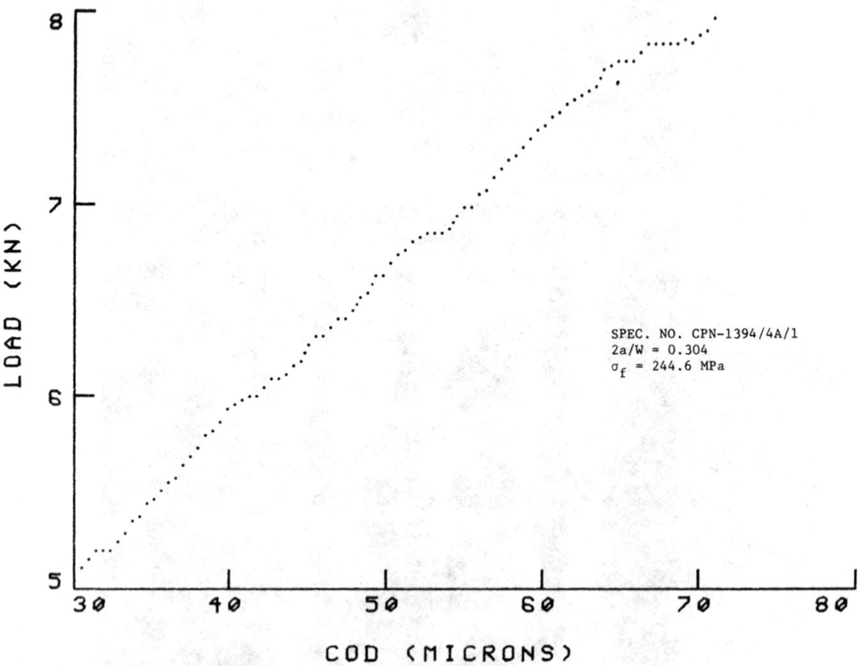

FIG. 6—*Detail of load-COD curve for* $[0/90]_{2s}$ *laminate (notch length = 7.62 mm) showing jumps in COD as load approaches its ultimate level.*

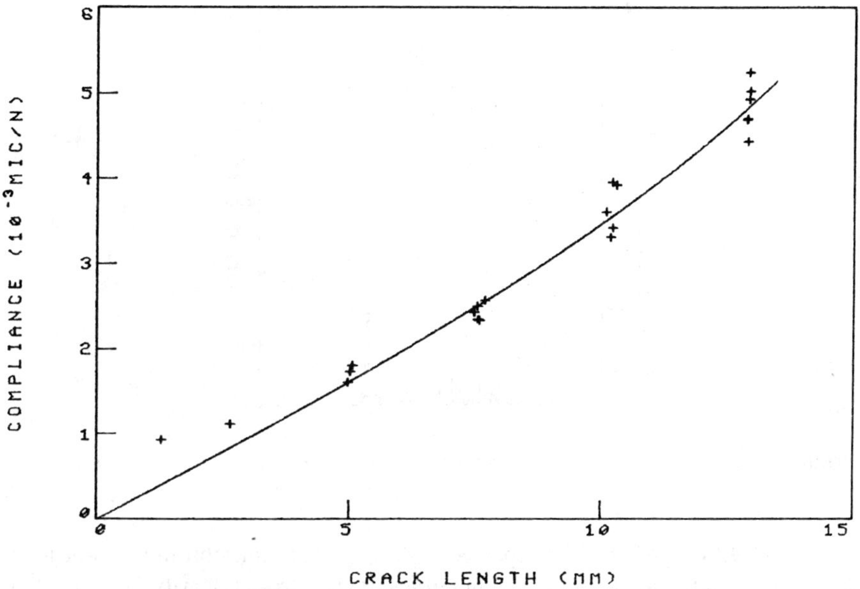

FIG. 7—*Local compliance (COD measured with IDG) versus notch length for unidirectional specimens. Compliance is highly sensitive to notch length.*

displacement gage (IDG)[3]. The load-COD curves are highly nonlinear, which results from fiber breakage and local matrix plastic deformation. Details of a representative load-COD curve for a $[0/90]_{2s}$ laminate are shown in Fig. 6 revealing jumps in COD as the load approaches its ultimate value. These jumps in COD could be directly related to fiber breakage through visual observations, for example, CCTV. Similar nonlinear load-COD curves and jumps in COD were recorded for all other laminates to differing degrees, depending upon lay-up. A significant effect of notch length in the elastic region has been recorded as well, as can be seen in the resulting "local" compliance shown in Fig. 7 (solid line represents predictions based on elastic stress analysis given in Ref *10*).

All these base-line results for the fracture behavior of boron/aluminum laminates indicate the need for a nondestructive inspection technique that is readily utilized and can depict the failure initiation and progression and the failure processes in real time. The following discussion presents representative AE results.

Accumulative Events and Counts

The most straightforward AE measurement is count-rate as a function of applied load and its correlation with the far-field load-displacement curves, for example,

[3]Details of the IDG technique and representative results are given in Ref *6* and are not repeated here for the sake of conciseness.

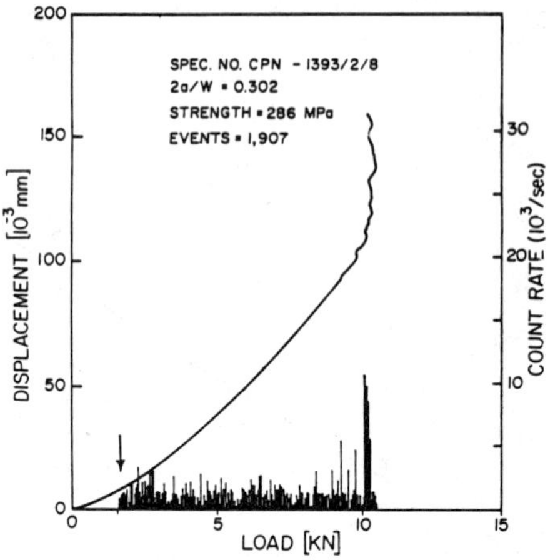

FIG. 8—*Far-field displacement and AE count rate as a function of load for* $[0/90]_{2s}$ *laminate.*

Figs. 8 and 9. The AE results repeatedly showed that emission initiates at load levels for which the load-displacement curves deviate from linearity, as indicated by the arrows in Figs. 8 and 9. Also, increased AE activity is associated with higher nonlinearity in load-displacement curves, that is, increased inelastic deformation with increasing load. In most cases, whenever jumps in displacement occur, a sudden surge in AE activity is recorded. Therefore, distinctly different count-rate results were recorded for each of the different laminates as they exhibit different deformation characteristics. It should be noted that the emission initiation load can be as low as 20% of ultimate load. In summary, the AE results can be easily and directly correlated with the deformation characteristics and can furnish

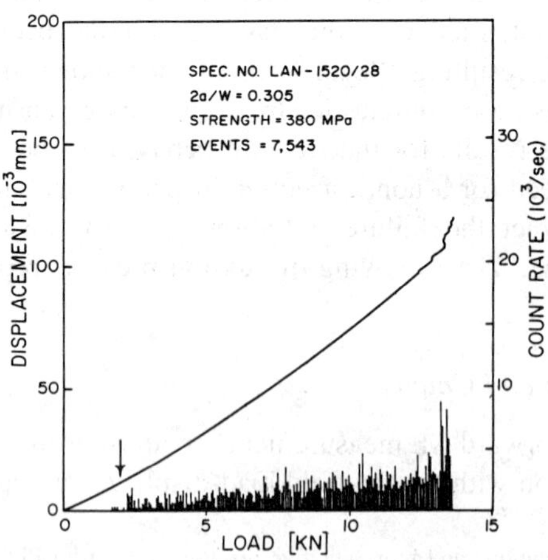

FIG. 9—*Far-field displacement and AE count rate as a function of load for* $[0/\pm45/0]_s$ *laminate.*

information on damage/deformation initiation at a relatively early stage of loading.

The potential of the AE technique for distinguishing among specimens with different notch sizes (or notch length-to-width ratios, $2a/W$) has been addressed as well. The results for different $2a/W$ ratios are shown in Figs. 10 and 11 in an events-versus-applied load format. It seems that for all boron/aluminum laminates studied, no distinction could be made among the various notch lengths. In other words, notch length does not affect either the emission initiation load or the rate of increase in emission with increasing load. This lack of sensitivity of AE to notch length is surprising considering the significant notch sensitivity of boron/aluminum laminates, Figs. 1 to 3 and 5 to 7. It seems, however, that the extensive matrix plastic deformation that takes place all along the specimen length (to be discussed later) emits significantly more AE than do the localized regions of the notch-tip. The accumulative emission generated from the matrix deformation and probably from the interfacial failure overwhelms the emission associated with the local damage growth at the tip of the artificial notch.

Location Distribution Histograms of Events

The location at which damage occurs is obtained by placing the two AE transducers a known distance apart. From the time difference between stimulation of each transducer by an AE event (namely, occurrence of damage) and *a priori*

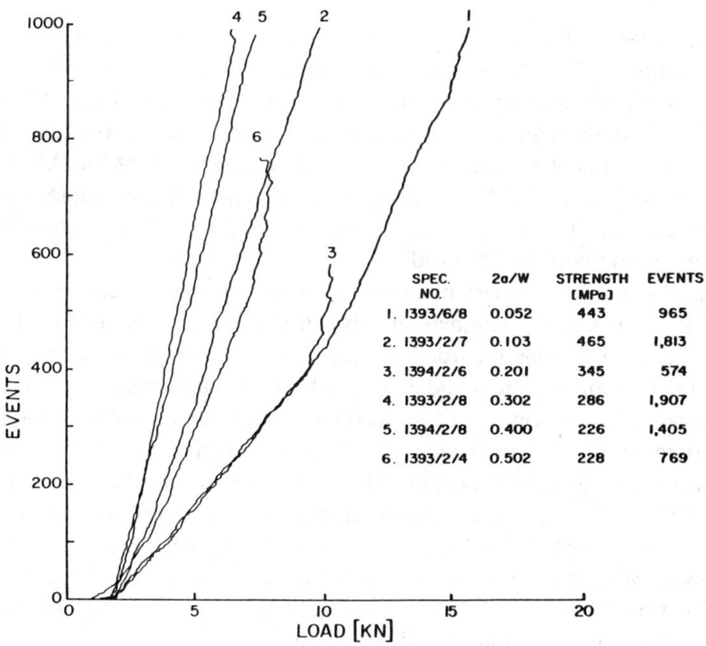

FIG. 10—*Accumulative events as a function of load for various notch lengths for* $[0/90]_{2s}$ *laminate.*

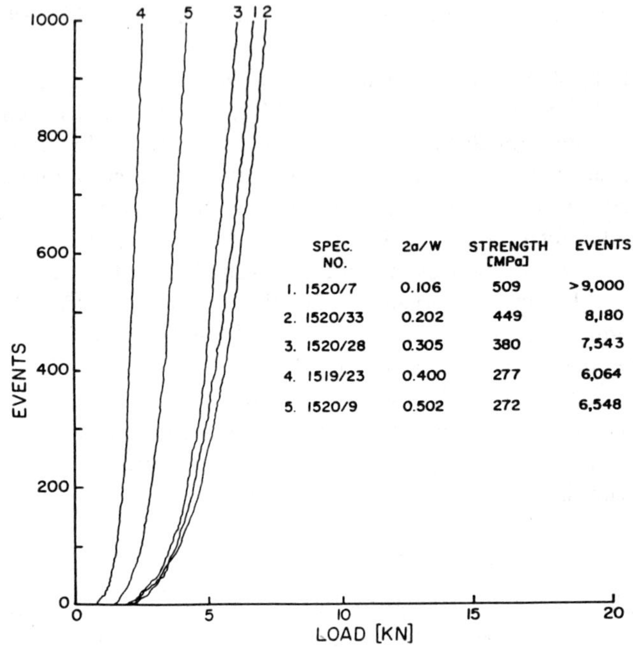

FIG. 11—*Accumulative events as a function of load for various notch lengths for* $[0/\pm 45/0]_s$ *laminate.*

knowledge of the surface wave speed of the subject material (usually obtained by proper calibration of the AE instrumentation), the location at which the AE event occurs can be determined in real time. With increasing load, AE events will emanate from throughout the specimen length, yielding location distribution histograms of events at various load levels. It is expected that the AE activity will be more pronounced at the location of stress raisers (for example, notches, impact damage, etc.) and thus existing damage could be detected and located and its progression could be tracked.

Location distribution histograms of events at various load levels for a notched ($2a/W = 0.3$) $[0/\pm 45/0]_s$ laminate are shown in Figs. 12*a* and *b*. For the sake of clarity, the distributions recorded at various load levels have been separated in the figure. The applied stress and associated accumulated number of AE events within the pre-selected window of 40 to 60 are given for each distribution. The results shown in Fig. 12 indicate that the artificially induced notch could be easily detected and located (Location 50 at the center of the specimen), that is, increased AE activity has been recorded at that location. Moreover, the existing notch could be located at a relatively low load level, within less than 20% of ultimate load. With increasing load, significant AE activity has been recorded also at Location 45, Fig. 12. This increased AE activity can be attributed to either damage progression along the 45° direction, as has been repeatedly observed visually through the CCTV, or it may be due to damage progression at the other notch tip and the failure of the AE instrumentation to accurately distinguish between two sources of emission (both notch tips) generating simultaneously.

The results shown in Fig. 12 also indicate that emission occurs practically throughout the specimen length, and increases with increasing load. It seems that this emission is due primarily to the significant matrix plastic deformation and interfacial failure typical to boron/aluminum laminates. In fact, the number of AE events occurring in the notch region (namely, Location 50) amounts to less than 20% of the total number of events occurring throughout loading. Obviously, this ratio will depend on window limits, specimen dimensions (being lower for longer specimens), and on laminate configuration, fabrication procedure, material quality, and constituent properties. Because of the relatively large amount of emission generated throughout the specimen length, no clear distinction could be made among various notch lengths in the events-versus-load curves shown in Figs. 10 and 11, even though the events recorded are generated within a relatively narrow, pre-selected window of 40 to 60.

Another important issue is to determine the smallest notch size the AE technique can detect and locate. From results for $[0/\pm45/0]_s$ laminate with various $2a/W$ values, such as those shown in Fig. 12, it has been determined that a damage size of $2a/W = 0.1$ could be easily located. For smaller notch sizes, determi-

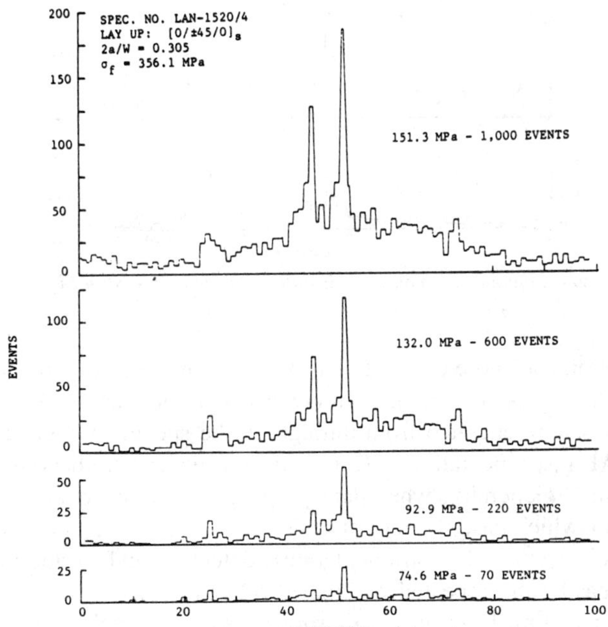

FIG. 12*a*—*Location distribution histograms of events for* $[0/\pm45/0]_s$ *laminate at various load levels. Notch (located at Location 50) can be easily detected and located, however, significant emission is detected throughout the specimen length and increases with increasing load.*

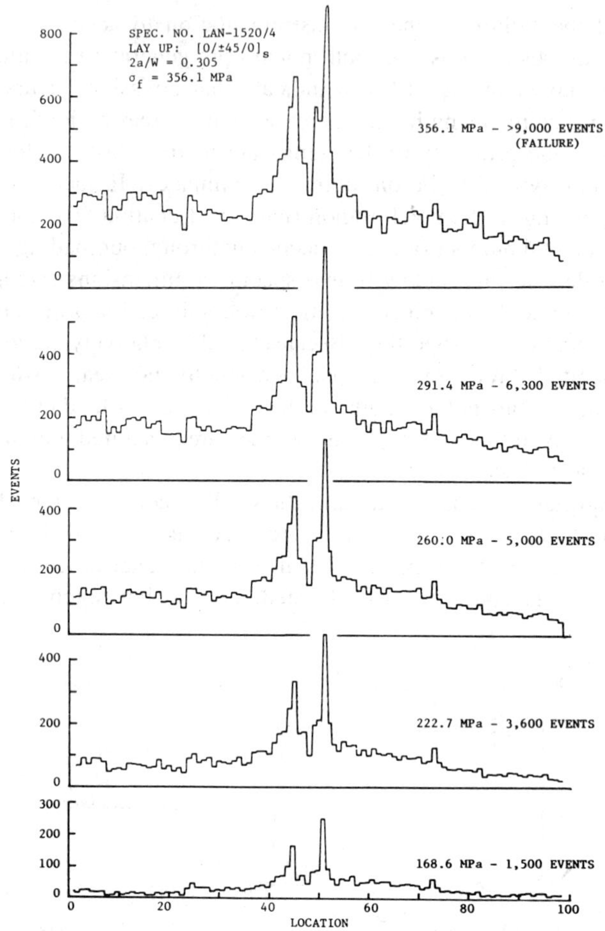

FIG. 12*b*—*Continued. (Note, scale is different from that used in Fig. 12*a).

nation of location was more difficult. The reason is that notch-tip damage growth for small notch lengths only occurs at relatively high load levels. At these levels, the rate of emission generated from throughout the specimen gage length is too high for the AE instrumentation to record, that is, the system becomes saturated or "locked-out." Generally, when damage progression occurs at relatively low load levels (for which case the rate of emission is sufficiently slow for accurate data acquisition by the AE instrumentation), detection and location of existing damage was easily and repeatedly obtained. On the other hand, when damage progression emanating from an existing notch occurs at higher load levels, the AE instrumentation fails to detect its location. Consequently, detection and location of the artificially induced notches in $[0]_8$, $[\pm 45]_{2s}$, $[0/90]_{2s}$, and $[0/\pm 45/90]_s$ laminates could not be made unless $2a/W \geq 0.2$.

Detection of Failure Mechanisms

Determination of the major failure mechanisms using AE can be based upon a variety of AE source intensity analyses such as events amplitude, duration, energy, counts per event, frequency spectrum, rise-time, etc. In this study, emphasis has been placed on analyzing the amplitude distribution histograms of events to determine the dominant failure mechanisms and processes in the various laminates.

A special concern in boron/aluminum laminates is the detection of fiber breakage and determining the load at which it occurs. For this purpose, as a first stage, individual 5.6-mil and 8.0-mil diameter fibers were loaded to failure. Ultimate strength and event amplitudes were recorded, Table 1. The results show that practically all fiber failures generated high-amplitude events of 95 dB and 98 dB for the two fiber types, respectively. The only exception that appears in Table 1 could have been caused by extraneous events generated just prior to the fiber breakage. The recorded amplitudes were highly reproducible despite the large scatter in strength, that is, Weibul shape parameter is 6.5 and 5.0 for the 5.6-mil and 8.0-mil fibers, respectively. It could be argued that the higher the strength of the fiber, the higher the amplitude (and energy) of the associated event. These event amplitudes approached the limits of the AE instrumentation dynamic range, which is from 0 dB to 100 dB. It could be expected that with a wider dynamic

TABLE 1—*Strength data and amplitude of acoustic emission events for single boron fibers.*

	5.6-mil Boron Fibers		8.0-mil Boron Fibers	
	Strength, MPa	Amplitude, dB	Strength, MPa	Amplitude, dB
	2140	96	1442	98
	2324	96	1715	98
	2497	96	2208	98
	2535	95	2391	98
	2584	96	2454	98
	2600	96	2524	98
	2667	95	3047	98
	3024	96	3058	98
	3027	95	3107	98
	3243	95	3262	98
	3259	95	3369	98
	3259	95	3379	98
	3454	96	3471	98
	3572	95	3606	98
	3627	96	3771	98
	3629	95	3891	98
	3739	38	3989	98
	3930	...		
	3951	95		
Average	3108		2981	
Standard deviation	563		927	

range, more accurate amplitude values would have been recorded and would most probably show a correlation between strength and amplitude. However, such an investigation was beyond the scope for this study. Here, the purpose was to determine the established assumption that fiber breakages do, in fact, result in high-amplitude events. It should also be added that the large scatter in fiber strength was recorded for fibers taken from different spools but manufactured during the same time period. This large scatter in fiber strength also indicates the need for a nondestructive test procedure that can easily detect and identify fiber failure in real time.

As a second stage, in order to identify fiber breakage within the composite, several unidirectional boron/aluminum specimens were loaded to a predetermined load level at which high-amplitude events were recorded, the specimens were removed from the testing fixture, the aluminum matrix was dissolved, and the number of broken fibers was recorded. A representative result is shown in Figs. 13 and 14, for which a unidirectional specimen with a notch length of $2a/W = 0.30$ has been loaded to approximately 40% of ultimate load. At this load level, the amplitude distribution histogram, Fig. 13 (shown for clarity on a logarithmic scale), includes several high-amplitude events, indicating the possible occurrence of broken fibers. The SEM photographs, Fig. 14, obtained after dissolving the outer layer of the aluminum matrix do reveal the existence of broken fibers. In summary, from the results shown in Table 1 and in Figs. 13 and 14, it could be concluded with confidence that the occurrence of high-amplitude events is associated with fiber breakage.

In the third stage, amplitude distribution histograms of events were recorded

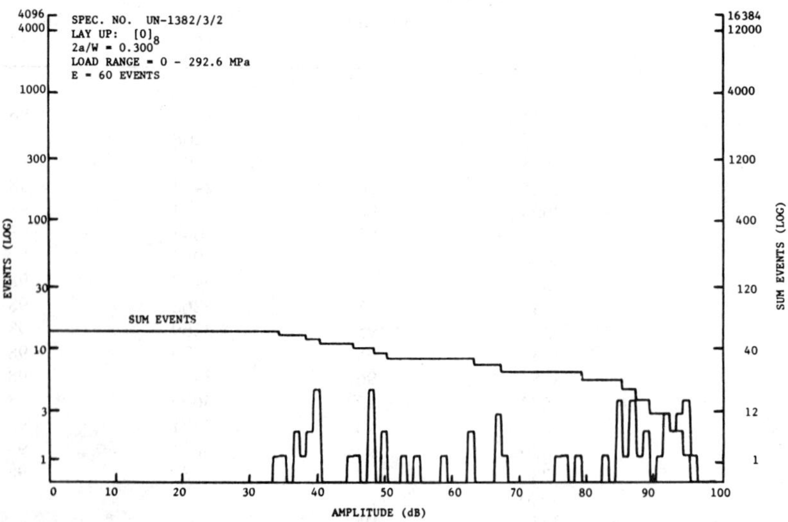

FIG. 13—*Amplitude distribution histogram and cumulative event amplitude distribution (CEAD) for unidirectional specimen loaded to approximately 40% of expected ultimate load, showing high-amplitude events.*

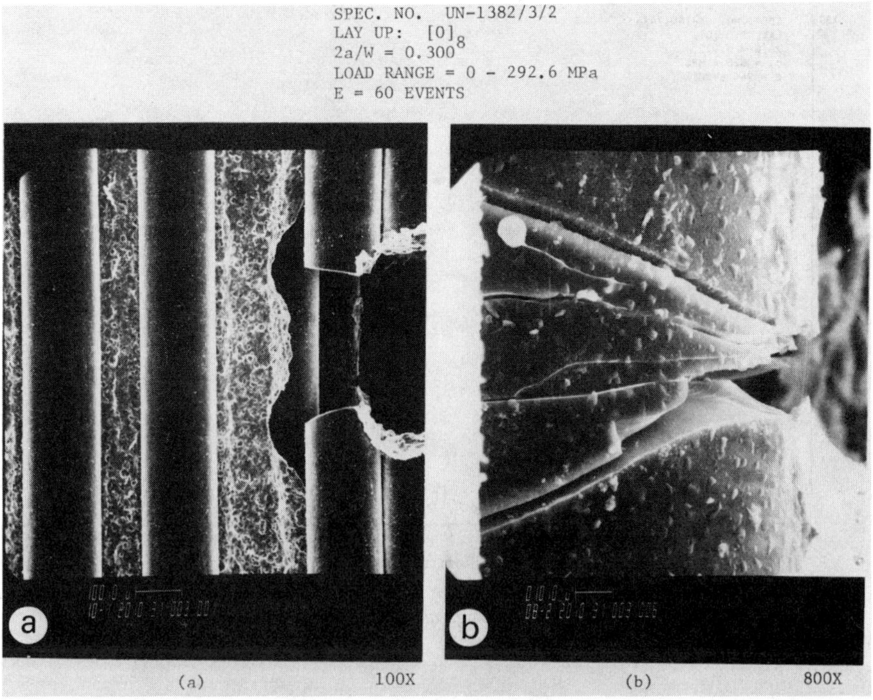

FIG. 14—*Scanning electron microscope photographs of broken fibers at notch-tip (after dissolving surface layer of aluminum matrix) for a unidirectional specimen loaded to 40% of expected ultimate load (same specimen as shown in Fig. 13): (a) notch-tip damage, broken fiber seen in third layer (×100) and (b) detail of broken fiber in third layer (×800).*

for all six boron/aluminum laminates with all six notch length-to-width ratios, Figs. 15 to 20. These histograms are for all the events accumulated during quasi-static loading to failure that occur within the location window of 40 to 60. It is of interest to note that different amplitude histograms were obtained for the different laminates, while within each given laminate configuration the amplitude histograms were consistent and independent of notch length. In other words, the results were highly reproducible, Fig. 21.

It should be noted that the amplitude distribution histograms of events shown in Figs. 15 to 20 indicate several event amplitudes below the AE system threshold, namely, 1 V (40 dB). The measured and actual amplitudes agree very well within the approximate range of 45 dB to 90 dB. However, for event amplitudes below 45 dB, the margin of error increases dramatically on the low side. Consequently, all the event amplitudes below 40 dB shown in the amplitude distribution histograms of events should be considered to be in the 40-dB to 45-dB range. The error between the actual and measured peak amplitude is due to the specific hardware and data processing design of the AE instrumentation, a technical explanation of which is beyond the scope of this paper. It should be noted, however, that for the purpose of this discussion the precise decibel level of the

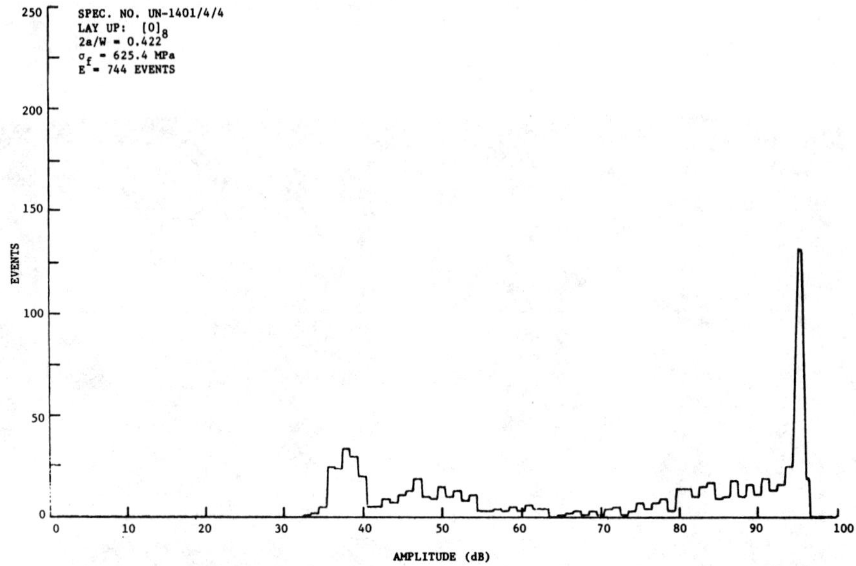

FIG. 15—*Amplitude distribution histogram of events accumulated throughout loading to failure for* $[0]_8$ *laminate showing primarily high-range amplitude events.*

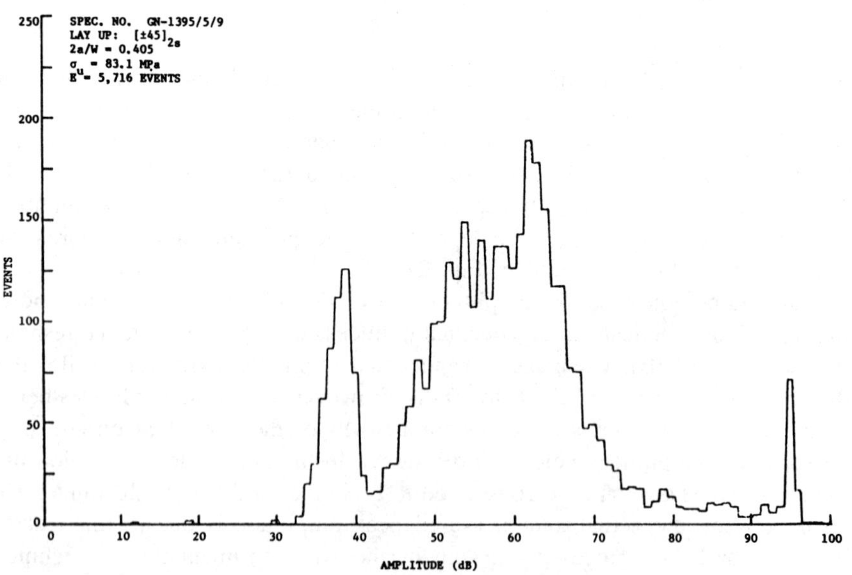

FIG. 16—*Amplitude distribution histogram of events accumulated throughout loading to failure for* $[\pm 45]_{2s}$ *laminate, showing primarily low and middle-range amplitude events.*

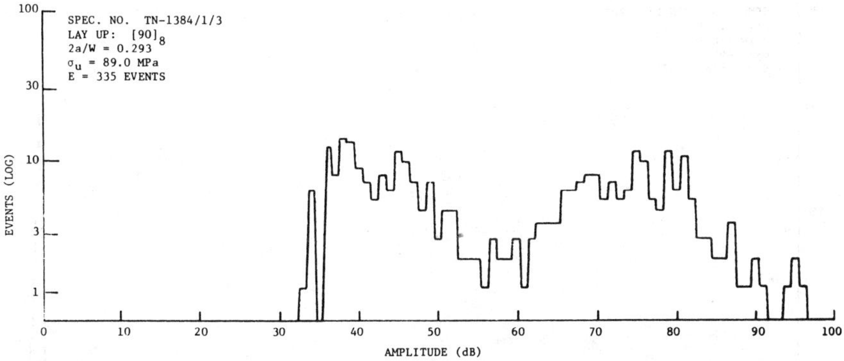

FIG. 17—*Amplitude distribution histogram of events accumulated throughout loading to failure for* $[90]_8$ *laminate, showing primarily low and middle-range amplitude events.*

event amplitudes is not critical. Rather, the purpose is to determine whether the distinct amplitude ranges can be associated with specific failure modes.

The amplitude distribution histogram of events for the $[0]_8$ specimen, Fig. 15, includes a large number of events in the 94-dB to 96-dB range that can be attributed to fiber breakage, as previously discussed. Some increased activity in the 40-dB range and a few events in the middle range, 50 dB to 70 dB, were recorded as well. Completely different amplitude distribution histograms were recorded for the $[\pm 45]_{2s}$ laminate, Fig. 16, for which a larger number of events were in the lower (40-dB) range and a significant amount of emission occurred in the middle (50-dB to 70-dB) amplitude range. Since significant matrix plastic deformation is expected to occur in the $[\pm 45]_{2s}$ laminate, it could be concluded that the middle range amplitude events are associated with this plastic deformation. This conclusion is also supported by a comparison among the AE results for all the laminates, as discussed later. It seems that the lower amplitude range is associated with either interfacial failure or significant shear deformation of the matrix along the fibers, or a combination thereof, depending upon the laminate configuration. The precise nature is difficult to distinguish due to the complex inter- and intra-ply states of stress.

In the $[90]_8$ specimens, Fig. 17 (shown for clarity on a logarithmic scale), a relatively large number of events occur in both the lower and middle-amplitude ranges. These two amplitude ranges seem to be associated with interfacial failure and matrix plastic deformation, respectively. These two failure modes are expected to occur in $[90]_8$ specimens. The high-amplitude events (94 dB to 96 dB) shown in Figs. 16 and 17 were generated when the specimens approached their ultimate tensile strength (UTS), which will be discussed later.

Based on the amplitude distribution histograms obtained for these laminate "building blocks," it should be expected that the amplitude distribution histograms for multidirectional laminates would include all three amplitude ranges, associated with interfacial failure or matrix shear deformation along the fibers,

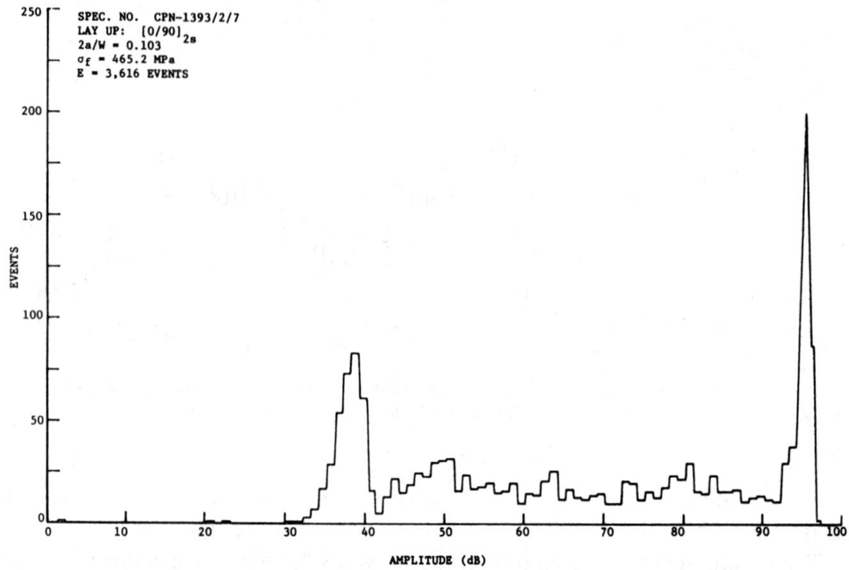

FIG. 18—*Amplitude distribution histogram of events accumulated throughout loading to failure for* $[0/90]_{2s}$ *laminate showing primarily low and high-range amplitude events.*

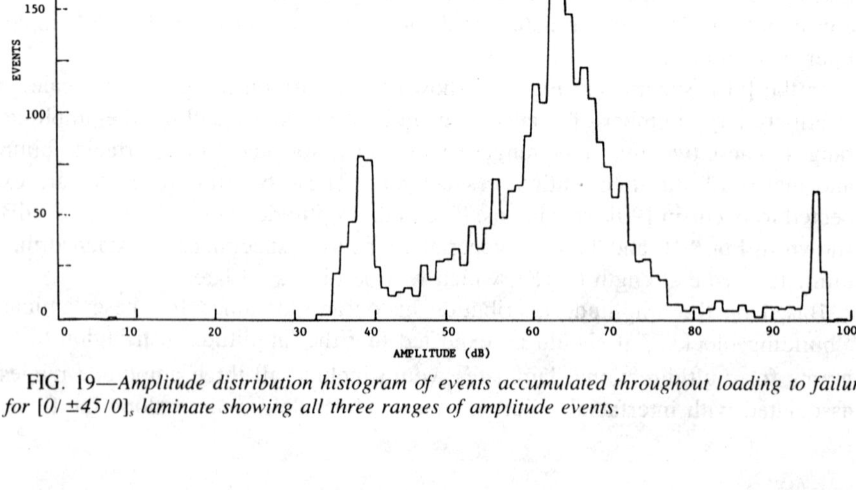

FIG. 19—*Amplitude distribution histogram of events accumulated throughout loading to failure for* $[0/\pm45/0]_s$ *laminate showing all three ranges of amplitude events.*

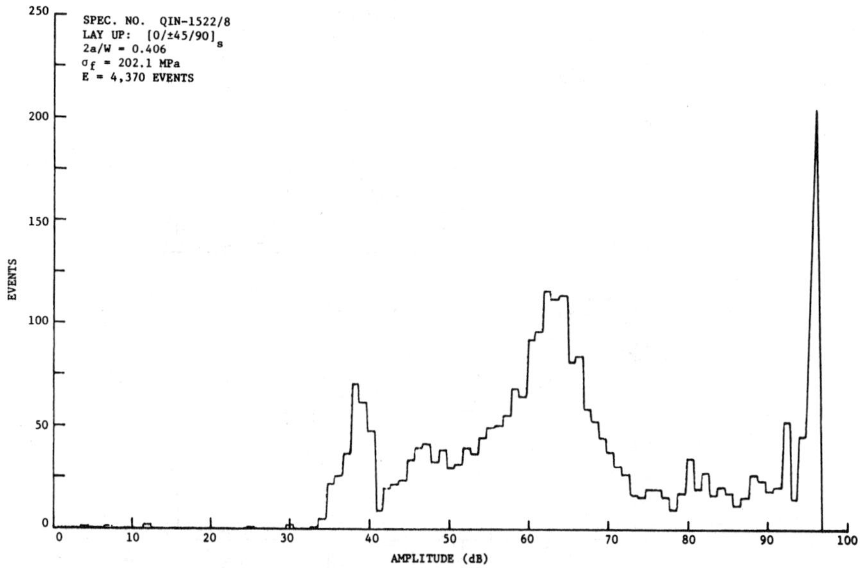

FIG. 20—*Amplitude distribution histogram of events accumulated throughout loading to failure for* $[0/\pm45/90]_s$ *laminate showing all three ranges of amplitude events.*

matrix plastic deformation, and fiber breakage. The results shown in Figs. 18 to 20 for $[0/90]_{2s}$, $[0/\pm45/0]_s$, and $[0/\pm45/90]_s$ laminates, respectively, confirm this. For the $[0/90]_{2s}$ laminate, Fig. 18, the amplitude distribution histogram is similar to that recorded for the $[0]_8$ specimen, Fig. 15, that is, a large number of high-amplitude events with limited activity in the middle-amplitude range. However, compared with the $[0]_8$ specimen, the amplitude distribution histogram of the $[0/90]_{2s}$ laminate also includes a larger number of low-amplitude events that might be contributed by the presence of the 90° plies. The results obtained for the other two multidirectional laminates, Figs. 19 and 20, clearly show all three amplitude ranges. As previously mentioned, the amplitude distribution histograms were highly reproducible and independent of notch length, Fig. 21.

The issue of identifying the precise failure modes associated with the lower (40-dB) and middle (50-dB to 70-dB) event amplitude ranges cannot be resolved conclusively at this stage. Based on a comparison between the amplitude distribution histograms shown in Figs. 15 to 18, it is plausible to associate the middle-range event amplitudes with matrix plastic deformation. Actually, a variety of mechanisms are associated with the inelastic deformation of the aluminum matrix, for example, the relative motion between adjacent layers, particulate fracture, microcracking, etc. No attempt was made in this study to distinguish among the various mechanisms, and for the sake of brevity they are generally defined here as plastic deformation. The low-amplitude events could also be due to a variety of failure mechanisms, for example, interfacial failure, extensive shear deformation of the matrix along the 0° and 45° fibers, matrix-matrix bond failure, etc.

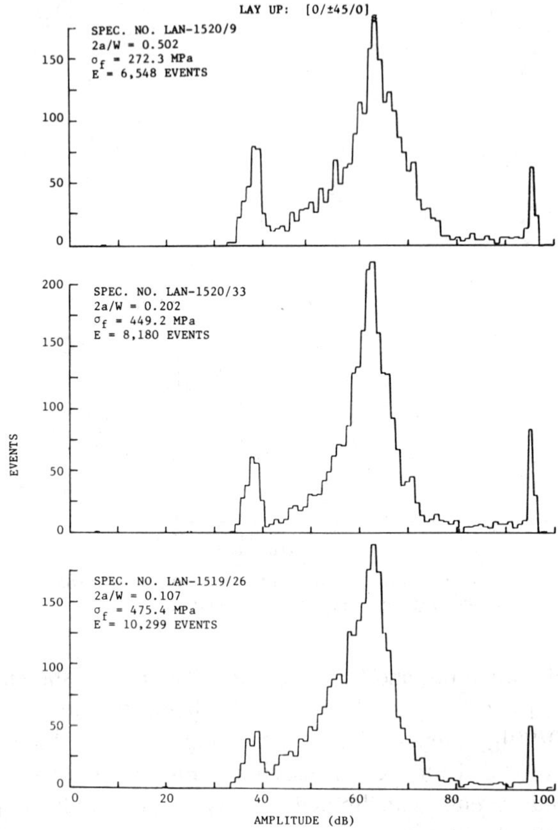

FIG. 21—*Amplitude distribution histograms of events accumulated throughout loading to failure for* $[0/\pm45/0]_s$ *laminate of various notch lengths showing high reproducibility of AE results.*

Based on the examinations of the fracture surface morphologies of multidirectional laminates, recalling that interfacial failures were noticed in the 90° and 45° plies, it seems that this failure mode dominates the low-amplitude events. The amplitude distribution histograms recorded for the $[90]_8$ specimens, Fig. 17, do correlate with this assumption. An additional study currently underway on selected unidirectional off-axis boron/aluminum monolayers may resolve this issue.

Virtually no emission has been recorded for the unreinforced aluminum 6061-F matrix. In other words, using the pre-selected AE instrumentation settings of threshold, gain, etc. as applied in this study, the neat matrix was practically silent. Similar results were also reported elsewhere, namely, Ref 2. It seems that the increased local plastic deformation, caused by the presence of the fibers, and the complex state of stress in the composite laminate generate significantly more emission than the unreinforced aluminum. The issue of correlating the number of high-amplitude events with the actual number of broken fibers was not ad-

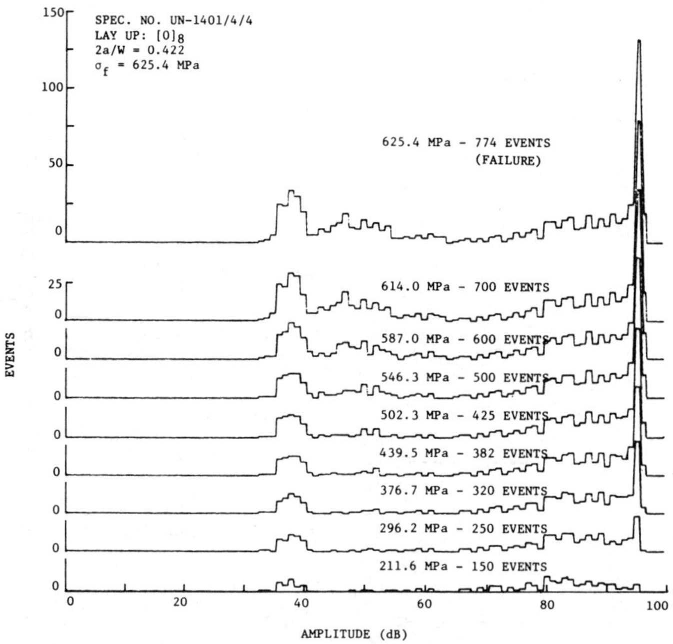

FIG. 22—*Amplitude distribution histograms of events for* $[0]_8$ *laminate at different load levels. High-amplitude events are detected at relatively low load levels.*

dressed in this study. Such attempts were made in Refs *3* and *5* with partial success and additional study into this issue is warranted.

Finally, an important issue is the applicability of the AE technique to monitor the failure processes previously discussed in real time. For this purpose, amplitude distribution histograms of events were recorded during loading at different load levels. Representative results for all laminates studied are shown in Figs. 22 to 27. The amplitude distribution histograms shown for the $[0]_8$ and $[0/90[_{2s}$ laminates, Figs. 22 and 23, respectively, indicate that high-amplitude events (namely, fiber breakage) occur at relatively low load levels, at approximately 20% of ultimate load. In most cases, fiber breakage is followed by low-amplitude events that seem to be associated with interfacial failure and whose number increases with the number of broken fibers. Similar results were repeatedly obtained for the $[0/\pm45/0]_s$ and $[0/\pm45/90]_s$ laminates, Figs. 24 and 25, respectively. High-amplitude events were recorded at approximately 20% of ultimate load, followed by an increased number of low (40-dB) and middle (50-dB to 70-dB) range amplitude events. For the $[\pm45]_{2s}$ and $[90]_8$ laminates, Figs. 26 and 27, respectively, events in the low and middle-amplitude ranges initiated first, while the high-amplitude events were detected at very high load levels, primarily when the load approached UTS. For the $[\pm45]_{2s}$ laminate, the number of high-amplitude events increased beyond the UTS when significant necking of the specimen occurs, resulting ultimately in the specimen fracture, at which stage fiber breakage

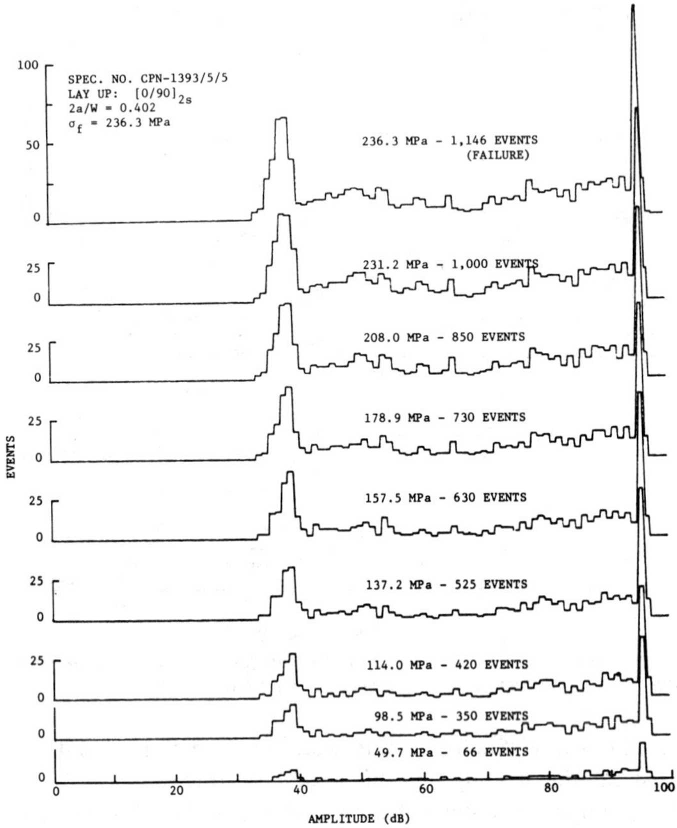

FIG. 23—*Amplitude distribution histograms of events for* $[0/90]_{2s}$ *laminate at different load levels. High-amplitude events are detected at relatively low load levels.*

occurs. For the $[90]_8$ laminate, only a few number of high-amplitude events were detected that occur at relatively high load levels and that are most probably associated with fiber splitting. Such fiber splitting has been observed through the SEM, for example, Fig. 4.

In summary, the representative results shown in Figs. 22 to 27 indicated that for boron/aluminum laminates containing 0° plies, fiber breakage may occur at relatively low load levels, followed by significant matrix plastic deformation. Both of these failure modes can be easily detected in real time through AE.

It should be emphasized that the AE results strongly depend on extrinsic (loading rate, gripping procedure, etc.) and intrinsic (laminate configuration, material system, etc.) parameters and on the AE instrumentation, for example, transducers, filters, data acquisition rate, etc. For example, using different types of fibers, or different aluminum alloys as matrix material may result in different AE characteristics, for example, Ref *2*. Also, the use of different AE instrumentation settings, transducers, etc. [*4*] will also yield different results.

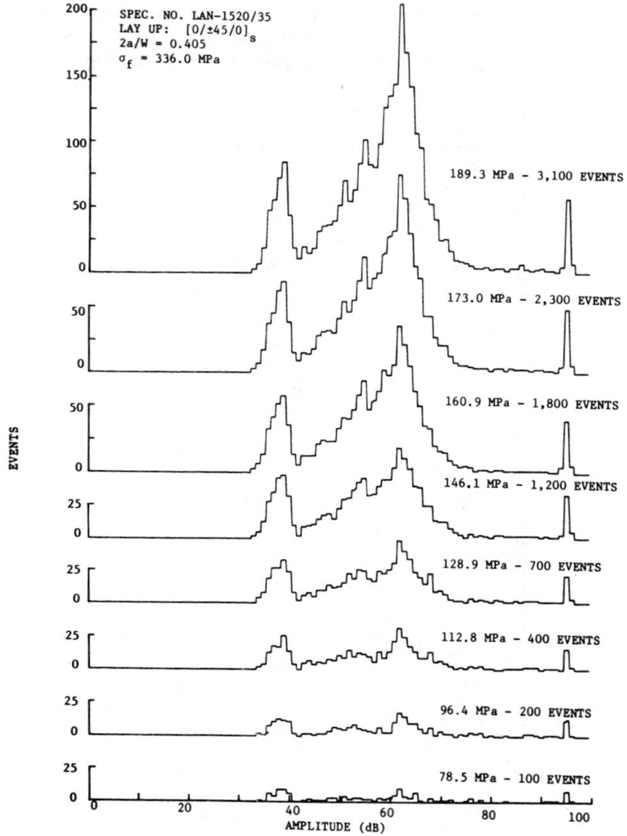

FIG. 24*a*—*Amplitude distribution histograms of events for* $[0/\pm45/0]_s$ *laminate at different load levels. High-amplitude events are detected at relatively low load levels.*

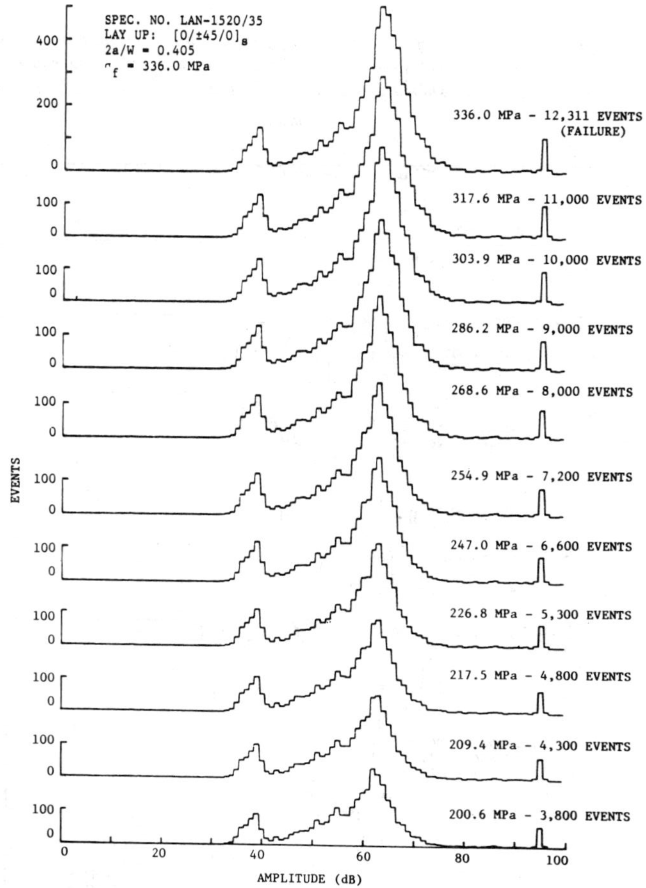

FIG. 24*b*—*Continued. (Note, scale is different from that used in Fig. 24*a*).*

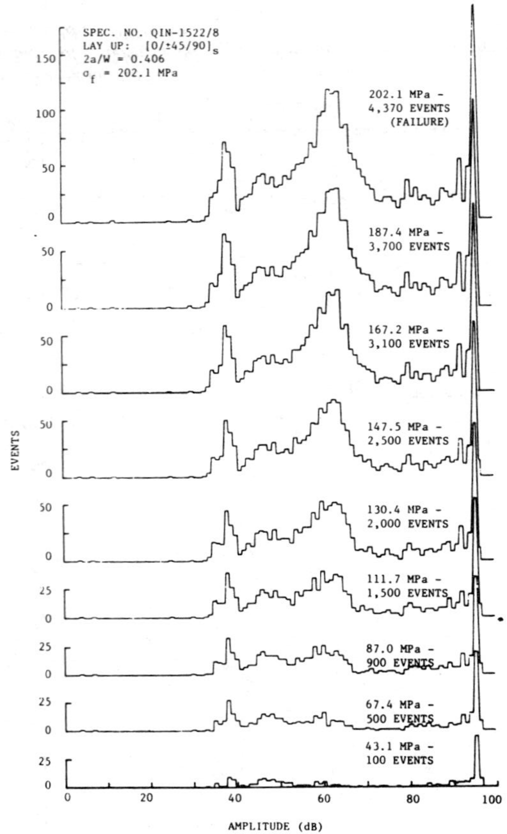

FIG. 25—*Amplitude distribution histograms of events for* $[0/\pm45/90]_s$ *laminate at different load levels. High-amplitude events are detected at relatively low load levels.*

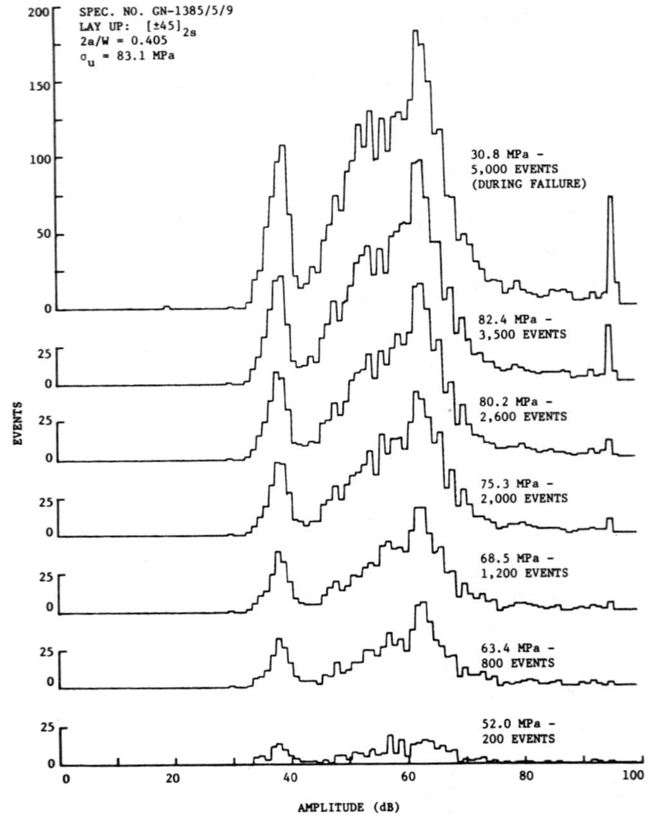

FIG. 26—*Amplitude distribution histograms of events for* $[\pm45]_{2s}$ *laminate at different load levels. High-amplitude events are detected at load levels approaching ultimate, preceded by low and middle-range amplitude events.*

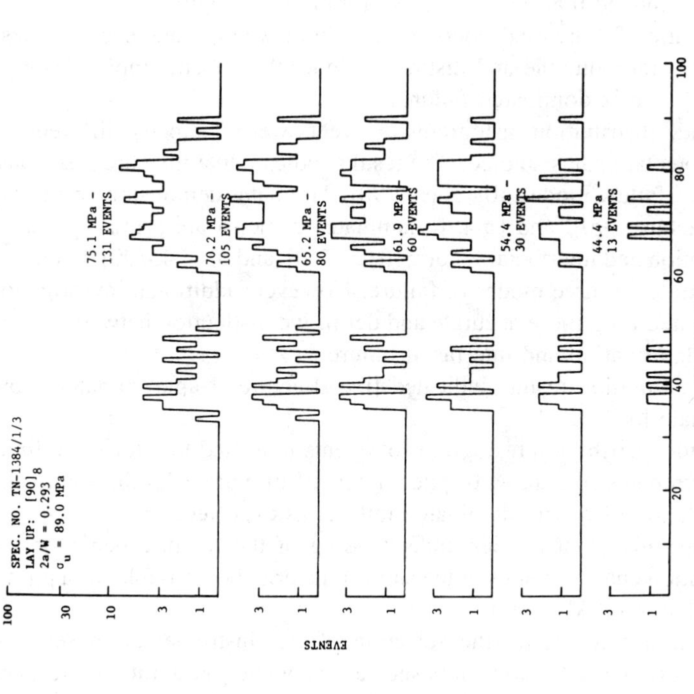

FIG. 27*a—Amplitude distribution histograms of events for* $[90]_8$ *laminate at different load levels. High-amplitude events are detected at load levels approaching ultimate, preceded by low and middle-range amplitude events.*

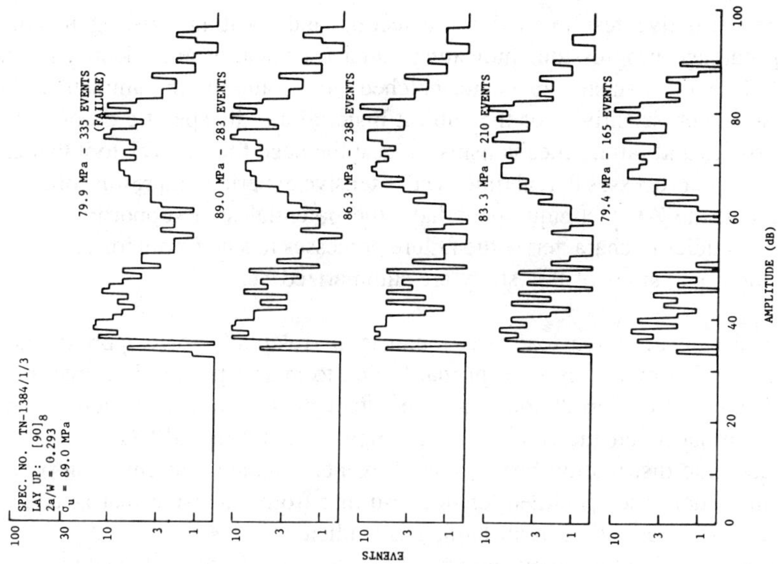

FIG. 27*b—Continued.*

Conclusions

This study addressed several issues regarding the viability of the AE technique as a nondestructive test method for detecting and locating existing notches, tracking damage progression, indicating damage criticality and identifying the major failure mechanisms in center-notched boron/aluminum laminates. The significant notch sensitivity of the subject material and its specific deformation characteristics and failure mechanisms suggest the need for an NDT tool that can identify failure processes in real time. The extensive experimental results obtained indicate that the AE technique does have the potential as a supporting tool in laboratory studies to characterize the failure processes in a new composite system. The major conclusions of this study are summarized here.

1. Emission initiation load was found to be relatively low, approximately 20% of ultimate strength, primarily due to matrix plastic deformation.
2. A qualitative correlation could be established between AE activity and deformation characteristics, for example, count rate and COD.
3. Location distribution histograms of events revealed that emission occurs throughout the specimen length, resulting from extensive matrix and interface deformation in all laminates studied.
4. Location distribution histograms of events could detect and locate existing notches provided that they exceeded a minimum length.
5. The fracture of individual fibers resulted in high-amplitude events. These results are reproducible and distinctly higher than event amplitudes generated by matrix dominated failures.
6. Amplitude distribution histograms of events were distinctly different for the various laminates studied. The results indicate that they are associated with three failure modes: $[0]_8$ and $[0/90]_{2s}$ laminates demonstrate primarily fiber fracture; $[90]_8$ and $[\pm45]_{2s}$ laminates demonstrate primarily matrix deformation and interfacial failure; $[0/\pm45/0]_s$ and $[0/\pm45/90]_s$ laminates demonstrate all three modes of failure. However, additional investigation is warranted for a more accurate and definitive distinction between matrix plastic deformation and interfacial failure.
7. Acoustic emission results indicated fiber fracture at approximately 20% of ultimate load.
8. Amplitude distribution histograms of events indicated that for multidirectional laminates containing 0° plies, fiber fracture precedes the significant accumulation of matrix-dominated failure, as expected.
9. It is imperative that a prior understanding of the fracture behavior and deformation characteristics of the subject material be available for a proper interpretation of AE results.
10. Certain issues regarding the selection of AE instrumentation settings, analyses of other AE parameters such as event energy, duration, etc., and primarily a clearer identification of the matrix-dominated failure mechanism in boron/aluminum laminates deserve additional investigation.

Acknowledgment

This work was supported by the Air Force Office of Scientific Research (OSR #79-0079). The authors wish to thank Mr. H. Perkinson for his help during the preliminary work and for performing the testing on individual fibers.

References

[*1*] Henneke, E. G., II, and Herring, H. W., in *Composite Reliability, ASTM STP 580*, American Society for Testing and Materials, Philadelphia, 1975, pp. 202–214.
[*2*] Grenis, A. F. and Levitt, A. P., in *Proceedings*, 1975 International Conference on Composite Materials (ICCM), Vol. 2, 1975, pp. 1058–1084.
[*3*] Grandemange, J. M. and Street, K. N., in *Proceedings*, 1975 International Conference on Composite Materials (ICCM), Vol. 2, 1975, pp. 1019–1050.
[*4*] Pipes, R. B., Ballintyn, N. J., Scott, W. R., and Carlyle, J. M., in *Composite Materials: Testing and Design (Fourth Conference) ASTM STP 617*, American Society for Testing and Materials, Philadelphia, 1977, pp. 153–169.
[*5*] Johnson, W. S., Biglow, C. A., and Bahei-El-Din, Y. A., "Experimental and Analytical Investigation of the Fracture Process of Boron/Aluminum Laminates Containing Notches," NASA Technical Paper 2187, National Aeronautics and Space Administration, Sept. 1983.
[*6*] Awerbuch, J. and Hahn, H. T., *Journal Composite Materials*, Vol. 12, 1978, pp. 222–237.
[*7*] Whitney, J. M. and Nuismer, R. J,. *Journal Composite Materials*, Vol. 8, 1974, pp. 253–265.
[*8*] Awerbuch, J. and Madhukar, M., *Journal of Reinforced Plastics and Composites*, Vol. 4, No. 1, 1985, pp. 3–159.
[*9*] Awerbuch, J. and Hahn, H. T., *Journal Composite Materials*, Vol. 13, 1979, pp. 82–107.
[*10*] Lekhnitskii, S. G., *Anisotropic Plates*, translated from the Second Russian Edition by S. W. Tsai and T. Cheron, Gordon and Breach, Science Publishers, Inc., New York, 1968.

P. L. N. Murthy[1] *and C. C. Chamis*[1]

Dynamic Stress Analysis of Smooth and Notched Fiber Composite Flexural Specimens

REFERENCE: Murthy, P. L. N. and Chamis, C. C., **"Dynamic Stress Analysis of Smooth and Notched Fiber Composite Flexural Specimens,"** *Composite Materials: Testing and Design (Seventh Conference), ASTM STP 893,* J. M. Whitney, Ed., American Society for Testing and Materials, Philadelphia, 1986, pp. 368–391.

ABSTRACT: A detailed analysis of the dynamic stress field in smooth and notched fiber composite (Charpy-type) specimens is reported in this paper. The analysis is performed with the aid of the direct transient response analysis solution sequence of MSC/NASTRAN. Three unidirectional composites were chosen for the study. They are S-glass/epoxy, Kevlar/epoxy, and T-300/epoxy composite systems. The specimens are subjected to an impact load that is modeled as a triangular impulse with a maximum of 8960 N (2000 lb) and a duration of 1 ms. The results are compared with those of static analysis of the specimens subjected to a peak load of 8960 N (2000 lb). For the geometry and type of materials studied, the static analysis results gave close conservative estimates for the dynamic stresses.

KEY WORDS: composite materials, fiber composites, stress wave propagation, flexural specimen, dynamic stress analysis, frequencies, mode shapes, transient response analysis, fracture toughness, smooth specimen, notched specimen, dynamic stress intensity, static stress intensity

Smooth and notched flexural specimen testing continues to be popular in the composites community for characterizing or qualifying fiber composites, or both. Reasons for the popularity are: (1) simplicity, (2) adaptability to adverse environments, (3) availability of simple equations for data reduction, and (4) availability of written ASTM standard testing procedures. In addition to these reasons, flexural specimen testing forces the material to respond like a structure by simultaneously subjecting it to tensile, compressive, and shear stresses. Furthermore, the notched flexural specimen (Charpy impact test specimen) testing enables easy determination of fracture toughness and impact resistance.

Fracture, in general, (be it static, quasi-static, or dynamic) is a dynamic event and, as such, is a very complex process. Fracture progression in a flexural

[1] NRC-NASA research associate and structures/composites engineer, respectively, National Aeronautics and Space Administration, Lewis Research Center, Cleveland, OH 44135.

specimen is controlled by the local dynamic stress field that is also complex. The local dynamic stress field is characterized by the stress waves that are normal, shear, flexural, and surface waves. Each of the dynamic stresses can initiate failure (or a defect) at a point or propagate this defect to fracture, or both. For a complete understanding of these phenomena, a detailed description of the local dynamic stress field is a prerequisite. This requires the use of a complex transient analysis involving direct time integration as opposed to the standard modal synthesis method such as those available in some general-purpose finite-element structural analysis codes. Solution 27 (direct transient response) of MSC/NASTRAN is one of the available tools for such analyses. The objective of the present paper is to report a detailed analysis of the dynamic stress field in smooth and notched Charpy specimens.

Although the notched Charpy test specimen has been used for years in testing metals and recently in testing composites, no analysis has been performed to determine the detailed dynamic stress state variation in the notch vicinity. In general, the physical problem of the notched Charpy test specimen is dynamic and nonlinear; solution of this problem is difficult. However, a good first-order approximation may be obtained by assuming linear behavior and a quasistatic load. Chamis[2] reports such an analysis that uses the static solution sequence of COSMIC NASTRAN. The main conclusions of Chamis are: (1) the stress state is biaxial, and (2) the Charpy test specimen is not suitable for assessing the impact resistance of nonmetallic fiber composites directly. In the present work, the effort is directed towards understanding the stress wave propagation and the attendant dynamic stress field in fiber composite smooth and notched flexural specimens.

Analysis

In this section, the specimen geometry, the finite-element idealization, the finite-element analysis method, and the composite systems analyzed are described.

Specimen Geometry

The geometry of the Charpy test specimen (ASTM Notched Bar Impact Testing of Metallic Materials (E 23-82) is shown in Fig. 1. As can be seen in this figure, the overall length of the specimen is 5.497 cm (2.164 in.) and the length between supports is 3.998 cm (1.574 in.). The specimen width is 1.001 cm (0.394 in.). The specimen unnotched depth is 1.001 cm (0.394 in.) and the depth at the notch is 0.800 cm (0.315 in.). The notch is 0.201 cm (0.079 in.) deep and has a 45° opening.

[2] Chamis, C. C., "Failure Mechanics of Fiber Composite Notched Charpy Specimens," NASA TM X-73462, National Aeronautics and Space Administration, Lewis Research Center, Cleveland, Sept. 1976.

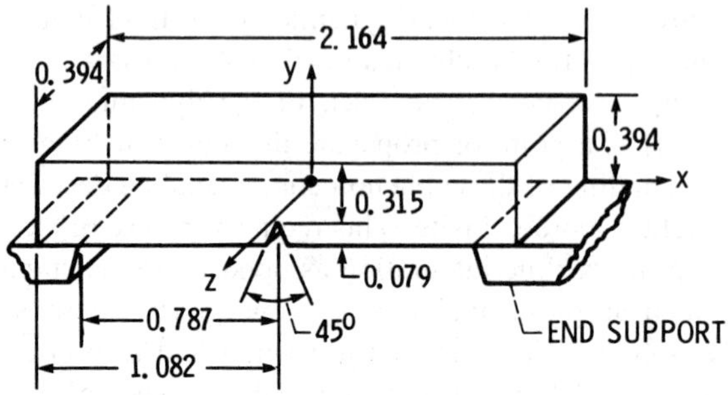

FIG. 1—*Geometry of ASTM Charpy test specimen. (All dimensions in inches; 1 in. = 2.54 cm.)*

Finite-Element Idealization

Two finite-element idealizations of Charpy test specimens—one without a notch (smooth specimen) and one with a notch—were selected for the present study. The details of the meshes are shown in Figs. 2 and 3. The entire specimen was modeled in order to properly represent the flat-end supports of the specimen and in order to avoid any effects on the local stresses (especially near the crack-tip) due to specified single-point displacement constraints that would be required if advantage was taken of symmetry. The material properties are assumed to be uniform, orthotropic, and obey a linear stress-strain law throughout the analysis. In addition, the specimen is assumed to be in a state of plane stress. The plane-stress assumption is also justified from the physics of the problem. The width restraints at the notch tip are negligible because of the very low value of the respective Poisson's ratio. For the present analysis, the plane-stress assumption implies that the stresses are permitted to vary along the specimen length and through the thickness but not across the width. This reduces the stresses to be calculated to three, two normal and one shear or a two-dimensional problem.

With these assumptions, plane-stress finite elements can be used to model the Charpy test specimen. For the smooth specimen, all the elements are quadrilateral. The notched specimen is modeled with both triangular elements and quadrilateral elements. The triangular elements are used as transition elements in the areas around the supports, the load application point, and the notch. These are the

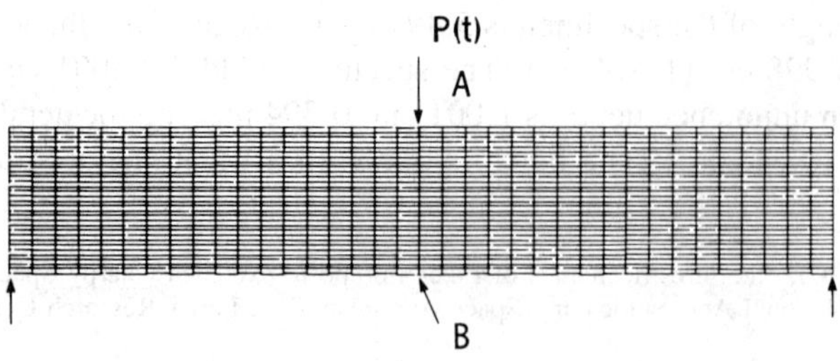

FIG. 2—*Finite-element idealization of smooth specimen.*

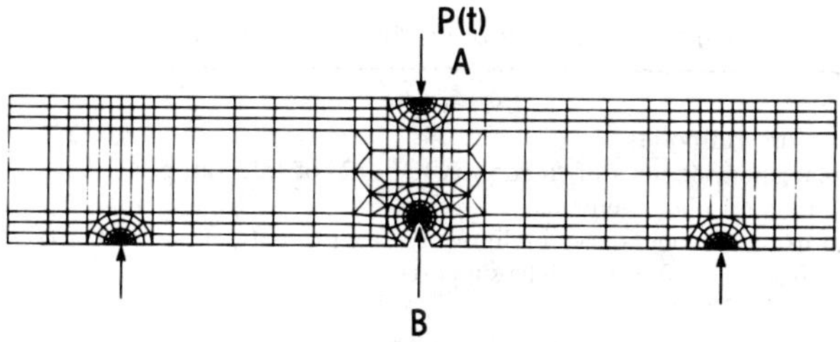

FIG. 3—*Finite-element idealization of notched Charpy specimen.*

regions where maximum stress concentrations are expected to occur and therefore are provided with a finer mesh. The boundary conditions prescribed are such that the node at the left support is constrained from x, y, and z displacements, and the node at the right support is constrained from y and z displacements. In addition, for the notched specimen, three nodes closest to the right support are constrained from displacement in the y direction in order to simulate flat support surface. The specimens are subjected to an impact loading. The form of impact is a triangular function with peak load of 8960 N (2000 lb) occurring at 500 μs of a total contact time of 1000 μs. The impact load is applied at the midspan top node in order to simulate a relatively small radius impactor.

The statistics of the finite-element representation for the smooth and notched specimens are shown in Table 1.

Finite-Element Analysis Method

The MSC/NASTRAN general-purpose structural-analysis finite-element computer program is used for the finite-element analysis. They are two-dimensional isoparametric elements. NASTRAN obtains the solution using a displacement formulation. This solution sequence employs a direct time integration scheme to obtain the transient response of a structure based upon the Newmark Beta method.[3] It provides stable results for the widest possible spectrum of practical problems without sacrificing either accuracy or efficiency. For complete details of the MSC/NASTRAN analysis see Footnote 4.[4] A brief description is given in the following paragraphs.

The differential equations of a linear structural problem may be written in the general matrix form

$$[p^2M + pB + K]\,\{u\} = \{F\} \tag{1}$$

where $p = d/dt$, M and K are the mass and stiffness matrices, B is associated

[3] Newmark, N. M., Journal American Society of Civil Engineers, Engineering Mechanics Division, EM-3, July 1959, pp. 67–94.

[4] MacNeal, R. H., "The NASTRAN Theoretical Manual, (Level 15.5)," MacNeal-Schwendler Corp., Los Angeles, Dec. 1972.

TABLE 1—*Statistics of the finite-element representation.*

SMOOTH SPECIMENS	
Number of nodes or grid points	1147
Number of displacement degrees of freedom (DOF) (2° of freedom per node)	2294
Number of quadrilateral plate elements	1080
DOF eliminated using the boundary conditions (2 from $u = 0$ and $v = 0$ at the left support and 1 from $v = 0$ at the right support)	3
Number of free DOF (2294 − 3)	2291
NOTCHED SPECIMENS	
Number of nodes or grid points	656
Number of displacement degrees of freedom (DOF)	1312
Number of quadrilateral plate elements	544
Number of triangular plate elements	82
Total number of elements (544 + 82)	626
DOF eliminated using the boundary conditions (2 from $u = 0$ and $v = 0$ at the left support and 3 from $v = 0$ for three nodes at the right support)	5
Number of free DOF (1312 − 5)	1307

with the damping matrix, u is the vector of displacement, and F is the load vector. The numerical integration is achieved by replacing p^2 and p by finite difference operators and then using explicit integration. The outputs of the transient analysis module include velocities, and accelerations as well as displacements. The output can be requested at even multiples of the integration time step. This feature affords some economy in output data preparation in cases where small time step is needed for greater accuracy.

Composite Systems Analyzed

Flexural (Charpy-type) test specimens made from three typical composite systems are analyzed. They are: T-300/epoxy, Kevlar/epoxy, and S-glass/epoxy composites. The specimens are all unidirectional composites with the fibers parallel to the length (x-axis, Fig. 1) of the specimen.

The plane stress-strain relationship (stiffness) coefficients required to input to NASTRAN are summarized in Table 2. These properties are obtained by using the resident data-bank in the composite mechanics computer code ICAN.[5] The

TABLE 2—*NASTRAN plane stress–strain relationships.*

Composite System	Mass density, $\times 10^{-4}$ (lb · s²/in.⁴)	Stress-Strain Coefficient			Orthotropy Ratios		
		G_{11}	$G_{12} = G_{21}$	G_{22}, $\times 10^6$ psi	G_{33}	G_{11}/G_{22}	G_{11}/G_{33}
S-glass/epoxy	1.813	8.7900	1.1700	3.2600	1.2700	2.70	6.92
Kevlar/epoxy	1.249	12.385	0.20082	0.57377	0.31635	21.59	39.15
T-300/epoxy	1.405	17.896	0.30275	1.1318	0.50871	15.81	35.18

NOTE—1 lb · $s^2/in.^4$ = 10.91 g · s^2/cm^4, and 1 mpsi = 6.894 GPa.

[5] Murthy, P. L. N. and Chamis, C. C., "ICAN: Integrated Compsoite Analyzer," TM-83700, National Aeronautics and Space Administration, Lewis Research Center, Cleveland, 1984.

relationships between the NASTRAN stiffness coefficients (Gs), and the usual engineering constants are

$$G_{11} = E_{\ell 11}/(1 - \nu_{\ell 12}\, \nu_{\ell 21}) \tag{2}$$

$$G_{12} = \nu_{\ell 21}\, G_{11} = \nu_{\ell 12}\, G_{22} = G_{21} \tag{3}$$

$$G_{22} = E_{\ell 22}/(1 - \nu_{\ell 12}\, \nu_{\ell 21}) \tag{4}$$

$$G_{33} = G_{\ell 12} \tag{5}$$

$$G_{13} = G_{23} = G_{31} = G_{32} = 0 \tag{6}$$

The notations in Eqs 2 to 5 are as follows: $E_{\ell 11}$ denotes the longitudinal modulus, $E_{\ell 22}$ is the transverse modulus, $G_{\ell 12}$ is the shear modulus, $\nu_{\ell 12}$ is the major Poisson's ratio, and $\nu_{\ell 21}$ is the minor Poisson's ratio. For an elastic material, the two Poisson's ratios are related by the well-known relationship

$$\nu_{\ell 21} = \nu_{\ell 12}\, E_{\ell 22}/E_{\ell 11} \tag{7}$$

Results and Discussion

Load Conditions

Two types of load conditions are used in obtaining the results. The first is a static loading where a force of 8960 N (2000 lb) is applied at the center of the specimen on the top surface. The second is an impact loading. This is modeled as a triangular pulse with a peak value of 8960 N (2000 lb) in the middle. The pulse is modeled to last for 1000 μs. The transient response is, however, obtained for three contact time periods (that is, 3 ms). A separate normal-modes analysis is used to determine the first five natural frequencies.

Natural Frequencies and Normal Modes

The NASTRAN normal-modes analysis module (Solution 3) is utilized to determine the first five natural frequencies and the associated mode shapes for the Charpy test specimens. They are shown in Fig. 4. The results are shown in Table 3 for S-glass/epoxy specimens. The frequencies are used to determine the time periods that aid in determining the integration time step for the transient analysis. A time step of 5 μs is chosen for the transient analysis based upon the time periods shown in Table 3.

Displacement and Stress Wave Propagation

The bulk-wave and shear-wave velocities are normally much higher compared to the flexural wave velocities. In order to capture the propagation characteristics of these waves, a much smaller time step of integration (0.1 μs) is chosen. This

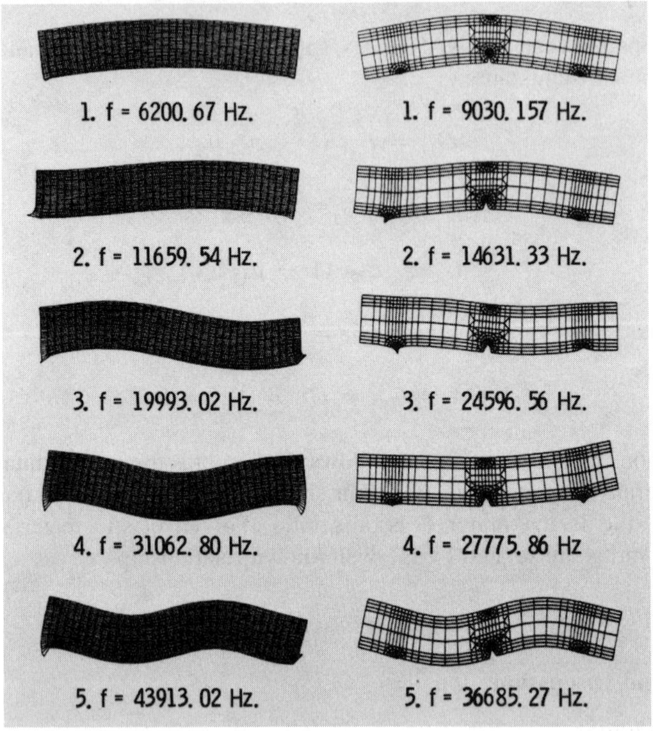

FIG. 4—*Natural frequencies and mode shapes for S-glass/epoxy smooth and notched flexural specimens.*

TABLE 3—*Natural frequencies and time periods for S-glass/epoxy smooth and notched specimens.*

Mode Number	Frequency, cycles/s	Time period, μs
	SMOOTH	
1	6200.67	161.3
2	11659.54	85.8
3	19993.02	50.2
4	31062.8	32.2
5	43913.02	22.8
	NOTCHED	
1	9030.16	110.7
2	14631.33	68.3
3	24596.56	40.7
4	27775.86	36.0
5	36685.27	27.3

requires 30 to 50 time steps for the wave front to reach the opposite face. The output of displacements and stress contours at various time intervals are saved and displayed graphically in Figs. 5 to 10. Figures 5 to 7 depict the dynamic displacement propagation in S-glass/epoxy, Kevlar/epoxy, and T-300/epoxy specimens. Two bulk-wave velocity parameters and one shear-wave velocity parameter are defined below to aid the following discussion of the displacement wave propagation results

$$C_{B11} = G_{11}/\rho \tag{8}$$

$$C_{B22} = G_{22}/\rho \tag{9}$$

$$C_{S12} = G_{12}/\rho \tag{10}$$

where ρ is the mass density of the material.

The computed values of C_{B11}, C_{B22}, and C_{S12} are shown in Table 4 for the three composite systems under study. The velocities are expressed in in./μs. The trend indicated by C_{B22} for the three materials (the transverse shock wave travels fastest in S-glass/epoxy and slowest in Kevlar/epoxy) is seen clearly in Figs. 5 to 7, specifically, the frames after 3 and 5 μs. A rough estimate of the normal wave velocity can be obtained by counting the number of elements that appear to be affected by the impact from Figs. 5 to 7. The normal wave velocity estimates

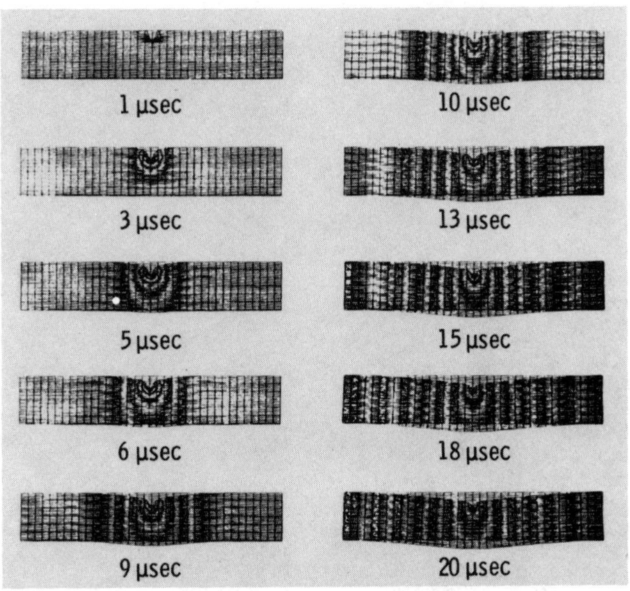

FIG. 5—*Dynamic displacement propagation in S-glass/epoxy composite flexural specimen. (Triangular impulse: 8960 N (2000 lb) maximum 1000 μs duration.)*

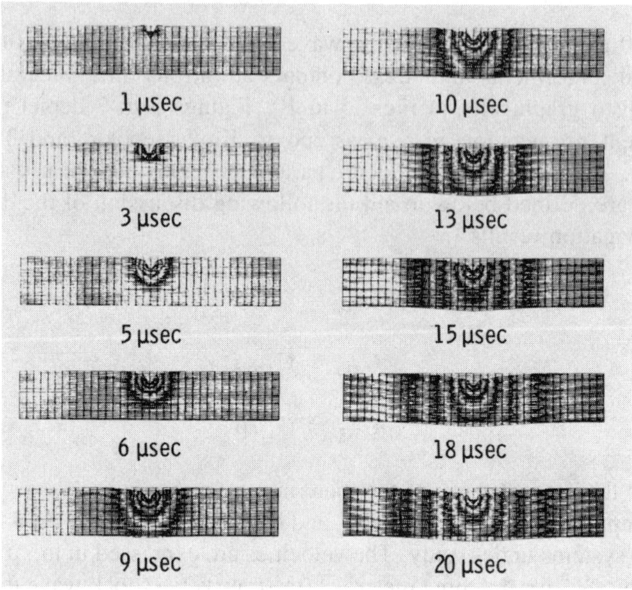

FIG. 6—*Dynamic displacement propagation in Kevlar/epoxy composite flexural specimen. (Triangular impulse: 8960 N (2000 lb) maximum 1000 μs duration.)*

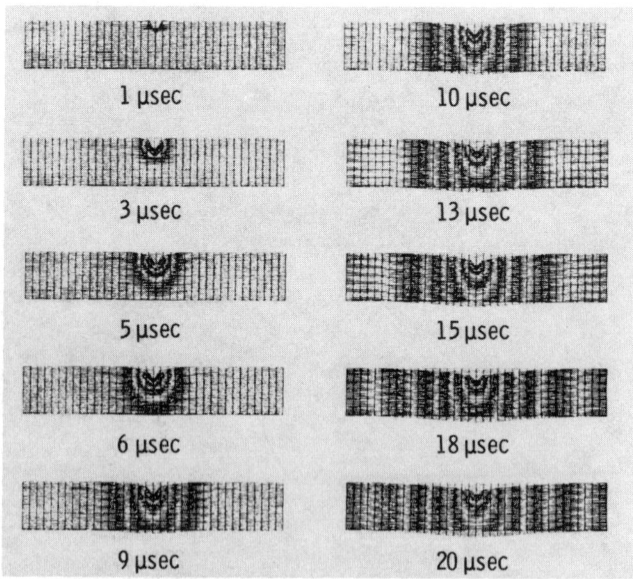

FIG. 7—*Dynamic displacement propagation in T-300/epoxy composite flexural specimen. (Triangular impulse: 8960 N (2000 lb) maximum 1000 μs duration.)*

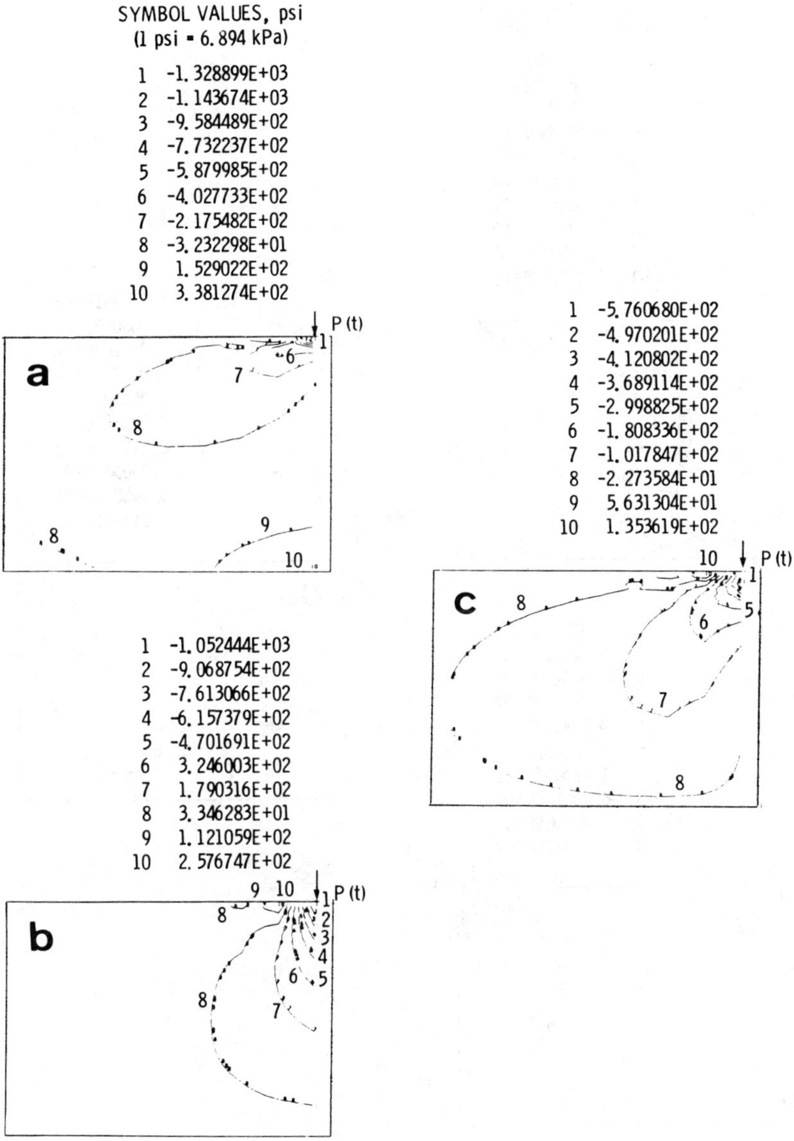

(*a*) σ_{11} contours.
(*b*) σ_{22} contours.
(*c*) σ_{12} contours.

FIG. 8—*Dynamic stress contours in S-glass/epoxy composite flexural specimen after 1% contact time.*

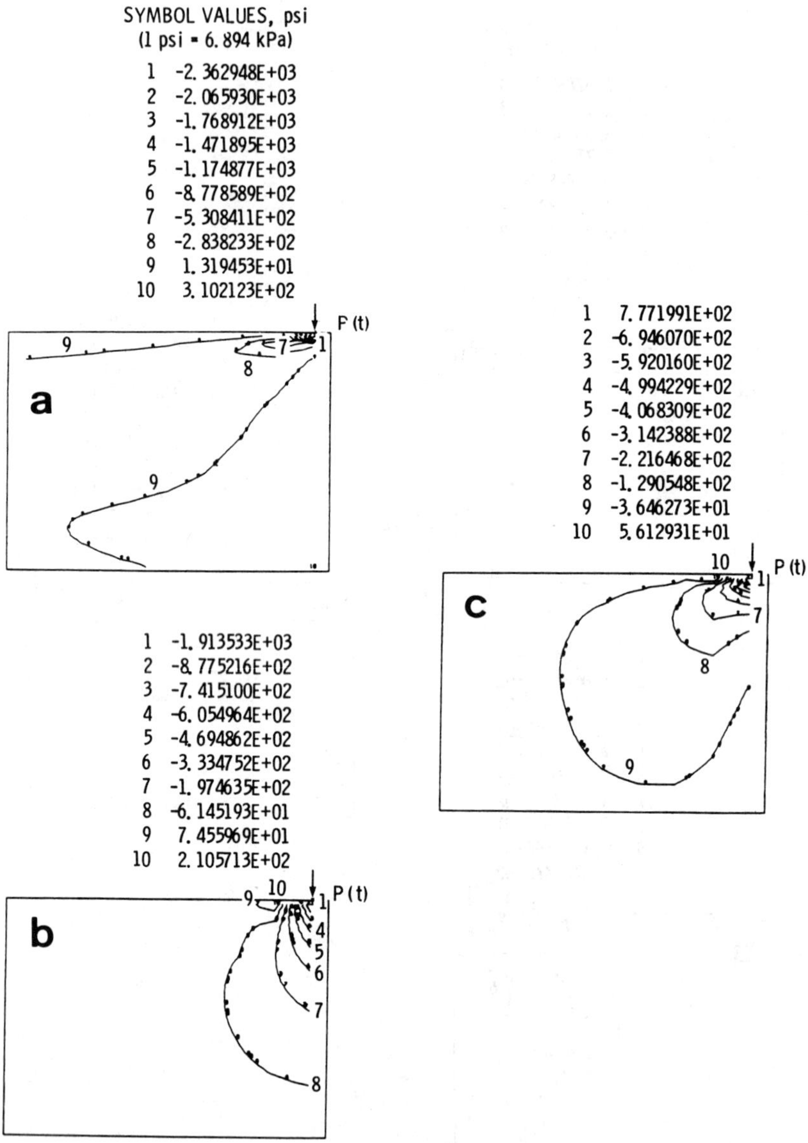

(a) σ_{11} contours.
(b) σ_{22} contours.
(c) σ_{12} contours.

FIG. 9—*Dynamic stress contours in Kevlar/epoxy composite flexural specimen after 1% of the contact time.*

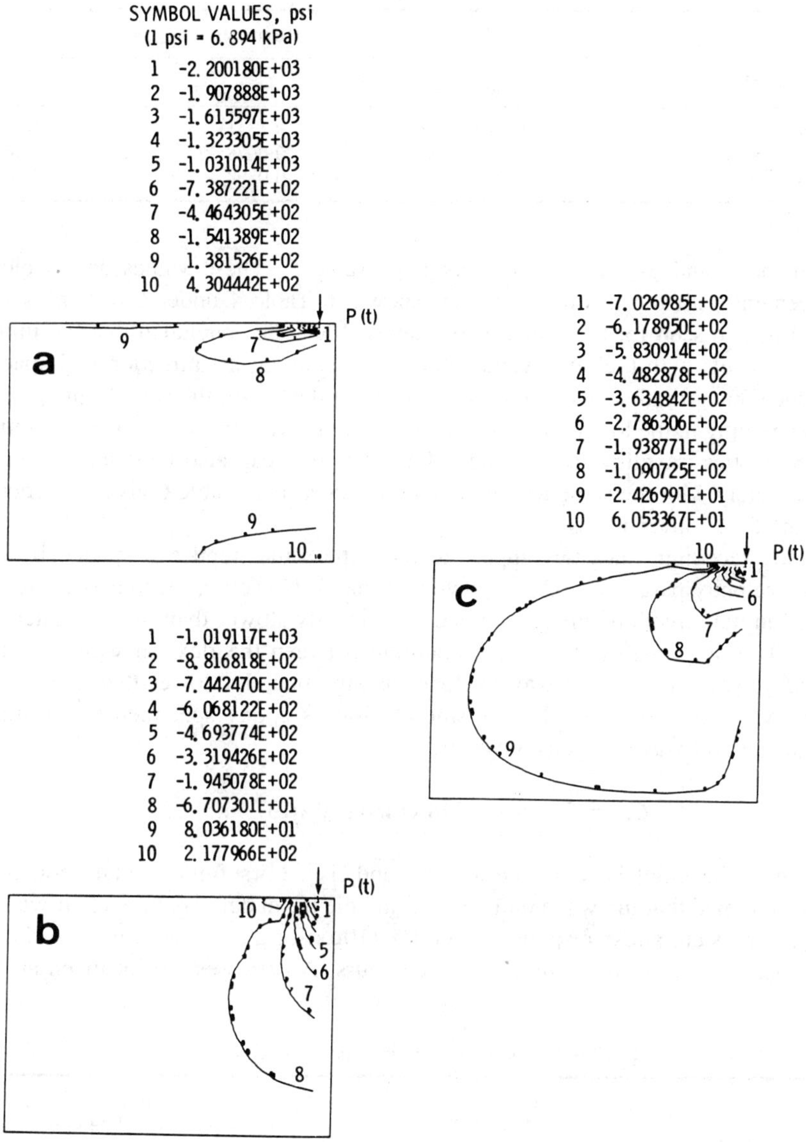

(*a*) σ_{11} contours.
(*b*) σ_{22} contours.
(*c*) σ_{12} contours.

FIG. 10—*Dynamic stress contours in T-300/epoxy composite flexural specimen after 1% contact time.*

TABLE 4—*Wave velocity parameters for the selected composite systems (in./μs; 1 in./μs = 2.54 cm/μs).*

	Parameter		
Material	C_{B11}	C_{B22}	C_{S12}
S-glass/epoxy	0.2201	0.1341	0.0837
Kevlar/epoxy	0.3149	0.0678	0.0503
T-300/epoxy	0.3569	0.0898	0.0602

from the 1 and 3-s frames are shown in Table 5. These values are in close agreement with the theoretical values shown in Table 4 under C_{B22}. The same trend is also seen for the 3 and 5 μs frames. It can be concluded that the initial shock travels with the bulk-wave velocity, C_{B22}, along the direction of impact.

Once the normal shock reaches the bottom of the beam, the wavefront appears to be moving in the longitudinal direction forming a flexural wave. The velocities of the waves traveling in the longitudinal direction can also be determined approximately with the same technique mentioned earlier. Table 6 gives the details for the frame after 13 μs.

The preceding velocities appear to have the same trend as depicted by the wave velocity parameter, C_{S12}, shown in Table 4. However, the waves along the longitudinal direction appear to move significantly slower than that indicated by C_{S12}. This is probably due to the coupling between the flexural wave and the shear wave. The flexural-wave velocity is significantly slower than the shear-wave velocity. For example, the smooth S-glass/epoxy specimen under study has a flexural-wave velocity given by

$$C_F = 2\ell f = 0.0348 \text{ cm}/\mu\text{s} \ (0.0137 \text{ in.}/\mu\text{s})$$

where ℓ is the length between the supports and f is the first fundamental frequency. (It is assumed that the wave number is 1 and the beam deflects into a half wave.)

The transient stress responses after 1% (10μs) of the contact time are shown in Figs. 8 to 10 in the form of stress contours. All stresses are localized at this

TABLE 5—*Normal wave velocity estimates after 1 and 3 μs.*

	Number of Elements		Velocity, in./μs (1 in. = 2.54 cm)		
			After		
Composite System	1 μs	3 μs	1 μs	3 μs	Average
S-glass/epoxy	10	25	0.1313	0.1094	0.1204
T-300/epoxy	7	16	0.0788	0.0700	0.0744
Kevlar/epoxy	5	12	0.0657	0.0520	0.0589

TABLE 6—*Normal wave velocity estimates after 13 μs.*

Composite System	Number of Elements	Velocity, in./μs (1 in. = 2.54 cm)
S-glass/epoxy	11	0.0510
T-300/epoxy	9	0.0416
Kevlar/epoxy	7	0.0320

early time. The longitudinal stress (σ_{11}) in the Kevlar/epoxy and T-300/epoxy composites is of about the same magnitude while that in the S-glass/epoxy is about half as much. The normal stress (σ_{22}) and shear stress (σ_{12}) are of about the same magnitude for all three composite systems. It appears from these stress results that, under the same impact conditions, the stress in the Kevlar/epoxy and the T-300/epoxy will reach fiber fracture stress magnitudes considerably earlier (about half the time) than in the S-glass/epoxy composite. Two impli-

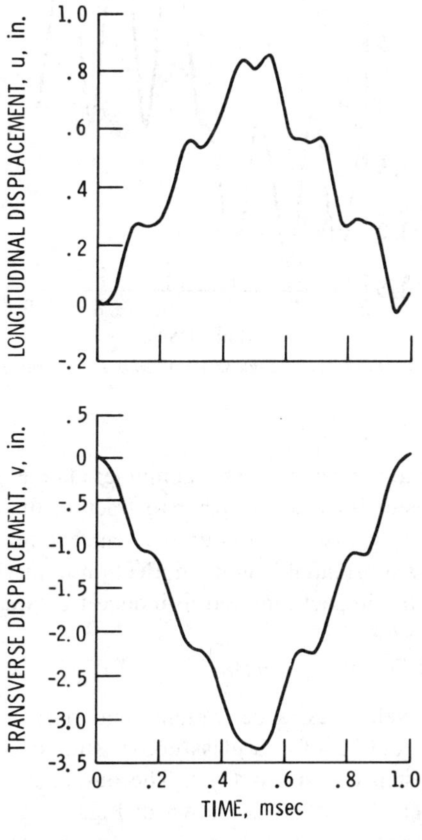

FIG. 11—*Displacement response of Point A for S-glass/epoxy smooth flexural specimen (1 in. = 2.54 cm).*

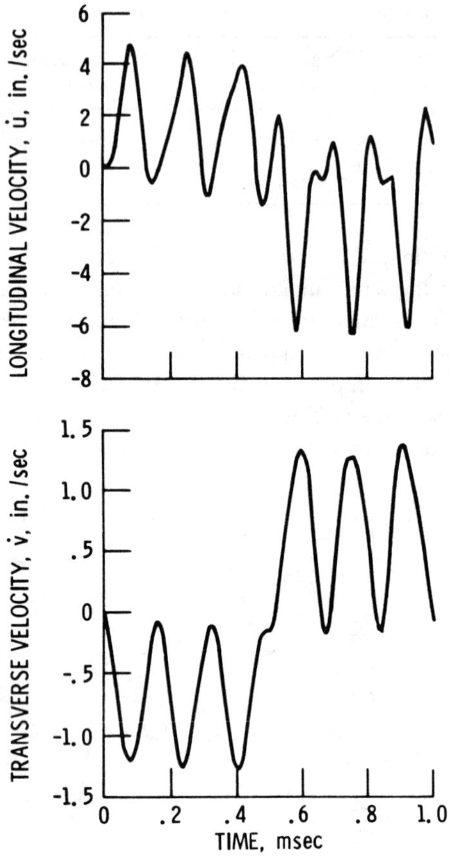

FIG. 12—*Velocity response of Point A for S-glass/epoxy smooth flexural specimen* (*1 in./s* = 2.54 cm/s).

cations follow relative to the same stress magnitude: (1) the S-glass/epoxy composite will sustain greater impact load prior to fracture than the T-300/epoxy, and (2) the rapid compressive stress built up will cause longitudinal compression failure accompanied by substantial bending deflection in the Kevlar/epoxy composite thus increasing the impact required to induce fracture.

Transient Response of Smooth Specimens

The displacements, velocities, accelerations, and stresses are obtained for a total contact time of 1000 μs for S-glass/epoxy smooth specimens. In these computations, a time step of 5 μs is used. The output is saved for every two time steps. Two points, A and B, as shown in Figs. 2 and 3 are selected for study. Point A is the load point and Point B is the opposite point at the bottom for smooth specimens and at the notch tip for notched specimens.

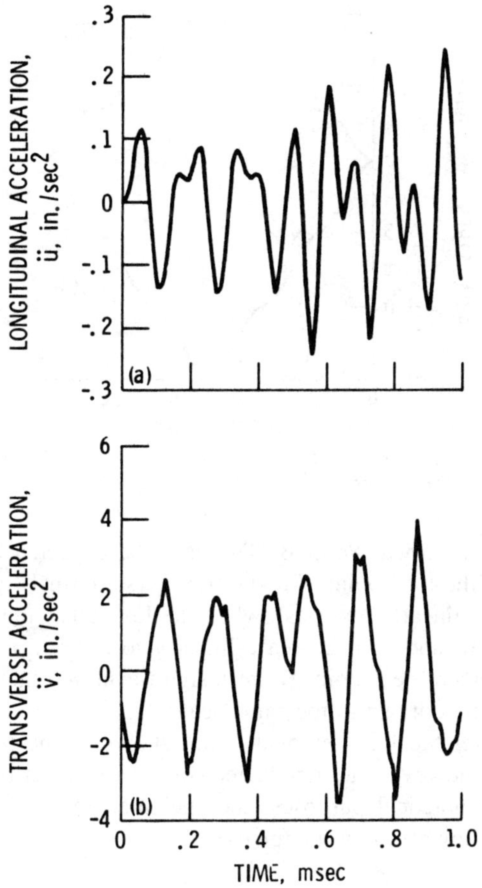

FIG. 13—*Acceleration response for S-glass/epoxy smooth flexural specimen (1 in./s^2 = 2.54 cm/s^2).*

The transient response results appear in Figs. 11 to 16. Figures 11 to 13 show the longitudinal (axial) and the transverse (bending/flexural) components of displacment, velocity, and acceleration of Point A plotted against time. The corresponding results for Point B are not shown as they are similar to Point A response both qualitatively and quantitatively. It can be concluded from these figures that the response is primarily in the first flexural mode. For example (from Fig. 11), the number of cycles in 1 ms is counted as a little over 6. From Table 3 the first fundamental time period is 161.3 μs, which implies that 6.2 (1000/161.3) cycles of response in first mode can be expected.

The stress response results are shown graphically in Figs. 14 to 16. Each figure contains the stress response of elements near Points A and B. The longitudinal stress behavior appears to be primarily the first flexural mode response. However, the magnitude of stress near Point A is about one and a half times greater than

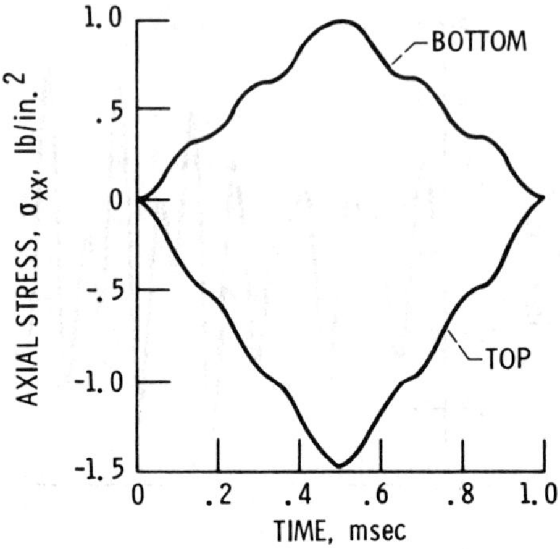

FIG. 14—*Axial stress, σ_{xx}, response of top and bottom elements near Points A and B for S-glass/epoxy smooth flexural specimen (1 psi = 6.894 kPa).*

the magnitude of stress near Point B. This is to be expected because of stress concentration near the load point, A. The transverse normal (σ_{22}) and the shear (σ_{12}) stress increase linearly to a peak when the load takes the maximum value of 8960 N (2000 lb) and then decrease linearly to a zero. The corresponding stresses for the element near Point B are insignificant and therefore can not be shown distinctly when drawn to the same scale.

The response after the load removal is not shown in the preceding figures. The computations, however, are conducted for three contact time periods. It is found that, for all practical purposes, the specimen remains static. A typical response curve is shown in Fig. 17 for Point A.

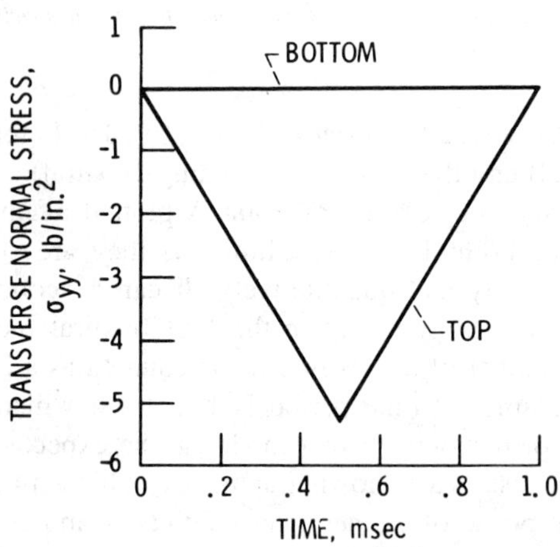

FIG. 15—*Transverse normal stress, σ_{yy}, response of top and bottom elements near Points A and B for S-glass/epoxy smooth flexural specimen (1 psi = 6.894 kPa).*

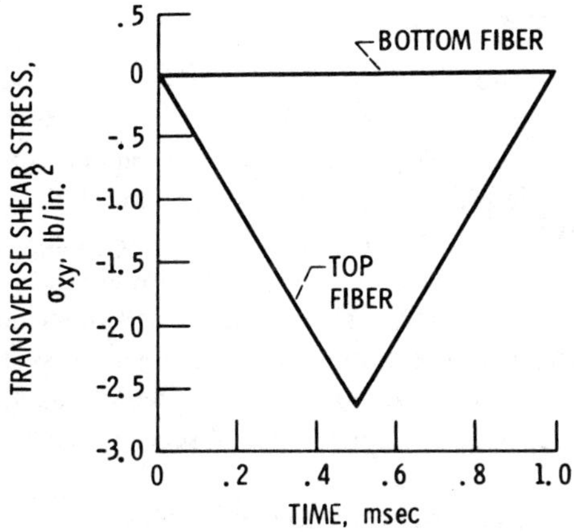

FIG. 16—*Transverse shear stress, σ_{xy}, response of top and bottom elements near Points A and B for S-glass/epoxy smooth flexural specimen (1 psi = 6.894 kPa).*

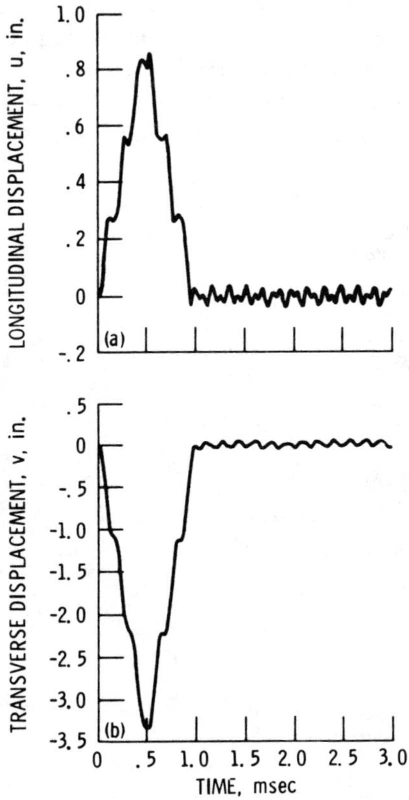

FIG. 17—*Displacement response of Point A for three contact time periods of S-glass/epoxy smooth flexural specimen (1 in. = 2.54 cm).*

Static and Dynamic Stress Contours

Smooth Specimens—The static stress contours under a load of 8960 N (2000 lb) and the dynamic stress contours at the peak load of 8960 N (2000 lb) are shown for a quarter span in Figs. 18 to 20. They appear to be identical. The peak values for the dynamic load case, however, are slightly lower than those for the static case. While the maxima for the longitudinal stress (σ_{11}) occur at the center of top and the bottom, the transverse normal stress and the transverse shear stress have the maxima almost adjacent to each other near the load application point. The transverse shear stress and the axial stress distribution is seen to approach the classical Euler-Bernoulli theory predictions away from the load application point. Steep stress gradients are observed near the load application point.

Notched Specimen—The results for notched S-glass/epoxy specimen near the crack tip are shown in Figs. 21 to 23. The static and dynamic stress contours at the peak load show similar trends as observed in the case of smooth specimens. Figure 24 shows the stress intensity near notch tip in relationship to the far field stresses. This distance from the notch tip is measured along the longitudinal direction towards the left support. It is seen that all the three stresses attain very high peaks, indicating severe local stress intensities near the notch tip. As the stress allowables for σ_{22} and σ_{12} are generally an order of magnitude lower than that for σ_{11}, one can expect a matrix-initiated failure followed by fiber fractures at this location.

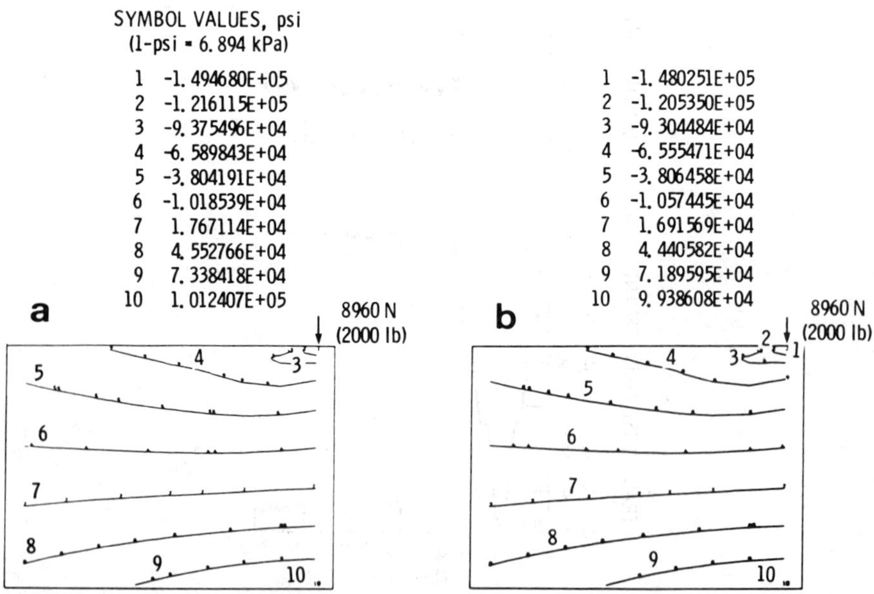

FIG. 18—*Comparison of static* (a) *and dynamic* (b) *axial stress,* σ_{xx}, *contours at peak load for S-glass/epoxy smooth flexural specimens.*

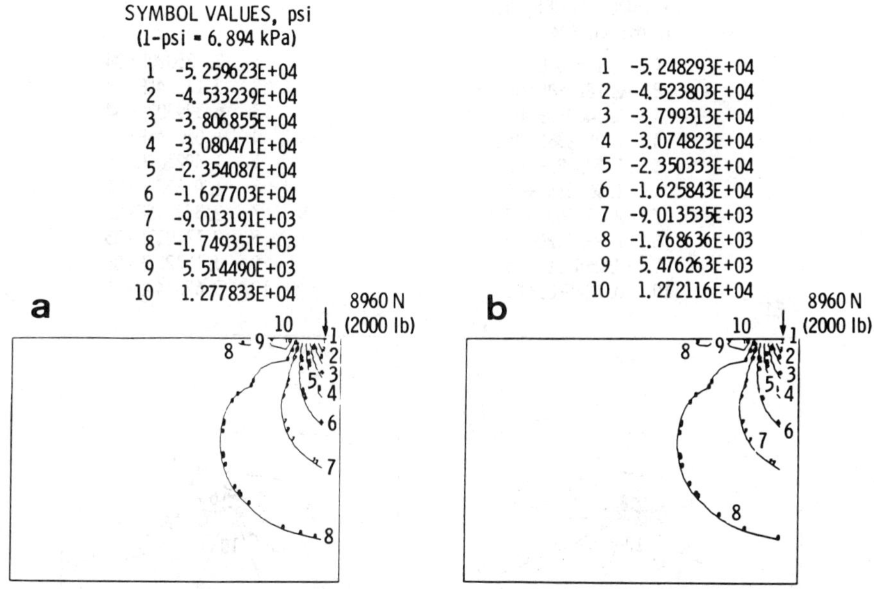

FIG. 19—*Comparison of static* (a) *and dynamic* (b) *transverse normal stress,* σ_{yy}, *contours at peak load for S-glass/epoxy smooth flexural specimen.*

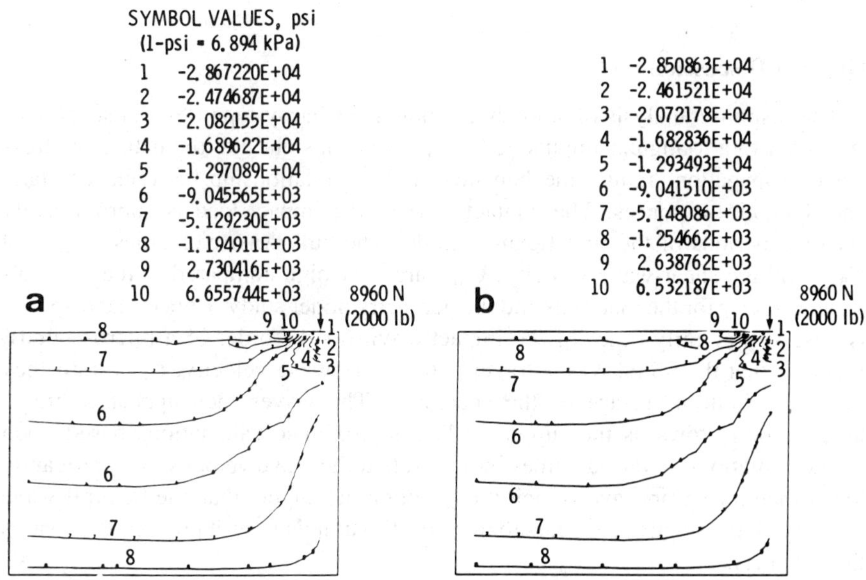

FIG. 20—*Comparison of static* (a) *and dynamic* (b) *transverse shear stress,* σ_{xy}, *contours at peak load for S-glass/epoxy smooth flexural specimen.*

SYMBOL VALUES, psi
(1-psi = 6.894 kPa)

Symbol	a	b
1	-2.175306E+04	-2.125020E+04
2	2.182584E+04	2.137141E+04
3	6.540473E+04	6.399302E+04
4	1.089836E+05	1.066146E+05
5	1.525625E+05	1.492362E+05
6	1.961414E+05	1.918579E+05
7	2.397203E+05	2.344795E+05
8	2.832992E+05	2.771011E+05
9	3.268781E+05	3.197227E+05
10	3.704570E+05	3.623443E+05

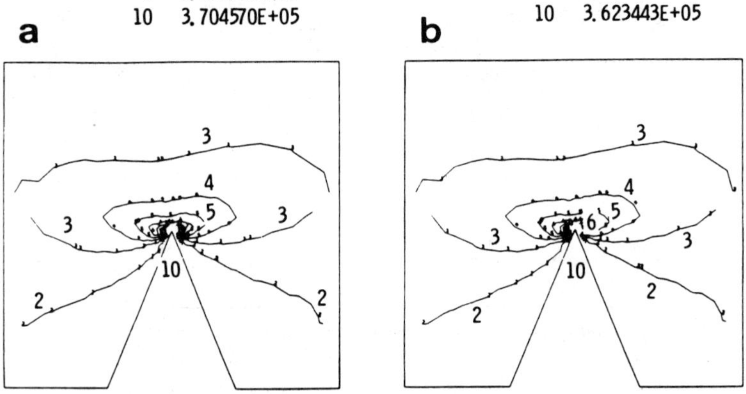

FIG. 21—*Comparison of static* (a) *and dynamic* (b) *axial stress*, σ_{xx}, *contours near the notch region at peak load for S-glass/epoxy notched flexural specimen.*

General Discussion

The response analysis of smooth and notched Charpy-type flexural specimens subjected to a triangular impulsive load provides insight into the nature of stress wave propagation, failure mechanisms, and the relationship between the static and dynamic responses. The contact time of the impact load is approximately six time periods of the first flexural mode. The bulk-wave velocities, C_{B11} and C_{B22}, and the shear-wave velocity, C_{S12}, are very high compared to the flexural-wave velocity for the materials and the geometry under study. Hence, the response is observed at early times of the impact event (or the order of a microsecond). It shows that the initial waves travel with a bulk-wave velocity, C_{B22}, until they reach the bottom surface of the specimen. The waves then appear to travel longitudinally towards the supports. The approximate calculations based upon the deformations at various times indicated that the wave velocity is significantly lower than the shear-wave velocity, C_{S12}, but much higher than the flexural-wave velocity. The wavefront appears to induce a flexural wave as it progresses towards the supports.

At longer times, the response is primarily in the first flexural mode. A com-

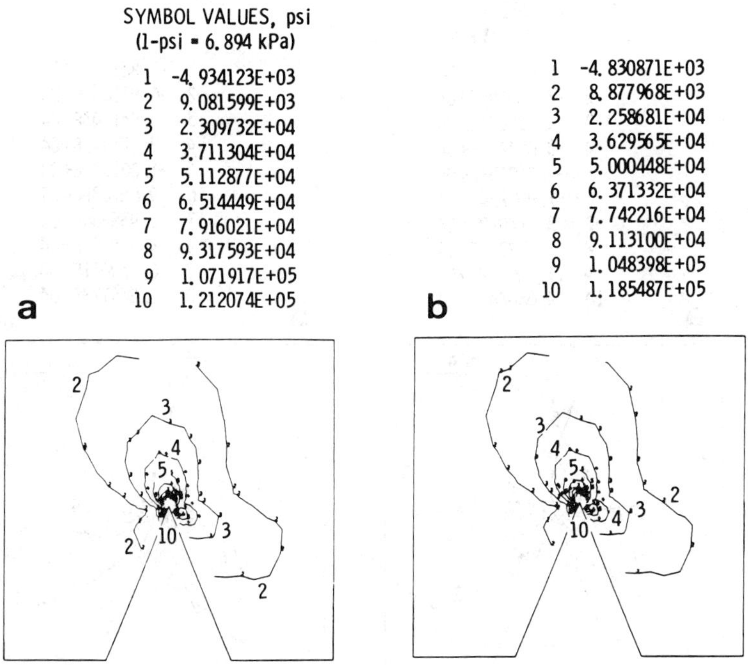

FIG. 22—*Comparison of static* (a) *and dynamic* (b) *transverse normal stress,* σ_{yy}, *contours near the notch region at peak load for S-glass/epoxy notched flexural specimen.*

parison with the static response indicates that little or no difference exists in the magnitude for stresses at the peak load. The static predictions, however, are on the conservative side.

The response of notched fiber composite specimens show severe stress intensities near the notch region indicating local failures in shear and transverse tension initially followed by fiber fractures since the longitudinal stresses approach fracture stress magnitudes.

The response after the load removal reduces to mere noise; the structure remains practically static.

Summary of Results

The results of the study on static, dynamic, and transient response of smooth and notched, fiber composite, Charpy specimens are listed here.

1. The wave propagation velocities can be estimated from the early time displacement propagation response; the estimates are in fair agreement with the theoretical predictions.

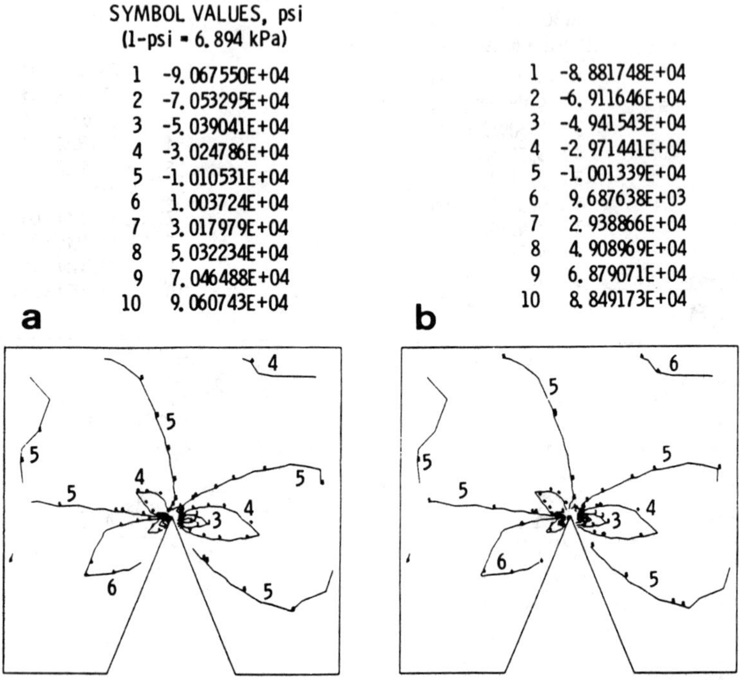

FIG. 23—*Comparison of static* (a) *and dynamic* (b) *transverse shear stress,* σ_{xy}, *contours near the notch region at peak load for S-glass/epoxy notched flexural specimen.*

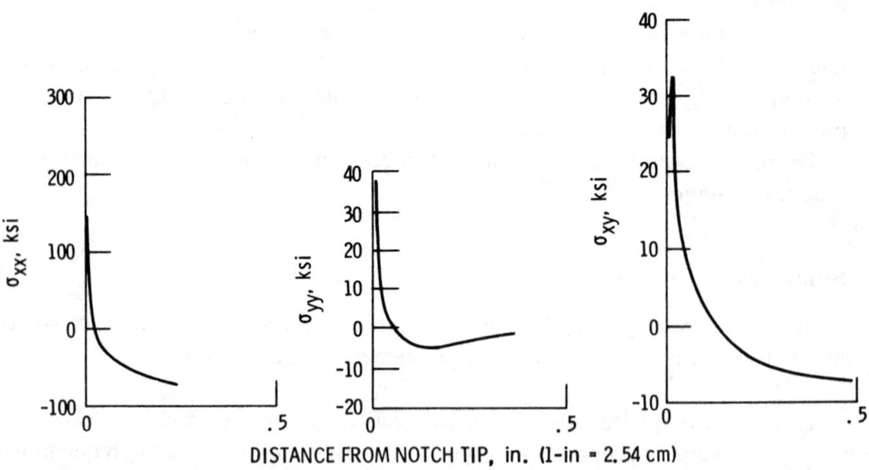

FIG. 24—*Stress intensity distribution near the notch tip for S-glass/epoxy notched flexural specimen (1 ksi = 6.894 MPa).*

2. The first five flexural mode shapes and frequencies show coupling effects from the thickness stretch and the transverse shear type modes; the frequencies are not integer multiples of the fundamental frequency.
3. The dynamic and static peak load stress contours are almost identical. The static peak load stress magnitudes are slightly higher. Stress predictions based upon a quasi-static approach lead to conservative estimates.
4. The transient displacement and velocity response appear to be primarily in the first flexural mode. The acceleration response shows contributions from higher modes.
5. The structure responds for the load duration time period only. The steady-state response after the load removal appears to be negligible.
6. The transverse normal and shear stresses vary linearly with time and follow the load path.
7. The notch-tip region develops severe stress concentrations compared to corresponding material strengths, and any of the three stresses could cause or initiate a failure.
8. Based upon the transient stress response, it appears that the failures in Charpy specimens are initiated at the notch tip by the shear stresses and the transverse tensile stresses followed by fiber fractures. However, the failures in smooth specimens are probably initiated by a combined stress state near the load application point followed by local interply delaminations.

Frank K. Ko[1]

Tensile Strength and Modulus of a Three-Dimensional Braid Composite

REFERENCE: Ko, F. K., **"Tensile Strength and Modulus of a Three-Dimensional Braid Composite,"** *Composite Materials: Testing and Design (Seventh Conference), ASTM STP 893*, J. M. Whitney, Ed., American Society for Testing and Materials, Philadelphia, 1986, pp. 392–403.

ABSTRACT: A geometric model for three-dimensional braid composite using the concept of average cosine is presented in this paper. In the form of contour diagrams, the fabric geometry and processing variables are related to the tensile strength and modulus of the composite. The tensile properties of the three-dimensional braid AS4-3K graphite/5208 epoxy-resin composite are compared to the state-of-the-art laminated unidirectional, woven and quasi-isotropic composites. A comparison between theoretical and experimental results are also made.

KEY WORDS: three-dimensional braid, composite materials, geometric model, average cosine, crimp, linear density, strength, modulus, fiber volume fraction, yarn orientation, processing variable, laminates, fully integrated fabric, fabric preform

As a result of the increasing demand for higher level of damage tolerance, through-the-thickness strength and interlaminar strength for structural composites, there is a revival of interest in the fully integrated (non-layered) net shape fabric preforms for the reinforcement of composites. One class of fibrous structure that has received particular attention is the three-dimensional braid formed by an irregular braiding process. The formation and the unique structural characteristics of the three-dimensional braid have been reviewed in several publications.[2–4] Basically, a three-dimensional braid is a three-dimensional fibrous assembly composed of orthogonally interlaced fiber bundles oriented in the bias direction with respect to the longitudinal axis of the structure. Because of its fully integrated nature (no layers), this preform virtually eliminates the problem of delamination. The freedom to distribute and orient the fibers coupled with an ability to form

[1] Associate professor, Fibrous Materials Research Center, Department of Materials Engineering, Drexel University, Philadelphia, PA 19104.

[2] Ko, F. K., "Three-Dimensional Fabrics for Composites," *Proceedings*, International Conference on Composite Materials, ICCM IV, Tokyo, 1982.

[3] U.S. Patent 4,312,261, 26 Jan. 1982.

[4] Ko, F. K. and Pastore, C. M., "Structure and Properties of an Integrated 3-D Fabric for Structural Composites," *Proceedings*, 2nd U.S./Japan Conference on Composites, 1983.

net structural shapes provide additional design flexibility. While these interesting design and structural features have been demonstrated in our laboratory (see Footnote 4), the thrust of our work has been to elucidate the structure-property relationship of the three-dimensional braid composites and to establish a basis for computer-aided design and manufacturing of the three-dimensional braid preform.

In a previous study by Ko et al (see Footnote 4), a geometric model was used to relate the tensile strength of the three-dimensional braid composite to the fabric geometry and processing variables. A theoretical relationship has been established between yarn orientation and the unit cell geometry, which is a function of yarn material property and processing variables such as the uniformity and frequency of the beat-up (compacting) motion. Since the three-dimensional braid formation process is not yet fully automated and a wide variety of yarns of different rigidity are being used, it is difficult to obtain a universal relationship for the yarn orientation and unit cell geometry. Accordingly, it is the objective of this study to develop a relationship between the average cosine of yarn orientation and the textile structural parameters such as crimp and linear density. With these geometric relationships, the volume fraction of the three-dimensional braid composite will be determined. In order to verify the prediction of the strength and modulus of the composite, the tensile properties of a three-dimensional braid AS4/5208 epoxy-resin composite were characterized. A comparison of the tensile properties was made between the three-dimensional braid composite and laminated composites.

A Geometric Model

For an idealized straight yarn segment in a unit cell of the three-dimensional braid as shown in Fig. 1, the yarn orientation is characterized by the angle, θ,

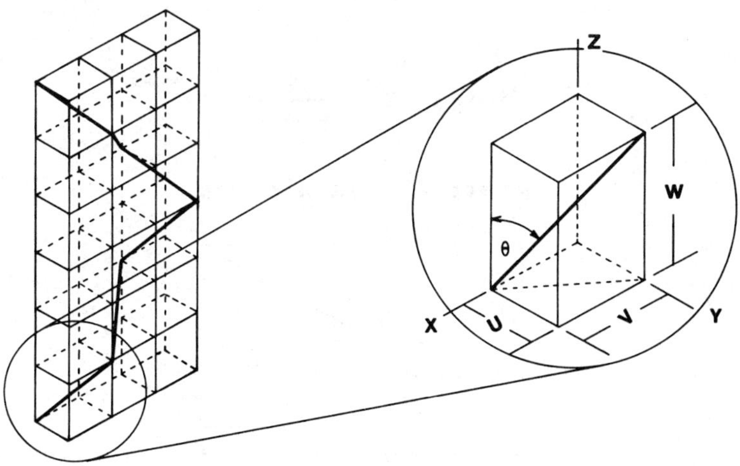

FIG. 1—*Idealized yarn segment in the unit cell.*

with respect to the Z axis and the angle, α, on the X-Y plane. In a given fabric, several imaginary planes of yarn segments may exist analogous to the laminates each assuming an orientation, θ_i. To obtain an average representation of yarn orientation, the average cosine is used herein, relating to yarn crimp, as shown in Fig. 2.

$$\text{Cos } \bar{\theta} = \frac{1}{1 + C} \tag{1}$$

where C is the crimp of the yarns in the fabric that can be considered the average fractional difference between the length of the yarn segment and the width of the unit cell. In a braiding process, the more often beat up takes place and more frequently the interlacing is made, the greater is the crimp in the yarn. Crimp may also be used to indicate the amount of geometric strain of the fabric preform. In Fig. 3, the yarn orientation angle is shown to increase with crimp. With the relationship between yarn orientation and yarn crimp established, we can proceed to relate yarn orientation to textile parameters such as yarn and fabric linear density, D_y and D_f, respectively. The linear density of the yarn or the fabric, by definition, is the mass per unit length of the structure that can be considered a measure of the average distribution of materials along the length of a structure. Since crimp is the fractional difference in length of the yarn segment in a unit cell, the relationship between crimp and linear density and thus a relationship between the average cosine of yarn angle and linear density can be established, as follows

$$C = \frac{D_f}{ND_y} - 1 \tag{2}$$

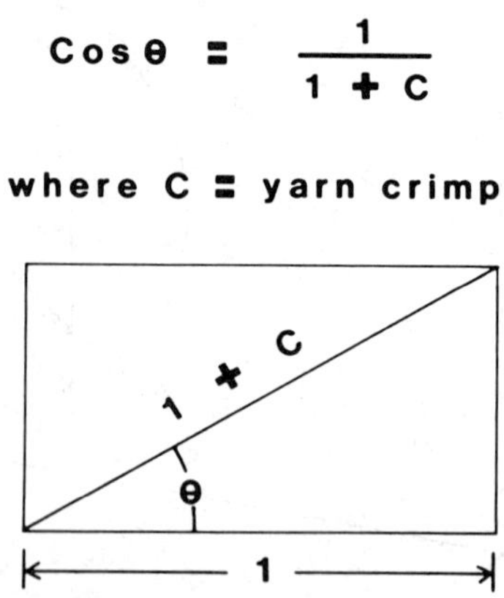

FIG. 2—*Relationship between yarn orientation and yarn crimp.*

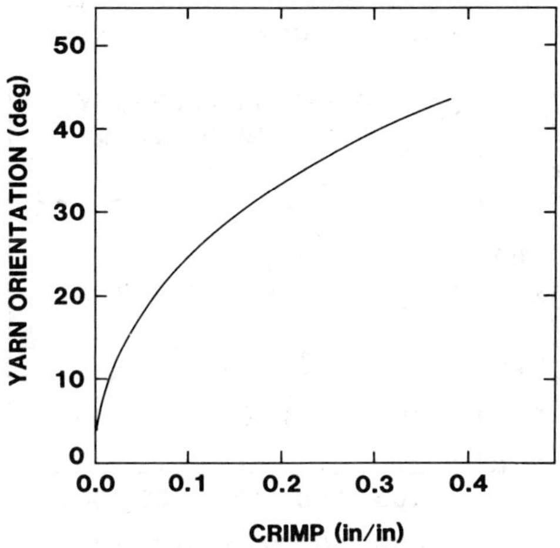

FIG. 3—*Relationship of yarn orientation and crimp in a three-dimensional braid.*

and

$$\text{Cos } \theta = ND_y/D_f \tag{3}$$

where

N = number of yarns in the fabric;
D_y = yarn linear density, denier or tex;
D_f = fabric linear density, denier or tex;
denier = mass (g) in 9000 m; and
tex = mass (g) in 1000 m.

For a given yarn of known linear density, as shown in Fig. 4, yarn orientation angle tends to decrease as the number of yarns in the fabric increases. For the same number of yarns in the fabric, yarn orientation angle decreases as the linear density of the fabric decreases. Since fabric linear density is influenced by the tightness of the beat-up process in fabric formation, this is consistent with our observation on the effect of processing parameters on the yarn orientation of the three-dimensional braid (see Footnote 4).

With the average angle of the yarn, one can also predict the tensile strength of the fabric for various structural parameters according to the following relationship

$$\sigma_f = \sigma_y \text{ Cos}^2\theta \tag{4}$$

where

σ_f = fabric strength, and
σ_y = yarn strength.

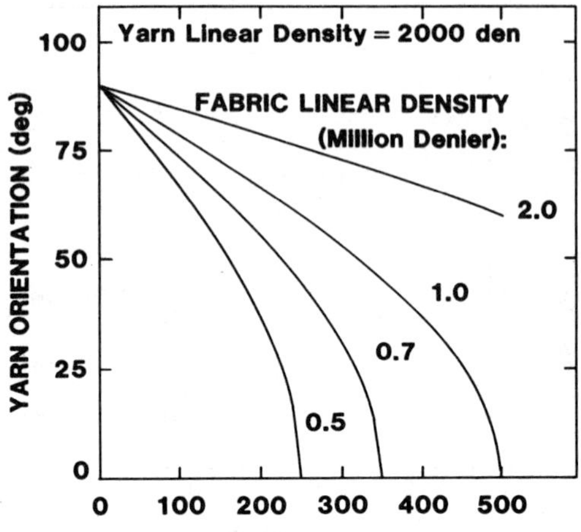

FIG. 4—*Relationship of orientation and number of yarns in a three-dimensional braid.*

The strength of the fabric preform, for a given yarn as shown in Fig. 5, tends to increase as the number of yarns in the fabric increases. The rate of increase is more rapid for fabrics with lower linear density. For the same number of yarns in the fabric, the strength of the fabric tends to increase as the linear density of the fabric decreases, since fabric linear density is directly related to yarn orientation angle.

In order to predict the properties of a composite, the volume fraction of the fabric was examined as a function of yarn and fabric linear density. According

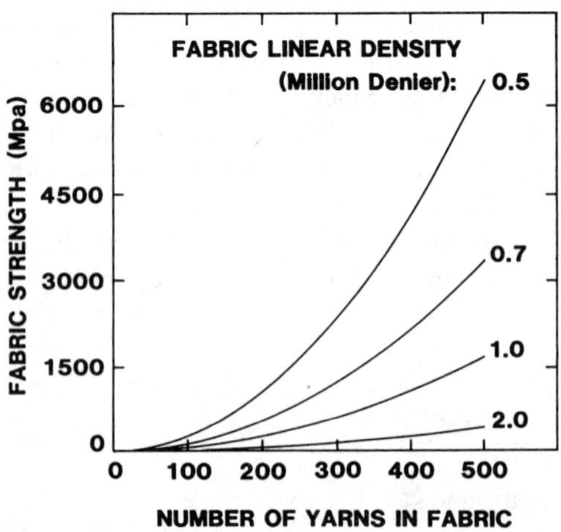

FIG. 5—*Prediction of fabric strength for AS4-3K graphite three-dimensional braid.*

to the definition of fiber volume fraction

$$V_f = NL_yA_y/L_fA_f \tag{5}$$

where

N = number of yarns in the fabric;
L_y = length of yarn in a given length of fabric, L_f;
A_y = cross-sectional area of yarn;
A_f = cross-sectional area of fabric;

and by substitution, one can show that

$$V_f = \left(\frac{D_f}{D_y}\right)\left(\frac{A_y}{A_f}\right) \tag{6}$$

For a net shape fabric that is formed to the dimension of the composite, the cross-sectional area of the fabric, A_f, may be considered equivalent to the cross-sectional area of the composite, A_c. From Fig. 6, it can be seen that fiber volume fraction decreases as the cross-sectional area of the fabric or the composite increases for a yarn of given linear density. For the same cross-sectional area, fiber volume fraction increases with fabric linear density. For example, for the AS4-3K graphite three-dimensional braid composite of 0.838 cm^2 (0.13 in.^{2}) cross-sectional area, the volume fraction increases from 0.3 to 0.75 as the fabric linear density increases from one-half million to one million. With a knowledge of the yarn and matrix properties, we can now predict the stress-strain properties of the three-dimensional braid composite using the equation of mixture for strength

$$\sigma_c = \sigma_f V_f + \sigma_m V_m \tag{7}$$

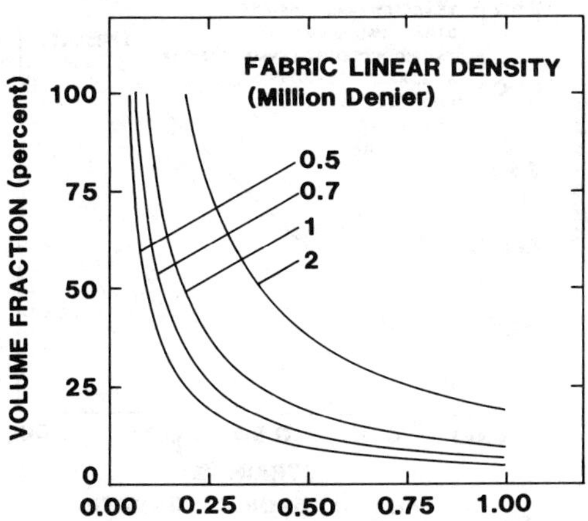

FIG. 6—*Theoretical fiber volume fraction of AS4-3K graphite/epoxy composite.*

For example, using the data listed below, we can obtain the stress-strain curve and the associated information as shown in Fig. 7.

Number of yarns in fabric, N	335
Yarn linear density, D_y	1970 denier
Fabric linear density, D_f	7.3×10^5 denier
Cross-sectional area, A_y	1.2×10^{-3} cm^2 (1.89×10^{-4} in.^{2})
Cross-sectional area of composite, A_c	0.593 cm^2 (0.092 in.^{2})
Modulus of matrix, E_m	4.5 GPa (0.65 Msi)
Stress-strain curve of yarn, $f(\epsilon_f)$	345 GPa (E_y = 50 Msi)
for example, $\sigma_y = E_y\epsilon_y$	$\epsilon_{y,u} = 1.53\%$

In order to verify the theoretical prediction, the tensile properties of a three-dimensional braid AS4/5208 graphite/epoxy composite were characterized.

Experimental

The verification of theoretical prediction and a valid assessment of the performance of the three-dimensional braid composite demands quality test data that were not available until recently. Through a technical exchange between our laboratory and the David Taylor Ship R&D Center, a composite of AS4 and 5208 epoxy resin was successfully prepared for this study. Specifically, the AS4-3K graphite three-dimensional braid preforms were prepared in our laboratory. A 1 × 1 construction was used to form the three-dimensional braid to the net dimension of 3.2 mm thick by 25.4 mm wide (¼ in. × 1 in.). The preforms were impregnated with a Narmco Co. 5208, 177°C epoxy resin using the resin film lamination technique in the David Taylor Ship R&D Center as described by

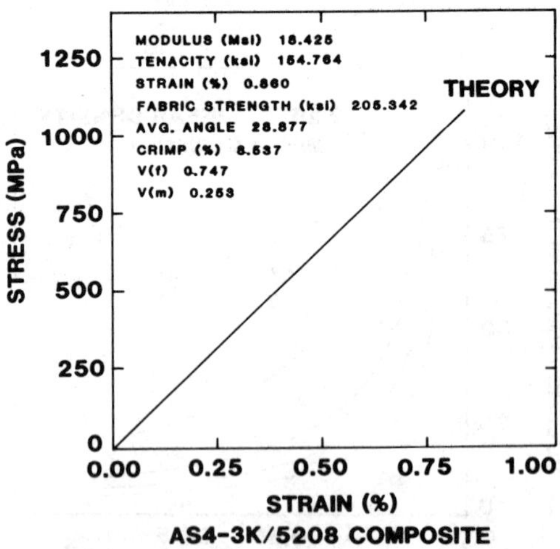

FIG. 7—*Theoretical stress-strain relationship of the three-dimensional braid composite.*

Macander et al.[5] The AS4/5208 composite system was selected such that a direct comparison with the similar laminated composites can be made.

Tension testing of the composites was carried out on a MTS hydraulic tension tester according to ASTM Test Method for Tensile Properties of Fiber-Resin Composites (D 303976) using a 152.4-mm (6-in.) gage length and 50.8-mm (2-in.) long fiberglass end tabs at a crosshead speed of 1.27 mm/min (0.05 in./min.). All the tension tests were carried out at the Naval Air Development Center (NADC) with the assistance of Dr. Jim Alper.

An observation of the failure pattern, as shown in Fig. 8, indicated that the three-dimensional braid composites failed by fiber breakage with the failure surface roughly perpendicular to the direction of loading.

A comparison of the tensile strength and modulus of the three-dimensional braid composite with laminates, as shown in Table 1, revealed that the strength and modulus of the three-dimensional braid composite were lower than that of

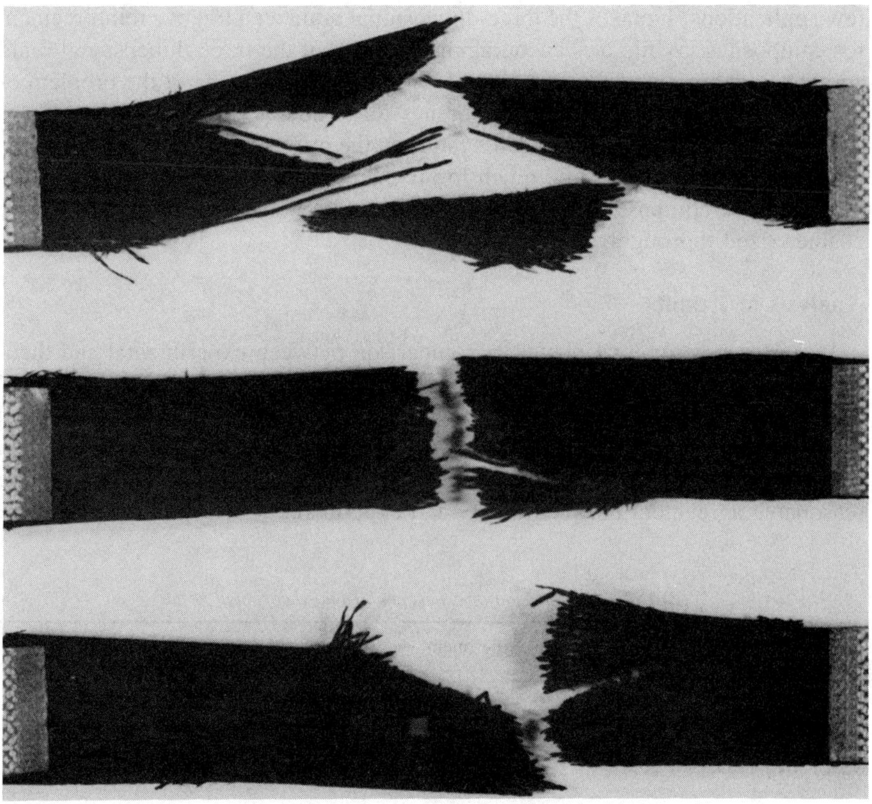

FIG. 8—*Failure behavior of three-dimensional braid composites.*

[5] Macander, A.B., Crane, R.M., and Camponeschi, E.T., Jr., "Fabrication and Mechanical Properties of Multidimensionally (X-D) Braided Composite Materials," in this volume.

TABLE 1—*Comparison of tensile properties of three-dimensional braid and laminated graphite/epoxy composites.*

	Three-Dimensional Braid AS/5208	Laminates[a]		
		Quasi-Isotropic C3K/5208	Woven AS/3501	Unidirectional AS/3501
Tensile strength (ksi)[b]	185	71.6	100	310
modulus (Msi)	15.3	7.7	10	21.5
Fiber volume (%)	67	65	62	62

[a] Based on Hercules Product Bulletin.
[b] 1 ksi = 6.9 MPa.

the unidirectional laminates as expected. However, comparing the laminated angle-ply and woven fabric composites, the three-dimensional braid composites had far superior properties. Since the unidirectional laminates are used in very few applications, it makes the three-dimensional braid an attractive reinforcement for composites. While it is encouraging to note that the three-dimensional braid composite attains satisfactory levels of tensile properties without the problem of delamination, there are still several issues to be addressed. For example, considering the high values of Poisson's ratio, on the order of 0.7 to 1, the transverse properties are expected to be much lower than the uniaxial properties. The behavior of the composite under static and dynamic compression should also be characterized thoroughly.

Analysis of Results

With the experimental results, a comparison between experimental and theoretical results can now be made as shown in Table 2. In general, the predicted strength and modulus are within 20% of the experimental results that are well below two standard deviations. The prediction of the geometric and physical parameters are within about 10% of the experimental observations. Despite the reasonable agreement between theory and experiments, the rupture strain of the

TABLE 2—*Comparison of experimental and theoretical results.*

	Experiment[a]	Theory	Difference
Tensile strength, ksi[b]	184.5 ± 25	154.8	−29.7
Tensile modulus, Msi	15.3 ± 2.3	18.4	+3.1
Rupture strain, %	1.2	0.86	−0.34
Fiber orientation, deg	21.1 ± 2.1	23	+1.86
Crimp, %	...	8.5	...
Fiber volume fraction	0.67 ± 0.03	0.75	+0.08

[a] Average of four replications.
[b] 1 ksi = 6.9 MPa.

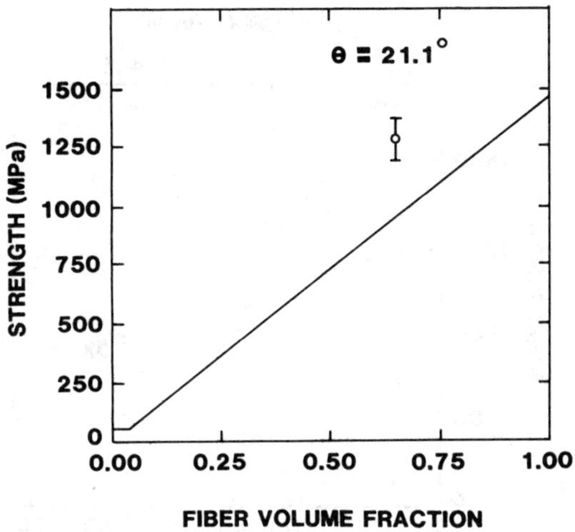

FIG. 9—*Theoretical strength prediction for AS4-3K graphite/epoxy composite.*

composite was well underestimated by the model. Considering the fact that the rupture strain of the yarn and the matrix are 1.53% and 1.3%, respectively, yarn geometry may contribute more to the total strain than is permitted by the model.

As a first approximation, we can use the model to examine the strength and modulus of the three-dimensional braid composites at various volume fractions and yarn orientations. In Figs. 9 and 10, the tensile strength and modulus are predicted as a function of volume fraction. It can be seen that very little contribution was received from the matrix, while the predicted modulus was within

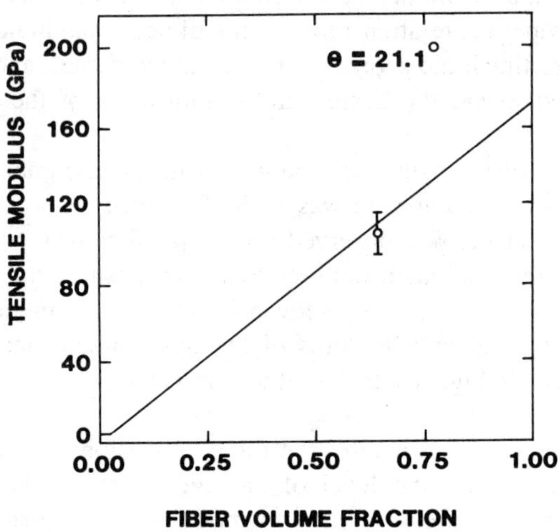

FIG. 10—*Theoretical modulus prediction for AS4-3K graphite/epoxy composite.*

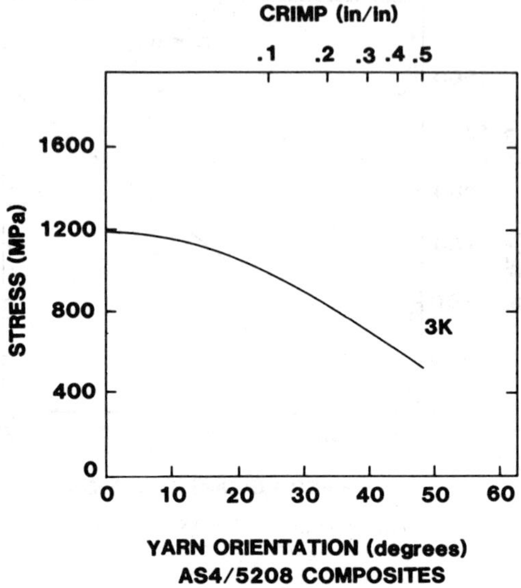

FIG. 11—*Prediction of strength as a function of yarn orientation.*

the error bound of the experimental observation, about 20% underestimation was observed for strength prediction. Using this geometric model, the strength and modulus of a three-dimensional braid composite can also be predicted, as shown in Figs. 11 and 12, for various yarn orientation.

Conclusions

A simple geometric model has been established to assist the study of the structure and properties relationship of three-dimensional braid composite. Contour diagrams relating braid geometry, processing variables, and tensile properties were constructed to aid the design and optimization of the three-dimensional braid preforms.

The prediction of strength and modulus of the three-dimensional braid AS4-3K/5208 epoxy-resin composite was with 20% error. However, greater discrepency, over 35% error, was observed for the prediction of strain.

Compared to the laminated unidirectional composites, the three-dimensional braid composite was 40% and 30% lower in strength and modulus, respectively. However, the strength and modulus of the three-dimensional braid composite was about twice as high as that of the laminated woven and quasi-isotropic composite.

With its fully integrated nature, the three-dimensional braid composite has a strong potential to raise the level of damage tolerance. With the more than satisfactory level of static strength and modulus, this three-dimensional braid composite provides room for trade-off and optimization of the properties in

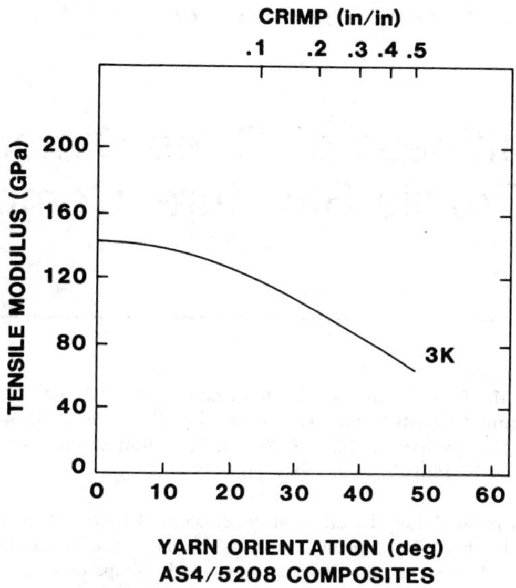

FIG. 12—*Prediction of modulus as a function of yarn orientation.*

different directions under different modes of deformation. When the dynamic and traverse static properties are fully characterized and optimized, the three-dimensional braid preform represents an attractive option for structural composites.

Acknowledgments

Part of this study was supported by the Office of Naval Research (ONR) under Grant No. N00014-83-K-0415 and by NADC. The assistance of Dr. Jim Alper, Lee Gause, Dennis Whyte, and Mary Williams were greatly appreciated.

Chang-Long Ma,[1] *Jenn-Ming Yang,*[1] *and Tsu-Wei Chou*[1]

Elastic Stiffness of Three-Dimensional Braided Textile Structural Composites

REFERENCE: Ma, C.-L., Yang, J.-M., and Chou, T.-W., **"Elastic Stiffness of Three-Dimensional Braided Textile Structural Composites,"** *Composite Materials: Testing and Design (Seventh Conference), ASTM STP 893,* J. M. Whitney, Ed., American Society for Testing and Materials, Philadelphia, 1986, pp. 404–421.

ABSTRACT: A methodology based upon the concept of fabric unit cell structure and an energy approach has been developed for studying the elastic properties of three-dimensional fabric composites. In general, the unit cell assumes the shape of a parallelepiped with its size determined by the weaving parameters. The yarns can assume the positions along the edges and diagonals in a unit cell, and each unit cell is centered around an "interlock" of yarns. To facilitate the analysis of complex yarn interactions, the unit cell is projected onto mutually orthogonal planes. Strain energies due to yarn axial extension, bending, and lateral compression are then considered in the two-dimensional plane weave structures. Closed-form expressions for axial elastic moduli and Poisson's ratios have been derived as functions of fiber volume fractions and fiber orientations. Numerical examples and experimental results of elastic properties for three-dimensional braided composites are presented.

KEY WORDS: composite materials, fabric composites, three-dimensional braided fabrics, elastic properties

Textile structural materials are gaining increasing technological importance as reinforcements for polymeric, metal, and ceramic matrix composites. In the case of two-dimensional woven fabric composites, a significant effort has been made by Ishikawa and Chou [*1–6*] to investigate the relationship between fabric structure and properties. They identified the various types of weaves by the material and geometrical patterns of repeats as well as the size of yarns. They also have established analytical models for analyzing the thermo-mechanical properties including the nonlinear stress-strain behavior.

The structure and properties of multidirectionally reinforced composites have recently been described by several authors [*7–13*]. These composites with multidirectional reinforcements are fabricated by a weaving or braiding technique. In comparing with two-dimensional composites based upon fabrics or unidirec-

[1]Visiting scholar, graduate student, and professor, respectively, Center for Composite Materials, and Mechanical and Aerospace Engineering Department, University of Delaware, Newark, DE 19716. Chang-Long Ma is on leave from the Department of Mechanical Engineering, Fushun Institute of Petroleum Technology, The People's Republic of China.

tional tapes, three-dimensional composites can be produced with considerable flexibility in their sizes along the length, width, and thickness directions. The integral nature of these composites provides improved stiffness and strength in the thickness direction, along which interlaminar failure often occurs in two-dimensional composites.

In spite of the high potential of three-dimensional composite applications in structural composites, there is a lack of basic understanding of their behavior. The purpose of this paper is to investigate the stiffness properties of three-dimensional composites. An analytical approach is adopted to model their constitutive behavior. The methodology of the analysis applies to three-dimensional structures in general. However, for the convenience of comparisons with experiments, three-dimensional Magnaweave [7] composites are considered in particular.

In general, a three-dimensional fiber arrangement can be considered as composed of two types of structures: the baseline structure formed by weaving fibers along three mutually orthogonal directions, and the interwoven or braided structure composed of fibers not orthogonal to one another. Three-dimensional composites can be "designed" to meet the performance requirements in different directions through suitable spatial arrangements of the reinforcing fibers. Figure 1 depicts the fiber spatial arrangements with respect to a rectangular Cartesian coordinate frame. The general features of the spatial arrangement can be better understood by focusing our attention on a typical location, for instance Point *d* in Fig. 1, where the interwoven or interlocking of fibers may take place.

Figure 2 depicts a "unit cell" volume with Point *d* as its geometric center. The size of the cell and hence, the parameters, p_a, p_b, and p_c, are determined by the fiber weaving pattern. The solid lines in Fig. 2 indicate all the possible fiber positions; diagonal fibers are denoted by Arabic numerals and baseline fibers by alphabets. The interaction among fibers at Point *d* depends upon the specific

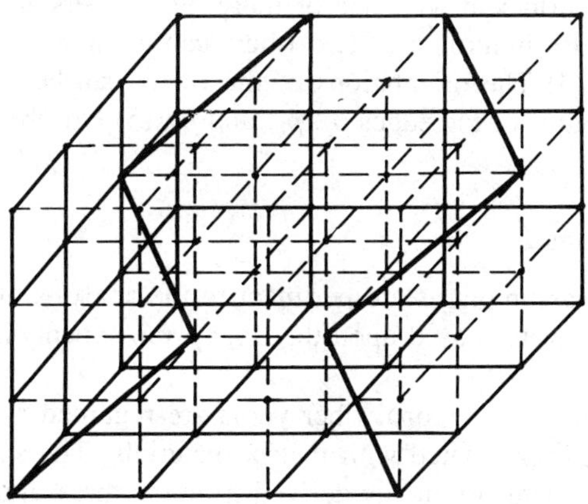

FIG. 1—*Schematic view of yarn positions in three dimension.*

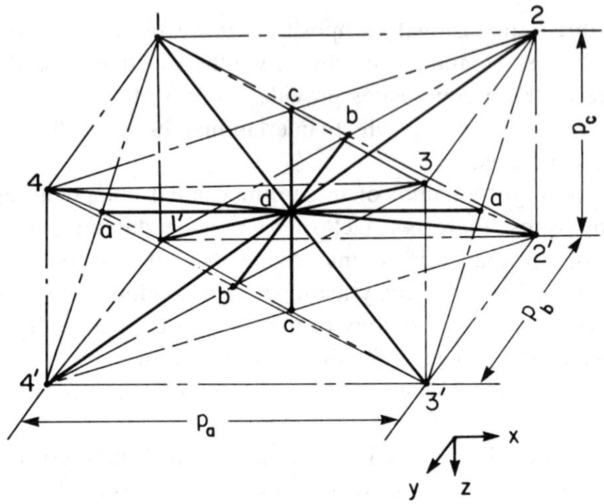

FIG. 2—*A "unit cell" structure for modeling three-dimensional woven composites. The baseline yarns are denoted by* aa, bb, *and* cc. *The diagonal yarns are 13', 42', 24', and 31'. The* d *denotes the center of the unit cell.*

weaving patterns. For instance, the baseline fibers *aa, bb,* and *cc* may be straight at the point of intersection such as in most of the cases of carbon/carbon composites [*14*], or may be curved and interlaced such as in the case of Ref *1*. It is also understood that the term "fiber" used here may represent a fiber yarn and thus, Point *d* in Fig. 2 represents a "region" of fiber intersection rather than a geometric point.

Some of the basic characteristics of a unit-cell are as follows. (1) The vertical edges such as 22′ and 44′ are assumed to be parallel to the direction of braiding in the three-dimensional braided and, hence, to the major axis of the composite specimen. (2) A specific fiber extending from 1 to *d* in Fig. 2 can further assume any one of the following four positions: 1′*d*, 2′*d*, 3′*d*, and 4′*d*. This fiber may or may not be interlocked with one or more other fibers at *d*. (3) A maximum of four fibers and a minimum of two fibers can assume the diagonal positions in the unit cell. (4) The orientations of the fibers can be varied by a suitable change of the lengths of the edges p_a, p_b, and p_c. (5) No fiber originates or ends in the unit cell.

Basic Assumptions

The following assumptions are pertinent to the analysis of three-dimensional woven fiber composites based upon the fiber arrangements in Fig. 2.

1. The baseline and diagonal fiber yarns are regarded as "composite rods" after being impregnated with matrix materials. The stiffness and strength of the composite are mainly derived from the three-dimensional composite rod structure.

2. The composite rods are homogeneous and linearly elastic, and have uniform circular cross-sections that do not become flattened under external loading.
3. The composite rods possess both tensile, compressive, and bending rigidities.
4. A jamming force exists at the region of contact between two interlocking composite rods. The rods are treated as either compressible or incompressible under the action of jamming forces.
5. The radius of curvature of a rod at the region of contact with another rod is equal to the sum of the yarn radii.

In order to facilitate the analysis, the three-dimensional diagonal composite rod structure of Fig. 2 is projected onto six mutually orthogonal planes along the $\pm x$, $\pm y$, and $\pm z$ directions. For instance, in the z-projections, the upper half of the unit cell (above Point d) is projected onto the $-z$ plane and the lower half is projected onto the $+z$ plane. In this way, each composite rod is only projected once along each coordinate axis. The baseline rod structure, on the other hand, is projected only along the $+x$, $+y$, and $+z$ directions. Figures 3*a* and *b* illustrate the projections of the baseline and diagonal rods, respectively, along the positive x, y, and z directions. It is further assumed that the composite rods in each projection can be regarded as a two-dimensional *plain weave* material. Figure 4 further illustrates the details of interactions of the baseline composite rods.

Several geometric parameters are given in Figs. 3 and 4. Here, ℓ_a, ℓ_b, and ℓ_c define the lengths of diagonals on the plane projections. θ_a, θ_b, and θ_c are the angles between ℓ_a, ℓ_b, ℓ_c, and p_c, p_b, p_a, respectively. Therefore, for instance, $\ell_a = \sqrt{p_b^2 + p_c^2}$, $\cos\theta_a = p_c/\ell_a$, and $\sin\theta_b = p_c/\ell_b$.

The volumes of the fibers in a given direction are defined as the projected length of the fibers multiplied by the fiber cross-sectional area. Thus, the projected relative fiber volume fractions along the x, y, and z directions are, respectively

$$V_{fx} = 5\, A_f/p_b p_c \tag{1}$$

$$V_{fy} = 5\, A_f/p_a p_c \tag{2}$$

$$V_{fz} = 5\, A_f/p_a p_b \tag{3}$$

Here, A_f denotes the composite rod cross-sectional area and $A_f = \pi D^2/4$, assuming circular rods of diameter D.

The Elastic Strain-Energy Approach

In considering the deformation of a unit cell, three types of elastic strain energies in the composite rods are taken into account. These include the strain energies due to bending (U_B), extension (U_E), and compression over the region

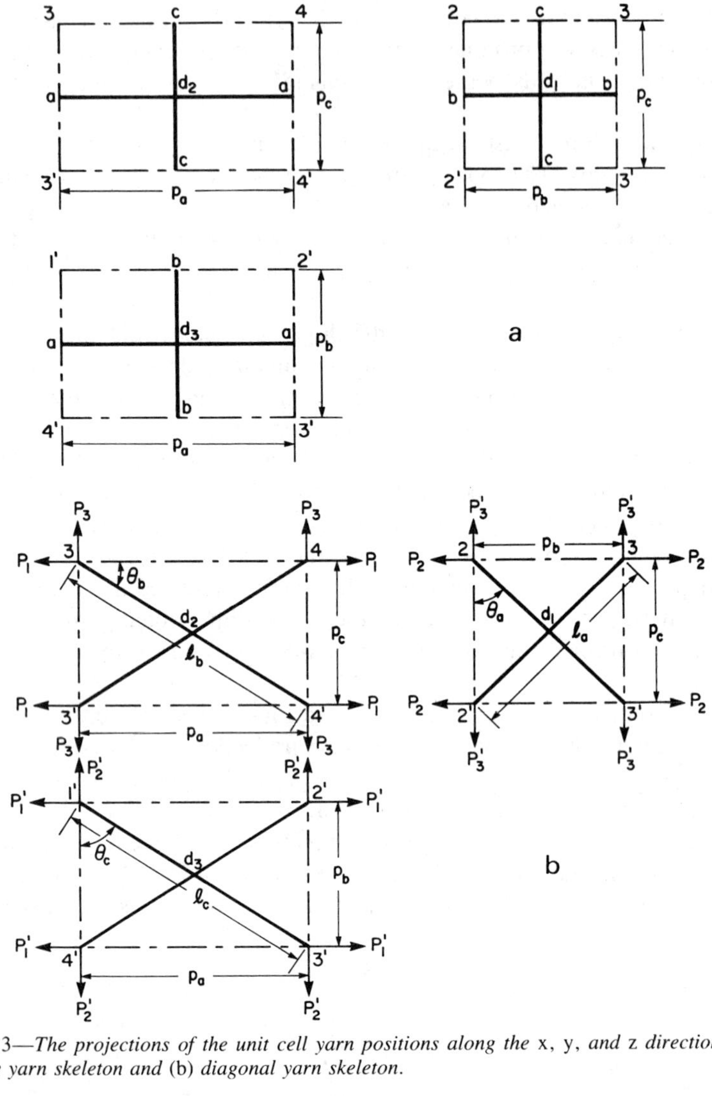

FIG. 3—*The projections of the unit cell yarn positions along the* x, y, *and* z *directions for* (a) *baseline yarn skeleton and* (b) *diagonal yarn skeleton.*

of fiber contact (U_C). The U_B per unit length of the rod is related to the externally applied bending moment (M) and the flexural rigidity (B) of the rod as

$$U_B = M^2/2B \tag{4}$$

U_E per unit length of the rod can be expressed in terms of the tensile force (T) and the tensile rigidity (λ) as

$$U_E = T^2/2\lambda \tag{5}$$

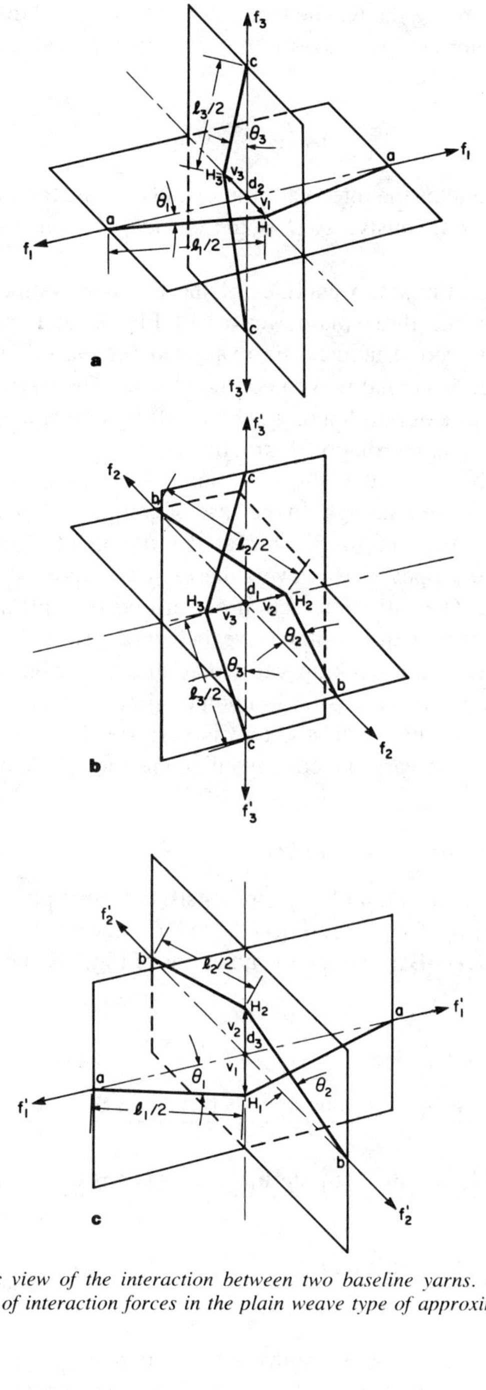

FIG. 4—*Schematic view of the interaction between two baseline yarns. Points* d_1, d_2, *and* d_3 *indicate the locations of interaction forces in the plain weave type of approximation.*

where $T = \lambda \epsilon_t$, ϵ_t being the tensile strain. The U_C is related to the contract force (V) between the interlocking rods, the rod diameter D, and the transverse compressive rigidity μ by

$$U_C = v^2 D/2\mu \tag{6}$$

where $V = \mu \epsilon_d$ denotes the inter-yarn force, and ϵ_d is the transverse compressive strain in the yarn. Obviously, $\epsilon_d D$ is the change of yarn diameter due to the inter-yarn force.

In the following, the strain energies of the baseline composite rod structure are considered for the three plain weaves of Fig 3*a* and the energies of the diagonal composite rod structures are examined for the six projections, which also assume two-dimensional woven configurations. The overall response of the unit cell structure to external loading is obtained by combining the responses of the baseline as well as the diagonal structures.

It should also be noted that the same analytical approach is valid for both tensile and compressive loadings. In the case of compressive loading, as opposed to tensile loading, the jamming force at the region of contact between two interlocking rods disappears. However, the elastic support of the matrix on the fiber crimp exists. Thus, the interaction of yarn crimp with matrix needs to be considered in analyzing the compressive behavior. This also implies that the contact force between interlocking yarns may diminish when the externally applied loading changes from tension to compression.

The analysis of the unit cell structure is performed by examining the strain energies of the baseline yarn skeleton and diagonal yarn skeleton in the following [*15–17*].

The Baseline Composite Rod Skeleton

The baseline yarn skeleton of Fig. 3*a* consists of three plain weaves. External forces are denoted by f_1, f_2, f_3, f'_1, f'_2, and f'_3 in Fig. 4.

aa-cc *Plain Weave*—Based upon the notations of Fig. 4*a*, the total strain energy is given by

$$\begin{aligned} U_T &= U_B + U_E + U_C \\ &= \frac{\ell_1^3 (f_1 \sin \theta_1 - v_1 \cos \theta_1)^2}{24B} + \frac{\ell_3^3 (f_3 \sin \theta_3 - v_3 \cos \theta_3)^2}{24B} \\ &\quad + \frac{\ell_1 (f_1 \cos \theta_1 + v_1 \sin \theta_1)^2}{2\lambda} + \frac{\ell_3 (f_3 \cos \theta_3 + v_3 \sin \theta_3)^2}{2\lambda} \\ &\quad + \frac{2v_1^2 D}{\mu} + \frac{2 v_3^2 D}{\mu} \end{aligned} \tag{7}$$

Here, ℓ_1, ℓ_2, and ℓ_3 are the yarn lengths between two neighboring interlocks (Fig. 4). It is assumed that all the baseline yarns have the same diameter, D, tensile rigidity, λ, flexural rigidity, B, and transverse compressive rigidity, μ. The

Castigliano's theorem requires that

$$\frac{\partial U_T}{\partial v_1} + \frac{\partial U_T}{\partial v_3} = 0 \tag{8}$$

Since $v_1 = v_3 = v$, Eq 8 can be used to solve for v, and we obtain

$$v = \frac{\sum_{i=1,3} \left[\frac{\ell_i^3 f_i \sin\theta_i \cos\theta_i}{12B} - \frac{\ell_i f_i \cos\theta_i \sin\theta_i}{\lambda} \right]}{\sum_{i=1,3} \left[\frac{\ell_i^3 \cos^2\theta_i}{12B} + \frac{\ell_i \sin^2\theta_i}{\lambda} \right] + \frac{8D}{\mu}} \tag{9}$$

The strain in the x direction is $\epsilon_1 = \delta p_a / p_a$. Thus,

$$\begin{aligned} \epsilon_1 p_a &= \delta p_a = \frac{\partial U_T}{\partial f_1} \\ &= \ell_1 f_1 \left(\frac{\ell_1^2 \sin^2\theta_1}{12B} + \frac{\cos^2\theta_1}{\lambda} \right) - v\, \ell_1 \sin\theta_1 \cos\theta_1 \left(\frac{\ell_1^2}{12B} - \frac{1}{\lambda} \right) \end{aligned} \tag{10}$$

Similarly, the strain in the z direction is $\epsilon_3 = \delta p_c / p_c$ and

$$\begin{aligned} \epsilon_3 p_c = \delta p_c = \frac{\partial U_T}{\partial f_3} &= \ell_3 f_3 \left(\frac{\ell_3^2 \sin^2\theta_3}{12B} + \frac{\cos^2\theta_3}{\lambda} \right) \\ &- v\, \ell_3 \cdot \sin\theta_3 \cos\theta_3 \left(\frac{\ell_3^2}{12B} - \frac{1}{\lambda} \right) \end{aligned} \tag{11}$$

Substituting v from Eq 9 into Eqs 10 and 11, we obtain

$$\epsilon_1 = M_1/N_1, \; \epsilon_3 = M_3/N_3 \tag{12}$$

where

$$\begin{aligned} M_1 = f_1 \Bigg\{ & \frac{\ell_1^3 \ell_3^3 \sin^2\theta_1 \cos^2\theta_3}{144 B^2} + \frac{\ell_1^4}{12 B \lambda} \\ & + \frac{\ell_1 \ell_3}{12} \left(\frac{\ell_1^2 \sin^2\theta_1 \sin^2\theta_3}{B \lambda} + \frac{\ell_3^2 \cos^2\theta_1 \cos^2\theta_3}{B \lambda} \right) \\ & + \frac{\ell_1 \ell_3 \cos^2\theta_1 \sin^2\theta_3}{\lambda^2} + 4 \ell_1 \left(\frac{\ell_1^2 \sin^2\theta_1}{12 B} + \frac{\cos^2\theta_1}{\lambda} \right) \\ & \times \left(\frac{2D}{\mu} \right) \Bigg\} - f_3 \ell_1 \ell_3 \sin\theta_1 \cos\theta_1 \sin\theta_3 \cos\theta_3 \\ & \times \left(\frac{\ell_1^2}{12 B} - \frac{1}{\lambda} \right) \left(\frac{\ell_3^2}{12 B} - \frac{1}{\lambda} \right) \end{aligned} \tag{13}$$

$$N_1 = p_a \sum_{i=1,3} \left(\frac{\ell_i^3 \cos^2 \theta_i}{12B} + \frac{\ell_i \sin^2 \theta_i}{\lambda} + \frac{8D}{\mu} \right) \tag{14}$$

$$\begin{aligned} M_3 = f_3 \Bigg\{ & \frac{\ell_3^3 \, \ell_1^3 \sin^2 \theta_3 \cos^2 \theta_1}{144 \, B^2} + \frac{\ell_3^4}{12 \, B \, \lambda} \\ & + \frac{\ell_1 \, \ell_3}{12} \left(\frac{\ell_3^2 \sin^2 \theta_3 \sin^2 \theta_1}{B \, \lambda} + \frac{\ell_1^2 \cos^2 \theta_3 \cos^2 \theta_1}{B \, \lambda} \right) \\ & + \frac{\ell_1 \, \ell_3 \cos^2 \theta_3 \sin^2 \theta_1}{\lambda^2} + 4 \, \ell_3 \left(\frac{\ell_3^2 \sin^2 \theta_3}{12 \, B} + \frac{\cos^2 \theta_3}{\lambda} \right) \\ & \times \left(\frac{2D}{\mu} \right) \Bigg\} - f_1 \, \ell_1 \, \ell_3 \sin \theta_3 \cos \theta_3 \sin \theta_1 \cos \theta_1 \\ & \times \left(\frac{\ell_3^2}{12 \, B} - \frac{1}{\lambda} \right) \left(\frac{\ell_1^2}{12 \, B} - \frac{1}{\lambda} \right) \end{aligned} \tag{15}$$

$$N_3 = p_c \sum_{i=1,3} \left(\frac{\ell_i^3 \cos^2 \theta_i}{12 \, B} + \frac{\ell_i \sin^2 \theta_i}{\lambda} + \frac{8D}{\mu} \right) \tag{16}$$

bb-cc *Plain Weave*—Based upon the notations of Fig 4*b* and following the same procedures as developed in the preceding section, the strains in the y and z directions are given, respectively, by

$$\begin{aligned} \epsilon_2 &= \frac{\delta \, p_b}{p_b} = \frac{1}{p_b} \frac{\partial \, U_T}{\partial \, f_2} = \frac{M_2}{N_2} \\ \epsilon_3 &= \frac{\delta \, p_c}{p_c} = \frac{1}{p_c} \frac{\partial \, U_T}{\partial \, f_3} = \frac{M'_3}{N'_3} \end{aligned} \tag{17}$$

where M_2 and N_2 are obtained by changing all the Subscripts 1 in Eqs 13 and 14 to 2. Similarly M'_3 and N'_3 are obtained by changing all the Subscripts 1 in Eq 15 for M_3 and Eq 16 for N_3, respectively, to 2.

aa-bb *Plain Weave*—Referring to the notation of Fig. 4*c*, the strains in the x and y directions are given, respectively, by

$$\begin{aligned} \epsilon_1 &= \frac{\delta \, p_a}{p_a} = \frac{1}{p_a} \frac{\partial \, U_T}{\partial \, f_1} = \frac{M'_1}{N'_1} \\ \epsilon_2 &= \frac{\delta \, p_b}{p_b} = \frac{1}{p_b} \frac{\partial \, U_T}{\partial \, f_2} = \frac{M'_2}{N'_2} \end{aligned} \tag{18}$$

where M'_1 and N'_1 are obtained by changing all the Subscripts 3 in Eqs 13 and

14 to 2. Similarly M'_2 and N'_2 are obtained by changing all the Subscripts 3 in Eq 15 for M'_2 and Eq 16 for N'_2, respectively, to 2.

The Diagonal Composite Rod Skeleton

Six projections in the *x*-*y*-*z* coordinate system are necessary to uniquely represent the interactions of yarns in Fig. 2. The following derivations are made based upon the three typical projections along the positive *x*, *y*, and *z* axis as shown in Fig. 3*b*. The external forces p_1, p_2, p_3, p'_1, p'_2, and p'_3 are defined in Fig. 3*b*. It is also assumed that all the diagonal rods have the same diameter, D_d, tensile rigidity, λ_d, flexural rigidity, B_d, and transverse compressive rigidity, μ_d. It is assumed in the following derivations that the yarns are transversely incompressible, that is, $\mu_d \rightarrow \infty$.

344'3' Plain Weave

$$U_T = \frac{\ell_b^3 (P_1 \sin\theta_b - P_3 \cos\theta_b)^2}{12 B_d} + \frac{\ell_b (P_1 \cos\theta_b + P_3 \sin\theta_b)^2}{\lambda_d} \tag{19}$$

The application of Castigliano's theorem leads to

$$\epsilon_1 = \frac{\delta p_a}{p_a} = \frac{1}{p_a}\frac{\partial U_T}{\partial P_1} = \frac{1}{p_a}\left\{\ell_b P_1 \left(\frac{\ell_b^2 \sin^2\theta_b}{6 B_d} + \frac{2\cos^2\theta_b}{\lambda_d}\right) - P_3 \ell_b \sin\theta_b \cos\theta_b \left(\frac{\ell_b^2}{6 B_d} - \frac{2}{\lambda_d}\right)\right\} \tag{20}$$

and

$$\epsilon_3 = \frac{\delta p_c}{p_c} = \frac{1}{p_c}\frac{\partial U_T}{\partial p_3} = \frac{1}{p_c}\left\{\ell_b P_3 \left(\frac{\ell_b^2 \cos^2\theta_b}{6 B_d} + \frac{2\sin^2\theta_b}{\lambda_d}\right) - P_1 \ell_b \sin\theta_b \cos\theta_b \cdot \left(\frac{\ell_b^2}{6 B_d} - \frac{2}{\lambda_d}\right)\right\} \tag{21}$$

233'2' Plain Weave

$$\epsilon_3 = \frac{1}{p_c}\left\{\ell_a P'_3 \left(\frac{\ell_a^2 \sin^2\theta_a}{6 B_d} + \frac{2\cos^2\theta_a}{\lambda_d}\right) - P_2 \ell_a \sin\theta_a \cos\theta_a \cdot \left(\frac{\ell_a^2}{6 B_d} - \frac{2}{\lambda_d}\right)\right\} \tag{22}$$

$$\epsilon_2 = \frac{1}{p_b}\left\{\ell_a P_2 \left(\frac{\ell_a^2 \cos^2\theta_a}{6 B_d} + \frac{2\sin^2\theta_a}{\lambda_d}\right) - P'_3 \ell_a \sin\theta_a \cos\theta_a \cdot \left(\frac{\ell_a^2}{6 B_d} - \frac{2}{\lambda_d}\right)\right\} \tag{23}$$

1'2'3'4' Plain Weave

$$\epsilon_2 = \frac{1}{p_b}\left\{\ell_c P'_2 \left(\frac{\ell_c^2 \sin^2\theta_c}{6 B_d} + \frac{2\cos^2\theta_c}{\lambda_d}\right) - P'_1 \ell_c \sin\theta_c \cos\theta_c \left(\frac{\ell_c^2}{6 B_d} - \frac{2}{\lambda_d}\right)\right\};$$

$$\epsilon_1 = \frac{1}{p_a}\left\{\ell_c P'_1 \left(\frac{\ell_c^2 \cos^2\theta_c}{6 B_d} + \frac{2\sin^2\theta_c}{\lambda_d}\right) - P'_2 \ell_c \sin\theta_c \cos\theta_c \cdot \left(\frac{\ell_c^2}{6 B_d} - \frac{2}{\lambda_d}\right)\right\} \quad (24)$$

Combinations of the Baseline and Diagonal Rod Skeletons

In this section, the combined responses of the baseline and diagonal rod skeletons to external loadings are examined. The focus is on the initial Young's moduli E_x, E_y, and E_z, and the Poisson's ratios υ_{31}, υ_{23}, and υ_{12}. First, uniaxial loadings are applied to determine the axial stiffness. Figure 5 indicates the resultant forces due to the baseline rod skeleton. The resultant force in the z direction, for instance, is

$$f_{3b} = f_3 + f'_3 = \epsilon_3 \left(\frac{N_3}{K_3} + \frac{N'_3}{K'_3}\right) \quad (25)$$

for $f_1 = f'_1 = f_2 = f'_2 = 0$ and

$$K_3 = \frac{\ell_3^3 \ell_1^3 \sin^2\theta_3 \cos^2\theta_1}{144 B^2} + \frac{\ell_3^4}{12 B \lambda} + \frac{\ell_3 \ell_1}{12}\left(\frac{\ell_3^2 \sin^2\theta_3 \sin^2\theta_1}{B\lambda} + \frac{\ell_1^2 \cos^2\theta_3 \cos^2\theta_1}{B\lambda}\right) + \frac{\ell_3 \ell_1 \cos^2\theta_3 \sin^2\theta_1}{\lambda^2} + 4\ell_3\left(\frac{\ell_3^2 \sin^2\theta_3}{12 B} + \frac{\cos^2\theta_3}{\lambda}\right) \times \left(\frac{D}{\mu} + \frac{D}{\mu}\right) \quad (26)$$

N_3 is defined in Eq 16. Also, N'_3 and K'_3 are obtained by changing Subscript 1 in Eqs 16 and 26, respectively, to 2.

In the case of the diagonal rod skeleton, the resultant forces are indicated in Fig. 5. The P_1, P'_1, P_2, P'_2, P_3, P'_3 are shown for three of the six projection

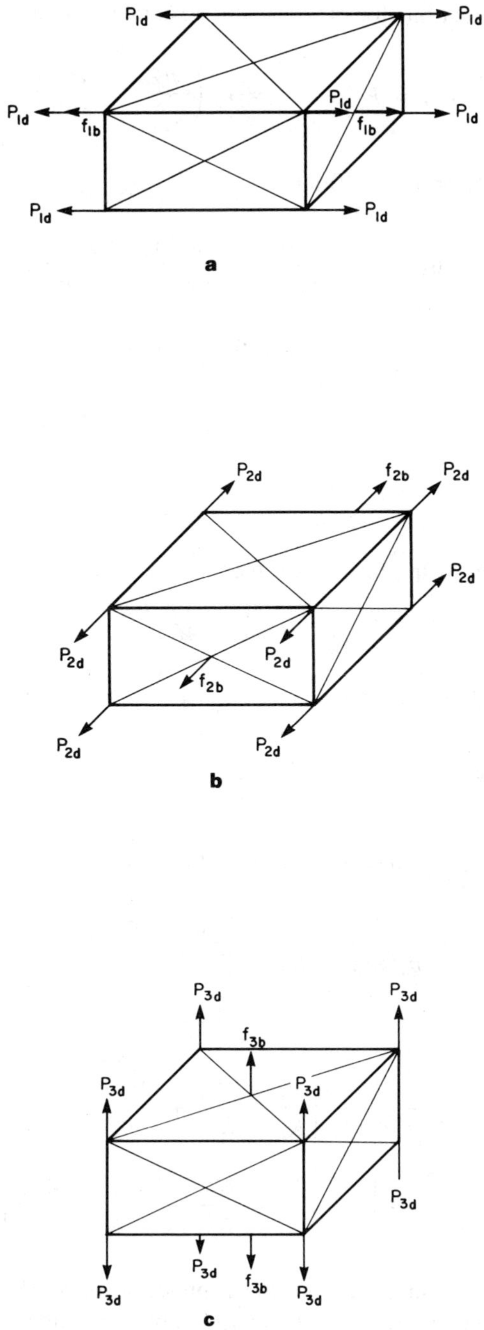

FIG. 5—*Combinations of forces acting on the baseline and diagonal yarn skeletons.*

planes. In the z direction, for example, the resultant force is

$$P_{3d} = P_3 + P'_3 = \epsilon_3 \left(\frac{p_c}{\ell_b \, \theta_3} + \frac{p_c}{\ell_a \, \theta'_3} \right) \tag{27}$$

for $P_1 = P_2 = P'_1 = P'_2 = 0$ and

$$\theta_3 = \frac{\ell_b^2 \cos^2 \theta_b}{6 \, B_d} + \frac{2 \sin^2 \theta_b}{\lambda_d} + \frac{8 \, D_d}{\ell_b \, \mu_d} \tag{28}$$

$$\theta'_3 = \frac{\ell_a^2 \sin^2 \theta_a}{6 \, B_d} + \frac{2 \cos^2 \theta_a}{\lambda_d} \tag{29}$$

Elastic Constants

Elastic Stiffness

The combination of the resultant forces of Fig. 5 in the z direction gives

$$F_z = f_{3b} + 4P_{3d} \tag{30}$$

and the corresponding axial stress is

$$\sigma_z = \frac{F_z}{p_a \, p_b \, V_{fz}} \tag{31}$$

Then, the axial Young's modulus in the z direction is

$$E_z = \frac{1}{p_a \, p_b \, V_{fz}} \left\{ \frac{N_3}{K_3} + \frac{N'_3}{K'_3} + \frac{4p_c}{\ell_b \, \theta_3} + \frac{4p_c}{\ell_a \, \theta'_3} \right\} \tag{32}$$

Similarly, we obtain

$$E_x = \frac{1}{p_b \, p_c \, V_{fx}} \left\{ \frac{N_1}{K_1} + \frac{N'_1}{K'_1} + \frac{4p_a}{\ell_c \, \theta_1} + \frac{4p_a}{\ell_b \, \theta'_1} \right\} \tag{33}$$

$$E_y = \frac{1}{p_a \, p_c \, V_{fy}} \left\{ \frac{N_2}{K_2} + \frac{N'_2}{K'_2} + \frac{4p_b}{\ell_a \, \theta_2} + \frac{4p_b}{\ell_c \, \theta'_2} \right\} \tag{34}$$

Here, K_2 and K_1 are obtained from the expression for K_3 (Eq 26) by changing Subscript 3 to 2 and 1, and 1 to 3, respectively. The same permutation can be applied to obtain K'_2 and K'_1 from K'_3, Q_2 and Q_1 from Q_3, and Q'_2 and Q'_1 from Q'_3. Also N_2 and N_1 are obtained from N_3 by changing Subscript 3 to 2 and 1, Subscript 1 to 1 and 3, and N'_2 and N'_1 are obtained from N'_3 by changing Subscript 3 to 2 and 1 and Subscript 1 to 1 and 3, respectively.

Poisson's Ratios

The derivations of the initial Poisson's ratios of the three-dimensional woven composites are outlined below. Following Fig. 5, we define, for instance

$$\nu_{31} = -\frac{\epsilon_1}{\epsilon_3} = \left[\frac{4p_c}{\ell_b\,\theta_3} + \frac{N_3}{K_3}\right] \Big/ \left[\frac{N_1}{W_1} + \frac{4p_a}{U_1}\right] \tag{35}$$

$$\nu_{23} = -\frac{\epsilon_3}{\epsilon_2} = \left[\frac{4p_b}{\ell_a\,\theta_2} + \frac{N_2}{K_2}\right] \Big/ \left[\frac{N'_3}{W'_3} + \frac{4p_c}{U_3}\right] \tag{36}$$

and

$$\nu_{12} = -\frac{\epsilon_2}{\epsilon_1} = \left[\frac{4p_a}{\ell_c\,\theta_1} + \frac{N'_1}{K'_1}\right] \Big/ \left[\frac{N'_2}{W'_2} + \frac{4p_b}{U_2}\right] \tag{37}$$

Here,

$$U_1 = \ell_b \sin\theta_b \cos\theta_b \left(\frac{\ell_b^2}{6\,B_d} - \frac{2}{\lambda_d}\right) \tag{38}$$

where U_2 and U_3 are obtained by changing Subscript b in Eq 38 to c and a, respectively. Also

$$W_1 = \ell_1\,\ell_3 \sin\theta_1 \cos\theta_1 \sin\theta_3 \cos\theta_3 \left(\frac{\ell_1^2}{12\,B} - \frac{1}{\lambda}\right)\left(\frac{\ell_3^2}{12\,B} - \frac{1}{\lambda}\right) \tag{39}$$

$$W'_2 = \ell_1\,\ell_2 \sin\theta_2 \cos\theta_2 \sin\theta_1 \cos\theta_1 \left(\frac{\ell_2^2}{12\,B} - \frac{1}{\lambda}\right)\left(\frac{\ell_1^2}{12\,B} - \frac{1}{\lambda}\right) \tag{40}$$

where W'_3 is obtained by changing Subscripts 1 and 2 in Eq 40 to 2 and 3, respectively.

In general, for $p_a \neq p_b \neq p_c$, the unit structure cell of the three-dimensional braided composites can be regarded as an orthotropic composite because it has three mutually orthogonal symmetric planes. So, there are nine independent constants, such as E_1, E_2, E_3, ν_{12}, ν_{23}, ν_{31}, G_{23}, G_{31}, and G_{12}.

Braided Composites

The theoretical modeling of three-dimensional composites just developed is applied to examine the elastic properties of three-dimensional braided composites, such as Magnaweave. The unit cell of a three-dimensional braided fabric without laid-in yarns consists of diagonal yarns only. The orientations of the yarns depends on the selection of the braiding motion parameter.

The elastic stiffness constants for three-dimensional braided composites, from Eqs 32–34, are

$$E_x = \frac{1}{p_b\, p_c\, V'_{fx}} \left\{ \frac{4p_a}{\ell_c\, \theta_1} + \frac{4p_a}{\ell_b\, \theta'_1} \right\} \tag{41}$$

$$E_y = \frac{1}{p_a\, p_c\, V'_{fy}} \left\{ \frac{4p_b}{\ell_a\, \theta_2} + \frac{4p_b}{\ell_c\, \theta'_2} \right\} \tag{42}$$

$$E_z = \frac{1}{p_a\, p_b \cdot V'_{fz}} \left\{ \frac{4p_c}{\ell_b\, \theta_3} + \frac{4p_c}{\ell_a\, \theta'_3} \right\} \tag{43}$$

The Poisson's ratios are, from Eqs 35 to 37

$$\nu_{12} = -\frac{\epsilon_2}{\epsilon_1} = \left[\frac{4p_a}{\ell_c\, \theta_1}\right] \Big/ \left[\frac{4p_b}{U_2}\right] \tag{44}$$

$$\nu_{23} = -\frac{\epsilon_3}{\epsilon_2} = \left[\frac{4p_b}{\ell_a\, \theta_2}\right] \Big/ \left[\frac{4p_c}{U_3}\right] \tag{45}$$

$$\nu_{31} = -\frac{\epsilon_1}{\epsilon_3} = \left[\frac{4p_c}{\ell_b\, \theta_3}\right] \Big/ \left[\frac{4p_a}{U_1}\right] \tag{46}$$

Results and Discussion

Numerical calculations are conducted for the case of graphite Magnaweave fabric in an epoxy matrix. The elastic material properties used in the calculations are: $E_f = 34 \times 10^6$ psi (AS4 Magnamite graphite fiber), $\nu_f = 0.22$, $E_m = 0.5 \times 10^6$ psi (epoxy matrix), and $\nu_m = 0.34$.

In order to discern the effect of fiber configuration on composite initial modulus, we also examine the case of "crimp exchange," namely, the change of shape of an interlock. This is equivalent to assume $\lambda_d = \mu_d \rightarrow \infty$ and the composite rods are inextensible and incompressible. However, as far as the initial modulus is concerned, the difference between the cases with and without this assumption is small.

To facilitate the calculation, the orientation angle between a yarn segment and the z-axis is defined as

$$\theta = \tan^{-1}\left(\sqrt{u^2 + v^2} \Big/ w\right) \tag{47}$$

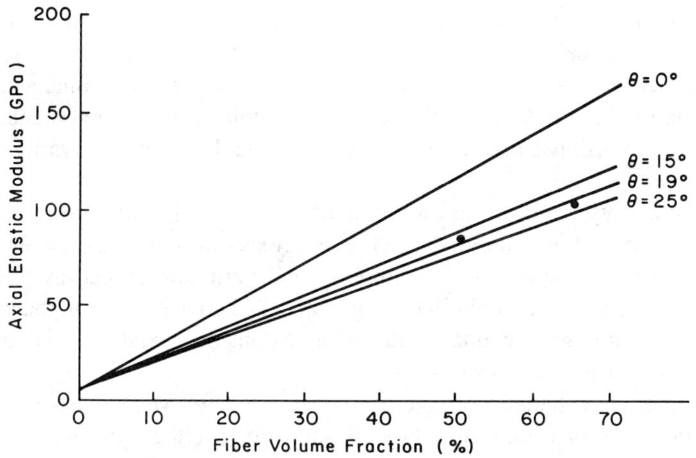

FIG. 6—*Predicted axial elastic tensile modulus of Magnaweave composites as a function of fiber volume fractions and fiber orientation angles. Solid circles indicate experimental results* [19].

Here, u, v, and w are dimensionless parameters obtained by normalizing the unit cell dimension, P_a, P_b, and P_c, respectively, by the yarn bundle diameter, d. The w is a function of the compacting action during fabric formation, while u and v are the yarn displacement resulting from the shedding motion.

According to Ko [*18*], the effect of density of yarn interlacing expressed in terms of the normalized unit cell height, w, on yarn orientation is significant. As w increases, the average yarn orientation angle decreases. An increase of the w value from 1 to 5 brings about the decrease of the yarn angle, θ, from 55° to 38°, 25°, 19°, and 15°, respectively, for w = 2, 3, 4, and 5. With the knowledge of the relationship between the yarn orientation and w, the approximated unit cell geometry can be obtained.

Numerical results demonstrating the predicted axial elastic moduli along the braiding direction as functions of the yarn orientation and fiber volume fraction are given in Fig. 6. The elastic moduli are sensitive to the geometrical braiding pattern; they increase as the yarn orientation angle decreases, that is, the fiber becomes more aligned with the tensile axis. The comparison of the present analysis with the limited experimental data available in the open literature shows good agreement.

Conclusions and Discussions

1. A methodology based upon the concept of fabric unit cell structure and an energy approach has been developed for studying the elastic properties of three-dimensional fabric composites in general. Fabric composites composed of either three sets of mutually orthogonal yarns or diagonally intersecting yarns are special cases of the present model.

2. In view of the complexity of the yarn configurations at their interlocking positions, the model is not intended to simulate each individual "lock" separately. The interactions among the yarns are dealt with in an approximate fashion by projecting the yarn positions onto six sets of mutually orthogonal planes. Within each two-dimensional projection, the interactions between two yarns are taken into account.

3. The energy approach is formulated based upon the interaction of yarns in two dimension, at an "interlock." The interlocks do not rotate under load due to symmetry considerations. The impregnated yarns are treated as "composite rods" in the form of "straight line" segments. The contributions to the composite elastic strain energy are due to the axial extension, bending, and transverse compression of the composite rods.

4. Numerical calculations have been performed for Magnaweave composites with four yarns in a unit cell. Results indicate that the axial tensile modulus (along the braiding direction) is sensitive to the geometrical braiding pattern.

5. The numerical results also demonstrate that for a fixed fiber volume fraction, the axial tensile modulus increases as the fiber becomes more aligned with the tensile axis. This implies for composites with a ductile matrix, the secant modulus will increase with the load.

6. It is known that in a three-dimensional braided fabric the braiding parameters (p_a, p_b, and p_c) are usually not fixed within a given specimen and they vary with location, for example, at locations at the center or near the edge of a specimen. Thus, the overall elastic property of a specimen should be considered as a volume average over several different θ_3 configurations.

7. The initial compressive modulus can be treated in the same manner as the tensile modulus. The lateral support of the matrix material on the fiber yarns against fiber micro-buckling needs to be considered at large compressive deformation.

8. The problems of elastic shear deformation and Poisson's effect of three-dimensional braided composites are being examined based upon the energy approach. The problems of nonlinear elastic deformation are of major practical significance and deserve careful examination.

9. The present approach is applicable to two-dimensional woven composites as having been demonstrated in Ref *20* for triaxial woven composites.

Acknowledgment

Tsu-Wei Chou wishes to thank the support of the Office of Naval Research (Contract No. N00014-83-K-0400). Dr. L. H. Peebles, Jr., is the Technical Monitor.

References

[*1*] Ishikawa, T. and Chou, T. W., *Journal of Composite Materials*, Vol. 16, Jan. 1982, pp. 2–19.

[2] Ishikawa, T. and Chou, T. W., *AIAA Journal*, American Institute of Aeronautics and Astronautics, 1983, p. 1714.

[3] Ishikawa, T. and Chou, T. W., *Journal of Materials Science*, Vol. 17, 1982, p. 3211.
[4] Ishikawa, T. and Chou, T. W., *Journal of Composite Materials*, Vol. 17, 1983, p. 92.
[5] Ishikawa, T. and Chou, T. W., *Journal of Materials Science*, Vol. 18, 1983, p. 2260.
[6] Ishikawa, T. and Chou, T. W., *Journal of Composites*, Vol. 17, 1983, p. 399.
[7] Ko, F. K., "Three-dimensional Fabrics for Composites—An Introduction to the Magnaweave Structure," K. Kawata and S. Umekawa, Eds., ICCM-IV, International Conference on Composite Materials, Tokyo, 1982, p. 1609.
[8] Florentine, R. F. in *Proceedings*, 37th Annual Conference, Reinforced Plastics/Composites Institute, The Society of Plastics Industry, Paper 10A, 10B, Jan. 1982, p. 1.
[9] U. S. Patent 4,312,261, 26 Jan. 1982.
[10] Lachman, W. L., "Multidirectionally Reinforced Carbon/Carbon Composites," *Proceedings*, International Conference on Composite Materials, Toronto, 16–20 April, 1978
[11] Halpin, J. C., Jerina, D., and Whitney, J. M., *Journal of Composite Materials*, Vol. 5, No. 36, Jan. 1971, p. 36.
[12] Jerina, K. L., "The Effective Moduli of 3-D Reinforced Fibrous Materials," Ph.D. thesis, Washington University, May 1974.
[13] Rosen, B. W., Chatterjee, S. N., and Kibler, J. J., in *Composite Materials: Testing and Design (Fourth Conference), ASTM STP 617*, American Society for Testing and Materials, Philadelphia, 1977, p. 243.
[14] Delmonte, J., *Technology of Carbon and Graphite Fiber Composites*, Van Nostrand Reinhold Company, New York, 1981.
[15] Hearle, J. W. S. and Shanahan, W. J., *Journal of the Textile Institute*, Vol. 69, 1978, p. 81.
[16] Hearle, J. W. S. and Shanahan, W. J., *Journal of the Textile Institute*, Vol. 69, 1978, p. 92.
[17] Leaf, G. in *Mechanics of Flexible Fibre Assemblies*, Sijthoff & Noordhoff, Alphen aan den Rijn, The Netherlands, 1980, p. 143.
[18] Ko, F. K., "Structure and Properties of an Integrated 3-D Fabric for Structural Composites," presented at the U. S.–Japan Composite Conference, July 1983.
[19] Florentine, R. A. in *Proceedings* 29th SAMPE Convention, Reno, Society for the Advancement of Materials and Process Engineering, 1984, p. 600.
[20] Ma, C. L. and Chou, T. W., unpublished results.

Aleksander B. Macander,[1] *Roger M. Crane,*[1] *and Eugene T. Camponeschi, Jr.*[1]

Fabrication and Mechanical Properties of Multidimensionally (X-D) Braided Composite Materials

REFERENCE; Macander, A. B., Crane, R. M., and Camponeschi, E. T., Jr., **"Fabrication and Mechanical Properties of Multidimensionally (X-D) Braided Composite Materials,"** *Composite Materials: Testing and Design (Seventh Conference), ASTM STP 893,* J. M. Whitney, Ed., American Society for Testing and Materials, Philadelphia, 1986, pp. 422–443.

ABSTRACT; This paper describes research concerning the resin impregnation and characterization of multidimensionally braided fiber-reinforced composite materials. These materials are an alternative to traditional laminated structures, having the potential for being more damage tolerant. Three graphite fiber systems were used in this investigation, and three processes were investigated for resin impregnation of the multidimensionally braided material using vacuum or pressure. Two were resin transfer techniques and the third was a resin film lamination technique. While all three methods are presented, the latter technique was chosen for impregnating the test specimens due to the consistently low void content and superior surface quality achieved by this method. Three variables having an important bearing on the performance of braided materials were investigated. These included the effect of braid pattern, tow size, and edge condition on the tensile, compressive, flexural, and interlaminar shear properties. The properties were obtained in the braid direction only. The cutting of the specimen edges substantially reduced both tensile and flexural strengths and moduli. Of the three braid patterns investigated (1 × 1, 1 × 1 × 1/2F and 3 × 1), the 3 × 1 braid showed superior tensile performance and the 1 × 1 × 1/2F pattern exhibited superior flexural properties. Variation in fiber tow size caused variations in tensile, flexural, and short-beam shear properties. The 12K tow size specimens exhibited the best performance. All braided composite materials in the uncut edge condition showed significant improvements in their short-beam shear strengths, being equal to or greater than unidirectional laminated composites. This latter characteristic may be one of several indicators that multidimensionally braided composites are inherently damage tolerant.

KEY WORDS: composite materials, multidimensional braiding, damage tolerant composites, resin impregnation, braid patterns, braid tows, graphite/epoxy composites, mechanical properties

A new class of fiber-reinforced composites is being developed that promises significant improvement in damage tolerance compared to conventional lami-

[1]Materials engineers, David Taylor Naval Ship Research and Development Center, Bethesda, MD 20084.

nates. The principal feature of this new material is the through-the-thickness network of reinforcing fibers that forms a multidimensionally (X-D) braided skeletal structure. The absence of planes susceptible to delamination in this material should result in a tough composite resistant to crack growth. It has been demonstrated that the X-D braiding process can generate both symmetric and asymmetric skeletal net shapes of varying thickness and contour. Two areas of investigation yet to be fully developed are methods to impregnate the X-D braided skeletal preforms and the evaluation of the mechanical properties resulting from the X-D braiding.

This paper will present the development of a methodology to impregnate and rigidize the X-D skeletal preforms with Narmco's 5208 structural epoxy resin. The X-D braided preforms were braided into flat panels using either Thornel T-300, Celion, or AS-4 graphite fibers. These preforms were braided using different tow sizes, braid patterns, and edge conditions. The tensile, compressive, flexural, and interlaminar shear properties are discussed as a function of the various construction variables. These properties were developed in the braid or axial direction only. An examination of the resulting failure surfaces is also presented.

X-D Braided Fiber Preforms

Examples of methodology used to produce the X-D braided materials were based on techniques that have been presented by Bluck [*1*], Florentine [*2*], and Ko [*3*].

In order to explain the terminology used for the braid pattern a brief description of the braiding loom will be given. The loom consists of a rectangular base plate, carrier tracks, fiber carriers, and an upper header plate. An example of such a machine is shown in Fig. 1.

Each carrier track has a regular series of slots along its own length. A cartesian array of slots is formed when the individual tracks with their slots are positioned adjacent to each other. The slots are filled with fiber or yarn carriers to which one end of a yarn is attached while the other end is attached to the header plate. By moving the yarn carriers in a prescribed manner, various preforms can be made. Figure 2 shows the pattern of travel for one yarn carrier whose braid pattern is described as 1×1. The first number in the braid pattern designation represents the number of spaces the yarn holder advanced in the x-direction. The second number represents subsequent spaces the yarn holder advanced in the y-direction. The resultant orientation of the yarn is thus the vector generated by the x-y motion of the yarn holder in the bed of the loom. The 3×1 braid pattern motion is similarly obtained. In the $1 \times 1 \times 1/2$F, 50% of the carriers do not participate in this carrier motion, remaining fixed in the axial direction.

Resin Impregnation of Preforms

The skeletal yarn preform that is braided to net shape must be impregnated with a suitable resin and rigidized. Three impregnation procedures were inves-

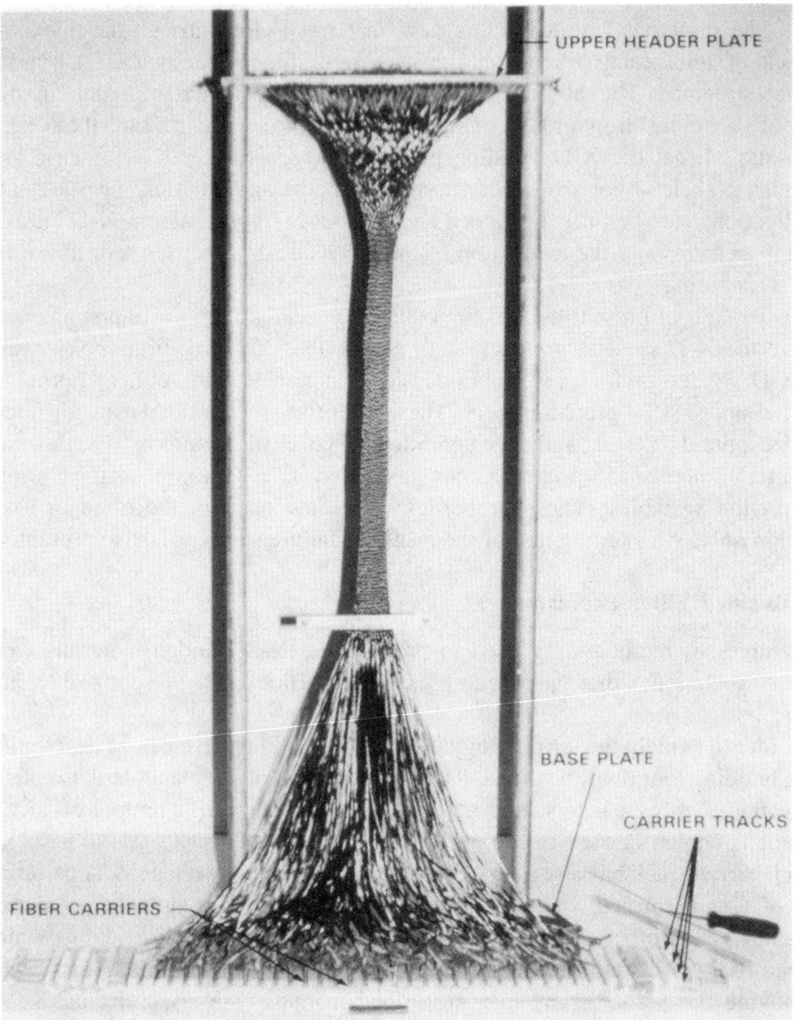

FIG. 1—*Semi-automatic loom for X-D braiding of skeletal yarn preforms.*

tigated. These included vacuum impregnation, pressure impregnation, and resin film lamination.

The vacuum and pressure impregnation techniques are very similar. In both cases the material to be impregnated is placed in a closed mold having a predetermined internal volume that controls the fiber and resin volume fractions of the finished part. The part thickness can be accurately obtained by using spacers such as 181 style glass cloth, and TX-1040 peel ply material. An example of this setup is shown in Fig. 3. The mold is designed to have both resin fill and bleed lines that indicate complete impregnation of the skeletal fiber preform. The

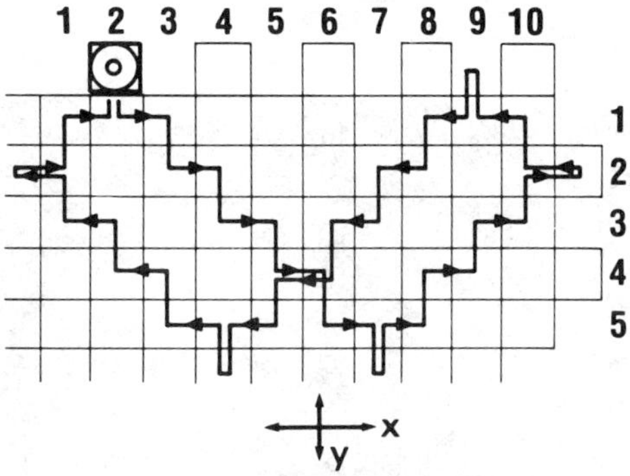

FIG. 2—*Schematic of the* x-y *motion of one yarn carrier in the Cartesian bed of a loom; carrier is following a 1 × 1 braid pattern.*

mold with the preform and spacers in place is then heated to 121°C (250°F). The resin is also heated in a container to reduce its viscosity. In the vacuum method, the mold is placed into an autoclave or vacuum oven that is evacuated to 762 mm (30 in.) mercury (Hg) and is preheated to 121°C (250°F). The container of fluid resin outside of the autoclave is connected to the input side of the mold with 6.4-mm (1/4-in.) copper tubing. A valve is located in the system to control resin flow. After all temperatures have stabilized, the valve is opened to allow resin to flow into the mold. The resin is drawn into the mold until there is a steady flow from the exit ports. The vacuum is then released and the recommended time/temperature cure cycle followed. With the pressure impregnation process, the same procedure is followed with the exception of the transfer of resin. The resin is pumped into an evacuated mold using 690 kPa (100 psi) air pressure. The resin is continuously pumped until a steady flow occurs at the exit ports. Finally, vacuum and pressure are released and the recommended time/temperature cycle is then followed.

It was found that although both methods were promising, quality in terms of

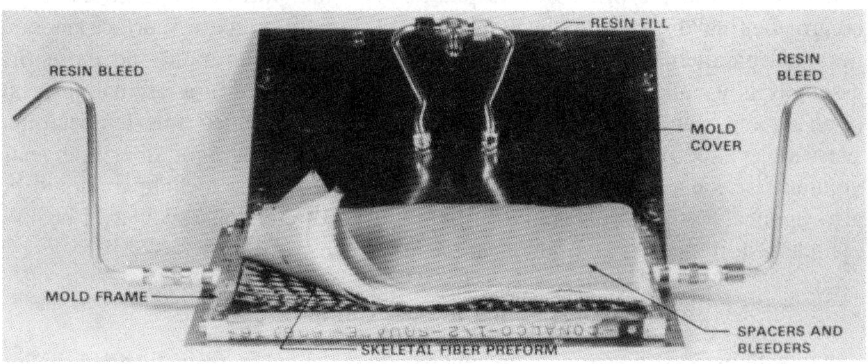

FIG. 3—*Vacuum and pressure resin impregnation techniques for X-D braided preforms.*

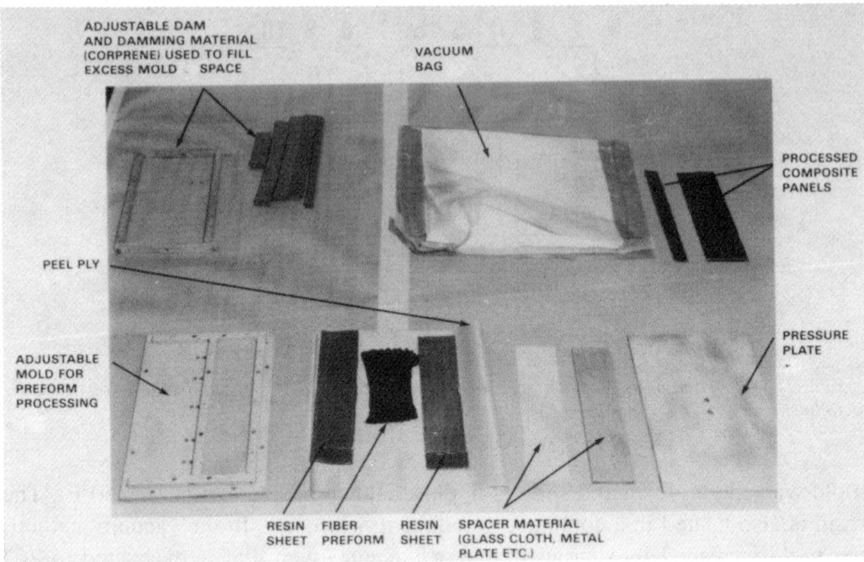

FIG. 4—*Impregnation technique for X-D braided preforms using a resin precast sheet and closed mold.*

controlling voids and porosity was not reproducible. With additional investigation a new method also using a closed mold was developed [*4*]. This technique illustrated in Fig. 4 uses 3.2-mm (1/8-in.) thick precast resin sheets to impregnate the preform. The dry preform is essentially sandwiched in the mold between two plies of the precast resin sheets. The mold is then placed in an autoclave that is heated to 121°C (250°F) and evacuated to 762 mm (30 in.) Hg. Air is removed from the preform and the melting resin allowing more positive wicking action of the resin into the preform. After the preform is saturated and the system is free of any entrapped air, the vacuum is released and the mold is removed from the autoclave. At this point in the process, the preform is in a B-staged condition. The next step in the process is to place spacer material in the mold cavity to control the final thickness of the preform panel. Finally, to control surface smoothness and parallelness, a pressure plate is placed over the mold and the entire assembly is vacuum bagged. The assembly is then placed in a laminating press or an autoclave and the preform cured using the recommended time/temperature/pressure cycle. This process produced panels of a consistent fiber and resin volume fraction and a void content of less than 1%. The technique should be also applicable to more complex fiber-reinforced skeletons such as I-beams, cylinders, foil shapes, and other structural sections.

Material used in the Investigation of X-D Composite Panels

All X-D braided preforms were impregnated using the resin film lamination technique previously described due to the quality of the impregnated preform. The resin system used was a Narmco Corp. 5208, 177°C epoxy resin. The X-D

TABLE 1—*Mechanical properties of the Celion, AS-4, and Thornel T-300 graphite fibers.*

Fiber Property	Fiber Type		
	Celion (6K and 12K)	AS-4 (3K, 6K, and 12K)	T-300, 30K
Tensile strength, GPa (ksi)	3.24 (470)	3.59 (520)	2.69 (390)
Tensile modulus, GPa (Msi)	234.4 (34.0)	234.4 (34.0)	220.6 (32.0)
Density, g/cm^3	1.77	1.80	1.77
% Elongation	1.4	1.53	1.2

braid characterization was conducted using three graphite fiber systems: Celion fiber produced by the Celanese Corp.; Thornel T-300 produced by the Union Carbide Co.; and AS-4 fibers produced by the Hercules Corp. The fiber properties are given in Table 1.

The Thornel T-300 X-D preforms were purchased from the Atlantic Research Corp., Alexandria, VA. The AS-4 and Celion preforms were obtained from the Philadelphia College of Textiles and Science, Philadelphia, PA. Table 2 summarizes pertinent information on these materials.

Mechanical Characterization

All X-D composite materials were characterized using ASTM tests for tensile, compression, flexural, and short-beam shear properties. A minimum of three specimens for each braid configuration was prepared and tested according to the following ASTM methods.

ASTM Test Method for Tensile Properties of Fiber-Resin Composites (D 3039-76)

ASTM Test Method for Compressive Properties of Unidirectional or Crossply Fiber-Resin Composites (D 3410-75)

ASTM Test Methods for Flexural Properties of Unreinforced and Reinforced Plastics and Electrical Insulating Materials (D 790-71)

ASTM Test Method for Apparent Interlaminar Shear Strength of Parallel Fiber Composites by Short Beam Method (D 2344-76)

Initial testing for compressive properties using the width given in ASTM D 3410-75 resulted in variations in strength of about 50%. Specimen width was varied from 6.4 mm (0.25 in.) to 25.4 mm (1.00 in.) in order to determine the optimum dimension. It then became evident that widths of 19 mm (0.75 in.) and 25.4 mm (1.00 in.) resulted in consistent strengths. These are the widths that were used in subsequent testing.

Results and Evaluation

Effect of Edge Condition and Braid Pattern

This phase of the evaluation consisted of characterizing X-D composites with a 1 × 1, 3 × 1, and 1 × 1 × 1/2F braid pattern made with T-300, 30K tow

TABLE 2—*X-D braided fiber preforms used in material evaluation.*

Graphite Fiber	Fiber Manufacturer	Tow[a] Size	X-D Braid[b] Pattern	Preform Size, mm (thickness × width × length)
Celion	Celanese Corp.	6K	1 × 1	3.2 × 25.4 × 254
		12K	1 × 1	3.2 × 25.4 × 254
Thornel T-300	Union Carbide Corp.	30K	1 × 1	3.2 × 152.4 × 254
				3.2 × 25.4 × 254
			1 × 1 × 1/2F[c]	3.2 × 63.5 × 254
				3.2 × 25.4 × 254
			3 × 1	3.2 × 76.2 × 254
				3.2 × 25.4 × 254
AS-4	Hercules Corp.	3K	1 × 1	3.2 × 25.4 × 254
		6K	1 × 1	3.2 × 25.4 × 254
		12K	1 × 1	3.2 × 25.4 × 254

[a] Tow size designation of 3K, 6K, 12K, and 30K indicates a filament count of 3000, 6000, 12 000, and 30 000 in a yarn bundle, respectively.
[b] All preforms were impregnated with Narmco 5208 epoxy resin.
[c] One half of the fiber in the 1 × 1 braid pattern were fixed in the axial direction.

fibers, impregnated with the Narmco 5208 epoxy resin. Specimens with both cut and uncut edges were tested since both edge configurations are possible for structural applications. Specimens with uncut edges were braided to the test width of 25.4 mm (1.0 in.).

The results of the braid pattern and edge condition characterization are given in Table 3. The most striking result is the variation in properties between specimens with cut and uncut edges. There is approximately a 50% reduction in tensile properties, and approximately a 40% reduction in flexural properties in specimens with cut edges compared to the specimens with uncut edges. The material exhibiting the best performance in the cut edge condition was the 1 × 1 × 1/2F braid configuration. The fixed fibers, running parallel to the edges of the specimen were thereby not cut when the specimens were cut. These specimens were the only ones in the cut edge condition having continuous fibers. As seen in Table 3, the 1 × 1 × 1/2F construction exhibited the highest strength and stiffness in the cut condition when tested in tension, compression, and flexure compared to the 1 × 1 and 3 × 1 braid pattern. Although short-beam shear tests were not conducted for any condition, the trend in performance for the 1 × 1 × 1/2F material should be maintained.

The most dominant factor affecting the performance of specimens with no fixed fibers, that is, the 1 × 1 and 3 × 1 constructions, is unquestionably the apparent fiber angle as measured from the loading axis. This becomes evident especially for those specimens having cut edges. Table 3 shows the 1 × 1 cut specimen to have an apparent fiber angle of ±20°, whereas the 3 × 1 specimens have an angle of ±12°. The difference in the angles would lead one to expect greater strength and stiffness in the 3 × 1 construction where the apparent fiber angle is the least off-axis. The performance of the three-braid configurations in the cut edge condition follows this premise and can be listed in order of decreasing performance as 1 × 1 × 1/2F, 3 × 1, and 1 × 1.

For the three-braid configurations in the uncut edge condition, the strength and stiffness in both tension and flexure is significantly higher for all braid patterns compared to the same properties obtained for cut specimens. The 3 × 1 and 1 × 1 × 1/2F have properties superior to the 1 × 1 braid pattern. This result is expected since the 1 × 1 construction has an apparent fiber angle that deviates the most from the axis of loading, ±20° as shown in Table 3. Considering the apparent fiber angle of ±15°, the 1 × 1 × 1/2F should have material properties that are inferior to the 3 × 1 braid that has an apparent fiber angle of ±12°. However, taking into account that one half of the fibers in the 1 × 1 × 1/2F are oriented in the load direction, it would be expected that the 1 × 1× 1/2F would have properties superior to the 3 × 1. The actual results are varied. The tensile strength and stiffness of the 3 × 1 are superior to the 1 × 1× 1/2F, whereas, the opposite is true for the flexural strength and stiffness. Actually, the flexural properties of all three-braid constructions are remarkably similar.

The physical appearance of the failure surfaces of the tension specimens for all braid patterns were very similar regardless of edge condition. A close-up view

TABLE 3—*X-D braided (Thornel T-300[a]/5208 epoxy) composite property data: 1 × 1, 3 × 1, and 1 × 1 × 1/2F braid patterns with uncut and cut edges.*

	Fiber Type and Braid Pattern					
Property[b]	T-300, 1 × 1 (uncut)	T-300, 1 × 1 (cut)	T-300, 3 × 1 (uncut)	T-300, 3 × 1 (cut)	T-300, 1 × 1 × 1/2F (uncut)	T-300, 1 × 1 × 1/2F (cut)
V_f, %	68	68	68	68	68	68
σ_{xt} MPa (ksi)	665.6 (96.5)	228.7 (33.2)	970.5 (140.8)	363.7 (52.7)	790.6 (114.7)	405.7 (68.9)
E_{xt} GPa (Msi)	97.8 (14.2)	50.5 (7.3)	126.4 (18.3)	76.4 (11.1)	117.4 (17.0)	82.4 (12.0)
σ_{xc} MPa (ksi)	...	179.5 (26.0)	...	226.4 (32.8)	...	385.4 (55.9)
E_{xc} GPa (Msi)	...	38.7 (5.6)	...	56.6 (8.2)	...	80.8 (11.7)
σ_{xf} MPa (ksi)	813.5 (118.0)	465.2 (67.5)	647.2 (93.9)	508.1 (73.3)	816.0 (118.3)	632.7 (91.8)
E_{xf} GPa (Msi)	77.5 (11.2)	34.1 (4.9)	85.4 (12.4)	54.9 (8.0)	86.4 (12.5)	60.8 (8.8)
ν_{xy}	0.875	1.36	0.566	0.806	0.986	0.667
Apparent fiber angle	±20°	±20°	±12°	±12°	±15°	±12°

[a]T-300 graphite yarn—30 000 tow.
[b]Tension and compression specimens were tabbed at grip ends.

of a fractured tension specimen with uncut edges is shown in Fig. 5. The failure appears to be a result of fiber breakage with some evidence of fiber pull-out. The failure surface was basically normal to the loading axis with a minimum of splintering. A compression specimen failure is shown in Fig. 6. It should be noted that these specimens had cut edges, and typically, failure occurred by shear-out of the braided yarn bundles and subsequent fiber buckling. Flexural specimens with uncut edges exhibited failure that manifested itself as resin cracks parallel to the surface tows on the tensile face. Figure 7 illustrates this characteristic flexural specimen failure.

It has been shown that differences in properties exist between X-D braided composite specimens with cut and uncut edges. Accordingly, care should be exercised in the interpretation and application of data that is reported. It could be postulated that the edge effect may be less pronounced in larger width panels. Thus, uncut specimen data may be more representative of the real mechanical capabilities of a large areal panel than cut edge coupon data.

FIG. 5—*Fractured tensile specimen, T-300/5208, 30K tow.*

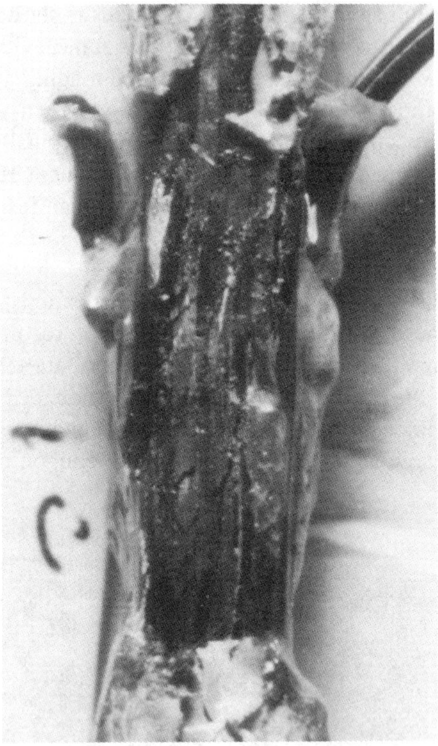

FIG. 6—*Fractured compression specimen, T-300/5208, 30K tow (view of cut edge showing tow debonds).*

Effect of Tow Size

Another variable that was also investigated, which plays a role in influencing material properties, is tow size of the yarn reinforcement. Three tow sizes were thus studied, namely, 3K, 6K, and 12K for the Hercules AS-4 graphite fiber and 6K and 12K for the Celanese Celion graphite fiber. All materials were braided to a 25-mm (1-in.) width in order that tests be conducted on specimens with uncut edges. As previously noted, the authors consider specimens with uncut edges as being more representative of the X-D braided properties of a large structure. Accordingly, tensile, flexural, and shear properties were developed for these three tow sizes and are given in Table 4. A discussion of the results follows in the succeeding sections.

Tensile Properties

The tensile data from Table 4 are presented graphically in Fig. 8. From the bar graph, it appears that for the three tow sizes investigated specimens with the 12K tow reinforcement have the highest ultimate tensile strengths. Tensile strength

FIG. 7—*Fractured flexural specimen, T-300/5208, 30K tow (surface view showing tow debonds).*

seems to increase with tow size for both the AS-4 and Celion fibers. Although not part of the initial tow study, tensile strength in Table 4 for the Thornel 300 material having a 30K tow appears to be comparatively lower than for the other tow sizes. Therefore, considering the differences in the various fibers used, with Thornel 300 having the lowest fiber strength as seen in Table 1, the initial implication that tensile strength was continuously increasing with tow size needs to be modified. It is thus proposed that the performance of the 12K tow specimens may suggest a certain optimum tow size. With the exception of the AS-4 (3K), Celion (6K), and Thornel 300 (30K) specimens, it is shown in Table 4 that the tensile moduli were consistently on the high side in the range of 110 GPa to 117 GPa (16 to 17 $\times$ 10^6 psi). The seemingly anomalous values of moduli for the 3K and 6K specimens may be attributable to the variation in apparent fiber angles. It is appreciated that these differences could have an adverse effect on tensile strength but not to the extent observed. For laminated composites the Tsai-Wu Tensor Failure Theory [*5*] only predicts a 23% reduction in strength in an AS-4 laminate when fiber orientation is changed from $\pm15°$ to $\pm20°$. The reduction shown in Table 4 for the 3K AS-4 material is 44% for tensile strength. For the X-D material, using the Tsai-Wu Theory as a first approximation, this large

TABLE 4—*X-D braided graphite/epoxy composite properties as a function of braid pattern [uncut specimens, 25.4 mm (1 in.) wide] including comparative data for a laminated fabric composite.*

	Fiber Type and Braid Pattern						
Property	AS-4, 3K 1 × 1	AS-4, 6K 1 × 1	Celion, 6K 1 × 1	AS-4, 12K 1 × 1	Celion, 12K 1 × 1	Thornel 300, 30K 1 × 1	Thornel 300, 8 Harness Satin Fabric
V_f, %	68	68	56	68	68	68	65
σ_{xt}, MPa (psi)	736.8 (106 870)	841.4 (122 040)	857.7 (124 400)	1067.2 (154 790)	1219.8 (176 910)	665.6 (96 530)	517.1 (75 000)
E_{xt}, GPa (Msi)	83.5 (12.1)	119.3 (17.3)	87.8 (12.7)	114.7 (16.6)	113.1 (16.4)	97.8 (14.2)	73.8 (10.7)
τ_{SBS}, MPA (psi)	114.8 (16 650)	126.0 (18 270)	71.4 (10 350)[b]	121.4 (17 600)	71.4 (10 350)[b]	...	69.0 (10 000)
ν_{xy}	0.945	1.051	0.968	0.980	0.874	0.875	0.045
σ_{xf}, MPa (psi)	885.3 (128 400)	739.8 (107 300)	...	1063.3 (154 210)	...	813.5 (117 990)	689.5 (100 000)
E_{xf}, GPa (Msi)	84.5 (12.3)	95.2 (13.8)	...	1385.2 (20.1)	...	77.5 (11.2)	65.5 (9.5)
Apparent fiber angle	±19°	±15°	±15°	±13°	±17.5°	±20°	0°

[a]Tensile specimens were tabbed with 1.6 mm (1/16 in.) thick 25.4 mm (1 in.) by 63.5 mm (2-1/2 in.) glass reinforced plastic tapered tabs at grip ends.
[b]Celion 6K and 12K specimens had cut edges for the short-beam shear tests only.

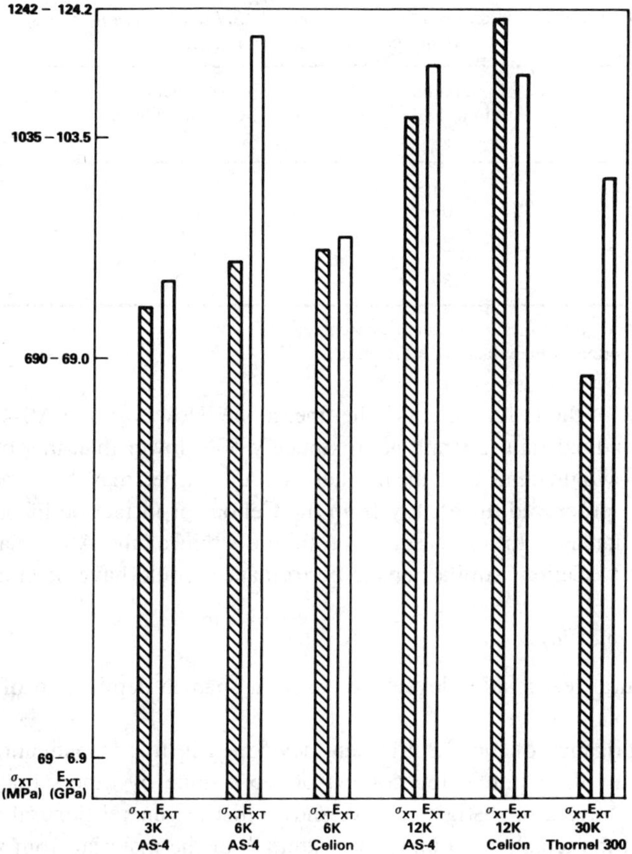

FIG. 8—*Graphical representation of tensile properties of AS-4, Celion, and Thornel 300 1 × 1 braided composites of various tow sizes.*

reduction in properties cannot solely be attributed to the greater apparent fiber angle. It is surmised that other variables not clearly identified play a significant role in this reduction.

Another phenomenon that may play a significant role in the performance of X-D braided composites is the axial effect created by the ''crowding'' of the braid yarns at the specimen edges. Although this effect appears to improve overall composite properties, the extent of the crowding effect may depend on tow size and may be more pronounced as the tow size increases. It is postulated that such is the case for the 12K tow specimens, where this edge effect may extend further into the specimen width than is the case for the 3K tow. As a result, the 12K material has superior properties.

Table 5 presents experimental values of tensile strain to failure in both the longitudinal and transverse directions for both AS-4 and Celion specimens. For the Celion specimens, it is seen that the longitudinal strain to failure is approx-

TABLE 5—*Strain to failure of the 1 × 1 braid AS-4 and Celion tension specimens (data obtained from strain gages).*

	Longitudinal Strain at Failure, %	Transverse Strain at Failure, %	Fiber Only[a] Strain, %
Celion 6K	1.04	1.08	1.4
Celion 12K	1.08	1.03	1.4
AS-4 3K	0.87	0.68	1.53
AS-4 6K	0.76	0.70	1.53
AS-4 12K	0.89	1.08	1.53

[a]Celion data reported in manufacturer's brochure;
AS-4 Data reported in manufacturer's brochure.

imately 75% of the failure strain of the fiber alone. However, the AS-4 specimen had a longitudinal failure strain approximately 50% lower than that of the fiber alone. This would tend to indicate that the AS-4 fiber may have been either mishandled, processed differently from the Celion, or in fact be inherently susceptible to damage. In any case, if the failure strain of the AS-4 material was comparable to Celion, similar tensile performance would have been expected.

Flexural Properties

The flexural test results developed, using a span to depth ratio of 16:1, are given in Table 4.

The significance of the flexural data developed in this investigation may be more meaningful if put in the context of work done by others for laminated composites. Several investigators have shown that in general flexural strength is greater than the tensile or compressive strength of the material from which it is made [*6–8*]. This behavior is explained to be a consequence of the nonuniform stress distribution through the thickness of the flexural specimen. In the flexural specimen, the maximum stress occurs only in the outermost fibers and thus the probability of a flaw to be located within this region of higher stress and thus contributing to failure is lower than it would be for the entire specimen. On the other hand, in tensile or compressive specimens, the entire cross-section of the specimen is assumed to experience an average stress. Thus, the probability of a flaw existing in this net section and contributing to failure and reduced strength is greater than for the flexural specimens.

In general, the data given in Table 4 tends to support the findings on flexural strength discussed by Bullock [*6*], Whitney [*7*], and Zweben [*8*]. The 3K specimen showed a flexural strength value that was 20% greater than its tensile strength. However, the flexural strength of the 6K specimen was 12% lower than its tensile strength. The 12K specimen had a flexural strength equal to its tensile strength.

Zweben [*8*] gives an extensive review of the flexural test method used in this investigation and cautions that the shorter span specimens recommended by

ASTM D 790-71 may actually give reduced flexural modulus values. He indicates that a laminate whose plies are uniformly spaced and oriented and whose tensile and compressive moduli are equal should have a flexural modulus whose value would equal the tensile modulus. Deviations from this may result from several sources. The two that are noted here are resin-rich surfaces and improper span-to-depth ratios of test specimens. Zweben recommends a ratio of 60 : 1 for measuring flexural modulus as this significantly minimizes the deflection contribution due to shear effects.

In this investigation, since a span-to-depth of 16 : 1 was used, one would expect the flexural modulus to be lower than the tensile modulus. Furthermore, additional reductions in modulus could also be expected as a result of resin-rich specimen surfaces. On examination, no such resin-rich surfaces existed. The resulting data from this investigation was found, however, to be inconsistent. Table 4 shows that for the 3K material, the value of the flexural modulus was equal to the value of the tensile modulus. The 6K flexural modulus was 20% less than the tensile modulus, whereas the 12K flexural modulus was 20% greater than the tensile modulus. Zweben [*8*], using a span-to-depth ratio of 16 : 1, reports a flexural modulus for a unidirectional Kevlar 49/polyester composite that is 35% lower than a modulus obtained with a span-to-depth of 60 : 1. This information has a bearing on the data of Table 4 as it suggests a higher value of flexural modulus for the AS-4 materials if a 60 : 1 test specimen was used. Zweben shows that as the span-to-depth ratio increases, the flexural modulus asymptotically approaches the tensile modulus. Furthermore, to resolve inconsistencies such as those shown in Table 4, Zweben recommends that a specimen with a 16 : 1 ratio be used for strength determinations and a specimen with a 60 : 1 ratio be used for flexural stiffness determinations.

The mode of failure was examined for each of the flexural specimens tested. Specimens of each tow size were sectioned, mounted, and polished. Photomicrographs of the failed sections are given in Fig. 9. In each specimen, the damage occurred on the compressive side of the specimen in the form of fiber debonds. Due to the nature of the braid pattern, the debonds were not planar and followed a pseudo-continuous path, parallel to the surface of the specimens. The number and lengths of the debonds varied from one braid pattern to another.

The 3K flexural specimens showed the greatest number of yarn debonds as shown in Fig. 9*a*. Debonds appear to originate at the loading face and continue to propagate, coalescing one plane below the surface normal to the loading nose in both directions. The debonds continue to progress into the thickness of the material, propagating outward from the loading area.

The 6K specimen showed similar damage behavior as illustrated in Fig. 9*b*. Here again, fiber debonds progress from the compressive surface for a distance of one tow thickness below. At this point, they seem to coalesce to form a pseudo-planar delamination. There is an additional debond that occurs one tow thickness below the previous debond. This additional debond seems to initiate from the previous debond at a position where a second tow transitions into the

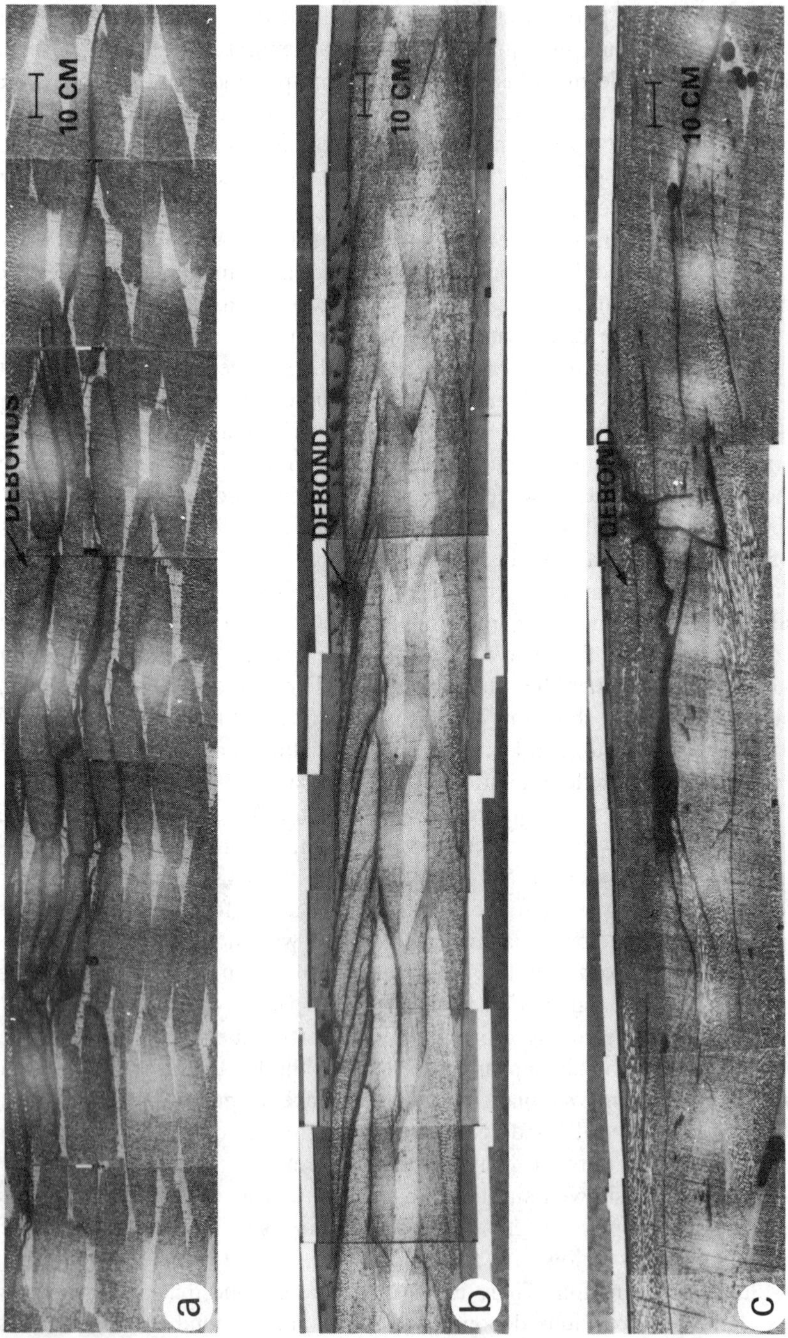

FIG. 9—*Photomicrographs of flexural specimen failure surfaces: 1 × 1 braid pattern, AS-4 graphite fiber; (a) 3K tow*, (b) 6K tow, and (c) *12K tow.*

thickness above. The severity of damage in this specimen is less than that contained in the 3K specimen.

The 12K specimen had the most damage as observed in Fig. 9*c*. Some debonds occurred but appeared to be contained and not allowed to progress to the extent that they did in the 3K and 6K specimens. In this instance, evidence of tow fracture is noticeable. These cracks progress normal to the surface to a depth approximately three tow thicknesses into the specimen.

From this cursory examination, it appears that the larger tow sizes are more capable of inhibiting debond growth than are the smaller tows. This may be the result of two phenomena. First, with the smaller tow, there appears to be more of a discontinuity between the tows caused by larger pockets of resin at the crossover region of the fibers. Second, the angles that the fibers appear to have as they progress through the material may exacerbate the debond propagation. The smaller tows appear to contain larger angular differences than the larger tows in the plane of the material. Also, the braid length may affect the angular transition of fibers in their travel through the specimen. These aspects have not been addressed adequately to obtain a comprehensive explanation of the damage progression.

Short-Beam Shear Properties

The short-beam shear properties that were developed in this investigation are gage marks indicative of the ability of X-D braided materials to resist in-plane shear loads and suppress damage growth. Conventional laminated composites have an inherent weakness that manifests itself in poor resistance to interlaminar shear. For instance this weakness usually results in extensive damage to the composite especially if subjected to low velocity impact. With the X-D braided composite, delamination as such does not exist due to the multidimensional, interconnecting network of fibers. What has been shown in the previous discussion on bending is a shear mode of failure where yarn tows experience debonding.

ASTM Method D 2344-76 was used to conduct the short-beam shear tests. The specimens had a span-to-depth ratio of 4 : 1 although the requirement for specimen width was modified to 25.4 mm (1 in.), that is, the uncut width of the braided composite, rather than the accepted width of 6.4 mm (1/4 in.). The results given in Table 4 show shear values for all AS-4 fiber tow sizes in the range from 110.2 to 124.1 MPa (16 to 18 ksi). By contrast, a typical laminated composite using graphite fabric has a short-beam shear of 69 MPa (10 ksi). Thus, it is seen that the AS-4 material represents an approximate improvement of 60 to 80% over the conventional fabric laminate, and in fact these improvements are equal to or exceed unidirectional laminate shear values.

These results are even more promising in light of the 25.4-mm (1-in.) specimen width used. Sattar and Kellogg [9] showed that width plays a factor in the stress distribution of the short-beam specimen, where shear strength varies across the width, being maximum at the edges and a minimum at the center. They also showed that the short-beam shear strength, as determined from beam theory, was a conservative estimate which becomes increasingly more conservative up to a width-to-depth ratio of 4 : 1. For the 25.4-mm (1-in.) wide specimens used in this investigation, the width-to-depth ratio was approximately 10 : 1. This choice was due to the fact that an uncut, full-width specimen was desired to avoid damage resulting from cut edges. The trend from Sattar and Kellogg's work would indicate that the short-beam shear strength determined in this investigation from beam theory is a conservative estimate.

In order to gain an insight into the fracture sustained by the X-D braided materials, tested short-beam shear specimens were suitably prepared and examined. Failure surfaces for the AS-4 braided 3K, 6K, and 12K tow specimens are shown in Fig. 10. It was found that all specimens sustained some form of damage on the tensile face. Specifically, the 3K tow specimens exhibited fiber fracture with cracks progressing along and into (normal) the tensile face of a specimen. Yarn debonds were found to exist, but growth appeared limited and did not extend deeper than a few tows into the thickness. Yarn debonds only occurred at tow/resin interfaces. The 6K specimens that carried a higher shear stress, had very few cracks along and into (normal) the surface on the tensile side. However, tow debonds within the body of the specimens appeared to be more numerous than for the 3K specimens. These debonds were not limited to the tow/matrix bond but also occurred within the individual tow. The damage region was larger and more severe. It would thus appear that the 6K material has a greater capacity for damage tolerance since it was able to develop a much higher failure load than the 3K material (a 26% improvement). This would tend to indicate that as the loading progressed, the damage that occurs is self-limiting and the specimen can continue to carry increasing loads up to a critical threshold value. The 12K tow specimens had a shear value somewhat lower than the 6K tow specimens and exhibited a damage condition very similar to the 3K specimens. Table 4 also gives values for the Celion 6K and 12K tow specimens that had cut edges. The lower shear values for these specimens compared to the AS-4 specimens with uncut edges is also illustrative of the edge effect discussed previously.

Summary

The multidimensionally braided preforms generated for this investigation have been impregnated with a typical structural epoxy resin and cured successfully. Processing problems commonly encountered by other investigators relating to impregnation and air entrapment have been resolved as documented by this paper.

The through-the-thickness or multidimensionally braided composite has performance characteristics that are superior to fabric laminated composites in the braid direction only. This is clearly evidenced by improved axial strength, axial stiffness, and short-beam shear properties as observed from Table 4.

The performance of the braided composite is a function of both tow size and

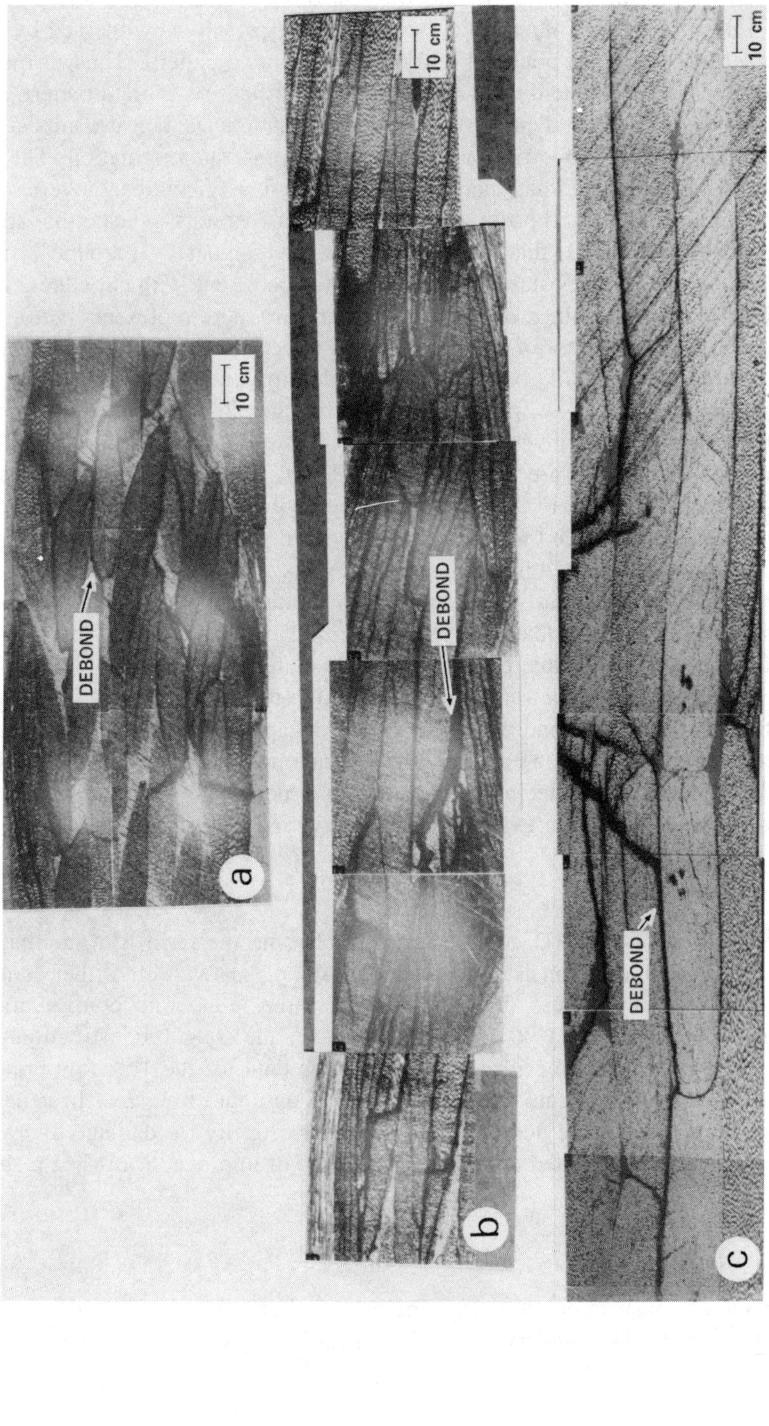

FIG. 10—*Photomicrographs of short-beam shear specimen failure surfaces: 1 × 1 braid pattern. AS-4 graphite fiber;* (a) *3K tow,* (b) *6K tow, and* (c) *12K tow.*

braid pattern. The investigation suggests that an optimum tow size may exist as evidenced by the superior tensile and flexural properties of the 12K tow composite. Likewise, an optimum braid pattern may be present as witnessed by the highest tensile properties obtained with the 3 × 1 braid pattern. Thus, it might not be unreasonable to meld the 12K tow size with the 3 × 1 braid pattern and expect tensile and flexural properties superior to laminated fabric composites, possibly approaching properties of unidirectional composite orientation. The investigation also showed that there is a definite edge effect that governs the behavior of all braided composite coupon specimens. Results indicate that specimens with fully braided, that is, uncut edges, have higher "all around" axial values including shear resistance as compared to specimens with cut edges. It is postulated that data obtained on uncut specimens are more representative of the areal strength and stiffness of full-size braided structural elements than are data from cut edge speicmens. Furthermore, although impact tests were not conducted on any specimens, short-beam shear data reported herein appears to support the impact findings of other investigators, such as Brown [*10*] and Gause [*11*], that X-D braided composites are more damage tolerant than laminated composites. This is further evidenced by the static test results of this study that indicate that failure surfaces are reduced in size since sequential ply failure no longer exists.

The information reported in this paper represents an effort to develop material property data that is currently very limited for multidimensionally braided composites. It should be noted that there remain significant technical areas that require attention. These areas of interest pertain to: (1) material properties including far more axial data as well as transverse and through-the-thickness behavior under both static and dynamic conditions; (2) analytical capability that could relate part geometry, braid orientation, and braider operation; (3) automated fabrication capability to realize full (technical and low cost) potential of the X-D braiding process; (4) nondestructive evaluation capability; and (5) repair capability.

Conclusions

Impregnation of the X-D composite material using the resin film lamination technique is capable of consistently producing flat panels with high fiber content and a void content of less than 1%. Of the three X-D braid configurations examined (1 × 1, 1 × 1 × 1/2F, and 3 × 1), the 3 × 1 construction has demonstrated the best over-all performance. The data for the 12K tow braided composite seems to indicate the existence of an optimum tow size. In general, the X-D composites have demonstrated a greater capacity for damage tolerance than conventional laminated composites by virtue of improved short-beam shear properties.

Acknowledgments

The authors wish to acknowledge the technical encouragement and financial support of Dr. H. H. Vanderveldt of the Naval Sea Systems Command.

References

[*1*] Bluck, R. M., "High Speed Bias Weaving and Braiding," U. S. Patent No. 3,426,804, 11 Feb. 1969.

[*2*] Florentine, R. A., "Apparatus for Weaving a Three-Dimensional Article," U. S. Patent No. 4,312,216, 26 Jan. 1982.

[*3*] Ko, F., "Tensile Evaluation of the Mechanical Properties of Magnaweave Composites," presented at the Annual Meeting of the American Helicopter Society, Philadelphia, 1983.

[*4*] Crane, R. M. and Macander, A. B., "Invention Disclosure for Resin Impregnation and Processing Technique for Rigidizing Net Shaped Fibrous Skeletal Composite Preforms," Navy Case No. 67962.

[*5*] Jones, R. M., *Mechanics of Composite Materials,* McGraw-Hill, Washington, DC, 1975, pp. 80–83.

[*6*] Bullock, R. E., *Journal of Composite Materials,* Vol. 8, 1974, pp. 200–206.

[*7*] Whitney, J. M., et al in *Composite Materials: Testing and Design (Third Conference), ASTM STP 546,* American Society for Testing and Materials, Philadelphia, 1974, pp. 30–46.

[*8*] Zweben, C., et al in *Composite Materials: Testing and Design (Fifth Conference), ASTM STP 674,* S. Tsai, Ed., American Society for Testing and Materials, Philadelphia, 1979, pp. 228–262.

[*9*] Sattar, S. A. and Kellogg, D. H. in *Composite Materials: Testing and Design, ASTM STP 460,* American Society for Testing and Materials, Philadelphia, 1969, pp. 62–71.

[*10*] Brown, R. T. "Application of Through-the-Thickness Braiding Process" to be presented at the Fiber Society/SAMPE Technical Program Philadelphia College of Textiles and Science, June 1984.

[*11*] Gause, L. W., "Mechanical Characterization of Magnaweave Braided Composites," *Mechanics of Composites Review,* Air Force Materials Laboratory, Oct. 1983.

Summary

The papers in this volume have been grouped into five major sections, each corresponding to a session title in the Seventh Conference on Composite Materials: Testing and Design. For convenience, the summary will be presented by major heading.

Structures

Toughness is a key issue in structural applications of composite materials. This is reflected in the four papers presented in this section. The effect of manufacturing and service-induced damage on the static and fatigue strength of aircraft composite structures is discussed by Garrett. Analysis of composite bolted repairs subjected to biaxial or shear loads is presented by Bohlmann, Renieri, and Riley. Process-induced defects, stress concentrations, and impact behavior are key issues in the characterization of T300/V-378A graphite/bismaleimide composites as performed by Ramkumar, Grimes, and Kong. Fracture toughness of thick-wall graphite/epoxy filament-wound composites is considered by Riddle and Beckwith. Both Mode I and Mode II energy release rates are determined for through-the-thickness cracks and delaminations.

Failure Mechanisms

Emphasis in this section is on matrix dominated failure modes. Mandell, Grande, Tsiang, and McGarry propose a single-fiber interfacial shear strength test in which the fiber is loaded in compression by a probe. Nicholls discusses a unique approach to determining the biaxial stress-strain behavior of neat resin specimens. Hahn and Williams identify transverse splitting, in addition to the usual modes of fiber failure and shear crippling, as a major failure mode for unidirectional composites subjected to compression loading. A microstructural analysis of laminates failed under uniaxial compression is performed by Kar, Herfert, and Kessler. Their investigation reveals that specimens tested at elevated temperature and high moisture content ($>1\%$) display different characteristics than specimens tested at room temperature with a lower moisture content.

Strength

Damage tolerance issues of stress concentrations and impact behavior are presented in this section. Lagace shows a laminate stacking sequence effect on the strength of graphite/epoxy composites containing circular holes. The stacking sequence effect is attributed to interlaminar stresses and ply constraints. The effect of laminate thickness on apparent fracture toughness is the subject of the paper by Harris and Morris. They consider through-the-thickness cracks in conjunction with three different test methods. Fracture toughness is shown to be a function of laminate thickness. Avva, Vala, and Jeyaseelan report on the combined effect of low-velocity impact and tension-tension fatigue loading on the residual strength of graphite/epoxy laminates. A residual strength model based on classical fracture mechanics is also presented.

Delamination

Delamination is the primary failure mode of interest in damage tolerance characterization of advanced composites. Low-velocity impact results in delamination. Delamination is also a manufacturing defect of key concern. Thus, this section is very timely. The seven papers included in this section represent a good cross section of current work in the area of delamination. Chai uses a double-cantilever-beam test to correlate neat resin fracture toughness with composite fracture toughness. His results indicate that a thin bondline resin specimen is required to simulate composite behavior. A symmetric double cracked-lap-shear specimen design for characterizing mixed-mode delamination is presented by Armanios, Rehfield, and Reddy. Less data scatter is obtained from the double crack-lap specimen as compared to the conventional single crack-lap specimen. Softening layers utilized as a mechanism for improving delamination resistance is the subject of the paper by Browning and Schwartz and the paper by Chan, Rogers, and Aker. Both of these papers demonstrate improved delamination resistance with the use of adhesive inner layers. In the paper by Browning and Schwartz a double-cantilever-beam specimen is utilized to measure Mode I behavior, while the edge delamination specimen, supplemented by a finite-element analysis, is utilized by Chan, Rogers, and Aker to characterize mixed-mode response. A unique design of the edge delamination specimen is presented by Soni and Kim. This specimen yields Mode II response with compression interlaminar normal stress for external tension loading, while resulting in mixed-mode response for external compression loading. The global-local laminate model is used to determine the interlaminar stresses. The initiation and propagation of delaminations in step-lap joints constructed of composites is presented by Ratwani and Kan. Joints with and without bondline flaws are considered. A contiuum analysis in conjunction with NASTRAN is utilized for determining interlaminar stresses. Mall and Johnson characterize the fracture behavior of adhesive bonded joints constructed of graphite/epoxy adherends. A double-cantilever-beam specimen is utilized to measure Mode I response and a cracked-lap-shear specimen for mixed-mode behavior. They find total strain-energy release rate to be the governing parameter in debond growth.

Analysis and Characterization

This section features papers on emerging materials such as three-dimensional woven and braided structures. A number of other topics are also included. Madhukar and Awerbuch demonstrate the use of acoustic emission for studying damage initiation and propagation in center-notched boron/aluminum laminates subjected to quasi-static tensile loading. In particular, they investigate the potential of the acoustic-emission technique as a means of tracking damage in metal-matrix composites. Dynamic stress analysis in notched and unnotched flexural specimens is the subject of the paper by Murthy and Chamis. Both a slow rate of loading and impact loading are considered. A triangular pulse (to simulate concentrated loads) is utilized in conjunction with a NASTRAN finite-element code to obtain numerical results. The last three papers involve three-dimensionally reinforced materials. Ko presents an analysis and characterization of a three-dimensional braided material. In the paper by Ma, Yang, and Chou, an extensive mathematical model applicable to three-dimensional woven composites is presented. The theoretical analysis is supplemented by experimental data on Magnaweave composites. Fabrication and characterization of multi-dimensionally braided composites is discussed in the paper by Macander, Crane, and Camponeschi. Three graphite fiber systems and three processes are investigated in this work.

James M. Whitney

Materials Research Engineer, Air Force Wright Aeronautical Laboratories, Wright-Patterson Air Force Base, OH; symposium chairman and editor.

Author Index

Subject Index

J

K

L

M

N

O

P

R